EMERGING SOLUTIONS FOR FUTURE MANUFACTURING SYSTEMS

IFIP – The International Federation for Information Processing

IFIP was founded in 1960 under the auspices of UNESCO, following the First World Computer Congress held in Paris the previous year. An umbrella organization for societies working in information processing, IFIP's aim is two-fold: to support information processing within its member countries and to encourage technology transfer to developing nations. As its mission statement clearly states,

> *IFIP's mission is to be the leading, truly international, apolitical organization which encourages and assists in the development, exploitation and application of information technology for the benefit of all people.*

IFIP is a non-profit making organization, run almost solely by 2500 volunteers. It operates through a number of technical committees, which organize events and publications. IFIP's events range from an international congress to local seminars, but the most important are:

- The IFIP World Computer Congress, held every second year;
- Open conferences;
- Working conferences.

The flagship event is the IFIP World Computer Congress, at which both invited and contributed papers are presented. Contributed papers are rigorously refereed and the rejection rate is high.

As with the Congress, participation in the open conferences is open to all and papers may be invited or submitted. Again, submitted papers are stringently refereed.

The working conferences are structured differently. They are usually run by a working group and attendance is small and by invitation only. Their purpose is to create an atmosphere conducive to innovation and development. Refereeing is less rigorous and papers are subjected to extensive group discussion.

Publications arising from IFIP events vary. The papers presented at the IFIP World Computer Congress and at open conferences are published as conference proceedings, while the results of the working conferences are often published as collections of selected and edited papers.

Any national society whose primary activity is in information may apply to become a full member of IFIP, although full membership is restricted to one society per country. Full members are entitled to vote at the annual General Assembly, National societies preferring a less committed involvement may apply for associate or corresponding membership. Associate members enjoy the same benefits as full members, but without voting rights. Corresponding members are not represented in IFIP bodies. Affiliated membership is open to non-national societies, and individual and honorary membership schemes are also offered.

EMERGING SOLUTIONS FOR FUTURE MANUFACTURING SYSTEMS

IFIP TC 5 / WG 5.5 Sixth IFIP International Conference on Information Technology for Balanced Automation Systems in Manufacturing and Services
27–29 September 2004, Vienna, Austria

Edited by

Luis M. Camarinha-Matos
New University of Lisbon, Portugal

Springer

Library of Congress Cataloging-in-Publication Data

A C.I.P. Catalogue record for this book is available from the
Library of Congress.

ISBN (HB) 0-387-22828-4 / e-ISBN 0-387-22829-2 Printed on acid-free paper.

Printed in the United States of America.

9 8 7 6 5 4 3 2 1 SPIN 11311560

springeronline.com

TABLE OF CONTENTS

TECHNICAL SPONSOR:

IFIP WG 5.5 COVE
Co-Operation infrastructure for Virtual Enterprises and electronic business

TECHNICAL CO-SPONSORS

OESTERREICHISCHE COMPUTER GESELLSCHAFT
AUSTRIAN COMPUTER SOCIETY

Holonic Manufacturing Systems

ORGANIZERS

OESTERREICHISCHE COMPUTER GESELLSCHAFT
AUSTRIAN COMPUTER SOCIETY

TU WIEN
TECHNISCHE UNIVERSITÄT WIEN
VIENNA UNIVERSITY OF TECHNOLOGY

ORGANIZATIONAL CO-SPONSORS

New University of Lisbon

UNINOVA
Instituto de Desenvolvimento de Novas Tecnologias

STEERING COMMITTEE
Luis M. Camarinha-Matos (PT) [SC chair]
Hamideh Afsarmanesh (NL)
Vladimir Marik (CZ)
Heinz-H. Erbe (DE)

Conference chairman: A Min Tjoa (AT)
Program chairman: Luis M. Camarinha-Matos (PT)

Track A co-chairs: Vladimir Marik (CZ), E. H. Van Leeuwen (AU)
Track B chair: Hamideh Afsarmanesh (NL)
Track C chair: Mauro Onori (SE)
Track D co-chairs: Luis Seabra Lopes (PT), Olga Stepankova (CZ)

BASYS'04

6th IFIP International Conference on Information Technology for Balanced Automation Systems in Manufacturing and Services

Vienna, AUSTRIA, 27-29 September 2004

REFEREES FROM THE PROGRAMME COMMITTEE

Track A: Multi-agent systems in Manufacturing
- □ J. Barata (PT)
- □ R. W. Brennan (CA)
- □ M. Deen (UK)
- □ M. Fletcher (UK)
- □ W.A. Gruver (CA)
- □ K. Hall (US)
- □ D. Kotak (CA)
- □ J. Lazansky (CZ)
- □ V. Marik (CZ)
- □ D. McFarlane (UK)
- □ J. Mueller (DE)
- □ E. Oliveira (P)
- □ M. Pechoucek (CZ)
- □ L. Sheremetov (MX)
- □ A. Smirnov (RUS)
- □ S. Tamura (J)
- □ P. Valckenaers (BE)
- □ E. H. Van Leeuwen (AU)

Track B: Networked enterprises
- □ H. Afsarmanesh (NL)
- □ P. Bernus [AU]
- □ L. M. Camarinha-Matos (PT)
- □ W. Cellary (PL)
- □ S. Dudstar (AT)
- □ H. Erbe (DE)
- □ C. Garita (CR)
- □ T. Goranson (US)
- □ T. Kaihara (JP)
- □ K. Kosanke (DE)
- □ G. Kovács (HU)
- □ A. Molina (MX)
- □ L. Nemes (AU)
- □ G. Olling (US)
- □ J. Pinho Sousa (PT)
- □ G. Putnik (PT)
- □ R. Rabelo (BR)
- □ W. Shen (CA)
- □ A Min Tjoa (AT)
- □ R. Wagner (AT)

Track C: Integrated design & assembly
- □ T. Arai (US)
- □ R. Bernhardt (DE)
- □ B. Lindberg (SE)
- □ R. Molfino (IT)
- □ D. Noe (SI)
- □ M. Onori (SW)
- □ S. Ratchev (UK)
- □ B. Raucent (BE)
- □ I. Rudas (HU)
- □ M. Santocchi (IT)
- □ G. Schreck (DE)
- □ R. Tuokko (FI)
- □ H. Van Brussel (BE)

Track D: Machine learning and data mining in industry
- □ M. Barata (PT)
- □ M. Botta (IT)
- □ M. Bohanec (SI)
- □ P. Brezany (AT)
- □ D. Mladenic (SI)
- □ H. Motoda (JP)
- □ S. Moyle (UK)
- □ J. Paralic (SK)
- □ T. Rauber (BR)
- □ J. Rauch (CZ)
- □ L. Seabra Lopes (PT)
- □ O. Stepankova (CZ)
- □ D. Wettschereck (UK)
- □ F. Zelezny (CZ)

OTHER REFEREES

António Abreu (PT)
João Rosas (PT)
F. Maturana (US)
P. Vrba (US)

FOREWORD

Agility and distribution

Industry and particularly the manufacturing sector have been facing difficult challenges in a context of socio-economic turbulence which is characterized by complexity as well as the speed of change in causal interconnections in the socio-economic environment. In order to respond to these challenges companies are forced to seek new technological and organizational solutions. Knowledge intensive approaches, distributed holonic and multi-agent systems, collaborative networks, data mining and machine learning, new approaches to distributed process modeling and supervision, and advanced coordination models are some of the example solution areas. Information technology plays a fundamental role in this process. But sustainable advances in industry also need to consider the human aspects, what led to the concept of "balanced automation systems" in an attempt to center the discussion on the balance between the technical aspects of automation and the human and social facets. Similar challenges are faced by the service sector. A continuous convergence between the areas of manufacturing and services has, in fact, been observed during the last decade.

In this context two main characteristics emerge as key properties of a modern automation system – ***agility*** *and* ***distribution****. Agility because systems need not only to be flexible in order to adjust to a number of a-priori defined scenarios, but rather must cope with unpredictability. Distribution in the sense that automation and business processes are becoming distributed and supported by collaborative networks. These networks can be observed at the inter-enterprise collaboration level, but also at the shop floor level where more and more control systems are designed as networks of autonomous and collaborative nodes. Multi-agent and holonic approaches play, naturally, a major role here. Advances in communications and ubiquitous computing, including the new wireless revolution, are fundamental enablers for these processes.*

In this context, the IFIP BASYS conferences were launched with the aim of promoting the discussion and sharing of experiences regarding approaches to achieve a proper balance between the technical aspects of automation and the human and social points of view. A series of successful BASYS conferences were held in Victoria, Brazil (1995), Lisbon, Portugal (1996), Prague, Czech Republic (1998), Berlin, Germany (2000), and Cancun, Mexico (2002). Following the IFIP vision, BASYS offers a forum for collaboration among different regions of the world.

This book includes the selected papers for the BASYS'04 conference that is held in Vienna, Austria, jointly organized by the Technical University of Vienna and the Austrian Computer Society. This 6th conference in the series addresses Information

Technology for Balanced Automation Systems in Manufacturing and Services. The main focus of this conference is to explore new challenges faced by the integration of Knowledge and Technology as major drivers for business changes, considering Product and Services Life Cycles.

The conference is organized in four main tracks, also reflected in the structure of the book:

- *Track A: Multi-agent and holonic systems in manufacturing, covering architectures, implementation solutions, simulation, collaborative and mobile approaches, intelligent systems and optimization.*
- *Track B: Networked Enterprises, covering infrastructures for networked enterprises, collaboration support platforms, performance, measurement approaches, modeling, and management of collaborative networks.*
- *Track C: Integrated design and assembly, covering new approaches for assembly systems design, configuration and simulation, sensors for assembly, and advanced applications.*
- *Track D: Machine learning and data mining in industry, covering case based reasoning, soft computing, machine learning in automation, data mining and decision making.*

Put together, these contributions offer important emerging solutions to support agility and distributed collaborative networks in future manufacturing and service support systems.

The editor,
Luis M. Camarinha-Matos, New University of Lisbon

KEYNOTE

1 NETWORKED RFID IN INDUSTRIAL CONTROL: CURRENT AND FUTURE

Duncan McFarlane
Centre for Distributed Automation and Control
Institute for Manufacturing
University of Cambridge, UK
dcm@eng.cam.ac.uk

This paper introduces the notion of networked Radio Frequency Technology (RFID) and reviews the work of the Auto ID Center in providing a low cost, global networked RFID solution. The paper then examines the role of networked RFID in changing the nature of industrial control systems operations. In particular the notions of connectedness, coordination and coherence are introduced as a means of describing different stages of adoption of RFID.

1. INTRODUCTION

1.1 Aims of the Paper

Radio Frequency Identification or RFID has sprung into prominence in the last five years with the promise of providing a relatively low cost means for connecting non electronic objects to an information network (refer to Finkenzeller (1999) for technical details). In particular, the manufacturing supply chain has been established as a key sector for a major deployment of this technology. This paper introduces the concept of networked RFID and discusses its role in the development of product driven industrial control. Firstly, however, we review some of the developments in RFID.

1.2 Developments in RFID

The concepts behind RFID were first discussed in the mid to late 1940's, following on from technical developments in radio communications in the 1930's and the development of radar during World War II (Landt *et al.*, 2001). An early published work exploring RFID is the landmark paper by Harry Stockman, "Communication by Means of Reflected Power" (Stockman, 1948). Stockman stated then that "Evidently, considerable research and development work has to be done before the remaining basic problems in reflected-power communication are solved, and before the field of useful applications is explored."

The 1950s were an era of exploration of RFID techniques – several technologies related to RFID were developed such as the long-range transponder systems of "identification, friend or foe" (IFF) for aircraft. A decade of further development of

RFID theory and applications followed, including the use of RFID by the U.S. Department of Agriculture for tracking the movement of cows. In the 1970's the very first commercial applications of the technology were deployed, and in the 1980's commercial exploitation of RFID technology started to increase, led initially by small companies.

In the 1990's, RFID became much more widely deployed. However, these deployments were in vertical application areas, which resulted in a number of different proprietary systems being developed by the different RFID solutions providers. Each of these systems had slightly different characteristics (primarily relating to price and performance) that made them suitable for different types of application. However, the different systems were incompatible with each other – e.g. tags from one vendor would not work with readers from another. This significantly limited adoption beyond the niche vertical application areas – the interoperability needed for more widespread adoption could not be achieved without a single standard interoperable specification for the operation of RFID systems. Such standardisation was also needed to drive down costs.

The drive towards standardisation started in the late 1990's.There were a number of standardisation efforts, but the two successful projects were:

(a) the ISO 18000 series of standards that essentially specify how an RFID system should communicate information between readers and tags

(b) the Auto-ID Centre specifications on all aspects of operation of an RFID asset-tracking system, which has subsequently been passed onto EAN.UCC (the custodians of the common barcode) for international standardisation

The next section focuses on the Auto ID Center and its developments.

1.3 Auto ID Center: 1999-2003

The Auto-ID Centre (Auto-ID Center, 2003) was a university-based organisation that was formed in 1999, initially by the MIT, the Uniform Code Council, Gillette and Procter and Gamble. The motivation of the Centre was to develop a system suitable for tracking consumer packaged goods as they pass through the supply chain in order to overcome problems of shrinkage and poor on-shelf-availability of some products. The requirements for RFID in the supply chain context are in stark contrast to those applications that preceded the centre as is illustrated in Table 1 from Hodges *et al.* (2003) where issues of volume, complexity and life differ markedly.

The Centre expanded, involving Cambridge University in 2000 and other universities in following years, and by October 2003 had over 100 member companies, all with a common interest in either supplying or deploying such a technology in their companies. Early on in the life of the Centre, it became clear that RFID would form a cornerstone of the technological solution, and along with the help of some end-user and technology companies, the Centre was instrumental in driving down the cost of RFID to a point where adoption started to become cost-effective in some application areas. Part of the solution to keeping costs down is a single-minded drive to reduce RFID tag complexity, and one approach to this advocated by the Auto-ID Centre is to store as little data about products as possible actually *on the tag*. Instead, this information is stored on an organisation's computer network, which is much more cost-effective.

Table 1 – RFID Application Characteristics (Hodges *et al.*, 2003)

	Tolling	Library	Asset	Baggage	Security	Supply Chain
Complexity of Information on Tag	M	L	H	L	L	L
Single or Multiple Applications for Each Tag	S	S	S	S	S	M
Volume of Tags	L	L	L	M	M	H
Expected Life of Tag	H	H	H	M	M	L

The specific aims of the centre were thus:

1. *Low Cost RFID solutions:* were developed by reducing the chip price on a tag, which was achieved by reducing amount of silicon required, which required the reduction of the information stored on chip to a serial number or ID only, with all other product information held on a networked data base.
2. *A Universal System*: in order to achieve business justification through multiple applications/companies standard specifications were proposed for tag/reader systems, and data management and communication systems.

The Auto ID Center's development work, now carried on by the Auto ID Labs in six locations, is described next.

2. THE ANATOMY OF NETWORKED RFID

As discussed earlier, the key to the recent RFID deployments has been the network connection of RFID tagged objects. We now discuss requirements for such a Networked RFID approach.

2.1 Networked RFID Requirements

A networked RFID system generally comprises the following elements:

1. A unique identification number which is assigned to a particular item.
2. An identity tag that is attached to the item with a chip capable of storing – *at a minimum* – the unique identification number. The tag is capable of communicating this number electronically.
3. Networked RFID readers and data processing systems that are capable of collecting signals from multiple tags at high speed (100s per second) and of pre-processing this data in order to eliminate duplications, redundancies and misreads.
4. One or more networked databases that store the product information.

With this approach, the cost of installing and maintaining such systems can be spread across several organizations while each is able to extract its own specific benefits from having uniquely identified items moving in, through and out of the organization's operations.

2.2 The EPC Network

The EPC Network is the Auto ID Center's specification for a Networked RFID system. The EPC Network consists of six fundamental technology components, which work together to bring about the vision of being able to identify any object anywhere automatically and uniquely. These are:

1) The Electronic Product Code (EPC)
2) Low-cost Tags and Readers
3) Filtering, Collection and Reporting
4) The Object Name Service (ONS)
5) The EPC Information Service (EPCIS)
6) Standardised vocabularies for communication

These six elements together form the core infrastructure of the EPC Network and provide the potential for automatic identification of any tagged product. Figure 1 illustrates a schematic of how the elements interface with each other for a toaster.

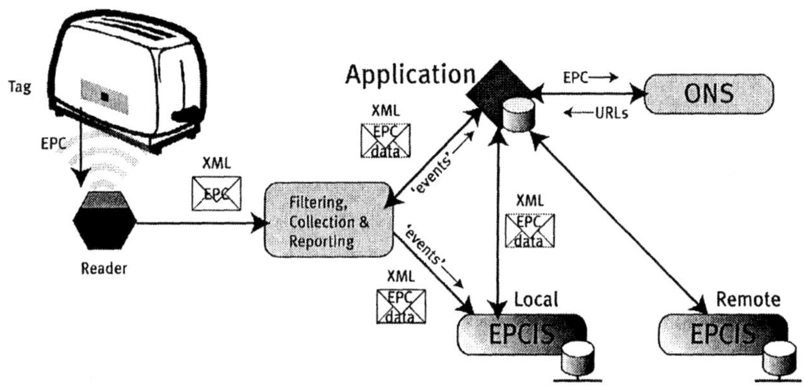

Figure 1 – Architecture of the EPC Network

We outline each component briefly below, and the reader is referred to Harrison (2004) for example for further details.

2.2.1 The Electronic Product Code (EPC)

The aim of the EPC is to provide a unique identifier for each object (Brock, 2001a). Designed from the outset for scalability and use with networked information systems, the EPC typically consists of three ranges of binary digits (bits) representing:

a) an EPC Manager (often the manufacturer company ID)
b) an object class (usually the product type or "SKU") and
c) a unique serial number for each instance of a product.

As well as being the lookup 'key' to access the information about the tagged object on the network, the EPC concept has also been an important factor in driving down the production costs of tags and readers (Sarma, 2001); by stipulating that the tag need only store the unique EPC identity number, it is possible to design tags with much lower on-board memory requirements, since the additional information about the tagged object can be stored in distributed networked databases, tied to the object via its EPC number.

2.2.2 Low-cost Tags and Readers

Radio Frequency Identification (RFID) is a key technology enabling automatic reading of multiple items simultaneously, without requiring manual scanning of each individual item. The reader emits radio waves of a particular frequency. When passive tags (called passive because they lack their own power supply) enter the range of a reader, their antenna absorb energy from the radio field, powering the microchip which stores the unique EPC identity code – and returning this information back to the reader via a modulation of the radio waves.

2.2.3 Filtering, Collection and Reporting ('Savant')

A widescale deployment of RFID tags and readers could potentially result in overloading of the information network (bandwidth and database storage capacity) with raw data from RFID readers. It is important to ensure that just significant data and 'events' are transmitted. These software 'events' contain information and are able to trigger processes in higher-level applications and information systems.

2.2.4 The Object Name Service (ONS)

The Object Name Service (ONS) is used to convert an EPC into a number of internet addresses where further information about a given object may be found. Currently, the ONS specification deals with a *static* implementation based on the Domain Name Service (DNS) which provides IP address lookup for the internet. Recognising that potentially several parties in the supply chain may also hold relevant data about an object, it is likely that static ONS will be augmented with a *dynamic* ONS counterpart, which is able to provide a lookup for many instances of a given product, pointing to the various other parties across the supply chain, which also hold information.

2.2.5 The EPC Information Service (EPCIS)

While the ONS points to various sources of information, it must be recognised that different companies will use different database vendors and different implementations and that there is currently great reluctance to share information between trading partners. However, in order to obtain maximum benefit from the EPC Network infrastructure, companies need to share some information in order to be able to respond in a more timely manner to the new data available, e.g. allowing manufacturers to adjust production rates to synchronise with actual real-time consumer demand detected by smart shelves with embedded readers.

2.2.6 Standardised vocabularies for communication

Having obtained the data via ONS and EPCIS, it is important that its interpretation is unambiguous and ideally self-describing. This is the role of standardised vocabularies. Approaches based on the Extensible Markup Language (XML) provide a way of marking up structured data for communication and exchange between diverse applications and different parties (refer to (Brock, 2001b) and (Floerkmeier *et al.*, 2003) for more details).

3. IMPACT OF NETWORKED RFID ON INDUSTRIAL CONTROL

Having established the structure and functionality of a networked RFID system, we now focus on its role in an industrial control environment. The first point to make is that although the networked RFID system is essentially and information providing Service, in an industrial control context it needs to be considered as part of a closed loop process (see Figure 2). In understanding the way in which RFID is introduced into the closed loop we find it helpful to consider three stages of integration:

1. **Connection:** the stage at which the physical integration of RFID data with the existing sensors used in the operation is achieved. The data at this stage is merely used for monitoring purposes and does not influence the resulting decisions or actions.
2. **Coordination:** the stage in which networked RFID data is exploited to provide an increased quality of product information in the closed loop which can enhance the decision making and execution processes.
3. **Coherence:** the availability of the increased quality of product information leads to a reeingeering of the decision making process and/or the physical operation being controlled.

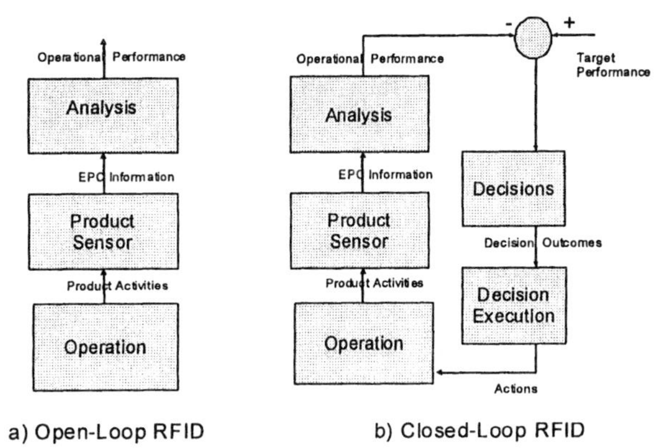

Figure 2 – Open Loops vs Closed Loops RFID

We will briefly discuss each of these stages, and comment on their relevance to ongoing developments in RFID-based industrial control.

3.1 Connection: RFID As An Additional Sensor in the Closed Loop

The most fundamental impact of the introduction of tagged products is that an additional sensor stream is introduced into the industrial control environment. Note that bar coding and other direct product inspection systems rarely play a role in industrial control environments owing to their difficulty in achieving reliable automation. Hence, typically, information as to the identity and movement of products is currently determined indirectly through the combination of proximity sensors and manual records.

The introduction of RFID enables a more accurate and automatable form of product monitoring, and can enable regular updating of production and order status, inventory levels etc.

In mid 2004, this stage of deployment represents the status quo in the commercial use of networked RFID. It is observed that many potential implementers are seeking simply to understand the issues and challenges in connecting RFID while work in establishing a business basis proceeds in parallel. Some comments on achieving RFID connections are provided in (Chang *et al.*, 2004).

3.2 Coordination: Quantifying Product Information Quality

The main value of the introduction of a networked RFID solution such as the EPC Network is in enhancing the quality of product information available to make decisions. By product information quality, we refer to properties or dimensions such as:

- *accuracy*: the precision and reliability associated with the collection of product information
- *completeness*: the amount of product information relevant for a given decision, that is available
- *timeliness*: the timeliness of the availability of product information

A qualitative assessment of different product information sources against these dimensions is given in Figure 3. In this diagram we distinguish between the stand alone and networked RFID solutions – the latter with direct data base access has the ability to provide a more complete level of information about a given item.

The coordination of networked RFID data raises a number of questions about the implementation of the system which are being addressed both academically and industrially at present:

- How should the RFID hardware be arranged to maximise the impact on the industrial control system?
- What are the other sensing issues, and how should the RFID data be best coordinated with these sensors to maximise the effectiveness of decisions made?
- How should the RFID data be filtered and prepared to be most effectively integrated?
- How can the impact of better product information on resulting decisions be qualified?

Any of the industrial developments being reported in the commercial press at present (RFID Journal, 2004) refer to the management of such issues, and academically, work has been performed to provide a theoretical framework for examining the role of information quality (McFarlane, 2003; McFarlane *et al.* 2003b) and its benefits, e.g. (Parlikad *et al.*, 2004).

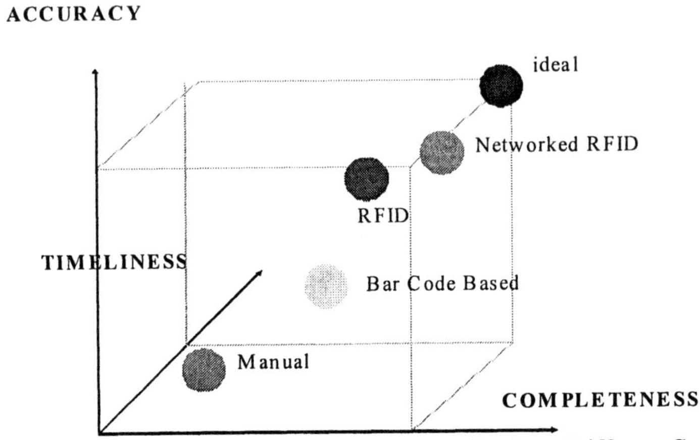

Figure 3 – Product Information Quality from Different Sources

3.3 Coherence: Networked RFID Supporting Product Intelligence

Many pundits have indicated that RFID may become a disruptive technology for the industrial supply chain, e.g. (Sheffi, 2004). The ready availability of high quality product data can not only enhance existing decision making processes in the supply chain (e.g. inventory management, quality control, shelf replenishment) but can lead to a radical rethinking of the nature of the decisions themselves and the resulting actions. For example, in (Wong *et al.*, 2002) the nature of retail shelf replenishment is examined in detail and in (Fletcher *et al.*, 2003) the role of RFID in developing a radical mass customised packaging environment is discussed. Essentially, a networked, RFID tagged object can play a rather different role in the operations it is subject to, compared to the way it is managed today.

In particular, the introduction of a networked RFID system can alter the role of a product from a purely *passive* one, to one in which a product – representing a section of a customer order – can *actively* influence its own production, distribution, storage, retail etc. We refer to this as an "intelligent product" – the notion and uses of intelligent products have also been reported in (Bajic *et al.*, 2002) and (Karkannian *et al.*, 2003). We formalise the concept of an intelligent product with the following working definition (McFarlane *et al.*, 2003a):

An intelligent product is a physical and information based representation of an item for retail which:

1. *possesses a unique identification*
2. *is capable of communicating effectively with its environment*
3. *can retain or store data about itself*
4. *deploys a language which can articulate its features and requirements for its production, usage, disposal etc...*
5. *is capable of participating in or making decisions relevant to its own destiny on a continuous basis*

The corresponding *intelligent product* for a soft drink can is illustrated in Figure 4 in which the physical can is connected to a network and thus to both information stored about it and also to a *decision making (software) agent* acting on its behalf. The concept of a software agent is important to the following discussion and is defined as:

A software agent is a distinct software process, which can reason independently, and can react to change induced upon it by other agents and its environment, and is able to cooperate with other agents.

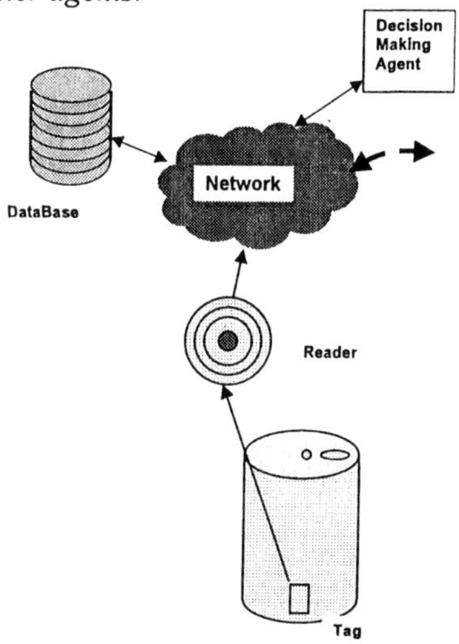

Figure 4 – "Intelligent drink can"

The intelligent product, defined here, is hence an extension of the product identification system provided by a networked RFID system – incorporating a software agent that is capable of supporting decisions made on behalf of the product.

The notion of software agents in the development of industrial control systems has been discussed for some time (see for example (Marik *et al.*, 2002; Deen, 2003) and the references therein). Software agents have been used to develop a radical set of future industrial control architectures in which disruption management, rapid reconfiguration and low cost customisation are the key drivers. The introduction of networked RFID, coupled to a software agent based industrial control environment can be seen to enable key elements of a radically new control system in which products as part of customer orders drive their own operations. The reader is referred to (McFarlane *et al.*, 2003a; Harrison *et al.*, 2004) for more details on this concept in the manufacturing domain.

4. SUMMARY

This paper introduced the networked RFID concept and summarised the key ways in which it can impact on industrial control systems. The interested reader is referred to the Cambridge Auto ID Labs activities for more details (Auto ID Labs, 2004).

5. ACKNOWLEDGEMENTS

The author would like to thank his colleagues at the Auto ID Labs at Cambridge and before that at the Cambridge Auto ID Center for their significant contributions to the developments that are summarised in this paper. The financial contributions of the large number of industrial sponsors of this work are also gratefully acknowledged.

6. REFERENCES

1. Auto-ID Centre, Auto ID Center website archive: http://archive.epcglobalinc. org/index.asp, 2003
2. Auto ID Labs @ Cambridge, website: www.autoidlabs.org/cambridge, 2004
3. Bajic E, Chaxel F. Holonic Manufacturing with Intelligent Objects. In Proceedings of 5th IFIP International Conference on Information Technology for Balanced automation systems In Manufacturing and Services, Cancun, Mexico, 2002.
4. Brock DL. Electronic Product Code™ (EPC™) – A Naming Scheme for Physical Objects. Auto-ID Center White Paper, 2001.
5. Brock DL, The Physical Markup Language (PML) – A Universal Language for Physical Objects. Auto-ID Center White Paper, 2001.
6. Chang Y, McFarlane D. Supply Chain Management Using AUTO-ID Technology – Preparing For Real Time, Item Level Supply Chain Management. In Evolution of Supply Chain Management: Symbiosis of Adaptive Value Networks and ICT, Kluwer Academic Publisher, USA, 2004.
7. Deen SM. (Ed). Agent Based Manufacturing: Advances In The Holonic Approach , Springer-Verlag Berlin Heidelberg, 2003.
8. Finkenzeller K. RFID Handbook. 1st edition, Wiley & Sons LTD, 1999.
9. Fletcher M, McFarlane D, Lucas A, Brusey J, Jarvis D. The Cambridge Packing Cell – A Holonic Enterprise Demonstrator. In Multi-Agent Systems and Applications III, LNAI 2691, Springer Verlag, Heidelberg, 2003, pp. 533-543.
10. Floerkemeier C, Anarkat D, Osinski T, Harrison M. PML Core Specification 1.0. Auto-ID Center White Paper, 2003.
11. Harrison M, McFarlane D, Parlikad A, Wong C Y. Information management in the product lifecycle – The role of networked RFID. Accepted for 2nd International Conference on Industrial Informatics, Berlin, 2004.
12. Harrison M. EPC Information Service – Data Model and Queries. Auto-ID Center White Paper, 2004.
13. Hodges S, McFarlane D. Radio frequency identification: technology, applications and impact. In Proceedings of the OECD Conference, Brussels, 2003.
14. Karkkainen M, et al. Intelligent products – a step towards a more effective project delivery chain. In Computers In Industry 50, Elsevier Science, 2003, pp. 141-151.
15. Landt J, Catlin B. Shrouds of Time: The history of RFID. Published by AIM, The Association for Automatic Identification and Data Capture Technologies, http://www.aimglobal.org/technologies/rfid/resources/shrouds_of_time.pdf, 2001.
16. Marik V, Stepankova O, Krautwurmova H, Luck M. (Eds.). Multi-Agent Systems and Applications II, LNAI 2322, Springer-Verlag, Berlin Heidelberg, 2002.
17. McFarlane D. Product Identity and Its Impact on Discrete Event Observability. In Proceedings of ECC, Cambridge, UK, 2003.
18. McFarlane D, Sarma S, Chirn J-L, Wong C Y, Ashton K. The Intelligent Product In Manufacturing Control And Management. Engineering Applications of Artificial Intelligence: special issue on Intelligent Manufacturing, Vol. 16, No. 4, 2003, pp. 365-376.
19. McFarlane D, Sheffi Y. The Impact of Automatic Identification on Supply Chain Operations. International Journal of Logistics Management, Vol. 14, No. 1, 2003, pp. 1-17
20. Parlikad A, McFarlane D. Investigating The Role Of Product Information In End-Of-Life Decision Making. Proceedings of 11th IFAC Symposium on Information Control Problems in Manufacturing, San Salvador, Brazil, 2004.
21. RFID Journal, website: www.rfidjournal.com, 2004.
22. Sarma S. Towards the 5¢ Tag. Auto ID Center White Paper MIT-AUTOID-WH-001, http://www.autoidcenter.org/research, 2001.
23. Sheffi Y. RFID and Innovation, In Proceedings of the MIT Summer School in Logistics and Operations Management, 2004.
24. Stockman H. Communication by Means of Reflected Power. In Proceedings of the IRE, 1948, pp. 1196-1204.
25. Wong CY, McFarlane D, Zahrudin A, et al. The Intelligent Product Driven Supply Chain. In Proceedings of IEEE International Conference on Systems, Man and Cybernetics, Hammamet, Tunisia, 2002.

PART A

MULTI-AGENT AND HOLONIC SYSTEMS IN MANUFACTURING

2 IMPLEMENTATION ISSUES WITH HOLONIC CONTROL DEVICE COMMUNICATION INTERFACES

Jason J. Scarlett[1], Robert W. Brennan[1], Francisco Maturana[2], Ken Hall[2], Vladimir Marik[3] and Douglas H. Norrie[1]
[1]*Department of Mechanical and Manufacturing Engineering*
University of Calgary, 2500 University Dr. N.W. Calgary, CANADA T2N 1N4
[2]*Rockwell Automation Advanced Technologies*
Allen Bradley Drive 1, Mayfield Heights, OH, USA, 44121
[3]*Rockwell Automation*
Americka 22, Praha, CZECH REPUBLIC, 120 00

This paper focuses on implementation issues at the interface between holonic control devices (HCDs) and agent-based systems. In particular, we look at a function block-based approach to communication that is applicable to existing IEC 61131-3 systems and emerging IEC 61499 systems.

1. INTRODUCTION

In this paper we focus on the physical holons or "holonic control devices" (HCDs) that reside at the lowest level of a holonic manufacturing system (HMS) (HMS, 2004). At this level, HCDs must have the capabilities of typical embedded control devices as well as the ability to function in the larger holonic system. In other words, HCDs must interface with the sensors and actuators of the physical processing equipment and provide the real-time control functions that implement and monitor the required sequence of operations; they must also communicate with other holons to negotiate and coordinate the execution of processing plans and recovery from abnormal operations.

Although there has been a considerable amount of progress towards developing collaborative problem solving systems at the planning and scheduling level and the physical device level of the manufacturing enterprise (McFarlane and Bussmann, 2000) there has been very little work on tying these worlds together. In other words, without an effective real-time interface between the information world (i.e., software agents) and the physical world (i.e., physical agents or holons), agents and machines will continue to exist and operate largely apart as they do today.

One of the main barriers is the very different approach to software development at these two levels. This is primarily because of the need to satisfy real-time requirements at the device level, but also because of the historical evolution of industrial control (e.g., ladder logic's relationship to relay wiring diagrams). Recent international standards efforts such as the International Electrotechnical Commission's IEC 61131-3 (Lewis, 1996) and IEC 61499 (IEC, 2000) standards have made progress in addressing the issues of open programming languages and distributed control models, however the issue of interfacing industrial control software to agent-based software remains.

A second area of concern is that of inter-holon communication. Within each HCD, the distributed intelligence that sets them apart from typical embedded controllers is enabled by software agents that are capable of communicating with other agents (and holons) through message passing. Although the approach to inter-agent communication is well established at the higher levels of the manufacturing enterprise by the services of agent platforms such as FIPA-OS (FIPA, 2004) and JADE (JADE, 2004), inter-agent communication at the device level becomes more problematic. On the software agent side, well-established communication protocols (e.g., Ethernet) are typically used. However, because of the more stringent requirements for latency, reliability and availability on the physical side, specialised communication protocols (e.g., CAN (Robert Bosch, 1991) and DeviceNet (DeviceNet, 2004)) are required.

In this paper, we investigate how the low-level control (LLC) and high-level control (HLC) domains can be interfaced. The LLC and HLC architecture proposed for this integration uses function blocks for the LLC domain and software agents for the HLC domain (Christensen, ????).

The paper begins with an introduction to two possible approaches to interfacing the agent and machine worlds. We then focus on the issues that arise when implementing these approaches. In particular, we look at the advantages and disadvantages of using existing programming approaches (IEC 61131-3) at the device level and discuss the potential advantages of an IEC 61499 based approach. As well, we investigate current approaches to implementing deterministic inter-holon communication at the device level and propose an alternative approach to this problem. We also investigate the requirements for integrating low-level control language with the agent level language and communication. The paper concludes with a summary of our experiences with the real-time interface problem as well as with our suggestions for further research in this area.

2. A LOW-LEVEL INTERFACE

In this section, we look at two possible approaches to interfacing the agent and machine worlds: (i) a data-table approach as illustrated in Figure 1(a), and (ii) a function block adapter approach as illustrated in Figure 1(b).

2.1 Data Tables

Given the architecture of a programmable logic controller (PLC), the first approach is arguably the most obvious since it takes advantage of the basic memory structure and execution model of common PLCs. For example, in Figure 1 a *data table* is used to allow "messages" to be passed between the agent world and the control world. During each PLC scan cycle, state information (e.g., input and output image table data and other addressable data) is written to a data table, which is then transformed to a format that is understandable to the agent system (e.g., FIPA Agent Communication Language (ACL) (FIPA, 2004)). As well, agent messages to the low-level control system are transformed to the appropriate data table format and read by the PLC (i.e., written to its RAM memory) during each PLC scan cycle.

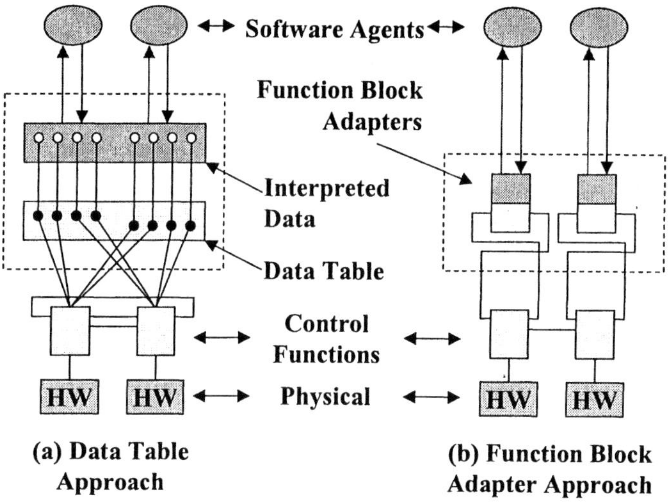

Figure 1 – A transformation interface

Although this approach is quite straight forward, it is very hardware and application dependent. For example, explicit knowledge of the PLC's addressing structure is required for this to work. As well, limitations on the amount of RAM available in the PLC for this type of data exchange may result in customisation of exactly what is read and written for each specific application.

For the remainder of section we will focus on the second approach, function block adapters, which was first proposed by Heverhagen and Tracht (2002) for IEC 61131-3 based systems. Given the "open systems" focus of the IEC 61131-3 industrial programming standard, this approach has the potential to overcome the drawbacks of the data table approach.

2.2 Function Block Adapters

Function block adapters were first proposed by Heverhagen and Tracht (2002) to provide a means of unambiguously expressing the interface mapping between IEC 61131-3 based control systems and object-oriented or agent-based software systems. To achieve this mapping, they propose a hybrid IEC 61131-3 function block, called a function block adapter (FBA) that expresses the mapping between IEC 61131-3 function blocks (Lewis, 1996) and Real-time Unified Modelling Language (RT-UML) capsules (please refer to Lyons (1998) for more information on RT-UML capsules, and Fletcher et al. (2001) for the relationship to IEC 61499 function blocks).

Given that the agent side of the system can be developed using a UML-based tool, it follows that an interface between the control software (e.g., IEC 61131-3 function blocks) and a RT-UML capsule is all that is needed for the transformation interface between the agent world and the control world.

As shown in Figure 2, Heverhagen and Tracht (2002) suggest that a hybrid IEC 61131-3 function block / RT-UML capsule can be used to map between the control world (i.e., the IEC 61131-3 function block, MyFB) and the object/agent world (i.e., the RT-UML capsule MyCapsule). The convention for IEC 61131-3 and IEC 61499 function blocks is that inputs are shown on the left and outputs are shown on the right. In Figure 2, MyFB can send messages to the object/agent system via outputs D, E, and F; messages are received from the object/agent system via inputs A, B, C. The black and white squares connecting MyCapsule and MyFBA represent the RT-UML ports.

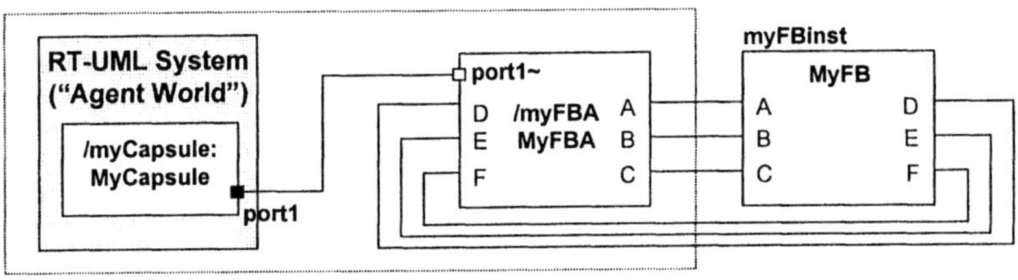

Figure 2 – IEC 61131-3 function block adapters
(from (Heverhagen and Tracht, 2002))

In order to unambiguously express the mapping between MyFB and MyCapsule, Heverhagen and Tracht proposed a simple FBA language. They note that the key to this working properly is that the interface should be simple: i.e., the interface should not specify what happens after a signal is translated and sent to a capsule or to a function block.

Figure 1(b) illustrates how we can now modify the transformation interface using function block adapters. In a more complex application however, multiple function block adapters may be used as well as multiple capsule interfaces on the agent side in order to reduce the complexity of the FBA interfaces.

Since IEC 61131-3 shares the same scan-based execution model with conventional PLC systems, the implementation of function block adapters is not as simple as Figures 1 and 2 imply. For example, Heverhagen and Tracht suggest two approaches: (i) with the FBA implemented on the object/agent side, and (ii) with the FBA split across both sides. In the next section, we investigate the use of IEC 61499 function blocks to implement FBA's. The FBA concept appears to be a closer fit with this model because of IEC 61499's event-based model and its use of service interface function blocks. This approach will be discussed in the next section.

3. IMPLEMENTATION ISSUES

In this section we summarise our experience implementing the second approach discussed in the previous section. We begin with a description of the IEC 61499 model and compare this with Heverhagen and Tracht's IEC 61131-3 approach. Next, we look at the issue of inter-object communication in a distributed real-time environment.

3.1 Function Block Adapter Implementation

On the surface, the IEC 61499 implementation of function block adapters appears to be very similar the IEC 61131-3 implementation as is illustrated in Figure 3.

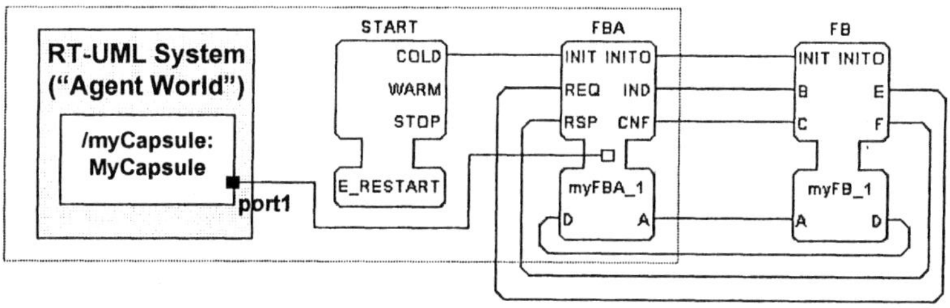

Figure 3 – IEC 61499 function block adapters

Comparing this with Figure 2 however, one can see that some of the interface is now implemented with IEC 61499 events (upper portion of the function blocks in Figure 3). In Figure 2, signals B, C, F and E are used to signal events. For example, a "true" value on B indicates that data is available to be read by input A; a "true" value on C indicates that MyFB has read the data on input A. As well, some additional information can be made available using the standard IEC 61499 protocols. For example, when MyFBA sends an event signal to MyFB's input B, it will set its QI input to "true" if data is available to be read on A; alternatively, it will set QI to "false" if there is no data available.

In order to illustrate this approach, we show the two basic forms of data transfer in Figures 4 and 5: agent or capsule initiated transfer and function block initiated transfer respectively.

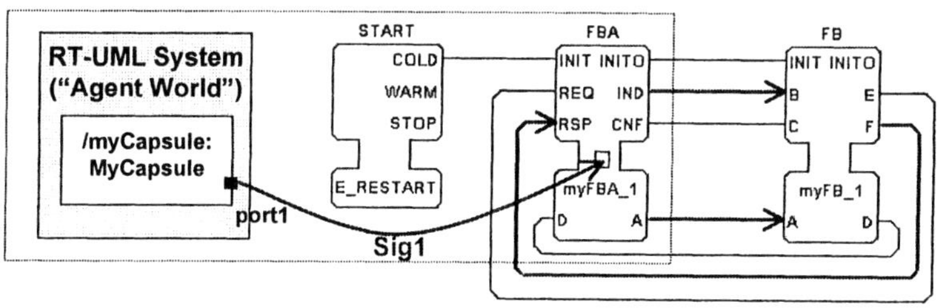

Figure 4 – Capsule initiated data transfer

In Figure 4, communication is initiated by a capsule (i.e., representing a software object or agent) in the "agent world". The capsule sends its data (i.e., "Sig1") via port1. This data is then made available on output "A" of the IEC 61499 function block adapter (i.e., "FBA"). FBA next indicates that data has been received and is available by initiating event "IND". "FB" then acknowledges receipt of the data by issuing event "F" (this is received on FBA's "RSP" event input). It should be noted that no message is sent to the capsule if communication is asynchronous.

Figure 5 illustrates synchronous communication that is initiated by the low-level control system. In this case, data is made available at output "D" of FB. When FB is ready to send this data to the higher-level agent system, it signals FBA with output event "E". This initiates an "REQ" event on FBA's input, which in turn results in the data being sent to the agent system (i.e., "Sig2"). In this case, the agent system acknowledges the transmission with "Sig3" via port1, allowing FBA to confirm to FB that its data was received (i.e., FBA issues a "CNF" event to FB).

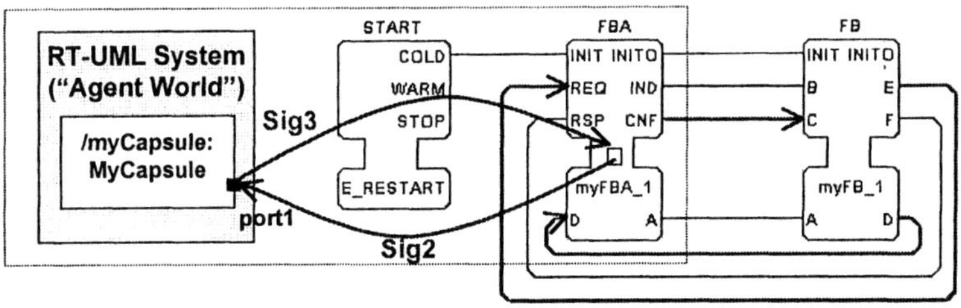

Figure 5 – Function Block initiated data transfer

As noted previously, the use of IEC 61499 event connections simplifies this approach. Arguably, the more significant difference in the implementation however, is that MyFBA is implemented as an IEC 61499 service interface function block (SIFB). As the name implies, interface function blocks provide services to the function block application. For example, resource initiated services such as a subscriber interface (to an Ethernet connection) or an analogue-to-digital converter interface can be implemented as a SIFB. Similarly, application initiated services such as a publisher (to an Ethernet connection) or a digital-to-analogue converter interface can be implemented as a SIFB.

As a result, the specialised hybrid function block / capsule (shown in the centre of Figure 2) is no longer required. For example, in the IEC 61499 implementation, the FBA shown in Figures 4 and 5 is a composite function block consisting of a FBA controller and a publisher/subscriber pair as shown in Figure 6.

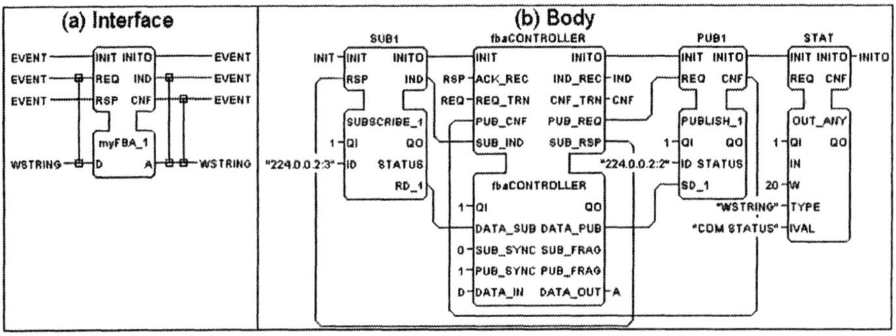

Figure 6 – Composite Function Block Adapter in IEC 61499

The FBA controller (fbaCONTROLLER) carries out the same basic functionality as the IEC 61131-3 FBA; the publisher/subscriber pair consists of two standard IEC 61499 SIFB's (SUB1 and PUB1) that in this case access Ethernet communication services. For agent-to-function block communication, the Ethernet protocol is sufficient in most cases. However, for function block-to-function block communication, a deterministic communications protocol is more appropriate as will be discussed in the next subsection.

3.2 Communication Protocols

Like other safety-critical systems, holonic systems at the device level inhabit an environment where incorrect operation can result in the harm of personnel and/or equipment (Storey, 1996). In a real-time distributed system, the overall integrity of the system is tightly linked to the integrity of the communication network. The suitability of a specific protocol for safety-critical applications must consider a wide range of issues such as redundancy, data validation, fault isolation, and timing. At the device level, or the level of inter-HCD communication, it is important to be able to guarantee the delivery of messages. As a result, a real-time embedded system protocol such as TTCAN (Marsh, 2003), FTT-CAN (Ferreira et al., 2001), TTP/C (Marsh, 2003), Byteflight (Kopetz, 2001), or FlexRay (Kopetz, 2001) is appropriate at this level.

Real-time protocols fall into two main categories: event-based and time-based protocols. Much of the discussion about choosing a protocol begins with the assumption that time-triggered protocols are the only ones suited to safety-critical applications. This assumption is based on the belief that time-triggered schemes are deterministic (higher degree of predictability) and event-based schemes are not (Claesson et al., 2003). For example, it is argued that it is not possible to predict the latency of event-based systems because of the uncertainties involved with arbitration. Another way to state this is that in an event-based system, the latency of messages changes depending on the volume of network traffic. This variation introduces a sense of uncertainty that some claim cannot be tolerated in a safety-critical environment. On the other hand, a purely time-triggered system will always have the same delivery delay times, bringing a sense of certainty to the network.

However, given the event-based model described in the previous section (i.e., IEC 61499), an event-based communication protocol would provide a closer match. Traditionally the uncertainty in message delivery makes time-triggered the preferred option. However, introducing a priority to an event-based system may be able to address the issue of uncertainty. The literature on safety-critical communication protocols does not include an event-based protocol that employs message priorities to deterministically describe the messaging delays. The authors are currently investigating an alternative approach to existing time-triggered protocols that uses dynamic priority setting (Scarlett et al., 2004). This approach appears very promising, resulting in a protocol that nicely matches the interface implementation described in the previous subsection.

4. CONCLUSIONS

In this paper we have presented two approaches to implementing the low-level interface between the information world (i.e., object/agent systems) and the physical world (i.e., PLC systems). The focus of our work has primarily been on the second approach, which involves the use of a special type of function block (a function block adapter or FBA) that allows unambiguous mapping between both sides. Given the event-based, distributed nature of the IEC 61499 model, this approach appears to be well suited to the notion of a FBA service. In this case, implementing a FBA in IEC 61499 does not require a hybrid function block as it does in IEC 61131-3; instead, the FBA can be thought of as a specific SIFB type.

Our current work in this area is focusing on refining the implementation of holonic control devices. In particular, we are focusing on the issue of inter-HCD communication as noted in section 3.2. Initial simulation results with our proposed event-based, dynamic priority communication protocol have indicated that the protocol is very flexible and result in real-time performance that is comparable to existing time-based protocols (Scarlett et al., 2004). We are now investigating a physical implementation of this communication protocol using the Systronix aJile Euroboard (SaJe, 2004) platform.

5. REFERENCES

1. Christensen, J. H., HMS/FB Architecture and its Implementation in S.M. Deen (ed.), Agent-Based Manufacturing. Berlin/Heidelberg: Springer-Verlag, 2003, pp. 53-87.
2. Claesson, V., Ekelin, C., Suri, N., (2003) "The event-triggered and time-triggered medium-access methods," In *Proceedings of the IEEE International Symposium on Object-Oriented Real-Time Distributed Computing* (ISORC'03).
3. DeviceNet (2004) ODVA Web Site, http://www.odva.org/.
4. M. Fletcher, R. W. Brennan, and D. H. Norrie, "Design and evaluation of real-time distributed manufacturing control systems using UML Capsules," 7th *International Conference on Object-oriented Information Systems*, Springer-Verlag, pp. 382-386, Calgary, 27-29 August, 2001.
5. Ferreira, Pedreiras, Almeida & Fonseca, "The FTT-CAN protocol for flexibility in safety-critical systems," *IEEE Micro*, pp. 81-92.
6. Foundation for Intelligent Physical Agents (2004) Web Site, http://www.fipa.org/.
7. Heverhagen, T. and Tracht, R. "Implementing function block adapters", *Lecture Notes in Infomatics*, Verlag, pp. 122-134, 2002.
8. IEC TC65/WG6 (2000) Voting Draft – Publicly Available Specification - Function Blocks for Industrial Process-measurement and Control Systems, Part 1-Architecture, International Electrotechnical Commission.
9. Java Agent Development Framework (2004) Web Site, http://sharon.cselt.it/projects/jade/.
10. Kopetz, H., (2001) "A comparison of TTP/C and FlexRay," TU Wien Research Report 2001/10.
11. Lewis, R. (1996) *Programming Industrial Control Systems using IEC 1131-3*, IEE.
12. A. Lyons, "UML for real-time overview," *Technical Report of ObjecTime Ltd*, 1998.
13. Marsh, D., (2003) "Network protocols compete for highway supremacy," *EDN Europe*, pp. 26-38.
14. McFarlane, D. C., and S. Bussman (2000) "Developments in Holonic Production Planning and Control", *International Journal of Production Planning and Control*.
15. Robert Bosch GmbH. (1991). Bosch CAN Specification version 2.0, Retrieved April 8, 2003 from http://www.can.bosch.com/docu/can2spec.pdf
16. SaJe, "Real time native exectution," http://www.systronicx.com/saje/index.html, 2004.
17. Scarlett, J.J., Brennan, R.W., and Norrie, D.H., "A proposed high-integrity communication interface for intelligent real-time control," Submitted to *Intelligent Manufacturing Systems Forum (IMS-Forum)*, 2004.
18. Storey, N. (1996) *Safety-critical Computer Systems*, Addison-Wesley.

3 MAKING A PERFECT 'GIN AND TONIC': MASS-CUSTOMISATION USING HOLONS

Martyn Fletcher
Agent Oriented Software Limited,
Institute for Manufacturing, Mill Lane, Cambridge CB2 1RX, UK.
Email: martyn.fletcher@agent-software.co.uk

Michal PĚCHOUČEK
Department of Cybernetics, Faculty of Electrical Engineering, Czech Technical University in Prague, Karlovo Namestri 13, 121 35 Prague 2, CZECH REPUBLIC.
Email: pechouc@labe.felk.cvut.cz

The paper presents a model of mass-customisation in manufacturing based on designing and deploying intelligent software agents. We illustrate how this mass-customisation would work in a novel scenario – making a perfect 'Gin and Tonic'. We also discuss some of the benefits this balanced approach can offer businesses in terms of pragmatic holonic software engineering within complex environments and a formal representation of holon operations to academia.

1. INTRODUCTION

The emergence of consumers needing specialised products and services tailored to their particular requirements has resulted in manufacturing companies having to exert greater control over how their product families are configured, presented and delivered. We focus on a particular domain of such personalisation of products, namely *mass-customisation* because it highlights the facilities needed by a manufacturing business to re-organise its shop-floor and supply chain. In the context of this paper, mass-customisation is the customisation and personalisation of manufactured products and services for individual customers at a mass production price. Currently available models for customising how a product can be configured and its presentation altered focus on ensuring that artefacts are manufactured with sufficient generality in a single organization and rely on a central configuration station (often manual) at the end of the production line that can refine the product appropriately. Yet this approach is not true mass-customisation as the factory still produces batches of products that are to be sold to specific retail outlets, which are then beholden to undertake focussed marketing efforts to sell the goods.

A finer-grain mass-customisation model will enable an individual person to issue a unique configuration, possibly via the Internet, of how they want their product to look and feel. Furthermore they do not want to wait long lead times for delivery. This type of mass-customisation is finding its way into factories of various manufacturing domains, such as the envisaged 5-day car or the responsive packing of personal grooming products. In both these environments, the customer selects how they want their intended purchase to be configured, for example in the case of a car purchase system, a user might specify "I want a car with a 3.2 litre engine, 6-speed manual gearbox, painted midnight blue and with a particular style of CD player installed". A key point concerning these existing models for mass-customisation is that they focus on the assembly of sub-components and that the user only has the capability to select which component they wish installed into their product. In this paper we propose a model of mass-customisation that offers the customer the capability to decide how a product is made based on the combination of non-discrete sub-components that can be assembled to meet the user's unique needs. An industrial example where such customisation would be of significant benefit is the process industry. Here batches of chemicals are combined and processed in specific ways to make a final chemical that suits the needs of the customer who placed the order.

Within the scope of this paper, we choose a more light-hearted case study, namely a small-scale manufacturing and robotic system that could be built into a 'themed' pub or cocktail bar. This system lets the customer select how they want a 'Gin and Tonic' drink be made for their personal taste. The drink is assembled with the customer selecting the type of glass, the volume of ice, the volume of Gin (of which they may be several varieties to choose from), and the proportion of Tonic water to be added. The finished drink would then be delivered to their table using a shuttle-based transportation system – ready for the person to enjoy!

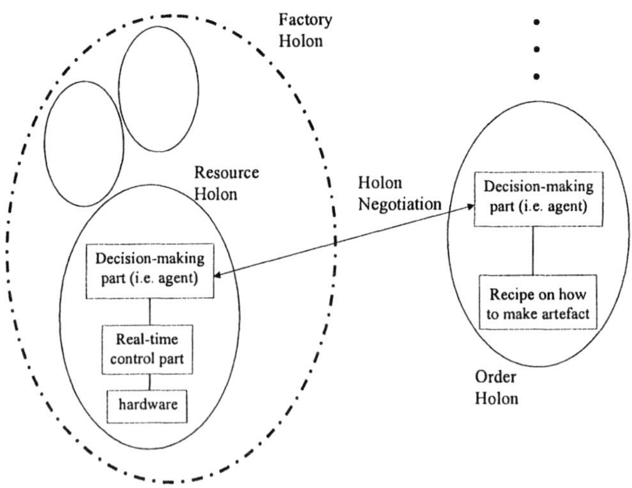

Figure 1 – Control of a Holonic System with Intelligent Software Agents

The technology we intend to use to construct the control system for this 'Gin and Tonic' maker is a new generation of Intelligent Manufacturing Systems (IMS) called *holonics*. Holonics uses intelligent software agents (Figure 1) to control how distributed and real-time processes are executed and coordinated. We will use the SMART formal framework for Agency and Autonomy of (Luck and d'Inverno, 2003) to design and deploy our agent-based holons. The paper is structured as follows. In section 2, we review relevant literature on holonic manufacturing for mass-customisation. Section 3 presents our case study for using mass-customisation, namely the perfect 'Gin and Tonic' maker environment. Section 4 presents a model of the holonic system to control the 'Gin and Tonic' maker. This model is based of the SMART approach for designing holons and their interactions in terms of intelligent software agents. Some conclusions are presented in section 5.

2. LITERATURE REVIEW

In traditional manufacturing environments (both at the internal factory level and at the entire supply chain management level), having customisation of product families and making low cost goods has been considered to be mutually exclusive. Mass production provides low cost artefacts but at the expense of uniformity. As (Davis, 1996) highlights, customisation of products was in only the realm of designers and craftsman. The expense generally made it the preserve of the rich. For example if you wanted a suit of clothes made, then you can either have an 'off the peg' suit but if you want clothes that are made to your specific body measurements and made from the desired material then you need a skilled tailor who is often rather costly. Today, new interactive technologies like the Internet, allow customers and retailers to interact with a manufacturing company to specify their unique requirements that are then to be manufactured by automated and robotic systems.

To clarify by an example, existing car assembly plants usually build batches of the same car, leave them on the car lot and try to sell them by aggressive marketing. In a factory geared to mass-customisation of discrete part assembly, people would select the exact specifications of their product, e.g. a car in terms of all configurable options (paint colour, leather seats etc). Then the entire production line, containing a variety of entities (e.g. assembly cells, inspection stations, automated guided vehicles and so on) would reconfigure themselves to build this specific product. The car can then be delivered to the person in a few days of asking. This reconfiguration of machines, re-planning, re-scheduling and handling faults are very difficult to achieve in current factories even if they are geared towards simple forms of 'option-based' mass-customisation. Examples of mass-customisation in the beverage industry are very limited: it is usually the case that the brewers decide how a mixed drink should look and then market this style. For instance Smirnoff mixes a given amount of vodka with a fixed volume of citrus juice and markets it under the name Smirnoff Ice™. Yet everyone is different and so someone might want a different mix of vodka and juice, which is rather difficult for large-scale brewers to make. Such mass-customisation must also operate within the scope of 21st century factories (or pubs) where customisation can occur not just at the assembly stage but also throughout the entire manufacturing process.

Holonic manufacturing systems are a particular variety of IMS based on the ideas of (Koestler, 1967) that many natural and man-made organisations are more flexible to changes when they are inhabited by stable intermediately entities. In a production context, these entities (called holons) need to act autonomously and cooperatively to ensure the overall organisation is more robust, responsive and efficient than today's manufacturing systems can offer. HMSs are recursive in their construction, with each holon having the option to contain sub-holons and combining real-time control with artificial intelligence to manage low-volume high-variety manufacturing processes. Also FIPA has provided templates for how agents should communicate and how multi-agent systems should be managed. A significant part of their standards effort has related to using the "Belief, Desire, Intention" (BDI) model of rational agents. Beliefs model the world state and are obtained from continuous, imprecise and incomplete perceptions. As the agent's specific purposes may change over time, it needs to know its own objectives and desires. When trying to achieve these goals, the agent must create a sequence of actions that cannot be changed as often as the environment changes. Thus the overall system needs to be committed (i.e. have an intention) to execute a certain sequence.

However it should be noted this architecture has received little attention in industry and is yet to prove itself in real-world HMS scenarios where mass-customisation demands that high quality user interfaces, system agility and robustness are paramount (Mařik, Fletcher and Pěchouček, 2002).

3. THE PERFECT 'GIN AND TONIC'

This section describes our case study of how holonic mass-customisation will operate in terms of a manufacturing environment to make and deliver a perfect 'Gin and Tonic' for each customer in a bar. The physical environment is characterised by:

- Customers sit on bar stools next to drinking stations on the bar. Each station has a touch-sensitive screen displaying an Internet web-page so that consumers can specify how their drink should be made (e.g. set relative proportions of Gin).
- At the drinking station, there is also a Radio Frequency Identification (RFID) reader that can read the identity of a tag embedded in the glass the consumer is drinking from. The station also has a sensor to detect how full the glass is, in order to make recommendations about when to purchase another drink.
- A MonTrack™ conveyor system runs the length of the bar upon which independent shuttles move along. These shuttles carry the consumer's drink through using a flexible fixture that can adapt to the size and shape of the glass being transported. A shuttle can stop at either of the two drink assembly cells in order that the drink can be made, or at any drinking station so that the appropriate customer can take their drink. Only when the glass reaches the consumer who ordered the drink will the glass be released. The shuttle can determine that it is at the correct drinking station because it also carries a RFID reader and stops when it reads a tagged glass that the customer is currently using. This means that a customer can move freely between drinking stations (say because a pretty girl at the other end of the bar invites him for a chat).
- There are two drink assembly cells, each with a docking station to firmly hold the shuttle. The first is dedicated to selecting the correct glass type from storage and placing ice into the glass. The cell can also pour any measure of two different types of Gin into the glass. The second cell has access to the same two bottles of Gin (which are located on a turntable) and can also pour from three bottles of specialist Gin. Only cell 2 can add Tonic water into the glass.

- To achieve this functionality, each cell has an anthropomorphic robot (possibly a Fanuc M6i) with a flexible end-effecter that can pick up and pour either the Gin or the Tonic water out of the correct bottles. Each bottle has a RFID tag on it and the end-effecter has a reader so the bottles can be placed anywhere inside the robot's working envelope and it can still determine the correct bottle.

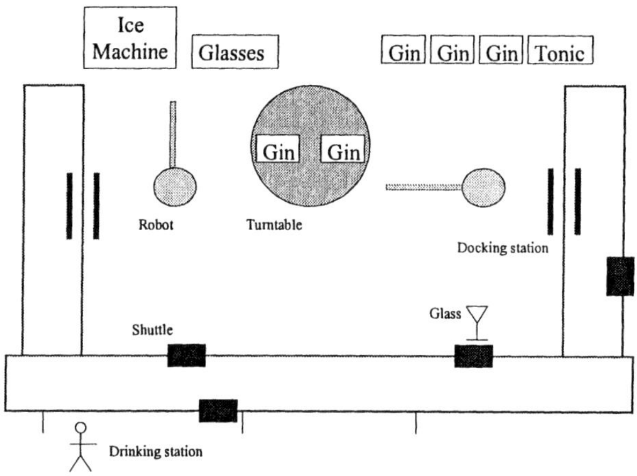

Figure 2 – Schematic Layout of Physical Environment

The layout of the perfect 'Gin and Tonic' environment is shown in Figure 2. The operations by the system for mass-customisation are:

- The consumer sits at a drinking station and specifies the configuration of their drink. If they are very thirsty then they may wish to indicate that they want the drink quickly and are willing to pay some more money for the privilege of speedy delivery. This amount of money is used by the holons in their negotiations and will be deducted from the consumer's credit card when the drink arrives. Agreeing the amount of money to be spent is needed because the consumer is not getting a 'standard' measure of Gin but rather the precise number of millilitres that he/she wants.
- An order holon (modelled using a software agent) is created to ensure that the drink is made correctly and delivered to the customer on time and to budget.
- The order holon interacts with the necessary resource holons in the system (also modelled as agents) to satisfy the goals within the recipe associated with making the drink. The generic recipe for making a perfect 'Gin and Tonic' is: (i) reserve the services of a shuttle to transport the drink around the system, (ii) select the correct type of glass and put it on top of the shuttle, (iii) add the correct volume of ice into the glass, (iv) add the correct type and volume of one or more Gins, and (v) add the correct amount of Tonic water.
- The drink is then delivered to correct drinking station where the customer is now sitting (maybe different from where he/she placed the order) using the RFID tags on the customer's glass for recognition.
- The information, in a local XML database, associated with the unique RFID tag attached to the glass is updated to reflect how the drink has been made and to

whom it belongs. Using this information, customer profiles can be created to better market the drinks and also to aid the bar's replenishment of used bottles.
- If the sensor at a drinking station determines that a drink is nearly finished then the customer is offered another drink (possibly at a promotional price).

We now demonstrate that the SMART (d'Inverno and Luck, 2001) approach can bring significant benefits to our modelling of the agent-based holons and their interactions in controlling the perfect 'Gin and Tonic' making environment.

4. SMART

4.1 Overview of SMART

The richness of the agent metaphor has lead to many different uses of the term and has caused a situation where there is no commonly accepted notion of what constitutes an agent. In response, Luck and d'Inverno have developed the SMART agent framework to unambiguously and precisely provide meanings for common agent concepts and terms. SMART enables alternative models of particular classes of agent-based system to be described, and provides a foundation for subsequent development of increasingly more refined agent-oriented concepts, such as holonics. The SMART approach does not exclude (through rigid definition) any particular class of agent. Rather it provides a means to relate different classes of components within an agent-oriented system, e.g. the holonic control system for our 'Gin and Tonic' making environment. The SMART process is as follows. Initially, the software designer must describe the physical environment and then, through increasingly detailed description, define the software components within the control system to manage this environment. These components are arranged into a four-tiered hierarchy comprising entities, objects, agents and autonomous agents (agents that established their own goals through motivations). These classes constitute SMART's view of the world. For our purposes, the aim of the SMART approach is to construct a formal framework for the components in the holonic control system and their interactions, using formal notation such as Z, which is independent of the agent architecture used to implement these agent-based holons. For an introduction to the Z formalism, readers are referred to www.zuser.org/z/

4.2 Designing Holons using SMART

As stated above, the SMART framework reflects the complex view of the world held by an agent-founded control system in terms of components of varying degrees of functionality. To formally model these components, a language like Z can be used so each component is represented as a schema and is included by other components. In Luck and d'Inverno's model, there are separate schemas for action, perception and state for each of the four component layers. In our refined model, we add a fifth layer to the component hierarchy, namely that of a holon because a holon refines the functionality of an autonomous agent in order to be cooperative and recursive. Hence there are Z schemas to represent *HolonAction*, *HolonPerception* and *HolonState* as shown in Figure 3. We refer interested readers to (Luck and d'Inverno, 2003) for a full description of how component schemas are defined. We focus on the new schemas using that style.

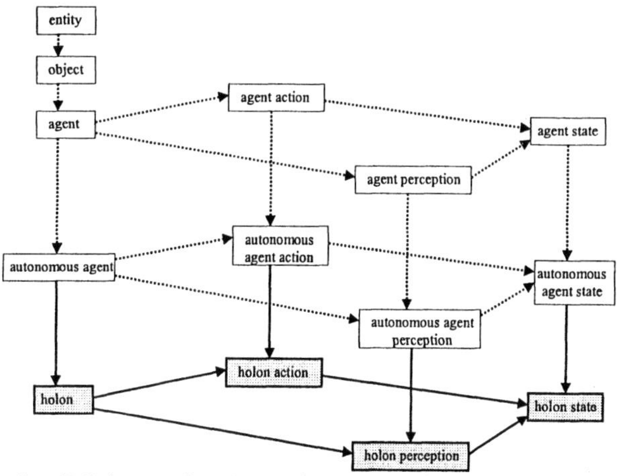

Figure 3 – Z Schemas in a Formal Framework for Modelling Holons

4.2.1 Holon Specification

We begin our specification of holons by introducing *roles*.

Definition: A role is a distinct entity, which contains a description of the relationship, and facilities that the participants in a team/sub-team (or holon / subordinate holon) relationship must provide. The role relationship is expressed in terms of the motivations and belief exchanges implied by the relationship. A role will lead to the autonomous generation of motivations by the holon and will impact the holon's behaviour and reasoning in order to address these motivations.

Like the other Z aspects of the formal model, the type of the role is described using a given set, [Role] as follows. The rows show how holons build upon the schemas of autonomous agents, agents, objects and entities. We have included columns for the order holon and a robot holon (an essential resource holon).

Schema	**Variable**	**Order**	**Robot**
holon	roles	{order management}	{material handling}
autonomous agent	motivations	{achievement, delivery}	{achievement, utilisation}
agent	goals	{acquire shuttle, get glass, get ice, get Gin, get Tonic}	{load glass, insert ice, poor Gin, poor Tonic}
object	capabilities	{interact with resources, use recipe}	{lift glass, hold bottle}
entity	attributes	{glass type, Gin volume, ice volume, Tonic volume}	{yellow, stationary, heavy}

A holon can now be defined.

Definition: A holon is an autonomous agent with a non-empty set of roles. It is specified simply as:

```
— Holon ————————————————————
| Autonomous Agent
| role : P  Role
|————————
| role != {}
|————————————————————————————
```

Where P is the Z syntax for a power set containing, in this case, the roles that any holon might perform. To illustrate these principles, consider a shuttle carrying a drink: the shuttle cannot be considered a holon because, while it may have the ability to determine its own motivations (such as wanting to take the optimal route or wanting to go to a repair in case of damage), it does not have the ability to define how its motivations fit into the roles of the overall system. In this respect, it relies on other holons (i.e. a Track Manager holon) for purposeful existence. However the robot is a holon because it has the need to be recursive and cooperative through a role, and has the ability to generate internal goals in order to satisfy a role. Suppose a role for the robot is material handling. In normal operations, the robot will generate motivations (and in turn create internal goals) for achievement (related to making drinks in the bar) and utilisation (related to ensuring it is working to maximise its throughput). These motivations can be decomposed, recursively, to motivations for each of six independently controlled joints/axes that give the robot its degrees or freedom, and these must be coordinated to make the drink. The robot will create motivations for its joints to make the requested drinks, but if it recognises that the schedule of operations is not optimal then it will generate the utilisation motivation to determine a better sequence of work. It could also recognise that if works for some long duration on a certain type of task then its performance could degrade and so it abandon this achievement motivation and generate a new motivation to compensate for this reduced performance. Such a robot is a holon because its motivations are not imposed, but are generated dynamically in response to its environment and roles.

4.2.2 Holon Perception, Action and State

Goals, motivations and roles are relevant to determining what a holon perceives in the environment, which can be independent of its roles etc. Therefore the schema below specifies a holon's perception as a modified version of what the underlying autonomous agent perceives schema to reflect these extensions. A holon will also have some mechanisms to determine its actions and behaviour with respect to the environment and its roles.

```
— HolonPerception ————————————————
Holon
AutonomousAgentPerception
holonperceives : P Role -> P Motivation -> P Goal -> Env -> View
——————
dom holonperceives ={roles}
————————————————————————
```

The action selection functions of a holon is a refinement of the AutonomousAgentAction scheme of an autonomous agent and one is produced every time a role is used to transfer knowledge (messages) among holons. The state of a holon is defined in terms of the state of an autonomous agent. Changes to this state are as a result of its roles, motivations, goals, perceptions and the environment. For brevity we have not included the HolonAction and HolonState schemes, we just point out that they have a similar structure (with addition of roles) to the schemes AutonomousAgentAction and AutonomousAgentState respectively. Finally we specify how a holon performs its next set of actions as a refinement of the

AutonomousAgentInteracts schema. These extensions to the agent hierarchy help us to formally define and describe how holons act autonomously, cooperatively and recursively in an unambiguous manner.

4.3 Applying the Formalism

Again consider our 'Gin and Tonic' environment. An order holon has a role that demands it gains the services of the resource holons (including the robot) and within which, the order holon has a motivation to make a drink for a consumer that satisfies his/her taste and incurs minimal cost. Meanwhile the robot has a role of making drinks and, within the scope of that role, has a motivation to produce as many drinks as possible per hour. Hence an *interaction* to achieve cooperative scheduling is needed to resolve this potential motivational conflict. An interaction is composed of two motivations and an interaction content that is designed to migrate the two holons from original states in their respective motivations to a pair of destination states. Therefore a dynamic holarchy (in other words a temporary coalition among holons with some prescribed organisation) is composed of a set of holons' motivations and their interactions. All interactions must only take places between two holons' motivations belonging to the same holarchy. Within the scope of such an interaction, the process may follow any style of agent-oriented collaboration metaphors, such as the phases of the classic Contract Net Protocol. From this starting point, we can observe that the Z formalism applied by the order holon during the Announcement phase of its MakeMeDrink motivation (interacting with the Robot holon's PourDrink motivation) is as shown below.

```
— Holon Interaction, Type: Contract Net ————————————

Holon: Robot; Motivation: PourDrink && Holon: Order; Motivation: MakeMeDrink

  — Phase1: Announce ————————————————————
    — Drink Specification ————————————

      Gin Type #1: 22 mili-litres
      Gin Type #2: 7 mili-litres
      Tonic: 73 mili-litres
      Ice: true
      Glass Type: tumbler
    ————————————————————————————————
    — Customer ——————————————————————
      Person Identity: Joe Bloggs
      Last Drink RFID Tag: 02453003218003628200273O
      Cost and Deadline: £1.73 @ Tuesday 9 December 2003, 12:45
    ————————————————————————————————
  ————
    — Process ———————————————————————
      for each Robot
        (request(:sender (:name Order) :receiver (set (:name Robot)))
        :content ((action (:name Robot) (pour_drink(drink_specification,customer))))
        :protocol fipa-call_for_proposal :language FIPA-SL :conversation-id order123)
    ————————————————————————————————
  ——————————————————————————————————
————————————————————————————————————
```

We can now exert balance and look at the case for pragmatic businesses to adopt the holonic vision. Merits of this agent-based holonic approach to mass-customisation include the opportunity for global optimisation of the customisation processes within the manufacturing business could be accomplished through this model. There are multiple criteria upon which a factory configuration can be judged and so optimised upon. For example, minimizing mean delivery time of a certain class of specially parts, maximizing the number of competing cells that can supply a part (i.e. giving alternatives if one cell cannot provide the requisite part), and minimizing the volume of parts stored. The agent-based holons provide means for such multi-criteria optimisation via mediation and so forth within their interactions Fault-tolerance and reliability are two criteria essential for any pragmatic mass-customisation. The dynamic agility that the holons' intelligent software agents have provides a solid foundation for the development of robust supply chains with supplier enterprises that can offer customised goods quickly to meet customer-specific orders. Moreover, by having holons use decentralised control, the system as a whole displays graceful degradation in the face of hardware failures, rather than complete collapse. This means that the time to deliver a customised product is kept short to maintain customer satisfaction. We now make some concluding statements.

5. CONCLUDING REMARKS

We have outlined the key features of a 'continuous' mass-customisation model for manufacturing using the holonic approach. The design and deployment of these holons is based on applying the SMART framework to build holons using intelligent software agents. We have also discussed several issues associated with how some typical holons will operate in a novel mass-customisation environment that makes a 'Gin and Tonic' to satisfy customers' unique requirements for their perfectly-combined drink. As businesses increasingly shift their emphasis towards high-variety low-volume production to meet the ever-changing demands of people for customized goods, management of the businesses resources via agent-based holons with distributed control are the logical consequence. Clearly there is incentive for businesses to introduce holonic and mass-customisation ideas onto their shop-floors, supply chains, and even into the odd pub, in order to meet the ever-growing demand for customised products. Moreover competition between businesses to manufacture such customized goods cost-effectively, balanced with the academic value provided from a formal Z-based framework, will make the arrival of holonics imminent. Future research will focus on evaluating the model, i.e. how scaleable it is.

6. REFERENCES

1. S. Davis, *Future Perfect, 10th anniversary edition*, Addison-Wesley, ISBN 020159045X, 1996.
2. M. d'Inverno and M. Luck, *Understanding Agent Systems*, Springer, 2001.
3. HMS Project, http://hms.ifw.uni-hannover.de/public/overview.html , 2003.
4. A. Koestler, *The Ghost in the Machine,* Arkana, 1967.
5. M. Luck and M. d'Inverno, Unifying Agent Systems, *Annals of Mathematics and Artificial Intelligence*, 37(1-2), 2003.
6. V. Mařik, M. Fletcher and M. Pěchouček, Holons and Agents: Recent Developments and Mutual Impacts, in *Multi-Agent Systems and Applications II*, Springer, 2002.

4 HOLONIC MANUFACTURING CONTROL: A PRACTICAL IMPLEMENTATION

Paulo Leitão[1], Francisco Casais[1], Francisco Restivo[2]
[1]*Polytechnic Institute of Bragança, Quinta Sta Apolónia, Apartado 134, 5301-857 Bragança, PORTUGAL, {pleitao,fcasais}@ipb.pt*
[2]*Faculty of Engineering of University of Porto, Rua Dr. Roberto Frias, 4200-465 Porto, PORTUGAL, fjr@fe.up.pt*

The ADACOR holonic architecture for manufacturing control addresses the agile reaction to unexpected disturbances at the shop floor level, by introducing supervisor entities in decentralised systems characterised by the self-organisation capabilities associated to each ADACOR holon. The result is an adaptive control architecture that balances dynamically between a more centralised structure and a more decentralised one, allowing the combination of global production optimisation with agile reaction to unexpected disturbances. The validation of the proposed architecture is required to verify the correctness and the applicability of its concepts. This paper describes the implementation of ADACOR concepts using multi-agent systems, especially through the use of the JADE agent development platform.

1. INTRODUCTION

Companies, to remain competitive, need to answer more closely to the customer demands, by improving their flexibility and agility while maintaining their productivity and quality. The traditional manufacturing control systems respond weakly to the emergent challenges faced by the manufacturing systems, given their poor capability to adapt with agility to unexpected internal disturbances and to external environment volatility. This weakness is mainly due to the rigidity of the current control architectures.

Several manufacturing control architectures using emergent paradigms and technologies, such as multi-agent and holonic manufacturing systems, have been proposed (see [1-5]). One of the proposed holonic architecture is the ADACOR (ADAptive holonic COntrol aRchitecture for distributed manufacturing systems) architecture [6], which addresses the agile reaction to disturbances at the shop floor level, increasing the agility and flexibility of the enterprise.

ADACOR architecture is build upon a set of autonomous and cooperative holons, each one being a representation of a manufacturing component that can be either a physical resource (numerical control machines, robots, pallets, etc.) or a logic entity (products, orders, etc.). The holon is a concept first introduced by Koestler [7] to represent the interactions in social organisations, later introduced in manufacturing by the HMS consortium (see http://hms.ifw.uni-hannover.de/).

A generic ADACOR holon comprises the Logical Control Device (LCD) and, if exists, the physical resource, capable to perform the manufacturing task. The LCD is responsible for regulating all activities related to the holon and comprises three main components: decision, communication and physical interface [8]. In ADACOR agents are used to implement the logical part of the holon, i.e. the LCD device.

ADACOR architecture defines four manufacturing holon classes: product, task, operational and supervisor. The product, task and operational holons are quite similar to the product, order and resource holons defined in PROSA reference architecture [2], while the supervisor holon presents characteristics noot found in the PROSA staff holon. The supervisor holon introduces coordination and global optimisation in decentralised control and is responsible for the formation and coordination of groups of holons.

The ADACOR adaptive production control approach is neither completely decentralised nor hierarchical, but balances between a more centralised approach to a more flat approach, passing through other intermediate forms of control [6], due to the self-organisation capability associated to each ADACOR holon, translated in the autonomy factor and in the propagation mechanisms [8]. For this purpose, ADACOR evolves in time between two alternative states: the stationary state, where the system control relies on supervisors and coordination levels to achieve global optimisation of the production process, and the transient state, triggered with the occurrence of disturbances and presenting a behaviour quite similar to the heterarchical architectures in terms of agility and adaptability.

The validation of these concepts requires their implementation and testing, to analyse their correctness and applicability. This paper describes the implementation of ADACOR concepts, at the Laboratory of Automation in the Polytechnic Institute of Bragança, Portugal, to verify their applicability and if the system works as specified, either in normal operation or in presence of disturbances.

Along the paper, the implementation of the behaviour of each ADACOR holon class, communication infra-structure, manufacturing ontology, decision-making mechanisms, graphical user interfaces, customisation of manufacturing holons and connection between the holonic control system and the physical manufacturing devices will be described.

2. AGENT DEVELOPMENT PLATFORM

The development of holonic manufacturing control systems based in the ADACOR architecture requires the previous implementation of their concepts in a prototype.

The ADACOR prototype uses agent technology to implement each holon. Multi-agent systems can be adequately developed using object-oriented languages, such as Java. However, the development of multi-agent systems requires the implementation of features usually not supported by programming languages, such as message transport, encoding and parsing, yellow and white pages services, ontologies for common understanding and agent life-cycle management services, which increases the programming effort. The use of agent development platforms which implement these features makes the development of agent-based applications easier and reduces the programming effort.

A significant set of platforms environments for agent development is available for commercial and scientific purposes, providing a variety of services and agent models, which differences reflect the philosophy and the target problems envisaged by the platform developers. Surveys of some agent development platforms can be found in [9-10].

The choice of an agent development platform obeyed to a set of criteria: to be an open source platform with good documentation and available support, ease to use, low programming effort, use of standards, features to support the management of agent communities like white pages and/or yellow pages and facilities to implement rule oriented programming.

The chosen platform was JADE (Java Agent Development Framework), provided by CSELT and available on http://jade.cselt.it/, because it responds better to the mentioned requirements. In fact, JADE simplifies the development of multi-agent systems by providing a set of system services and agents in compliance with the FIPA (Foundation for Intelligent Physical Agents) specifications: naming, yellow-page, message transport and parsing services, and a library of FIPA interaction protocols [11]. JADE uses the concept of *behaviours* to model concurrent tasks in agent programming and all agent communication is performed through message passing, using FIPA-ACL as the agent communication language. JADE provides the FIPA SL (Semantic Language) content language and the agent management ontology, as well as the support for user-defined content languages, which can be implemented, registered, and automatically used by the agents.

The agent platform provides a Graphical User Interface (GUI) for the remote management, allowing to monitor and control the status of agents, for example to stop and re-start agents, and also a set of graphical tools to support the debugging phase, usually quite complex in distributed systems, such as the Dummy, Sniffer and Introspector agents.

JADE offers also an easy and full integration with other useful tools, such as JESS (Java Expert System Shell) and Protegé 2000 (for knowledge based system development and management), and provides other features such as an active mailing list to support technical problems.

3. GENERAL ADACOR IMPLEMENTATION

In this section, the main issues related to the implementation of the ADACOR architecture prototype using the JADE framework will be described.

3.1 Internal Architecture of an ADACOR Holon

An ADACOR holon is a simple Java class that extends the **Agent** class provided by the JADE framework, inheriting its basic functionalities, such as registration services, remote management and sending/receiving ACL messages [11]. These basic functionalities were extended with features that represent the specific behaviour of the ADACOR holon, like the *HandleReceiveMessages*, *AllocateTask* and many other behaviours.

The start-up of an ADACOR holon comprises its initialisation (read the configuration files and load the behaviours) and its registration according to the initial organisational structure, defined by the configuration files, and is followed by the actual start-up of the holon's components, i.e. the communication, decision and physical interface components.

The behaviour of each ADACOR holon uses multi-threading programming, over the concept of JADE's behaviour, allowing to execute several actions in parallel. The behaviours launched at the start-up and those which can be invoked afterwards are provided in the form of Java classes.

The communication between holons is done over the Ethernet network, using TCP/IP protocol and is asynchronous, i.e. the holon that sends a message continues the execution of its tasks without the need to wait for the response. The messages specified in the ADACOR architecture are encoded using the FIPA-ACL agent communication language, the content of the messages being formatted according to the FIPA-SL0 language. The meaning of the message content is standardised according to the ADACOR ontology.

One of the holon's concurrent tasks (behaviour) waits continuously for the arrival of messages using the *block()* method to block the behaviour until a message arrives. The arrival of a message triggers a new behaviour to handle the message (the *HandleReceiveMessages* behaviour), thus implementing an asynchronous communication mechanism over the JADE platform.

Each supervisor holon has embodied a DF (Directory Facilitator) that provides yellow pages functionalities, allowing to locate holons within its group by their capabilities.

3.2 ADACOR Ontology

ADACOR defines its own manufacturing control ontology, expressed in an object-oriented frame-based manner, as recommended by FIPA Ontology Service Recommendations (see http://www.fipa.org/). This recommendation refers to the development of classes describing concepts and predicates, and their registration as a part of the application ontology, allowing a practical and fast way of creating an ontology with an immediate underlying implementation.

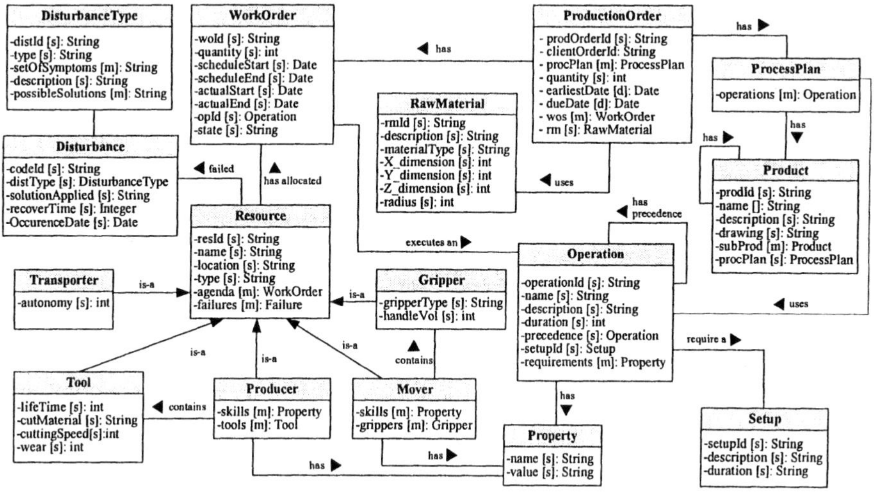

Figure 1 – ADACOR Ontology for Manufacturing Control

The manufacturing control ontology used in ADACOR is based in the definition of a taxonomy of manufacturing components, which contributes to the formalisation and understanding of the manufacturing control problem. These components are mapped in a set of objects, illustrated in the Figure 1, which defines the vocabulary used by distributed entities over the ADACOR platform, and indicates the concepts (classes), the predicates (relation between the classes), the terms (attributes of each class), and the meaning of each term (type of each attribute).

The ADACOR ontology was translated to Java classes according to the JADE guidelines that follow the FIPA specifications for the development of ontologies. The main class of the ADACOR ontology describes the concepts and predicates defined in the ontology, indicating its ontological role. Each ontological role is characterised by a name and a structure defined in terms of a number of slots that represent the attributes of the concept or predicate.

Instances of the ontological roles can be conveniently represented inside an agent as instances of application-specific Java classes each one r epresenting a role. The methods defined in each individual class used to describe each concept or predicate, allow to handle the data related to the object.

3.3 Decision-Making Mechanisms

The ADACOR decision component, illustrated in Figure 2, uses declarative and procedural approaches to represent knowledge and to regulate the holons behaviour. The knowledge base of each ADACOR holon is dependent of its type, objectives, skills and behaviour.

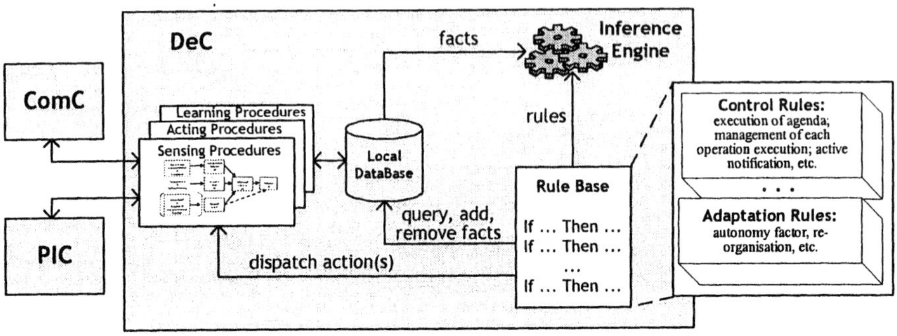

Figure 2 – Decision Component Architecture

The central element in the decision component is the rule-based system, which applies declarative knowledge, expressed in a set of rules. The advantage of this type of knowledge-based system is related to the simple and very comprehensive way to represent the reasoning capability of one holon. The simplicity and the associated high abstraction level of this approach compensates the typical weaknesses of these systems to handle incomplete, incorrect and uncertain information, or to implement complex systems, that require a large number of rules and can become very slow.

The rule-base system uses the JESS tool, which is a rule oriented programming infrastructure (Java based and JADE compatible) developed using the CLIPS (C Language Integrated Production System) language and uses the Rete algorithm as inference engine [12]. JESS handles data structures, functions and rules, requiring the use of a *clp* file to store the application knowledge base.

Each ADACOR holon class has its own *clp* file containing its knowledge base. The decision mechanisms that are common to all the ADACOR holons classes, such as the active notification, are placed in a special common *clp* file. The local database stores the short-term memory, i.e. the facts that represent the current state of the holon at a particular moment.

The set of rules defined in the knowledge base represents the behaviour of the ADACOR holon. As an example, the behaviour rule illustrated in Figure 3 is defined in the implementation of the task holon knowledge base.

```
(defrule Transport    "Will start the transport of the part"
   ?fact1 <- (Transport)
   ?fact2 <- (executing (jobInExecution ?wo))
   ?fact3 <- (WorkOrder (woID ?wo)(state ?state)(resName ?res)
             (location ?location)(precedence ?precedent))
=>
   (retract ?fact1 ?fact3)
   (assert (WorkOrder (woID ?wo)(state TRANSPORT)(resName ?res)
           (location ?location)(precedence ?precedent)))
   (ExecuteTransport ?wo ?*transport-path*)
)
```

Figure 3 – Invoking JADE Procedures from the JESS Environment

This rule has three conditions: the first condition is satisfied if the fact **Transport** is true; the second condition determines the name of the work order that is currently in execution; at last the third condition gets the information related to the operation in execution. When the three conditions are satisfied, the rule is selected and three actions are executed: first, the facts *fact1* and *fact3* are removed from the knowledge base; then the state of the work order is changed to **Transport**; finally, a behaviour in the JADE environment (linked to a simple Java class) that will be in charge to execute the transportation of the part is triggered.

Another type of connection supports the invocation of JESS commands from the JADE environment. This is done by introducing commands lines embedded in the Java program. In Figure 4, the extract of code illustrates the assertion of a new fact in the knowledge base, in this case a new work order.

```
...
decEngine.executeCommand("(assert (WorkOrder (woID " + value1 + ")
     (state " + value2 + ")(precedence " + value3 + "))))");
 ...
```

Figure 4 – Invoking JESS Commands from the JADE Environment

ADACOR holons also use procedural knowledge to represent knowledge. This type of knowledge is embodied in procedures, which are triggered as actions by some rules, each one being responsible for the execution of a particular set of actions. The scheduling algorithm is an example of this type of knowledge representation. Some other procedures are related to the acquisition of information by handling the arrival of messages from other holons or getting local information through the access to the physical manufacturing resource.

3.4 Graphical User Interfaces

In the ADACOR prototype, the operational and supervisor holons have graphical user interfaces to support the interaction with the user, illustrated in Figure 5.

The graphical user interface for the operational holons allows to visualise the local schedule, using a Gantt chart to show the work orders executed by the resource, to configure some operational holon parameters, such as the scheduler type or the activation of the autonomy factor, and to display statistical information related to the resource performance, such as the degree of utilisation, the number of work orders executed and the number of work orders delayed.

The graphical user interface for the supervisor holon allows to visualise the global schedule, using a Gantt chart to show the work orders executed by each lower-level resource, to display the resources under its coordination domain and their characteristics, and to configure some holon parameters, such as the schedule algorithm.

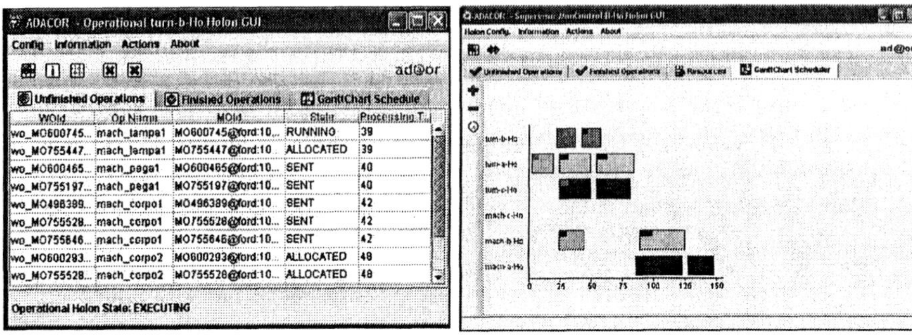

Figure 5 – Graphical User Interface of an Operational and a Supervisor Holon

Each ADACOR holon uses log files to store the relevant data during its life cycle, to support posterior analysis or to execute backup procedures in case of holon crash. In case of operational holons, this log file stores information about rejected, cancelled, failed and executed work orders. The task holon uses a log file to store the relevant information associated to the execution of the production order and to store the statistical report about the performance of the execution of the production order, such as the manufacturing lead time, tardiness, processing time and idle time.

4. APPLICATION SPECIFIC COMPONENTS

The holonification of the manufacturing components requires the configuration and customisation of the ADACOR holons and the development of wrappers that allow the connection between the control system and the physical devices.

4.1 Configuration Files of ADACOR Holons

The characteristics of each manufacturing holon are configured using a XML (eXtensible Markup Language)-based configuration file.

In case of the product holon it is necessary to introduce the product data model and the process plan. The description of the product structure is represented by a list of objects formatted according to a data structure, which includes the name of the sub-part, the number of parts necessary to produce the product and the estimated time to produce the part. The process plan defines the sequence of operations that must be executed to produce the product, containing a list of operations formatted according to an appropriated data structure, which mainly describes the name of the operation, a brief description about the operation, the estimated time to execute the operation, the reference to the part, a list of requirements associated to the operation, and the name of the operations that have precedence over this operation.

The characteristics of each resource are mapped in a XML-based configuration file that will be read by the operational holon. The data structure represents the resource model, which attributes reflect mainly the type of resource, the list of skills that the resource possesses and its location.

The organisational structure of the factory plant is defined in a XML-based configuration file that comprises the information related to the cell organisation and to the shop-floor layout. This organisational structure describes the possible manufacturing cells and associated coordinator entities, which will be converted into supervisor holons. With this organisational structure XML-based file, the operational holons can find their supervisor holons and their auxiliary resources, and the supervisor holons can find the list of holons that are in its coordination domain.

4.2 Physical Resource Interfaces

The implementation of operational holons that represent physical manufacturing resources requires the development of wrapper interfaces, supporting the integration of those resources. In the ADACOR architecture, the virtual resource concept was introduced to make transparent the intra-holon interaction [13].

The development of a virtual resource for each manufacturing device encompasses the implementation of the services at the server side, which will be invoked on the client side (PIC component from the operational holon). The client ignores the details of this implementation and each virtual resource can be re-used by other similar resources or holonic control applications.

Leitão et al. [13] describe the implementation of two different virtual resources to integrate two different automation resources, a PLC and an industrial robot. These two virtual resources implement the same services so that from the client side, whatever the resource is accessed, the invocation made is unique.

Here the virtual resource for an ABB IRB1400 load/unload industrial robot is briefly described, by the illustration of the implementation of the *read* service, as showed in the Figure 6.

```
public int read(String var,String type) {
  String[] vname=new String[1];
  vname[0]=var;
  short[] progNo=new short[1];
  progNo[0]=0;
  short varvalue=0;
  ...
  try {
     varvalue=h.s4ProgramNumVarRead(vname,progNo);
  }catch(IOException ioe) {System.out.println("Problem: " + ioe);}
  return ((new Short(varvalue)).intValue());
}
```

Figure 6 – *Read* Service for the ABB IRB 14000 Virtual Resource

The services provided by the virtual resource were developed using the RobComm ActiveX supplied by ABB [14], and accessed through TCP/IP. The major problem was the access to ActiveX from a Java program, since the ActiveX components are adequate to be manipulated by Windows-based programming environments. To overcome this problem, it was used the Jintegra tool (see http://j-integra.intrinsyc.com/) to convert the ActiveX component into a Java package [13].

The platform used to support the client-server interaction was CORBA. The analysis of the experimental implementation of the resource integration, by comparing the performances of CORBA, RMI and RMI-IIOP, is described in [13].

5. AUXILIARY TOOLS

During the implementation of the ADACOR prototype, a set of auxiliary tools was developed to support the configuration, operation and debugging of manufacturing control applications, providing functionalities to configure products and to supervise the factory plant in an integrated and global view.

As the global visualisation of all activities in the factory plant is difficult due to the use of multiple graphical user interfaces, the ADACOR Factory Plan Supervisor (AFPS) is used to monitor the production activities in the factory plant. Its graphical user interface is represented in Figure 7.

This tool allows to visualise the production process by enabling the visualisation of the manufacturing resources present in the factory plant, indicating their state and characteristics. In this tool, the visualisation of the transport resource has animation capabilities to help understanding of the material flow in the factory plant, indicating the direction of the movement and its actual load.

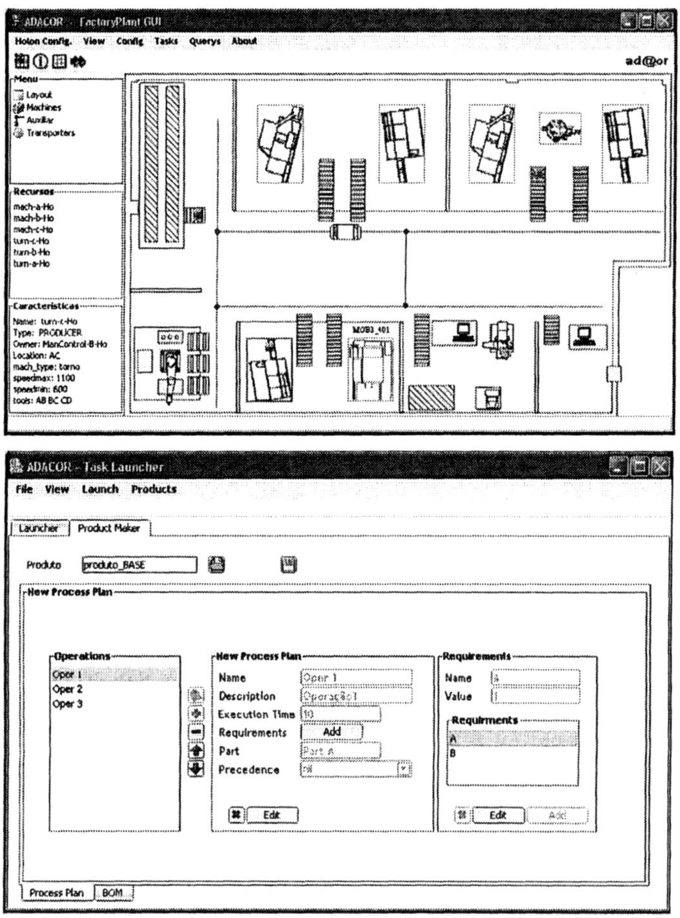

Figure 7 – Graphical User Interface of the Auxiliary ADACOR Agents

The ADACOR Product Manager (APM) agent, which graphical user interface is also represented in the Figure 7, allows defining new products in the system, by introducing the structure of the product and the process plan that defines the sequence of operations to execute the product. It also allows launching individual or pre-defined sequences of production orders to the factory plant, making easier the execution of experimental tests.

6. CONCLUSIONS

This paper described the implementation of the ADACOR holonic manufacturing control architecture concepts into a prototype.

Firstly, the implementation showed the capability of ADACOR architecture to represent and control a real environment. An application to a flexible manufacturing system was completed as a part of the doctoral thesis of an of the authors. The experience gained during the prototype implementation, debugging and testing, allowed proving essentially the applicability of the ADACOR concepts and the merits of the ADACOR collaborative/holonic approach.

The use of agent technology to implement the holonic manufacturing control prototype brings some important benefits: the software necessary to develop the application is simpler to write, to debug and to maintain, due to the smaller size of each distributed component. The use of Java language contributes for the platform independency, which is mandatory in manufacturing environment, due to its heterogeneous environment.

The use of JADE agent development tool brings several advantages in the development of holonic and multi-agent systems, such as the reduction of the development time and complexity. For this fact contributes the good documentation, efficient technical support and the set of functionalities provided by the platform that simplifies the development of multi-agent systems, such as the communication infra-structure, yellow and white pages and debugging tools.

7. REFERENCES

1. Parunak, H. Van Dyke, Baker A. and Clark S. The AARIA Agent Architecture: from Manufacturing Requirements to Agent-based System Design, Workshop on Agent-based Manufacturing, 1998.

2. Van Brussel, H., Wyns J., Valckenaers P., Bongaerts L. and Peeters P. Reference Architecture for Holonic Manufacturing Systems: PROSA, Computers In Industry, vol. 37, 1998, pp. 255-274

3. Brennan, R., Balasubramanian, S. and Norrie, D., A dynamic control architecture for metamorphic control of advanced manufacturing systems. Proceedings of the International Symposium on Intelligent Systems and Advanced Manufacturing, 1997, pp. 213-223.

4. --: Special Issue on Industrial Applications of Multi-Agent and Holonic Systems. Journal of Applied Systems Studies, 2(1),2001.

5. Deen S.M. (ed.): Agent-Based Manufacturing: Advances in the Holonic Approach. Springer Verlag Berlin Heidelberg, 2003.

6. Leitão P. and Restivo F.: Holonic Adaptive Production Control Systems. Proceedings of the 28th Annual Conference of the IEEE Industrial Electronics Society, Sevilla, Spain, 2002, pp. 2968-2973.

7. Koestler, A.: The Ghost in the Machine. Arkana Books, London, 1969.

8. Leitão P. and Restivo F.: Agent-based Holonic Production Control. Proceedings of the 3rd International Workshop on Industrial Applications of Holonic and Multi-Agent Systems (HoloMAS), Aix en Provence, France, 2-6 September, 2002, pp. 589-593.

9. Vrba P.: Java-Based Agent Platform Evaluation. In V. Marík, J. Müller and M. Pechoucek (eds), Multi-Agent Systems and Applications III, Volume 2691 of LNAI, Springer-Verlag, 2003, pp. 47-58.

10. Barata, J., Camarinha-Matos, L., Boissier, R., Leitão, P., Restivo, F. and Raddadi, M.: Integrated and Distributed Manufacturing, a Multi-agent Perspective. Proceedings of 3rd Workshop on European Scientific and Industrial Collaboration, Enschede, Netherlands, 27-29 June 2001, 145-156.

11. Bellifemine F., Caire G., Trucco T. and Rimassa G.: JADE Programmer's Guide. 2002.

12. Friedman-Hill E.J.: JESS, The Java Expert System Shell. Sandia National Laboratories, 1999.

13. Leitão P., Boissier R., Casais F. and Restivo F.: Integration of Automation Resources in Holonic Manufacturing Applications. In V. Marík, D. McFarlane e P. Valckenaers (eds), Holonic and Multi-Agent Systems for Manufacturing, Volume 2744 of LNAI, Springer-Verlag, 2003, pp. 35-46

14. --: RobComm User's Guide, version 3.0/3. ABB Flexible Automation, 1999.

5

CONTINGENCIES-BASED RECONFIGURATION OF HOLONIC CONTROL DEVICES

Scott Olsen, Jason J. Scarlett, Robert W. Brennan, and Douglas H. Norrie
Department of Mechanical and Manufacturing Engineering
University of Calgary, 2500 University Dr. N.W. Calgary, CANADA T2N 1N4

In this paper, we propose a dynamic approach to programmable logic controller (PLC) reconfiguration that is based on the IEC 61499 standard for distributed, real-time control systems. With this form of reconfiguration control, contingencies are made for all possible changes that may occur. We illustrate this approach with a simple system configuration that uses Rockwell Automation's Function Block Development Kit (FBDK) for the software implementation and Dallas Semiconductor's Tiny InterNet Interface (TINI) for the hardware implementation.

1. INTRODUCTION

By definition, "holons" contain both an information processing part and a physical part (Gruver et al., 2001). Moreover, "holonic systems" are essentially adaptive agent-machine systems. As a result, it is not surprising that the Holonic Manufacturing Systems Consortium (2004) has a work group devoted to these software/hardware devices or Holonic Control Devices (HCD). This paper focuses on the "adaptive" aspect of these agent-machine systems. In particular, we investigate how emerging software and hardware technologies can be taken advantage of to create systems at the device level that can dynamically reconfigure themselves in response to changes in the manufacturing environment (e.g., device malfunctions or the addition and/or removal of equipment).

Reconfiguration of conventional industrial controllers such as PLCs (programmable logic controllers) involves a process of first editing the control software offline while the system is running, then committing the change to the running control program. When the change is committed, severe disruptions and instability can occur as a result of high coupling between elements of the control software and inconsistent real-time synchronization. For example, a change to an output statement can cause a chain of unanticipated events to occur throughout a ladder logic program as a result of high coupling between various rungs in the program; a change to a PID (proportional/integral/derivative) function block can result in instability when process or control values are not properly synchronized.

In this paper, we propose a dynamic approach to PLC reconfiguration that is based on the IEC 61499 standard (IEC, 2000) for distributed, real-time control systems. With this form of reconfiguration control, contingencies are made for all possible changes that may occur. In other words, alternate configurations are pre-programmed based on the system designer's understanding of the current configuration, possible faults that may occur, and possible means of recovery. This approach uses pre-defined reconfiguration tables that, in the event of a device failure, allow the affected portions of an application to be moved to different devices by selecting an appropriate reconfiguration table.

The paper begins with an overview of our contingencies-based reconfiguration model. In order to implement this approach, we develop an IEC 61499 based reconfiguration management service that allows function block applications to be reconfigured dynamically (i.e., the management services ensure that the HCD is properly synchronised during reconfiguration). This reconfiguration manager also serves as an interface to higher-level software to enable intelligent reconfiguration (e.g., the use of multiagent techniques to allow the system to reconfigure automatically in response to change) (Brennan and Norrie, 2002).

Next, we illustrate this approach with a simple system configuration that uses Rockwell Automation's Function Block Development Kit (FBDK) (Christensen, 2004) for the software implementation and Dallas Semiconductor's Tiny InterNet Interface (TINI) (Loomis, 2001) for the hardware implementation. The paper concludes with a discussion of on how the reconfiguration process can be managed in a resource-constrained environment (i.e., such as on the TINI board), the limitations of Java for real-time distributed control, and the possibilities for intelligent reconfiguration in this type of system.

2. DESIGNING FOR RECONFIGURATION

2.1 A Contingencies Approaches to Reconfiguration

With this form of reconfiguration control, contingencies are made for all possible changes that may occur. In other words, alternate configurations are pre-programmed based on the system designer's understanding of the current configuration, possible faults that may occur, and possible means of recovery.

This approach uses a library of pre-defined configurations as shown in Figure 1. For example, in the event of a device failure, the affected portions of an application could be moved to different devices by selecting an appropriate configuration. As well, this detailed representation of the function block interconnections would allow higher-level agents to access the information required to make a smooth transition from one configuration to another, thus enabling dynamic reconfiguration.

2.2 An IEC 61499 Based Reconfiguration Management Service

An IEC 61499 based "reconfiguration manager" was developed to address the need for an interface between upper level agents (e.g., scheduling, fault monitoring, configuration) and lower control applications for dynamic reconfiguration of distributed systems. With respect to the manufacturing control system as a whole, external agents responsible for failure monitoring and system wide reconfiguration interact with each other as well as with the reconfiguration manager and the system to be controlled.

The reconfiguration manager must have certain capabilities in order to effectively link the higher-level agents to the lower level machine control. The reconfiguration manager running on the controller must be able to receive messages from agents in order to apply the appropriate control application to the controlled process, as shown in Figure 1. For example, scheduling agents (SA) in charge of scheduling parts to be processed and machine agents responsible for monitoring the status of the processing equipment may send information to a configuration agent (CA) that makes the decision of which control application should be implemented on the controller. Through the I/O capabilities of the controller, the control application may control and/or monitor a process that is accomplishing a specific task.

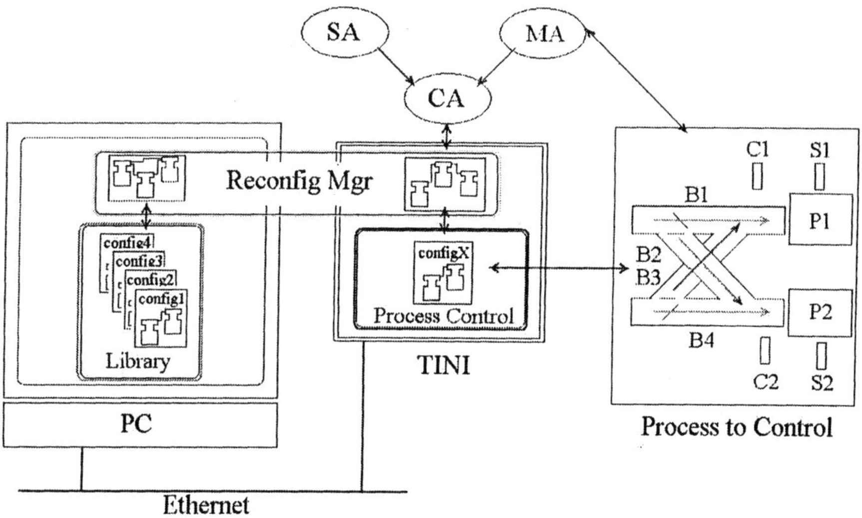

Figure 1 – The reconfiguration manager implementation

The basic function of the reconfiguration manager is to load and unload new control applications on the controller hardware at the request of outside agents. During the transition period when a control application is unloaded and a new one is being loaded, the reconfiguration manager must have the ability to maintain a transitional state for the controlled process to remain in. It should be noted that the control applications must have the built-in ability to save the state information of the controlled process, and also retrieve that state information for the next control application if necessary. These features allow for continuous smooth and stable performance of the controlled process.

Loading a new control application requires access to the library where all of the contingent control applications are stored. Since the control application only defines how function blocks are connected together and the values of any external variables, there must be access to the library where the class files are stored. These libraries may each be located on different hardware platforms. In order to be effective, the reconfiguration manager must perform its function under real-time constraints.

The reconfiguration manager is a distributed application running concurrently on the embedded control platform (i.e., the TINI platform) and the PC as shown in Figure 1. Communication between the PC and TINI is through an Ethernet connection. In the next section, we provide further details on the prototyping environment as well as some basic experiments that were conducted to test this approach to reconfiguration control.

3. EXPERIMENTS

3.1 The Prototyping Environment

The prototyping environment that is currently being used for our experiments with IEC 61499 consists of function block application software and a Java-based control platform.

The software component of our prototyping environment, Function Block Development Kit (FBDK) (Christensen, 2004), was developed by Rockwell Automation primarily for testing IEC 61499 concepts in a simulated environment. For our experiments, a MANAGER function block is used to provide the IEC 61499 device management services on our controller platforms. For example, this function block provides services to allow function block instances to be created, deleted, started, killed, etc. and also provides information on function block status (e.g., READY, INVALID_STATE, OVERFLOW, etc.). This allows distributed applications to be developed using FBDK and run (and managed) on a network of controllers (rather than just simulated on a single PC).

The hardware component of our prototyping environment consists of a Dallas Semiconductors Tiny InterNet Interface (TINI) (Loomis, 2001) board running on a Taylec TutorIO prototyping board. The TINI board includes a DS80C390 microcontroller (an Intel 8051 derivative) that supports JDK 1.2 (Java Development Kit) applications and also supports several forms of I/O such as discrete and analogue I/O, serial, Ethernet, 1-Wire and Controller Area Network (CAN). In order to experiment with the TINI board, the TutorIO board provides interfaces to the I/O (e.g., a two-line LCD, LEDs, etc.).

3.2 The Test Scenarios

The experimental set-up is composed of a conveyor belt system for transporting parts to a manufacturing process. As parts arrive at each process, they are counted so that when a certain number of them have arrived, the conveyor is stopped and the process is applied to a batch of parts. When the process has been completed, a signal is sent to start the conveyor again.

This particular conceptual manufacturing system was chosen so that it would be flexible enough to allow for various configurations but not overly complex. A less complex system allows for the focus to be placed on demonstrating the key concepts associated with the reconfiguration process rather than on the control application itself. Also, due to the limitations of the TINI, particularly with its relatively slow application loading times, a less complicated control application is desirable.

To compare the results of the reconfiguration manager and control applications running the TINI, a similar system was implemented entirely on the PC. A virtual device was set up on the PC with FBDK to simulate the TINI in order to run a modified version of the reconfiguration manager and the control applications. The inputs and outputs from the Taylec board were simulated with virtual lights and buttons that appear on the PC screen. The simulated versions of the reconfiguration manager and the control applications running on the "virtual TINI" have identical functionality as the TINI-based applications. This allows a reasonable comparison of the two systems and a meaningful assessment of the prototyping environment.

This part of the experiments was motivated by the limitations of the TINI platform. For example, one of the limitations is a result of a limit on the number of Publisher/Subscriber pairs a TINI can support. This is a result of the number of threads that a TINI can support. Since PCs typically have much higher processing capabilities in comparison to embedded platforms, the PC comparison provides a lower bound on these latencies. In terms of real-time performance, the PC may not possess superior capabilities. However, in this case, the limitations relate more to processor resource limitations which are rapidly being overcome by new platforms (discussed in section 4).

The three test scenarios reported in this paper are illustrated in Figure 2. The initial configuration, illustrated in Figure 2(a), involves a single conveyor (B1) that transports parts to a holding area where a process (P1) is applied to a batch of parts. The parts are counted by sensor C1 as they pass along the conveyor belt. When the process has been completed, sensor S1 sends a signal indicating that the P1 is ready for a new batch of parts.

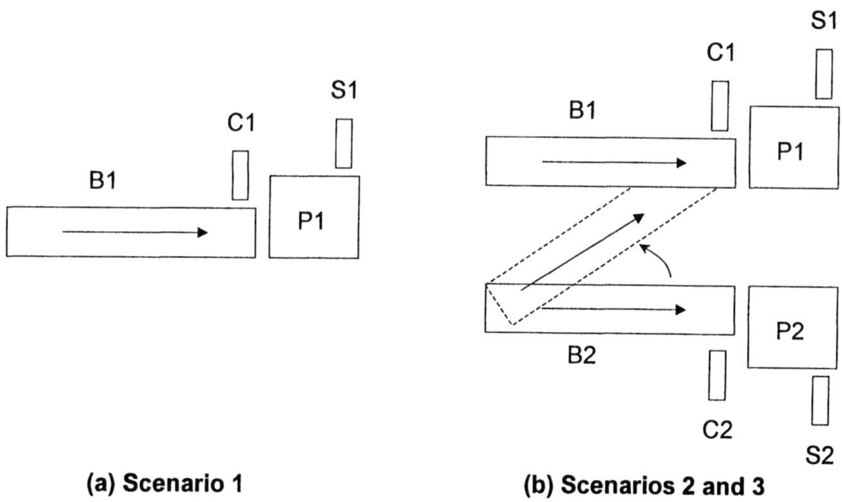

Figure 2 – The test scenarios

The second scenario, shown in Figure 2(b), depicts the situation where a second process (P2) is added. This contingency may be required in the event of additional equipment is introduced to the system or in the event that the controller for the second process has failed.

Finally, the third scenario, also illustrated in Figure 2(b), involves the case where a failure at P2, or in the event of rescheduling by a higher-level agent (e.g., to balance the two lines shown in Figure 4(b)). In this case, the conveyor feeding P2 is moved to feed P1, or a third conveyor is added to divert parts to P1 (shown by the dashed conveyor in Figure 2(b).

3.3 Experimental Results

In order to quantify the real-time performance of the reconfiguration manager, the time to load and run different applications was measured. Loading times were measured for applications running on the TINI and for the simulated 'virtual TINI' system to allow for a reasonable assessment of the performance on the prototyping platform. Two different TINI boards were used to perform the reconfiguration experiments, with nearly identical results from each one.

In Figure 3 we show the experimental results for the TINI platform. For these results, the application launch time represents the time to kill the application running on the TINI platform, FTP the contingent control application to the TINI platform, and then launch this new application. In order to compare the three scenarios described in the previous section, we calculate an application complexity metric that is based on application size (in KB), number of function block connections, and number of function blocks. This metric is normalized with respect to the initial scenario (i.e., this scenario has a complexity of 1). For example, scenario 1 has an application size s_1 = 5.02 KB and it uses c_1 = 27 connections to connect b_1 = 11 function blocks. As a result, scenario 3 (s_3 = 7.91 KB, c_3 = 44, b_3 = 20) has a relative complexity of 1.67: i.e.,

Complexity of scenario 3 = $(s_3 / s_1 + c_3 / c_1 + b_3 / b_1)/3 = 1.67$

Figure 4 shows the results for the 'virtual TINI' system noted previously. For these experiments, all of the same steps noted above were performed, however the control application was run on a 400 MHz, Pentium II processor.

Comparing the launch times to the relative complexity of the control applications in Figures 3 and 4, it is apparent that there is a correlation between launch time and application size. The launch time increases significantly with the increase in application size. Unfortunately, the launch times on the TINI platform are too slow for practical applications. This is partially a result of the increased processing requirements placed on the TINI platform by the reconfiguration manager process (shown in Figure 1). However, the overall performance of is still quite slow without this process running. For example, scenario 1 takes 85 seconds to launch and scenario 3 takes 102 seconds to launch on the TINI without the reconfiguration manager running.

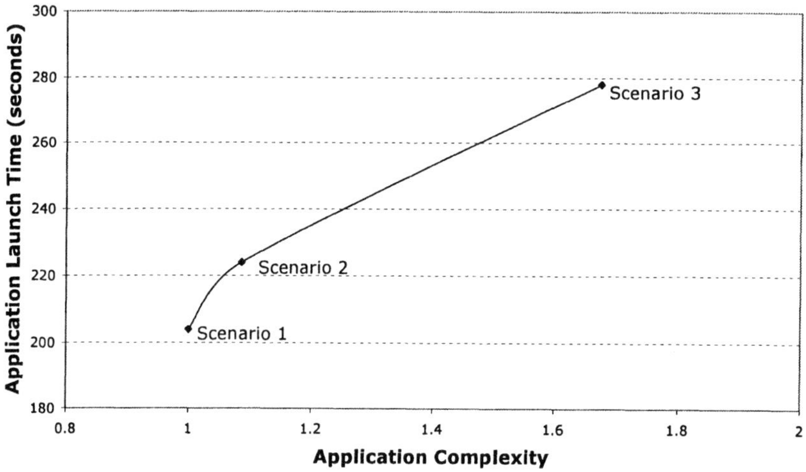

Figure 3 – Application launch time for the TINI platform

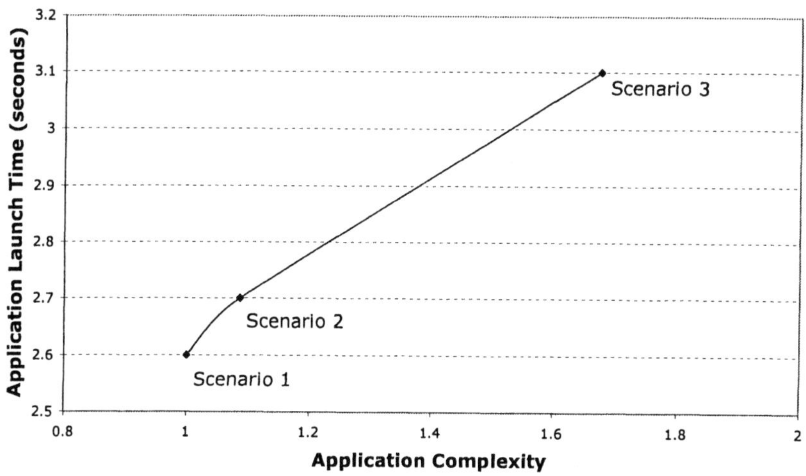

Figure 4 – Application launch time for the PC platform

4. CONCLUSIONS

It appears from the results reported in the previous section that the main limitation of this reconfiguration approach lies in the application launching process as illustrated in Figure 3. In particular, when reconfiguration is required, the existing control application is killed, a new application is loaded, and finally the new application is launched. Once the control application is launched, the TINI platform's real-time performance is quite acceptable (please refer to Rumpl et al., (2001) for more information on the TINI's run-time performance).

Reconfiguration should not necessarily require a complete replacement of the existing application however. For example, the changes required to move from a control application for scenario 1 to one for scenario 2 or 3 should only require the addition of a small number of function blocks and/or the "re-wiring" of variable and event connections. Unfortunately, the TINI Java Virtual Machine (JVM) supports a limited implementation of JDK 1.2, and as a result, does not support this form of reconfiguration. In particular, the TINI JVM does not support dynamic class loading and serialization.

In order to address this implementation issue, the authors are currently investigating alternative embedded Java-based platforms such as the Systronix aJile Euroboard (or SaJe) (SaJe, 2004) and the ImSys Simple Network Application Platform (or SNAP) (ImSys, 2004). Our work in this area is focusing on both dynamic and intelligent reconfiguration issues. For example, given the increased memory and processing speed of these platforms, we will be able to more closely investigate the timing issues associated with reconfiguration.

As well, the "re-wiring" approach noted above will be further investigated. The IEC 61499 standard's support of XML descriptions is particularly promising in this area (IEC, 2000). This format allows unambiguous specifications to be written for function block applications that can be used for initial system configuration as well as subsequent reconfiguration. In other words, changes in an application's configuration can be specified using a well-formed XML document; the reconfiguration manager's job is then to parse the XML description and to make the necessary changes to the application (e.g., changing connections, adding/removing function blocks, etc.). The anticipated advantages of this approach are that it should overcome the application loading issues described in this paper and also open the door for more intelligent approaches to reconfiguration (e.g., using higher level configuration agents to reason about new configurations rather than relying on pre-defined contingencies).

5. REFERENCES

1. Brennan, R.W. and Norrie, D.H. "Managing fault monitoring and recovery in distributed real-time control systems," *5th IEEE/IFIP International Conference on Information Technology for Balanced Automation Systems in Manufacturing and Services*, Cancun, Mexico, pp. 247-254, 2002.
2. Christensen, J.H. *Function Block Development Kit*, holobloc.com, 2004.
3. Gruver, W., Kotak, D., van Leeuwen, E., Norrie, D. (2001) "Holonic manufacturing systems – phase 2", *the International IMS Project Forum 2001*, Ascona, Switzerland.
4. HMS, *Holonic Manufacturing Systems Consortium Web Site*, http://hms.ifw.uni-hannover.de/, 2004.
5. IEC TC65/WG6 (2000) Voting Draft – Publicly Available Specification - Function Blocks for Industrial Process-measurement and Control Systems, Part 1-Architecture, International Electrotechnical Commission.
6. ImSys, *Web Site*, http://www.imsys.se, 2004.
7. Loomis, D. *The TINI Specification and Developer's Guide*, Pearson, 2001.
8. Systronix, *Web Site*, http://www.systronix.com, 2004.

6 THE MABE MIDDLEWARE

Alois Reitbauer[1], Alessandro Battino[2], Bart Saint Germain[3], Anthony Karageorgos[4], Nikolay Mehandjiev[5], Paul Valckenaers[3],

[1]*Profactor Produktionsforschungs GesmbH, Wehrgrabengasse 1-5, A-4400 Steyr, AUSTRIA*
alois.reitbauer@profactor.at
[2]*Institute of Production Engineering and Machine Tools (IFW) - University of Hannover, Schlosswender Strasse 5,30159 Hannover, GERMANY*
battino@ifw.uni.hannover.de
[3]*K.U.Leuven – P.M.A. Celestijnenlaan 300B, B-3001 Leuven, BELGIUM*
{paul.valckenaers, bart.saintGermain}@kuleuven.ac.be
[4]*Dept. of Communications and Computer Engineering, Univ. of Thessaly, 37 Glavani - 28th October Str, 382 21 Volos – GREECE, karageorgos@acm.org*
[5]*Dept. of Computation, UMIST, Sackville Str. M60 1QD, Manchester, UK*
mehandjiev@acm.org

The research project MaBE aims at the development of agent-based middleware supporting cooperation in open business environments, based on real-world case studies concerning virtual enterprises. These case studies provide us with a library of scenarios of inter- and intra- organisational cooperation, which impose certain requirements to the construction of effective agent-based middleware platforms supporting such cooperation. However, existing FIPA compliant, agent construction frameworks are insufficient to implement such agent-based middleware, as they cannot meet certain requirements imposed by the case studies. For example, the requirements involving dynamic handling of multiple and evolving ontologies, security and trust issues as well coordination in open environments cannot be directly addressed by current agent development tools and further research is required before building the appropriate middleware functionality supporting them. Since these issues are perceived as key enablers for open business collaboration, this paper provides an overview of the research undertaken to address them.

1. INTRODUCTION

The Virtual Enterprise (VE) paradigm (Cam, 2001) describes the development of mechanisms to support collaboration of existing business entities in a distributed environment. Multi-agent systems (MAS) are considered to be a suitable approach for modelling various cooperation scenarios within a Virtual Enterprise. Existing agent platforms like for example JADE (Jade, 2004), however, lack built-in support for some important requirements stemming from these scenarios.

The EU funded research project MaBE is developing a middleware system based upon a FIPA compliant agent platform (FIPA, 2002a) for supporting the development of collaborative business environments. MaBE focuses on extended and virtual enterprises and the applications prototyped within the project reside in the sector of productive industry and logistics Within this remit MaBE has focused on several important areas where further research is necessary to provide effective agent-based support to a number of real-world collaboration scenarios within Virtual Enterprises.

This paper provides an overview on the main research issues addressed. Because of the apparent diversity of research issues, the aim is to provide a holistic broad-brush picture of the set of issues rather than to go into a considerable level of depth exploring an individual issue.

The rest of the paper is organised as follows. In the next section we will present an exemplary industrial use case showing the requirements needed for collaboration within an Extended and Virtual Enterprise (Cam, 2001). Section 3 presents a reference model describing different forms of collaboration depending on organizational structures. Based on this reference model a set of required extensions to current multi-agent systems is described. Section 4 presents Coordination mechanisms for Supply Networks followed by a discussion of trust and security aspects in Section 5. Section 6 provides an overview of issues regarding communication semantics that have to be addressed for enabling cooperation.

2. INDUSTRIAL USE CASE

The solution described below is motivated by requirements from real world industrial use cases. As an example, the scenario of a company performing hardening processes is presented. The company consists of nine SME-organized as a Virtual Enterprise where new companies are dynamically joining and leaving the group and new processes are continuously introduced as needed.

The main objective for forming a Virtual Enterprise was the optimized usage of information as well as physical resources. However this possibility cannot be fully exploited due to the amount and the complexity of the information stored (e.g. part-treatment programs, quality control data, production plans). In the following the main impediments for collaboration within one company as well across companies are presented.

Within a single company, the main concerns are related with the production coordination. At present, the definition of batches and the scheduling are activities carried out manually and therefore implying big efforts. In order to reach a higher level of flexibility, there is the need for a resource coordination system allowing production planning, which takes into account the actual workload of the facilities, as well as real-time data about breakdowns and other disturbances.

Interaction between the companies induces mechanisms for communication (flow of information) as well as logistics (flow of material). These mechanisms make use of services currently available within the virtual enterprise but additional services not yet available are needed. Choosing the structure of a Virtual Enterprise it is expected to enable all involved entities to act as if they belonged to just one company, exploiting at the same time the logistic advantage constituted by a distribution of the factories in the territory. The transport of the parts (logistic service) can be carried out with VE means, with means of the customer or through logistic providers. The parts received from one customer can have as a destination the same customer of origin or another customer or even another company of the Virtual Enterprise (in case of distributed processes). The resulting number of possible interactions can only be efficiently coordinated with a system supporting automatic selection of services. Since such a system would connect different actors of the supply chain, topics like trust and security have to be addressed. It is, in fact, fundamental to provide a sufficient visibility in order to permit a services discovery, but without revealing reserved information or compromising the security of the communication. Moreover, to allow the interoperability between companies working in different business fields, a common language or at least a common frame of reference is required. For instance the data about a transport can be described in different ways inside the Logistic Provider and inside the Virtual Enterprise; and new concepts can be used when a new transport unit is introduced.

3. REFERENCE MODEL FOR COOPERATION

To provide an overall framework for analysing and understanding the case study, for modelling and implementing cooperation scenarios and for handling the requirements implied there, it is necessary to employ a theoretical reference model of collaboration within a Virtual Enterprise. The model on Figure 1, based on (Cam, 2001), was found to fit the scenario described above, and to resonate well with authors' experience. The model is based on different types of collaboration as follows.

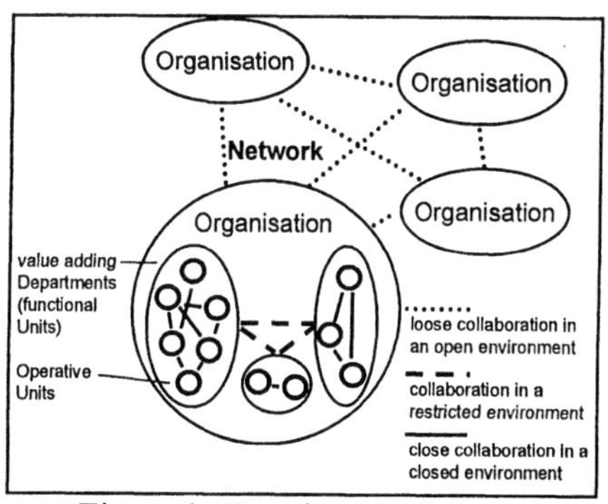

Figure 1 - Levels of Collaboration

Organisations at the lowest level are built on operative units representing atomic resources. These units cooperate among each other forming functional units also called departments or groups. Interactions between these functional units represent the collaboration taking place at enterprise or intra-organisational level. Networks of organisations are the next level of collaboration. Inter-organisational cooperation can further be divided into cooperation of a fixed number of partners or cooperation in an open environment where partners permanently can appear and disappear. Figure 1 presents a graphical representation of the different level of collaboration.

4. SUPPLY NETWORK COORDINATION

As presented in the use case above coordination not only implies coordination of all orders and resources meaning production logistics but also information logistics. A satisfactory solution has to provide monitoring and short term forecasting abilities as well as decision taking mechanisms. While Section 2 presented a static view of various forms of collaboration, a dynamic view on the system is required for resource coordination. The Supply Network paradigm (SAI, 2004) is seen as fitting well in here.

An entity in a supply network is defined by the services it provides to the environment. Combination of entities can either be established by establishing relations or by aggregation. An important factor concerning the established structure especially in open collaboration environments is that it is by no means fixed. The structure will dynamically change while the environment is changing.

Both in intra- and inter-organizational collaboration, autonomy and specificity of an entity has to be respected. As shown in Section 3 there is no single level of autonomy and specificity evolved in the collaboration scenarios. While in intra-organisational collaboration there is only single goal to be achieved among participating entities this is not the case in inter-organisational collaboration. Production entities in the system will be striving towards different and often also conflicting goal. Still, decisions made by one entity have to be respected by the other entities. Consequently a coordination mechanism must be able to define win-win solutions for all involved entities.

For the definition of a supply chain architecture the PROSA (HAD, 2003) model has been chosen. PROSA allows the modeling of holonic system structures using three main building blocks called *holons*.

- Product holons provide the process knowledge of a product, meaning all possible production steps combination and their order.
- Order holons manage one order (customer order, production order, a aggregation of orders, etc) throughout all production stages.
- Resource holons represent a material or immaterial resource within a supply network.

In addition to the architectural building blocks a mechanism for coordination of these entities is required. Open environments impose due to their dynamically changing structure great complexity regarding coordination. Centralised approaches are not applicable, as they require the perception of the whole environment. This however is either not or almost impossible in open environments.

The coordination mechanism presented here is based on the insights from the behaviour of ants searching for food. The main advantage of such an ant-based design is that individuals are not exposed to the complexity and dynamics of the global situation.

There are three different types of ants and respectively three stages of the resource coordination process each one fulfilling one dedicated purpose.

Feasibility ants are responsible for discover the topology of the supply network; finding out all respective services there can be accessed. This information can be used by explorer ants, which are responsible for exploring possible solutions through the supply network for a specific order. After the selection of a possible candidate solution found by one of the explorers, intention ants are used to spread the order agent's current intention along all resources involved in the orders respective objective. Based on these declared intentions, service providers can build a dynamic load forecast, which can be used by explorer ants to make their performance estimate reliable.

All information produced during each of the three phases has only a limited period for which it is valid. After this period, the information vanishes if it is not produced again. This procedure represents an analogy to the evaporation of pheromones – the analogy of information – used in ant communication. As a consequence of this evaporation of information each of the phases above has to be executed cyclically. This approach keeps information in the system always up-to-date and additionally and even more important makes the system robust to changes.

5. TRUST AND SECURITY

Current agent systems lack proper support for security and other trust building mechanisms (Pos, 2000). A trusted environment is however a vital point in business collaborations. Current standardisation activities within FIPA (FIPA, 2004) deal with the implementation of basic security mechanisms for message based security. Security mechanisms however are only one factor of trust. Reputation and confidence building mechanisms are additional factors that have to be addressed. As not all of these factors are relevant for any kind of collaboration – a clarification on this topic is needed.

Requirements for trust issues depend on the form of collaboration. We will now relate them to the model presented in Section 3, concentrating on where ICT support is needed.

Collaboration within departments happens within a closed system and in a trusted environment. An established legal framework forms the basis for confidence in the actions of the parties involved in collaborations. This confidence is supported by striving towards a common goal and by the fact that the acting parties are employees of the same company, bound by contracts.

Inter-department collaboration happens on a regular basis but less frequent than within a department. The collaboration scenarios are well defined and all actions take place in a closed, trusted environment as well. Trust however declines due to reduced personal contact in collaborations. Security at this level addresses the protection of information and resources based on organisational roles as well as on the context of collaboration.

In *inter-organisational collaboration* the environment is insecure, more dynamic, and in general not trusted. Security for collaboration scenarios within an organisation is completely controlled by the organisation itself. In an open environment a part of this control is handed over to the other involved parties. Mechanisms are needed which allow communication scenarios, in which not all parts of the infrastructure are controlled by the organisation itself.

Contracts between companies are as powerful as contracts regulating inter-organisational activities. However they more difficult to enforce as independent third parties are required for enforcement. As companies might have conflicting or competing goals, confidence in the actions of others is not present per se. The establishment of reputation also takes longer as interactions occur on an infrequent, business need driven basis.

In inter-organisational collaboration different scenarios of doing business exist. Collaboration based on strictly defined processes with known business partners represents the simplest form of inter-organisational collaboration. This can be extended to a fully open dynamic organisational network, where processes are not fixed, and collaboration takes place with unknown partners.

A different view on the roles of the importance of trust related issues are the states collaboration goes through in its lifecycle. Following (Cam, 2001) collaboration has four states. The initial state is the creation state, followed by the operation of the virtual organisation. During operation the organisation is suspect to evolve by changing its structure. The dissolution of the virtual organisation is then the last stage. The evolution stage can however be eliminated in closed and fixed collaborations. Confidentiality and data origin identification play a vital role in all stages. Authorization issues as well as reputation and confidence enforcing mechanisms are key requirements for the execution phase. The creation and respectively evolution phases are responsible for setting up and defining reputation and confidence mechanisms.

Due to the specific requirements to authorization mechanisms in collaborative environments (Edwards, 1996) that cannot be coped with using standard mechanisms like access control lists (ACLs) or access control matrixes (ACMs) as described in (Lampson, 1974). Although there is a lot of work available on reputation and confidence building mechanisms, there is no ready to use solution available for business applications. Agent-oriented systems however offer a very attractive way for implementing these features.

By applying the mediator pattern (Gamma, 1994) agent communications can be intercepted and trust related checks can be performed based on interaction between agents. This mechanism is for example also used for systems monitoring the access of medical databases (Liu, 2000) or access control in web service environments (Sko, 2003).

6. ONTOLOGY MANAGEMENT

To meet the semantic interoperation requirements at the tactical level in the presented case study, it is preferable to use ontologies, which are formal representations of the meaning of a set of concepts and their relationships in a domain of discourse. The use of ontologies is currently a fundamental part of agent behaviour (FIPA 2002) and it is in principle consistent with other approaches such as those used in Semantic Web and Enterprise Application Integration. However, the use of ontologies within the multi-agent systems community is biased towards semantic interoperation, leaving two technical research issues to the developers of business support systems such as MaBE: ontology mapping and dynamic ontologies.

6.1 Mapping of multiple ontologies

The need for ontology mapping stems from the fact that different views of the same reality may exist between collaborating partners. To achieve common understanding, there are three main approaches:

- Using current standards, (i.e. XML-based representations) all systems can communicate using the least common denominator of all ontologies. This causes the loss of information and respectively reduces the communication abilities making deep integration impossible.
- Create a common "world model" integrating all sub-models of all involved systems. The model created using this approach is undoubtedly very complex. Additionally this approach cannot be taken in open environments where not all models will be known at design time.
- Finally, mappings on-demand can be created between the ontologies of interested parties. This would keep the size of the ontologies low and minimise information loss. The mapping process would be distributed throughout the collaborative environment.

We have taken the Partially Shared Views approach (Lee, 1990) to collaborative ontology mapping on-demand, and created appropriate agent negotiation mechanisms implementing this within MaBE.

6.2. Dynamic handling of evolving ontologies

Interoperation at tactical level of the case study requires that ontologies may change at run time. The middleware platform should therefore provide facilities for handling evolving ontological concepts at run-time using one the following approaches:

a. *Using an external Ontology Server* is a promising approach to realising dynamic ontology management. The server acts as a run-time middleware, which administers the knowledge base containing various ontologies. It includes powerful reasoning mechanisms (inference engines) which deliver answers to knowledge base queries submitted by the client applications. Applications using the ontology server are easily adaptable to new data and information by changing the underlying ontologies and knowledge bases.
b. *Using Ontology Agents as mediators* is the preferred mode of implementing an Ontology Server in agent-based systems since numerous complex interactions with agents will be required. The Ontology Agents will provide translation, mapping and interoperation services, and act as mediator to agent interactions involving unknown ontological elements. They will also be responsible for storing and manipulating ontology evolution on run-time. So other agents will not be concerned with ontology management, but focused on the usage of ontologies to provide application functionality.
c. *Providing direct access to evolving ontologies to each agent.* Although appealing, the approach of having ontology agents being responsible for all ontology management operations is likely to result to communication explosion and to inefficient infrastructure. Therefore, our view is that Ontology agents should be responsible for storing and manipulating evolving ontologies but once any changes or conflicts have been resolved, for example through negotiation of ontological concepts, then individual agents should be able to use the updated ontology versions. This would require appropriate interaction protocols for ontology management and versioning.

Each of these approaches is more suitable for a particular style of collaboration from Section 3, for example the final one is tuned to the needs of intra-departmental collaboration whilst the first one is appropriate at inter-organisational level. The

MaBE middleware will thus have to provide an integrated approach to dynamic ontologies based on a combination of these approaches. An initial draft of this approach has been published elsewhere (Carp, 2004). Experiments are currently under way to determine the optimum system configuration with respect to the optimal integration of the three approaches.

7. CONCLUSION

The MaBE middleware aims satisfy a set of currently unfulfilled requirements for supporting business collaboration identified from a business collaboration reference model and representative case studies: the requirements of coordination, trust and security, and ontology management. In this paper the approaches followed to meet the above challenges are outlined: Coordination is based on a simple and powerful modelling framework – PROSA – which is combined with nature inspired coordination mechanisms designed for dynamic environments. Trust and security is based on mechanisms specifically designed for the characteristics of multi-agent systems supporting business applications. Finally, an integral approach to ontology mapping and dynamic ontology handling of evolving ontologies will be extending the ontology mechanisms of the underlying JADE kernel to ensure semantic interoperation of involved parties at all levels of collaboration.

ACKNOWLDEGEMENTS. The work was partially funded by the European Commission under the GROWTH project MaBE (Multi-agent Business Environment), project no. GRD1-2001-40496.

9. REFERENCES

1. Camarinha-Matos, L., Afsarmanesh, H.: Virtual Enterprise Modeling and Support Infrastructures: Applying Multi-agent System Approaches. Lecture Notes in Artificial Intelligence, Vol. 2086. Springer Verlag Berlin Heidelberg (2001) 335 -364.
2. Carpenter M., Gledson, A. and Mehandjiev, N., Support for Dynamic Ontologies in Open Business Systems. To appear in Proceedings of the AAMAS'04 AOIS Workshop. 2004.
3. Poslad S and Calisti M. Towards improved trust and security in FIPA agent platforms. Autonomous Agents 2000 Workshop on Deception, fraud, and trust in agent societies, Barcelona, June 2000.
4. FIPA: FIPA Abstract Architecture Specification, SC00001
5. FIPA: FIPA Ontology Service Specification, XC0086.
6. FIPA: FIPA Agent Message Security Object Proposal: FIPA Security TC Working Document (2004).
7. Edwards, W.: Policies and Roles in Collaborative Applications. ACM Conference on Computer Supported Cooperative Work (CSCW'96), Cambridge, Mass., (1996) 11-20.
8. Gamma E., Helm R., Johnson R., Vlissides J.: Design Patterns – Elements of Reusable Object-Oriented Code. Addison-Wesley Professional Computing Series. (1996).
9. JADE: http://jade.tilab.com.; Telecom Italia Labs.
10. Lampson, B.: Protection. ACM Operating Systems Rev. 8, 1 (1974), 18-24.
11. Jintae Lee and Thomas W. Malone. Partially shared views: A scheme for communicating among groups that use different type hierarchies. *ACM Transactions on IS*, 8(1) 1990.
12. Liu D., Law K. Wiederhold G.: CHAOS: And Active Security Mediation System. Proceedings of the 12th International Conference on Advanced Information Systems Engineering. Lecture Notes in Computer Science, Springer-Verlag: London, UK (2000) 232 – 246.
13. Skogsrund H. Benatallah B. Casati F.: Model-Driven Trust Negotiation for Web Services, IEEE Internet Computing – Vol (7) No (6) (2003) 45 – 52
14. Hadeli, Valckenaers, Zamrescu, Van Brussel, Saint Germain., Holvoet, Steegmans:. Self-organising in multi agent coordination and control using stigmergy. ESOA. (2003).
15. Saint Germain, B., Valckenaers, P., Verstraet, P., Bochmann, O., Van Brussel, H.: Supply network control, an engineering perspective. Accepted at 11th IFAC Symposium on Information Control Problems in Manufacturing, Salvador, Brasil, (2004).

7 AGENT-BASED SIMULATION: MAST CASE STUDY

Vladimír Mařík
Department of Cybernetics, Czech Technical University in Prague, Czech Republic & Rockwell Automation Research Center, Pekařská 10a, 155 00 Prague, CZECH REPUBLIC, marik@labe.felk.cvut.cz

Pavel Vrba
Rockwell Automation Research Center, Pekařská 10a, 155 00 Prague, CZECH REPUBLIC pvrba@ra.rockwell.com

Martyn Fletcher
Agent Oriented Software Limited, PO Box 318, Cambridge CB4 1QJ, UK martyn.fletcher@agent-software.co.uk

Summarizing all the specific features of the simulation systems needed for agent-based systems, the paper documents that the simulation tools of this kind do represent rather complex development environment for agent-based systems than one-purpose simulation software obvious in the case of "classical" centralized systems. The agent-oriented simulation tools explore and combine methods of real-time emulation, qualitative simulation, testing and diagnostic algorithms with classical methods of both the discrete and continuous simulation approaches, techniques of advanced visualization and run-time interfacing. The MAST simulation tool and the detailed description of its extension for the Cambridge Packing Cell Testbed are used as a case study.

1. INTRODUCTION

In the case of multi-agent systems, under the term "simulation" we understand processes which are – in comparison to "classical" centralized systems – more complex and include more tasks than just single simulation in the classical meaning. There are many quite specific requirements and expectations put on simulation of the agent-based systems:

a) First of all, we expect that the qualitative evaluation of emergent behavior of an agent-based system will be provided.

b) Agent-based system can be simulated only by another agent-based system – the centralized approach is not adequate. It is necessary to stress, that the existing simulation tools like Matlab or Arena are not sufficient for this purpose.

c) The simulation of both the controlled process and the agent-control system has to be provided. Because of the direct reusability of the agent-control algorithms, the agent-control part is, as a matter of fact, emulated as well. In such a way the agent-based simulation is usually organized as the interaction of two emulations.

d) In the case of the agent-based simulation, there are two kinds of interfaces expected, namely a nice and instructive human-machine visualization interface, and – as a rule – a machine-machine runtime interface between the agent-control system and the controlled process/manufacturing equipment (either emulated or physically connected – see below).

The process of developing and implementing an agent-based system relies on several phases and widely explores the simulation principles of diverse nature. This is especially true in the case of systems without any central element where unexpected emergent behavior can appear. The following stages of the design process based on simulation can be gathered like this:

(1) Identification of agents: The design of each agent-based system starts from a thorough analysis of (i) the system to be controlled or manufacturing facility to be deployed and (ii) the control/manufacturing requirements, constraints, and hardware/software available. The result of this analysis is the first specification of *agent classes* (types) to be introduced. This specification is based on the application and its ontology knowledge. The usual design principle – following the object-oriented methodology – is that each device, or each segment of the transportation path or each workcell is represented by an agent. In very up-to-date systems, also the product itself or semi-product can be considered as agent able to negotiate its own processing/assembling with the agents of the manufacturing environment.

(2) Implementation/Instantiation of agent classes from the agent type library. This library is either developed (step 1), or re-used (if already available). Particular agents are created as instances of the definitions in the agent type library. Furthermore, the implementation of communication links among these agent instances is established within the framework of initialization from these generic agent classes (for instance agents are given the names of their partners for cooperation). In such a way, the first prototype of an agent-based control/manufacturing system (or similarly, a supply-chain management system) is designed.

(3) Own Simulation: Behavior of a complex agent-based system is rather emergent than deterministic (Steels, 1994). The decision-making knowledge stored locally in the agents along with the patterns of inter-agent interactions result in an aggregate global behavior of the system, which cannot be precisely predicted in advance. Yet the direct experimental testing of the global behavior with the physical manufacturing/control environment being involved is not only extremely expensive, but non-realistic as well. Simulation is the only way out. For this purpose, it is necessary to have:

- A good model of the controlled process or the manufacturing facility or the virtual enterprise. This model must depict all the entities within the factory/enterprise and their interfaces to the external world.

- A good simulation tool for running the model of controlled process/manufacturing facility/virtual enterprise to provide the emulation of the physical manufacturing/control environment. Standard simulation tools like e.g. Matlab, Arena, Grasp, Silk, AnyLogic etc. can be used for these purposes.

- A suitable agent runtime environment for running the agents – reused from phases 1 and 2 – and for modeling their interactions. On the basis of the results of agent platforms comparison (Vrba, 2003a), the JADE platform as open-source or JACK (Howden, *et al.*, 2001) as a commercial tool can be recommended.

- System integration strategies developed and implemented in the form of sub-system interfaces. It is necessary to have the following two *run-time interfaces*: (i) an interface between the agent-control and the process emulation and (ii) an interface to link the agent-control with the physical manufacturing equipment. In the ideal case these two interfaces should be compatible (or identical at best) to enable the designer to switch from simulation/emulation system to the physical manufacturing/control system or virtual enterprise as appropriate.

- HMI (human-machine interfaces) for all the phases of the system design and simulation.

4. **Implementation of the target control /manufacturing system**: In this stage, the target control, manufacturing, production management or supply-chain system is re-implemented into the (real-time) running code. This implementation usually relies on ladder logic, structured text or function blocks at the lowest level of control. However, the higher-level control – carried out by the agents – is almost entirely reused, i.e. the same agents used in the simulation phase (3) are used also for the physical control. For instance in the ExPlanTech production planning MAS system (Říha, *et al.*, 2001), there was 70% of the agent code reused from the simulation prototype. Therefore, the choice of the multi-agent platform in the phases (3) and (4) is critical – it is advised to operate with the same agent platform.

The simulation phase is much more crucial for the development process of agent-based systems than it has been for the development of "classical" centralized systems. It enables besides others:

- to **predict the behavior of the system as a whole**. The fact there is no central unit in the agent-based system represents a critical barrier in a wider applicability of the agent-oriented ideas. The simulation runs help to understand the system behavior and to detect the patterns of emergent behavior. Considering that the behavior of a MAS is emergent, to ensure that all types of possible behavior were explored/covered by simulation still remains a painful problem (the situation is similar to that of system testing).

- to **predict and test the optimal scenario for the agent-based system** development

- to select **the most optimal negotiation framework and strategy** for individual units in the system

- to **directly link the simulation with real-life manufacturing/control processes**. That means that whereas a part of the agent-based system is engaged fully in the real-life activities, the remaining part can be just simulated. The shift of the borderline between the simulated (in more precise terms "emulated") and the real part of the system can be carried out in a quite smooth way. This would help to speed-up the initial "commissioning" process significantly.

The requirements on simulation tools or platforms for MAS-oriented solutions call for new types of simulation systems (simulation platforms) with embedded MAS principles. One of pioneering systems of this kind, presented in this paper, is the MAST simulation tool being developed by Rockwell Automation (Vrba, 2003b). Another example of such a system is the agent-based control and simulation of the scaled-down form of chilled water system for the US-Navy ships (Maturana, *at al.*, 2004).

2. MAST – Manufacturing Agent Simulation Tool

The development of the MAST modeling and simulation tool started about three years ago as one of the pilot projects of Rockwell Automation Research Center in Prague aimed at the investigation of holonic systems, i.e. the exploitation of multi-agent technologies in manufacturing control. The original idea was to implement the agent-based solution for material handling domain, particularly the transportation of materials/products between various manufacturing cells on the factory shop-floor using conveyors and AGVs. The attention has been paid mainly to the identification of agents (see Section 1, phase 1), i.e. the definition of agent types for basic components, like manufacturing cell, conveyor, diverter, AGV, etc., and the specification of communication protocols and scenarios used for the inter-agent cooperation (Vrba, 2003b).

To be able to test and validate the agent functionality without being connected to the physical manufacturing equipment, the simulation, or more precisely the emulation of the manufacturing environment had to be implemented as well (see Section 1, phase 3). Basically, simulated movement of virtual products triggers virtual sensors that send signals to appropriate agents while the control actions taken by agents are propagated back to the simulation through virtual actuators. The simulation part of the MAST tool is tightly linked with the GUI (HMI-interface) that provides the visualization of the simulation and allows the user to interact with it.

The agent behavior is aimed at the transportation of products among user-requested manufacturing cells. The agents cooperate with each other via message sending in order to find the optimal routes through the system – a cost based model is applied where each conveyor provides a transportation at predefined cost. The work cells are interconnected via a network of conveyors and diverters (switching components) that route transported products through the conveyor network following the optimal, i.e. least-cost routes. Main stress is put on the failure detection and recovery – the user can simulate a failure of any component and trace the reaction of agents looking for another delivery routes while avoiding the broken component. It is important to say that there is no central control element – the decision making and control processes are distributed over the agents that work autonomously and use message sending and on-line service discovery for mutual collaboration. The *plug-and-operate* approach is thoroughly applied for system integration allowing to add/delete/change any component/agent through the GUI on the fly.

For the implementation we decided to use the JAVA language because of the variety of JAVA-based FIPA-compliant agent development tools being available today, most of them as open sources. Originally, the development started with the FIPA-OS agent platform but due to the performance and memory consumption issues we selected the JADE platform instead. The main advantage of using the object oriented language – JAVA in this case – is that the description of a particular agent type is represented by a single JAVA class containing the general specification of agent's attributes and the set of rules according to which the agent behaves. Such a class is then used to create as many instances as required (see Section 1, phase 2) without the need to program the behavior of each agent instance individually.

3. A PPLICATION OF MAST TO CAMBRIDGE PACKING CELL

3.1 The Cambridge Packing Cell Overview

Although there is a number of academic and industrial research organizations active in the holonic/agent-based control field, there are very few holonic systems deployed in real factories making real products today. Main reasons for this situation are the higher investments needed to implement the agent-based manufacturing system and also a number of research issues that remain to be resolved like the appearance of unpredictable, emergent behavior of a community of agents (see Section 1) or missing framework for evaluation of the holonic system's performance and applicability (Fletcher, *at al.*, 2003a).

To give the opportunity to evaluate different holonic design and development strategies, the holonic packing cell has been constructed in the Center for Distributed Automation and Control (CDAC) at the Cambridge University's Institute for Manufacturing. This lab provides a physical testbed for experiments with agile and intelligent manufacturing with focus on two particular areas: (i) Automatic Identification (Auto-ID) systems and (ii) agent-based control systems (Fletcher, *at al.*, 2003b).

The automatic identification is an emerging technology designed to uniquely identify a specific product in a supply chain. It replaces the bar code with an Electronic Product Code (EPC) embedded in a radio frequency identification (RFID) tag comprised of a silicon chip and antennae. The EPC numbers, usually in form of 96 bit code, are read wirelessly via high frequency radio waves – the RFID readers then pass the information to a computer or an application system (MES/ERP).

In the Cambridge packing cell, the RFID tags are attached to Gillette personal grooming items (razors, shaving gel, deodorant and shaving foam) and also to boxes to which these items are packed. The orders are placed by a user that can select any three out of the four Gillette product types to be packed into two types of gift boxes. The layout of the packing cell is given in Figure 1. It consists of three conveyor loops (Montech track) to transport the shuttles with boxes – the navigation of shuttles into and out of the loops is controlled by two, independently operating gates that are provided with EPC codes of passing boxes from the RFID readers. Shuttles are held at two docking stations so that the robot (Fanuc M6i) can pick and place items into boxes. The items are held in a storage unit in four vertical slots (each for a particular type of the Gillette item); the items are picked up by the robot from the bottom of the slot and, in the case of unpacking operation, picked from the box and placed to the top of the slot.

For controlling all the operations of the packing cell, the agent-based control system has been implemented (Fletcher, *at al.*, 2003a). It comprises of the following *resource agent* classes with number of instances shown in brackets: Robot (1), Storage (1), Docking Station (2), Gate (2), Track (1), Box Manager (1) and Production Manager (1). Additionally, there are the *order agents* associated with particular gift box orders (one agent per one order) and *product agents* representing all available boxes in the system. The processing of particular order starts with the negotiation between the order agent and product agents to select the appropriate box in which the items will be packed. The product agent representing selected box then uses its own intelligence and cooperation with resource agents to determine how best to be packed: (a) the order agent queries the storage unit if it is able to provide the requested items, (b) the order agent negotiates with docking stations to reserve a processing slot, (c) there is a negotiation with the gates to ensure a proper routing to docking stations using the information from RFID readers, (d) once at the docking station, the product agent requests the robot to pack the items into the box (this includes a negotiation between the robot and storage unit).

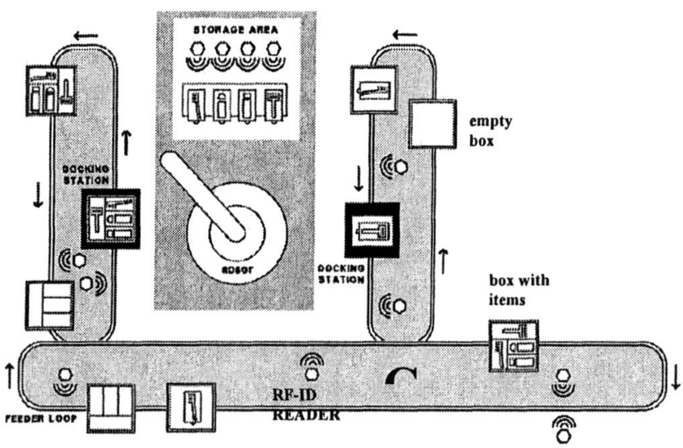

Figure 1 – Layout of the Cambridge packing cell

The agent control system – implemented in JAVA and using the JACK Intelligent Agents platform – is running on a Personal Computer. To provide the agents with an access to sensor and actuator parameters of the cell, the *blackboard* mechanism is introduced. The blackboard, running also on a PC, holds the synchronized copy of the data registers of an Omron PLC (connected via Ethernet) related to the sensors and actuators data. The agents can read from and write to the blackboard and thus observe the current status of the cell and perform desired control actions.

Recently, the CDAC testbed has been considerably extended with new manufacturing equipment. It includes mainly a second robot and storage area to increase the flexibility of the system allowing to (i) process more boxes concurrently, (ii) choose the appropriate place where the box will be packed based on availability of requested items in storages and (iii) simulate the failure conditions. New shelving storage system is used to hold the shuttle trays with both the empty and packed boxes as well as with the raw items that can be used to feed the storage areas. The shelving storage system is operated by a gantry robot that transports the trays with boxes or raw items between the particular shelves and the new docking station in the feeder loop where the tray is placed onto the waiting shuttle.

It is obvious, that such a substantial hardware extension of the lab along with the new manufacturing scenarios being introduced requires a new agent-based control system to be deployed. It has been recognized, that the agent-based solution for the material handling tasks used in the MAST tool can be easily extended to provide the graphical simulation of the CDAC lab (see Figure 2 for a screenshot) and, eventually, to be directly used for the physical control of the packing cell. This paper reports on the primary results achieved in realizing the CDAC-related extensions of the MAST tool. Particularly, newly implemented agents, like the RFID reader, Docking Station, Robot, etc. are described and the issues regarding the use of MAST for the physical control of the cell are discussed.

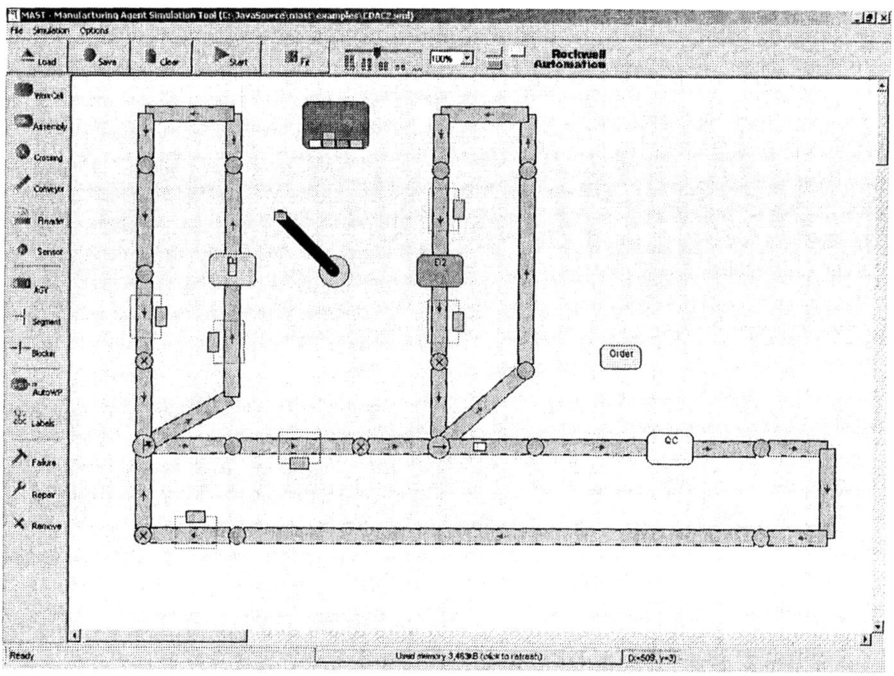

Figure 2 – Screenshot of MAST simulating Cambridge packing cell

3.2 The RFID reader agent

One of the major aim of the CDAC testbed is to demonstrate that the agent control can be integrated with the Auto-ID infrastructure. Physically, the lab is equipped with the RFID readers that send information about read EPC codes via Ethernet to a central server called Savant. The scanning period of the reader is quite high (e.g. 100 Hz) and within each scan all RFID tags that are in range of the reader are read at once. This results in large amount of redundant information generated by each reader that is sent to Savant. So the main purpose of Savant is to provide the EPC data filtering and storage in form of records containing the EPC number, name of the reader where it was read, timestamp and the flag if the RFID tag entered or left the reader. Such an information can be obtained from Savant by any client application using the SOAP protocol.

In an agent-based solution that integrates the Auto-ID technology it is reasonable to have a mediator agent that provides the EPC data to the other agents via standard agent communication protocols. To be able to easily add the Auto-ID support in the MAST tool, we decided not to use the Savant server – signals from emulated RFID readers are sent directly to the associated RFID reader agents (using a standard JAVA method call) that are doing the filtration locally. We have rather focused on implementing the mechanism of providing the EPC data to other agents. It is based on the FIPA **subscribe-inform** protocol which support has been embedded into all agent classes in MAST (by inheriting from general `MASTagent` class). Generally, this mechanism allows any agent (let say A) to subscribe for being informed by other agent (B) each time a particular event happens at the B-agent. In the case of the RFID reader agent, there are two events about which the reader informs the subscribed agents: (i) the product with an RFID tag entered the range of the reader and (ii) the product leaved the reader's range. In both these cases the reader agent sends the *fipa-inform* message including the EPC code(s) of the product to all agents that previously subscribed for one of these events. For subscription the *fipa-subscribe* message is sent to reader agent containing the `workPieceIN` string for the former or the `workPieceOUT` string for the letter type of the event (`workPiece` keyword is used in MAST to represent a product).

3.3 The Gate agent

The gate agent is responsible for routing the shuttles around the track. The gate is located at the crossing point of two conveyor loops (see Figure 1) and switches the shuttle coming from one of the two input tracks to one of the two output tracks in order to let the shuttle to remain in the current loop or to be rerouted to the other loop. How the particular shuttle will be switched obviously relates to the target docking station to which the box carried by the shuttle should be transported.

For the implementation of the gate agent in MAST we took the advantage of the existing diverter agent and the concept of least-cost product routing. Basically, the diverter holds an up-to-date routing table that contains the names of the work cells (docking stations) that are reachable using the diverter's output conveyors (tracks). These routing tables are determined by mutual cooperation of diverters, conveyors and workcells that exchange knowledge about reachable destinations along with costs of the delivery in a back-propagation manner (for further details see (Vrba, 2003b)). Once the product enters the diverter, the diverter agent searches the routing table for the product's destination workcell name in order to find out which of the output conveyors will direct the product to its destination following the least-cost route.

The modification for the Cambridge packing cell lies in the integration of the Auto-ID technology. The gate agent cooperates with the RFID readers using previously described subscription mechanism to get the EPC data of products incoming from input tracks to the gate. The issue is, that the EPC code, i.e. the ID of the product (represented as 24 characters string) does not directly contain the name of the destination docking station. To resolve this, the gate agent queries the product agent, that is supposed to be registered in the agent platform under a name that equals to the ID of the product (box). In the case that the EPC data sent by the reader contains more IDs (both box and items in it are marked with the RFID tags), the gate agent have to contact the yellow pages services in the agent platform to determine, which of the IDs relates to the existing product agent. The product agent then informs the gate agent about its destination so that the gate can properly navigate it.

3.4 The sensor agent

The sensors are used to detect the presence of shuttles at particular places on the track. The associated sensor agents support the same subscription mechanism as RFID readers to inform the other agents. Sensors are used particularly in combination with gates to allow only one shuttle moving through the gate at the same time. For this purpose, there are usually two sensors in front of the gate (at input tracks) and two sensors at the exit from the gate (at output tracks). The input sensors are closed by default, which means that the shuttle is stopped by the sensor in front of the gate. Once the gate performs a switch-operation for selected waiting shuttle, the sensor agent that blocks this shuttle is informed (the subscribe-inform mechanism is used again) – the sensor opens and the shuttle enters the gate. The exit of the shuttle is detected by one of the output sensors that inform the gate so that another waiting shuttle (if there is some) can be processed.

3.5 The robot and storage area agents

The robot agent has been implemented to control the packing/unpacking operations performed by the Fanuc M6i robot. The packing operation starts by receiving a request from the product agent when the shuttle carrying the box reaches the docking station. The message sent to robot includes the specification of required items (e.g. two gels and one razor). The robot agent starts the negotiation with the storage area agent to get the index of the slot from which the item of given type can be picked up (there are four vertical slots each for one type of the Gillette item). If the item is present at the bottom of the slot, the robot picks it and places it into the box; if not present, the robot continues with the other required item. When all items are processed (either packed to box or missing in the storage), the robot agent informs the product agent that the operation has finished.

3.6 The order and product agents

The order agents are responsible for processing the user orders for packing the customized gift boxes. The box can contain any combination of three Gillette grooming products choosing from four different product types (razor, shaving gel, foam, deodorant). Automatically generated order agent (one per order) starts the negotiation with product agents that represent available boxes in the system (either empty or already packed). If there is an empty box of requested type available and the associated product agent is willing to cooperate (i.e. is not currently "working for" another order agent), it is committed the processing of the order.

The product agent first contacts the yellow-pages services to obtain the list of storage areas. In the following contract-net protocol the storages are giving their bids in terms of how many of requested items they can provide; they also include name of the robot that operates the storage and the names of the docking stations. The product agent selects the storage offering the best bid and starts another negotiation with the docking station agents to reserve a processing slot (slots previously reserved for other product agents along with the priority of the order are considered). The product agent then changes its destination to selected docking station – any time the shuttle carrying a box associated with the product agent enters the input RFID reader of the gate, the gate contacts the product agent to get the name of its destination for proper navigation (see Section 3.3).

Once the processing of the box is finished by the robot (see Section 3.5), the product agent informs the order agent about the state of the order while releasing the shuttle from the docking station to return to the main feeder loop (see Figure 1). If the order is not fully completed, e.g. because there were not enough items in the storage, the product agent restarts processing of the order, i.e. contacts the storages again to obtain remaining items. Such a flexible behavior allows (i) to distribute the packing operation over several robots according to the current availability of items in storages, (ii) the product agent to put off the completion of the order until the missing items are inserted into a storage (e.g. by unpacking some boxes or by transporting raw items from shelving storage) and (iii) to integrate new storages and robots (possibly also new track loops) at runtime without the need to make any changes to program code of existing agent classes.

3.7 Using MAST agents for the physical control of the CDAC lab

As mentioned in Section 3.1, the idea is to use to MAST tool for the physical control of the Cambridge packing cell. However, there are still some issues that need to be resolved in order to reuse the MAST agents in the implementation of the target control system (see Section 1, phase 4). Particularly, the mechanism of accessing the sensor and actuator parameters have to be modified such that the MAST-agents will use the same interface for accessing either the simulation-part of the MAST tool (i.e. emulated hardware) or the physical manufacturing equipment of the lab.

In the current implementation, the virtual sensors and actuators of the simulation engine of MAST are directly linked with appropriate agents via standard JAVA method calls. Similar approach as the blackboard mechanism described in Section 3.1 will be used instead. The sensor and actuator data will be shared through the tags (data table) of the ControlLogix PLC using the prototype JAVA API. This API allows to directly access the data table, i.e. to read and write the tags remotely via Ethernet link from any JAVA program running on a PC. For example, in the case of the sensor component (Sect. 3.4), there is a sensor detecting presence of a shuttle and an actuator used to stop/release the shuttle. For each sensor there will be two tags in the PLC's data table distinguished by the name of the sensor (e.g. for sensor `s1` there will be tags `s1_sensor` and `s1_actuator`). As the simulation moves virtual shuttles around the track and the shuttle arrives at the sensor `s1`, simulation sets the `s1_sensor` tag in the data table to `true`. It is scanned by the appropriate sensor agent for changes to which the agent reacts, in this case by informing the subscribed agents. The shuttle is stopped if the `s1_actuator` tag is set to `stop`. When the sensor agent decides to release the shuttle, it changes `s1_actuator` to `go` value to which the simulation reacts by releasing the shuttle.

It is obvious that this mechanism allows to simply connect the agents with the physical hardware of the lab instead of the emulation-part of the MAST tool. The only thing needed is to link the physical sensors and actuators with the PLC (through its I/O interface) and store their values under the appropriate names.

Another issue is how to get the EPC data from the readers to the RFID reader agents (Section 3.2). The most convenient solution is to implement a JAVA driver for the physical RFID reader to receive unfiltered EPC data via the Ethernet directly from the reader and do the filtration locally in the RFID reader agent. The other way is to use the existing Savant server solution (see Section 3.1), i.e. to equip the RFID reader agent with the ability to receive already filtered EPC data from the Savant.

4. CONCLUSION

The main idea presented in this paper is that the simulation of agent-based system requires substantially different class of simulation systems and tools in comparison to "classical" centralized systems. The simulation systems applicable in the field of multi-agent system

a) have to be designed as agent-based systems as they have to – as a substantial part of their activities – **emulate** the behavior of the real MAS system. The off-line simulation mode is expected to be complemented or replaced by a real-time control of the physical real-life agent-based system or its part.

b) are expected to carry out – in the off-line mode – the emulation with the goal to achieve the **qualitative simulation** (in the sense of AI terminology) of the behavior of the MAS system as a whole with the stress to capabilities to detect the types/classes (in the optimum case all the types/classes) of potential emergent behavior. That's why, the models of agents should be mainly strongly knowledge-intensive ones suitable for qualitative simulation purposes. The knowledge-oriented analysis of behavior of each type of agents is a very important part of the simulation system design.

c) are explored like **testing, evaluation or diagnostic tools** of the agent-based system, especially during the period of the system design. The testing should start from different initial conditions, under different failures of various components – but nobody can confirm that all the potential states of patterns of behavior will be covered by the series of experimental simulation runs. Simulation – similarly to the case of software testing – cannot be considered as a complete evaluation of the system.

d) can use – for the emulation of the controlled process/physical manufacturing environment – existing standard discrete or continuous simulation tools (or their combination if both discrete and continuous processes have to be simulated concurrently)

e) should be equipped by an efficient **visualization module** as the main "output" of the simulation processes is the "movie" showing the behavior of the system

f) should **run in real time** and be equipped by a **run-time interface** to the real-life control hardware to enable the shift of the borderline between the simulation and real-time real-life control.

Describing the MAST tool and its extension for the Cambridge testbed, we have documented that all the features mentioned above are really needed. Especially the analysis, development and implementation of the RFID agents as well as the ideas behind the implementation of the real-time interface do represent the main technical contribution of this paper.

Summarizing all the specific features of the simulation systems needed for agent-based systems, we can conclude: **The simulation tools of this kind do represent rather complex development environments for agent-based systems than one-purpose simulation vehicles.**

5. REFERENCES

1. Fletcher M, McFarlane D, Thorne A, Jarvis D, Lucas A. Evaluating a Holonic Packing Cell. In Holonic And Multi-Agent Systems for Manufacturing, LNAI 2744, Springer Verlag, Heidelberg, 2003, pp. 246-257.
2. Fletcher M, McFarlane D, Lucas A, Brusey J, Jarvis D. The Cambridge Packing Cell – A Holonic Enterprise Demonstrator. In. Multi-Agent Systems and Applications III, LNAI 2691, Springer Verlag, Heidelberg, 2003, pp. 533-543.
3. Howden N. et al. JACK Intelligent Agents – Summary of an Agent Infrastructure. In Proceedings of IEEE International Conference on Autonomous Agents, Montreal, 2001.
4. Maturana F, Staron R, Hall K, Tichý P, Šlechta P, Mařík V. An Intelligent Agent Validation Architecture for Distributed Manufacturing Organizations. In Proceedings of the 6th IFIP International Conference on Information Technology for Balanced Automation Systems in Manufacturing and Services, Vienna, Austria, 2004.
5. Říha A, Pěchouček M, Vokřínek J, Mařík V. ExPlanTech: Exploitation of Agent-based Technology in Production Planning. In: Multi-Agent Systems and Applications II, LNAI No. 2322, Springer Verlag, Heidelberg, 2002, pp. 308-322.
6. Steels L. A Case Study in the Behavior-oriented Design of Autonomous Agents. In proceedings of the Third International Conference on Simulation of Adaptive Behavior, August 1994, Brighton, UK, pp. 445-452
7. Vrba P. JAVA-Based Agent Platforms Evaluation. In: Holonic and Multi-Agent Systems for Manufacturing, LNAI 2744, Springer Verlag, Berlin Heidelberg, 2003, pp. 47-58.
8. Vrba P. MAST: Manufacturing Agent Simulation Tool. In proceedings of IEEE Conference on Emerging Technologies and Factory Automation, September 2003, Lisbon, Portugal, Volume 1, pp. 282-287.

8

AGENT-BASED ARCHITECTURE FOR INFORMATION HANDLING IN AUTOMATION SYSTEMS

Teppo Pirttioja[1], Ilkka Seilonen[2], Pekka Appelqvist[1], Aarne Halme[1], Kari Koskinen[2]
Helsinki University of Technology
[1]Automation Technology Laboratory
[2]Information and Computer Systems in Automation
[1]teppo.pirttioja@hut.fi, www.automation.hut.fi
[2]ilkka.seilonen@hut.fi, www.automationit.hut.fi
FINLAND

This paper studies issues concerning the application of cooperative information agents to information handling in automation systems. The suggested approach utilizes agent-based layer as an extension to ordinary automation system, thus offering new functions without the need of replacing the existing a utomation system. The proposed architecture u ses a gent-based cooperation methods to enable flexible integration of heterogeneous and distributed data sources and functional or spatial hierarchical division for data abstraction and information filtering. In this case, information agents use BDI-model based manager and data handling modules for information processing. The approach is described with real-life inspired test scenario.

1. INTRODUCTION

Within the rise of the total complexity and the vast amount of acquired information in the automation systems there is a growing need for more powerful design methodologies and techniques. These methodologies should enable easier searching, combination and filtering of information to support end users decision-making. This paper discusses the potential match between properties that information agents provide and generic requirements in automation domain. This paper presents an agent-based architecture to take an advance of information agent technology for automation information handling problems. Agent-based information handling in automation is new research area as previously the application of agents has been focused to control functions.

This paper is outlined as follows: In Chapter 2 information solutions in automation system are represented together with a short review of information agents in other application domains. The suggested architecture is presented in Chapter 3 and the internal design is discussed in Chapter 4. The test scenario is represented in Chapter 5. Finally the conclusions and open questions are discussed in Chapter 6.

2. AUTOMATION AND INFORMATION AGENTS

First of all, the trend in information systems in automation is towards generic solutions and open architectures, i.e., the ability to combine different vendors' solutions is preferred (Tommila et al., 2001). Secondly, generic distributed information systems have evolved strongly and they have a number of properties that are also desired in the automation domain. Such properties are maintainability, openness, reliability, scalability, and easy connection between different resources (Tanenbaum and Steen, 2002). These issues motivate the further development of information systems in automation context.

2.1 Information Systems in Automation

The emphasis in information processing solutions in automation has traditionally been on reliability and solutions have typically been stand-alone (Tommila et al., 2001). In systems level, present day solutions are mainly based on OPC standard (OPC, 2004). This standard provides reliable real-time data access for individual variables and their continually changing numerical values. Also mechanisms for alarm events and access for history data is offered. In the instrumentation level the latest fieldbus standards provide all-digital two-way communication between the devices with modest support of control application design and implementation. Currently, the trend in the instrumentation level is towards more intelligent devices, which produce more and more diagnostic and monitoring information. Unfortunately this useful information is usually provided in vendor specific format.

In addition, spatially distributed instrumentation produce a huge amount of pure numerical data, which has to be processed in real-time, as it is available only in certain time window. Within information handling in automation the emphasis has traditionally been on enabling raw data exchange between distributed resources, while the semantic meaning of data has got little attention.

2.2 Agent-Based Information Systems

Efficient operation in knowledge intensive business needs right information, in the right place, and in right time. As the capacity of data storage and communication bandwidth are getting cheaper the information overload for the human operator has clearly emerged (Knowles, 1999). Agents as a design methodology and implementation tool might provide means to handle this (Ferber, 1999; Jennings, 2000; Luck, 2003, 2004; Tropos 2004). Recently, multi-agent technology has matured up to an industrial standard (FIPA, 2004). Some systematic engineering and documentation methods are also proposed (Luck, 2004, Tropos, 2004).

Lately interest has been in information agents communicating with meaningful messages based on shared ontology. Generally an information agent is a computational software entity that gathers and integrates information from heterogeneous and distributed data sources. One potential way to program these agents is to use BDI-model (Belief-Desire-Intention, see Rao and Georgeff, 1995; Ferber, 1999), and define information handling tasks as goals that agents try to achieve with searching, filtering and combining information. Furthermore, a variety of brokering techniques have been developed to match service requesters and service providers (Klush et al., 2003), including the use of semantics (Nodine el al., 2003).

2.3 Possibilities of Information Agents in Automation

In the automation domain there is a clear need for systematic design method for information handling in an environment that is distributed and dynamically changing by nature. On the one hand, agent-based architectures for automation control functions have been proposed by a number of researchers (Cockburn and Jennings, 1995; Parunak, 1999, Marik et al., 2002; Seilonen et al., 2002), and an industrial demonstration has also been presented (Jennings and Bussmann, 2003). Most of these architectures locate agents to a separate layer on the top of the physical automation system. On the other hand, applications of information agents in other application domains have number of aspects in common with automation applications, e.g., searching information from heterogeneous data sourçes.

Furthermore, using ontologies to define semantic meaning of messages has been studied, e.g., (OWL, 2004), and it is argued that these technologies could be useful in manufacturing applications (Obitko and Marik, 2003). In automation applications it could be valuable to apply an approach, where planning and execution are interleaved, as this is argued to support adaptation to changing environmental situations (desJardins el al., 1999).

Although a large number of useful technologies for effective information processing have been proposed, no combining architecture has been presented yet. Therefore, this paper introduces an architecture that integrates various above mentioned useful technologies.

3. AGENT-BASED INFORMATION SYSTEM ARCHITECTURE IN AUTOMATION

The purpose of information agents in the context of automation systems is to provide an additional intelligent and active information access layer, which will enable more easy and efficient utilization of information for human users. The information agents are intelligent in the sense that they handle information access goals. When an information access goal needs to be fulfilled the information agents will cooperatively find a way to provide that information if possible. The activeness of the information agents means that they can take initiative in information access. They can start cooperative information access operations themselves, if they just are aware of users interests.

To be able to use ordinary agent software development tools our information agents were situated to a separate layer on top of existing automation hardware and software, this is illustrated in Figure 1. This design follows our previous work with controlling agents (Seilonen et al., 2003).

3.1 Roles of Information Agents

The information agents form a cooperative agent society, where each agent has a certain role. We have defined these roles in the case of an automation application. Currently, this architecture includes agents operating in the following roles: *Client-*, *Information-*, *Process-*, and *Wrapper Agent* (e.g., see Figure 1). There the *Client Agent* provides human user an interface, and translates human understandable queries to agent communication language. *Information Agent* decomposes information queries to subqueries. *Process Agent* is responsible of certain spatially or functionally divided area of the total manufacturing process and *Process Agents* are arranged in hierarchically form to support information abstraction. *Wrapper Agent* is used to access the information stored in legacy information systems.

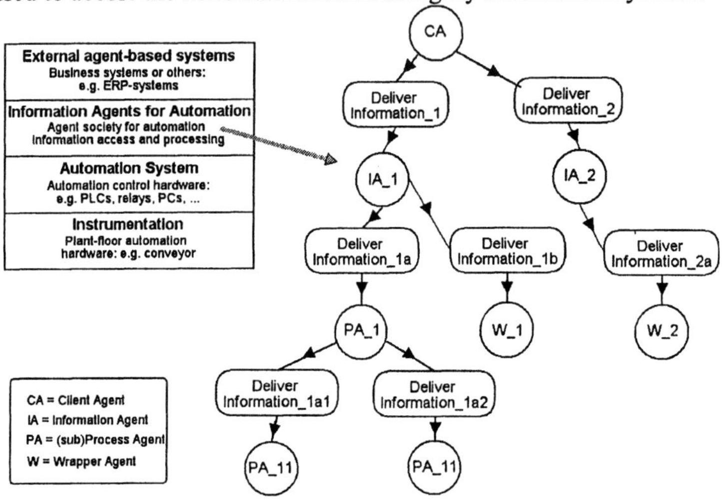

Figure 1 – an example of hierarchical setup of information agents

In our FIPA standard based approach agents register their services to the Directory Facilitator (DF) (FIPA, 2004). Although, agents register mainly that they are providing information, interest of certain information may be register also. Such information could be alarms or operation mode changes.

3.2 Functions of Information Agents

Potential functions for information agents in automation context include: diverse monitoring and diagnostic functions, advise the operator in the selection of operation mode, filtering relevant alarms from enormous number of alarms based on operational states, etc. Generally, functions that are decentralized by nature and benefit from agent cooperation are suitable for information agents. Especially, information agents could monitor actively in the device level and share information about emerging problems.

4. DISTRIBUTED INFORMATION PROCESSING BASED ON DOMAIN ONTOLOGY

The information agents need an internal design, which will give them the particular capabilities they need in their goal-oriented and cooperative information access operations. The BDI agent model is very suitable as a basic architecture for agents with goal-oriented operation. In addition to this, the information agents utilize FIPA interaction protocols as cooperation mechanisms and ontologies as data modeling technique. In this architecture, the domain ontology is the foundation for defining various models describing the process for the agents. The design of information agents should also enable the use of other information processing methods, such as principal component analysis or statistical methods.

4.1 BDI-model based Information Processing

The manager module controls the planning and execution of individual information access and processing tasks inside one agent. When information delivery goal is received the manager partitions it to a number of subgoals that match up to specific atomic operations. The execution of these operations is then conducted by individually information processing modules, illustrated in Figure 2.

First of all, the manager operation depends on the used interaction protocol, which specifies how and when to respond to the information delivery goal. Then the message content specifies what information processing modules are needed to fulfill the information delivery goal. If there is no internal module that can fulfill certain subgoal the manager tries to find out if some other agent is able to deliver this partial information. In this architecture the role of the manager is to decide which modules are used for information processing but not how.

4.2 Modules for Information Content Processing

The actual information processing is performed by modules, which are specialized to certain functions (see Figure 2). With information input modules these operations correspond to units of information read from particular information source (database, file, or conversation with other agents). Information processing module provides filtering, reasoning, and computational operations. Information output modules are used to distribute the processed information to other agents with conversations. Data exchange between different modules is based on domain specific ontology.

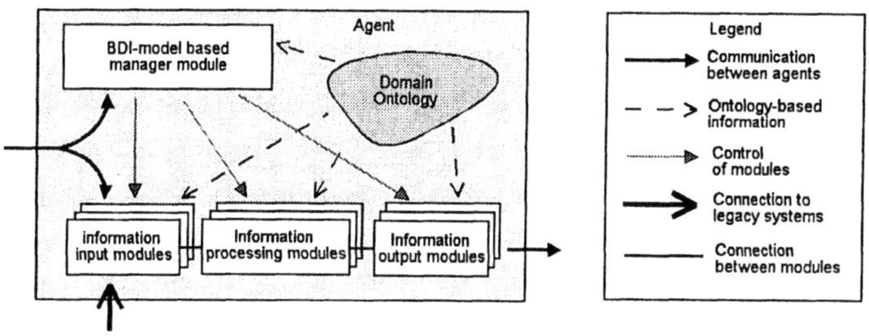

Figure 2 - Internal design of BDI-model based information agents

Using separated modules for different information processing functionalities is argued to enable easy system updates in the future. New modules with new features may be constructed and added to the existing ones without a need to program the operation of the manager entirely again.

4.3 Engineering Viewpoint

General agent platform services (communication and BDI-model agents) may be packed up with information agent services (matchmaking, and information delivery ontology) to form a basis for an engineer to build up an agent application. On top of that, at least conceptually, there is a layer providing configurable tools for automation information processing. The configuration files describing the particular application are left to the uppermost and distinct layer. In engineering viewpoint the idea is to provide user the possibility to program the operation of these agents with intuitive concepts and leave the description of the physical configuration of the production environment to a separated process models.

5. APPLICATION OF INFORMATION AGENTS IN PAPER MAKING

Our test case consists of preliminary implementation of information agents, which have connection to legacy information systems containing real production data. Generally the properties of pulp and paper are difficult to measure as the instrumentation that is used to measure important process quantities (e.g., pH, consistency, brightness) is subject to fouling and drifting (Leiviskä, 1999). Because the direct detection of malfunctions is problematic, most important measurements are crosschecked with physically doubled instrumentation and laboratory measurements may be used to verify the long-term stability.

5.1 Wrappers for Fusion of Measurement Information

The goal of our first test scenario was to produce integrated information about the operational condition of physical instrumentation to process operator. Initially this information was available in different user interfaces, and it was stored in three separate data sources using different data formats. In this scenario the *Information Agent* uses domain ontology to find out the different information types and DF to search responsible *Wrapper Agent* for each of these information types. As the *Wrapper Agents* are programmed to answer information queries in common presentation format, specified in the domain ontology, the fusion of information is straightforward. Sequence diagram for this scenario is shown in Figure 3.

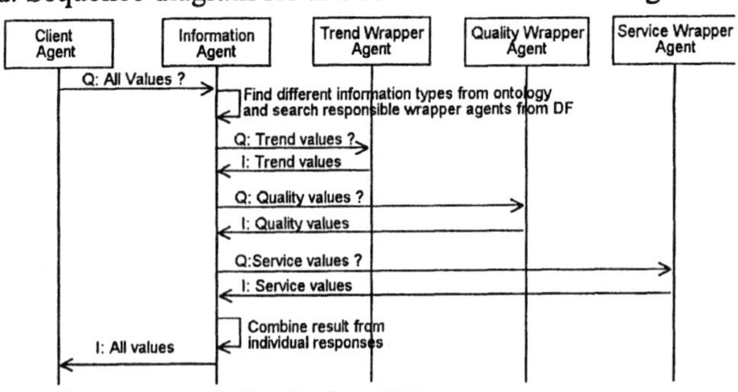

Figure 3 - Sequence diagram for fusion function

In the first test scenario the selection of FIPA query interaction protocol is reasonable as the requested data is available immediately in the data sources. In addition, the FIPA standard suggests the use of query-ref when the initiator agent wants some information that another agent knows. (FIPA, 2004)

5.2 Wrappers for Active Condition Monitoring

The second test scenario concerns the validation of online measurements. Uncertain on-line measurements are automatically compared by the agents to the exact laboratory measurements. As these laboratory measurements are available in certain time intervals, it was decided to use *Wrapper Agents* actively supervise the appearance of new measurements. When new laboratory measurement is available in legacy information system the *Wrapper* informs this to *Information Agent,* which may then use case-specific algorithm to find out if device is malfunctioning.

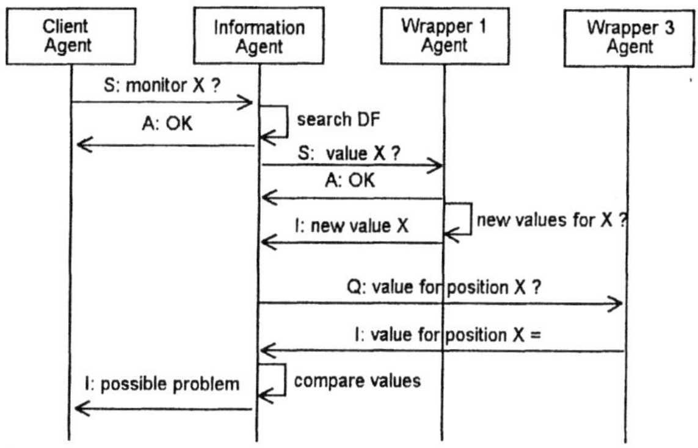

Figure 4 - Sequence diagram of monitoring function

In this scenario the use of FIPA subscription interaction protocol is justified by the fact that process operator is interested only in changes in monitoring status. These changes are possible only in times when new laboratory measurements are fed to the system. Furthermore, the FIPA standard suggests the use of subscription protocol when the initiator agent wants another agent to notify continuously about changes in the specified information. (FIPA, 2004)

6. CONCLUSIONS

In this paper an agent-based architecture for automation information handling is presented. The motivation for this architecture is the possible match between features that information agent technology tools provide and the needs of information processing in the automation context. This architecture gives the possibility to use systematic agent-oriented software engineering methods to construct automation information functions, and further realize the designs with latest agent platform implementations. Furthermore, our first two implemented test scenarios are targeted to help process operators to monitor the functionality of process devices.

In the future, the presented information agent architecture may be used to combine agents for automatic control and agents for information processing. Future research focuses on designing more functions based on this architecture and testing these by implementing them. Furthermore, using formal methods to describe the automation domain ontology for reasoning purposes and for the interaction between the agents is under further development.

7. REFERENCES

1. Cockburn D and Jennings NR. "ARCHON: A distributed artificial intelligence system for industrial applications". In Foundations of Distributed Artificial Intelligence, O'Hare GMP, Jennings NR, ed.: Wiley & Sons, 1995.
2. desJardins ME, Durfee HD, Ortiz CL Jr., Wolverton MJ, "A Survey of Research in Distributed, Continual Planning". AI Magazine, 4, 13-- 22, 1999.
3. Ferber J. "Multi-agent systems: An introduction to distributed artificial intelligence". Addison-Wesley, 1999.
4. FIPA. The Foundation for Intelligent Physical Agents, http://www.fipa.org, 2004.
5. Jennings NR. "On agent-based software engineering". Artificial Intelligence 117, 277–296, 2000.
6. Jennings NR and Bussmann S. "Agent-based Control Systems". IEEE Control Systems Magazine, 2003.
7. Knowles C. "Just-in-time information," Proceedings of the Second International Conference on the Practical Application of Knowledge Management, 1999.
8. Klush M, Omicini A, Ossowski S, Laamanen H. "Cooperative Information Agents VII", Lecture Notes on Artificial Intelligence 2782, Springer, 2003.
9. Leiviskä K. "Process Control", Vol 14 of the Papermaking Science and Technology Series, TAPPI, ISBN: 952-5216-14-4, 1999.
10. Luck M, McBurney P, Preist C. "Agent Technology: Enabling Next Generation Computing". AgentLink, ISBN 0854 327886, 2003.
11. Luck M, Ashri R, D'Inverno M. "Agent-Based Software Development". Artech House Publishers, ISBN: 1580536050, 2004.
12. Marík V, Fletcher M, Pechoucek M. "Holons & Agents: Recent Developments and Mutual Impacts". In Multi-Agent Systems and Applications II, Marik V, Stepankova O, Krautwurmova H, Luck M, ed. Springer-Verlag, Germany, pp. 323-335, 2002.
13. Nodine M, Ngu AHH, Cassandra A, Bohrer WG. "Scalable Semantic Brokering over Dynamic Heterogeneous Data Sources in InfoSleuth™", IEEE Transactions on Knowledge and Data Engineering, Vol. 15, No. 5, 2003.
14. Obitko M and Marík V. "Adding OWL Semantics to Ontologies", Holonic and Multi-Agent Systems for Manufacturing (HoloMAS), pp. 189-200, 2003.
15. OPC. OPC Foundation, http://www.opcfoundation.org/, 2004.
16. OWL. Web Ontology Language, http://www.w3.org/TR/owl-ref/, 2004.
17. Parunak DH. "Industrial and practical applications of DAI". In Multiagent systems, Weiss G ed. Cambridge, MA, USA, pp. 377-421, 1999.
18. Rao AS and Georgeff MP. "BDI agents: From theory to practice". Tech. Rep. 56, Australian Artificial Intelligence Institute, Melbourne, Australia, 1995.
19. Seilonen I, Pirttioja T, Appelqvist P. "Agent Technology and Process Automation". 10th Finnish Artificial Intelligence Conference (STeP 2002), Oulu, Finland, December 16-17, 2002.
20. Seilonen I, Pirttioja T, Appelqvist P, Halme A, Koskinen K. "Cooperating Subprocess Agents in Process Automation". 1st International Conference on Applications of Holonic and Multi-Agent Systems (HoloMAS 2003), Prague, Czech Republic, September 1-3, 2003.
21. Tanenbaum AS and Steen MV. "Distributed Systems: Principles and Paradigms". Prentice Hall, ISBN: 0-13-088893-1, 2002.
22. Tommila T, Ventä O, Koskinen K. "Next generation industrial automation – needs and opportunities". Automation Technology Review 2001, VTT Automation, 34-41, 2001.
23. Tropos. Agent software development methodology, http://www.troposproject.org/, 2004.

9 AN INTELLIGENT AGENT VALIDATION ARCHITECTURE FOR DISTRIBUTED MANUFACTURING ORGANIZATIONS

Francisco P. Maturana
Raymond Staron
Kenwood Hall
Rockwell Automation, Mayfield Heights, OH, USA
{fpmaturana, rjstaron, khhall}@ra.rockwell.com

Pavel Tichý
Petr Šlechta
Vladimír Mařík
Rockwell Automation Research Center, Prague, CZECH REPUBLIC
{ptichy, pslechta, vmarik}@ra.rockwell.com

In this paper, we focus on validation of Multi-Agent System (MAS) behavior. We describe the simulation architecture and the system design methodology to accomplish the appropriate agent behavior for controlling a real-life automation system. The architecture is explained in the context of an industrial-sized water cooling system. Nevertheless, it is intended to operate in a wide spectrum of control domains. In general, after the design of the control system is accomplished, a set of validation procedures takes place. The current needs are to validate both the control and the agent levels as integrated parts. Hence there is a need to establish a general architecture and methodology for easing the commissioning process of the control solution.

1. INTRODUCTION

Distributed systems such as manufacturing, supply chains, service industry, and information infrastructures require a flexible structure for the integration of their components to fulfill the market requirements of this century. Solutions to such requirements can be found in Intelligent Agent technology which provides an appropriate framework to integrate knowledge with efficient production actions (Brooks, 1986) (Wooldridge and Jennings, 1995) (Shen, *et. al.*, 2001).

The validation of agent behavior is not only a local to-the-agent issue but a global issue, where the interaction of the agents and associated latencies should also be modeled as part of the system. This requirement introduces an interesting complexity into the design of the architecture.

Distributed organizations emerges as a result of the dynamic interactions of its intelligent components, which can be human or artificial (intelligent agents or holons), or a hybrid (Christensen, 1994) (Mařík, *et. al.*, 2001). To validate the task sequences and interactions of the agents, it is required to understand the agent context from multiple views. We describe the gluing technology to integrate the pieces of the organization from the device level into upper enterprise levels. We focus on the validation infrastructure which is based on a Simulation Development Environment (SDE). The SDE is merged with the agent and control systems. The validation system allows the designer of the agents to verify the feasibility of the agents' actions prior to final commissioning of the system. This is a contribution to the system architecture to improve the design and performance of the future agent-based organization. The agent behavior can be refined exhaustively prior to its final deployment, without ad-hoc investments or complicated equipment in-the-loop. This infrastructure is synchronized with controllers to mimic the real-time operations in order to obtain good representations of the events occurring in the real world. We demonstrate the new infrastructure on an industrial-sized cooling system.

2. AGENT ARCHITECTURE FOR CONTROL

In the past, development of Agent architectures was focused on experimental systems of reduced scale (Maturana, *et. al.*, 2002) (Chiu, *et. al.*, 2001) (Tichý, *et. al.*, 2002). In those experiments, the foundation architecture for highly distributed control agents was established. Step by step, the new requirements were introduced into the extensions of the automation controllers to enable the creation of distributed intelligence in control.

We anticipate that agents will be distributed among multiple automation devices or Programmable Logic Controllers (PLCs) and therefore an agent infrastructure is needed to fit well the manufacturing environment, information networks, and enterprises in general. Each agent represents a physical process or machine or device and coordinates its operations with other agents. The MAS architecture is organized according to the following characteristics:

- Autonomy: Each agent makes its own decisions and is responsible for carrying out its decisions toward successful completion.
- Cooperation: Agents combine their capabilities into collaboration groups (clusters) to adapt and respond to diverse events and goals.
- Communication: Agents share a common language to enable interoperation.
- Fault tolerance: Agents possess the capability to detect equipment failures and to isolate failures.

2.1 Automation Architecture

In agent-based control, the controllers have an agent infrastructure for enabling the component-level intelligence. With this, it is possible to distribute the intelligence among multiple controllers using different agent sizes and populations. In this architecture, controllers of various sizes and capacities can be deployed. Different network connectivity can be used to exploit the distributed intelligence dimension that is added by the agents.

Regardless of the network topology (e.g., backbone or ring), the relationship among the agents is kept loosely coupled. There are dynamic interactions among the agents occurring during the decision-making process. These dynamic interactions establish logical relationships among the agents temporarily. The agents are designed based on FIPA specifications (http:\\www.fipa.org) and ContractNet protocol (Smith, 1980) to create and coordinate their activities throughout logical links. To enable the agent-based automation architecture, it was required to modify the controller's firmware. A common software infrastructure is shared among the different controllers.

The application software represents the physical components and processes of the facility under control by the agents. Each agent represents a physical device such as a valve, water service, heat load, etc. After the agent is created, it is ready to begin operations by carrying out initialization procedures (capability registration) and waiting for external messages or events from the control systems.

The agents contact each other within and outside the controllers via Job Description Language (FIPA/JDL) messages. FIPA/JDL is used by the agents to represent planning, commitment, and execution phases during the task negotiation. Information is encoded as a sequence of hierarchical actions with precedence constraints. JDL is also used to encode plan templates. A plan template is a representation of the agent behavior as parametric scripts. A parametric script has entry variables whose values are set during the planning process. Moreover, the script has associations with internal-to-agent functions which are executed to fulfill local decisions.

When an agent accepts a request, an instance of a plan template is created to record values emerging during the planning process. Requests are propagated throughout the organization using the Contract Net protocol. The requests visit multiple agents and negotiation clusters are formed.

For inter-organization conversations, the agents emit messages outside their organization via wrapping JDL messages inside FIPA envelops. This implementation includes Directory Facilitators (DFs) functionality to be FIPA compliant. A DF performs capability registration and matchmaking. For each capability request, a DF provides a list of agents that coincide with the requested capability. For instance, an overheating component requests cold water from its water service. This is a cooling process capability.

2.2 Intelligent Agent Architecture

The agents are goal-oriented entities. They organize the system capabilities around system missions. There are agents exclusively programmed to emit missions. Other agents are programmed to handle the mission requirements and the execution control. This type of distributed responsibility is easily handled using the agent programming methodology. Information is fractioned into small pieces and each of these is associated with separate agents. Importantly, agents are divided according to class types. Thus, information is encapsulated under a class type as a template to be used by the derived instances of that class type.

Goals emerge dynamically and these are agreed upon by the agents throughout negotiation. For instance, an agent that detects a water leakage in a pipe of the cooling system establishes a goal to isolate the problem. The agent then informs adjacent agents to evaluate the problem according to their views and borrowed data. This is the origin of a group based goal, which is to isolate the leaking pipes, in spite of the cost of operation. This action exceeds the pre-assigned priorities. Isolation is the highest possible priority.

The architecture of an agent has four components (Tichý, *et. al.*, 2002); planner, execution control, diagnostics, and equipment model. Important part is the execution control component that acts as control proxy, which translates committed plans into execution control actions. These actions are synchronized with the control logic programs. It also monitors events from the control logic and translate them into response-context events to be processed by the planner component.

Both the controller infrastructure and agent architecture facilitate the creation of agents for controlling the physical system. Thus, the control engineer and system engineer can experiment with distributed intelligence and control in a flexible manner. However, the puzzle is incomplete from the solution validation point of view. In general, after the automation system has been modeled in software, it is tested on physical pilot systems until a fine tuning of the control system is achieved. But with the introduction of the agent software, this operation becomes more difficult because the system has a larger number of control variables to be tested and stabilized. Therefore, physical testing of such a system becomes impractical. Hence, there is a need to incorporate a validation system to help the solution modeling process, with a minimum of manual operation and pre configuration cycles.

3. SIMULATION ARCHITECTURE

The general tendency in validation systems for automation control is to build physical prototypes or scaled down models of the real system. This practice is ideal from the accuracy of the observations that are extracted from the operations of the system during validation. Nevertheless, it is also practical to develop simulated models to enable extensive validation process.

Information is organized under the agent scope. This information relates to the transactions that occur during the planning process and during the execution of the plans. Agents enable the construction of more advanced strategies for controlling the system (according to the emergent behavior perspective). Advanced control strategies imply physical changes into the pilot facility, which adds cost and process uncertainty. From the predictive side of the spectrum, more advanced strategies allow for proactive diagnostics. But this requires the equipment to produce specific signatures that are generally obtained after a certain number of service hours. In simulation this can be done efficiently. Therefore, the obvious conclusion is to pursue an integrated architecture that includes all three elements: (1) control, (2) agents, and (3) simulation.

The main components of the validation system are: (1) agent/control software, (2) SDE, (3) soft controller, and (4) simulation. Figure 1 shows these components.

- *Agent/control software*: This component represents the agent and control software creation. Other publications describe more details about this component. This component produces three files types: (a) Agent object code (Agents.o): executable agent code to be placed in RAM of the controller; (b) Ladder logic code (.L5k): control programs written in ladder logic; and (c) Tag symbol topology (.xml): This represents the inputs and output variables of the field devices.

- *SDE*: This component takes the tag symbol topology and the simulation library to help the user match the control and simulation variables. The variable matching is a critical task that is generally performed manually in very separate contexts and by different people. Commonly, the tag symbols from control do not match the symbols from the simulation models. Another component of this architecture is a tool for importing the symbols from either source (control or simulation) into the other. In this manner, symbols from one source can be made available into the other for ensuring 100% correspondence among the symbols. This component produces an association file which is used to create a proxy. The proxy synchronizes the controllers and simulator clocks and also does data exchange.
- *Soft controller*: This component is an exact emulation of a hardware based controller. It allows for the creation of multiple controllers inside a single chassis as well as communication cards such as Ethernet/IP and ControlNet. This component is intended to contain agents and control programs, i.e., the behavior of the real multi-agent system is emulated in this environment (Mařík, *et. al.*, 2004).
- *Simulation*: This component represents the simulation environment. In this, the application-domain process is modeled using user-preferred techniques and languages. The fundamental idea is to deploy Commercial-Off-The-Shelf (COTS) simulation packages (e.g., Matlab, SolidWorks, Arena, etc.). COTS simulations are more practical from the industrial world point of view. Based on the majority of the cases observed, the usage of commercial simulators is more constructive than writing ad-hoc simulations.

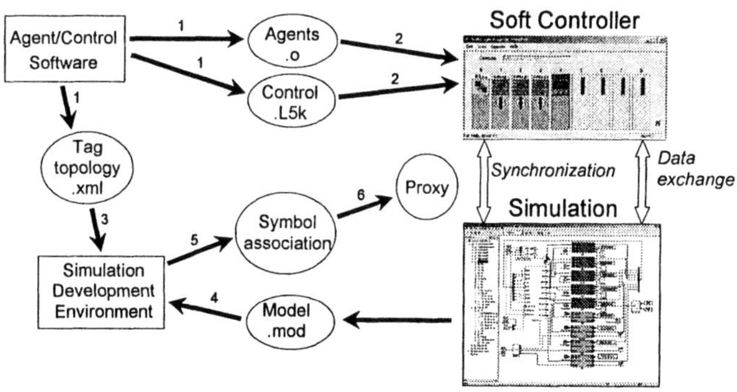

Figure 1 – Simulation system architecture

The simulation architecture adds many degrees of freedom to the design and validation of the agent-based automation system. With this, multiple strategies can be treated as equal and tested using a single computer without incurring into additional investment. Reusability of the infrastructure is a very relevant attribute. The following sections focus on the system design methodology.

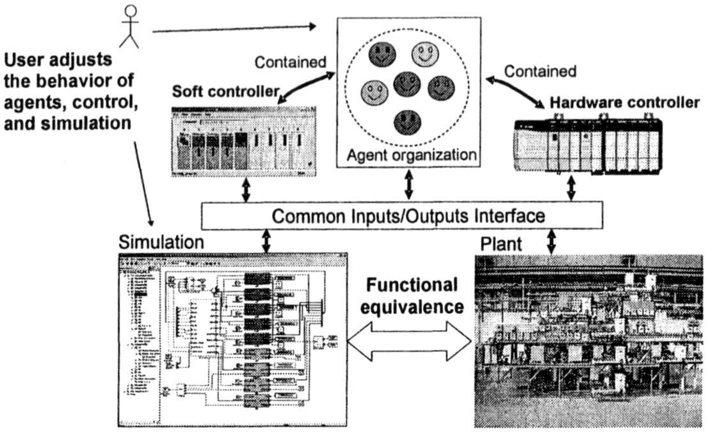

Figure 2 – Software equivalence

The design and validation process is as follows:

1) Users create a first prototype of the agent and control code to suit the characteristics of the physical plant;
2) A simulation engineer creates a simulation model of the physical plant;
3) Users create some desired agent strategies for fulfilling advanced control;
4) Agents and control software are downloaded into the soft controller;
5) Users match the tag topologies and create the synchronization proxy;
6) The integrated system is executed and observed to verify if the desired behavior is fulfilled correctly by the agents and control programs;
7) Users modify the behaviors by changing simulation, control, or agents;
8) Repeat Step (6). If users add more components or if users change the input/output configuration of the initial components, then repeat Step (5);
9) All the desired behaviors have been fulfilled to complete satisfaction; and
10) Software is transported into the physical plant and industrial controllers for final commissioning.

Another important aspect of this methodology is the common input/output (I/O) interface. Both the simulation and the physical plant expose the same set of I/O signals. Therefore, the agent software that interacted with the simulation will see no difference when connected to the physical equipment, because the interconnection has been done through a common I/O set. Nevertheless, at the hardware level, it is expected that some changes will occur regarding the characteristics of the equipment. It is understood that simulation can be very accurate in some cases, but it is still an idealization of the real situation. Nonetheless, these proposed changes are considerably lower than those occurring in a conventional commissioning process, yet from the lab into the pilot facilities.

4. SYSTEM MODELING

Figure 3 shows the cooling system under study. This cooling system is water based and it is currently used at a Navy site to mimic the cooling system of the DDG-51 destroyer class ship of the US Navy. This system is used for evaluation of advanced auxiliary machinery concepts. The cooling system is a reconfigurable fluid system platform with component-level intelligence. It includes the plumbing, controls and communications, and electrical components that mimic the real-life operations.

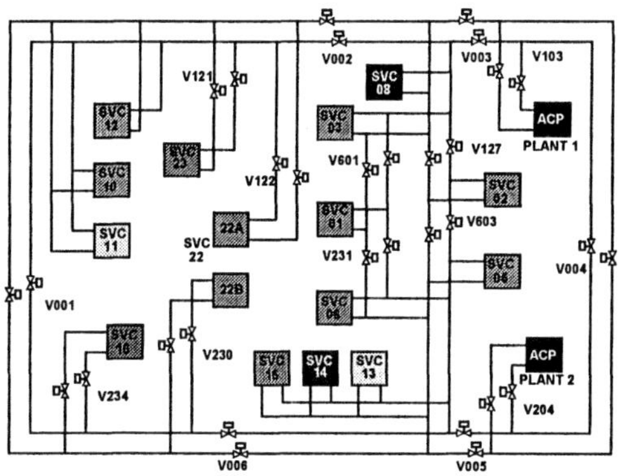

Figure 3 – Cooling system

Immersion heaters provide stimuli for each service (SVC boxes in Figure 3) so as to model actual heat transfer. Essentially, there are 3 subsystems, plants, mains and services. There is one plant per zone (i.e., currently 2 plants: ACP boxes). There are two types of services, vital (14) and non-vital (2). While in operation, under normal conditions, the cooling system is segregated in two zones to maintain the cold water from each source separate. These two zones increase the survivability of the system in case of damage occurring on one side.

The water from the cooling plants (named ACP plants) should never be mixed. Cooling flow is controlled by each service using a local flow circuit. As more services demand cooling the relative demand on the plants is increased. Under low load conditions, it is possible for one ACP to handle all the loads. However, high loading conditions will require that non-vital or low priority loads be shed from the cooling loop until a future time at which time the heat load and water distribution could be balanced again.

The ACP plants were modeled as a single agent each, which included pipes, valves, pumps, an expansion tank, and water-level, pressure, flow and temperature sensors. The main circulation piping is partitioned among 'T' pipe sections, i.e., passive agents. Load agents include a heat generator and a temperature sensor. Water Services agents include valves and flow sensors. There are standalone valves in the main circulation loop for the supply and return lines. This partitioning gives us a total of 68 agents.

5. RESULTS

The results will be presented in terms of the specific models (control, tags, and simulation) of the cooling system. There is no specific target result that can be easily pinpointed from this work but the capability to integrate agent control and simulation for validation purposes.

The following describes the system's topology. In Figure 4, the mapping of the physical I/O tags as a fragment of the tag symbol topology. The symbol information was automatically extracted from the agent/control models in XML format (refer to Section 3).

```
<Component name="SVC03">
<Tags>
<TAG name="SVC03allReqClose" host="slx1"
    access="r" type="boolean" value="0" />
<TAG name="SVC03allReqOpen" host="slx1"
    access="r" type="boolean" value="0" />
<TAG name="SVC03anyFailedToOpen" host="slx1"
    access="r" type="boolean" value="0" />
<TAG name="SVC03anyIsIsolatingLeak" host="slx1"
    access="r" type="boolean" value="0" />
<TAG name="SVC03reqClose" host="slx1" access="r"
    type="boolean" value="0" />
<TAG name="SVC03valveOK" host="slx1" access="r"
    type="boolean" value="0" />
<TAG name="SVC03waitingForRepair" host="slx1"
    access="r" type="boolean" value="0" />
</Tags>
<Component/>
```

Figure 4 – Cooling service I/O topology

An agent is a component (e.g., SVC03) with a set of tag elements. Each element identifies the name of the tag (e.g., 'SVC03allReqClose'), the name of the controller that contains the tag (e.g., 'slx1'), an access attribute which is 'r' for reading or 'w' for writing, depending on whether the simulation reads or writes into the variable, and a value type and an initial value. Figure 4 only shows a fragment of the tag topology for one service, we have approximately 2000 tags for the whole system.

Next, we explain a use case that was based on a Matlab/Simulink simulation of the cooling system. The simulation is a qualitative model, which includes water flow dynamics and heat transfer simulation for each of the components. Figure 5 shows a partial model of one of the cooling regions. It has five loads (SVC05, SVC06, SVC13, SVC14, and SVC15). Each simulation sub-model has I/O symbols that are imported to the SDE for subsequent matching.

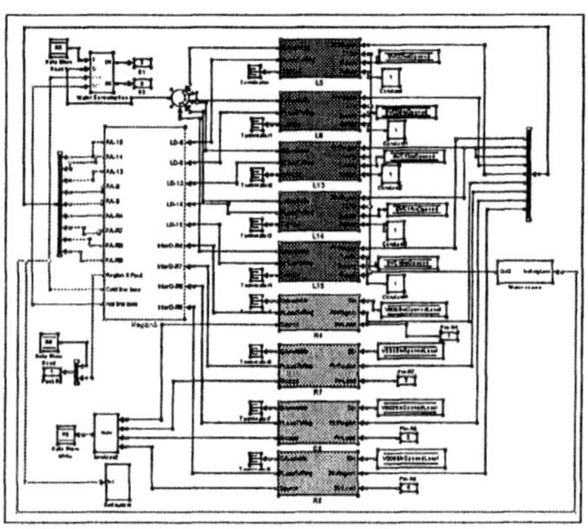

Figure 5 – Simulation sub-model (Simulink view)

After completing the matching of the symbols and proxy configuration, the system was executed to observe its behavior. The experiment shown in Figure 6 consists of emitting a system mission request into the cooling system. The request is to provide cooling under cruise conditions. The cooling system agents tried a configuration by emitting a series of sub-requests to different sections of the cooling system. The initial attempt (see the left part of Figure 6) failed because there was a problem in the water route discovery process. This experiment also failed for other missions such as cooling in 'battle' and 'in-port' modes.

In this experiment, we demonstrated the capability to observe and debug the system's behaviors using a simulation system, real agents and formal control algorithms. After deducing the probable causes of the error, the agent and control code was modified and next experiment was executed. The right part of Figure 6 shows the results. Now, the mission request went through some additional layers marking a successful completion.

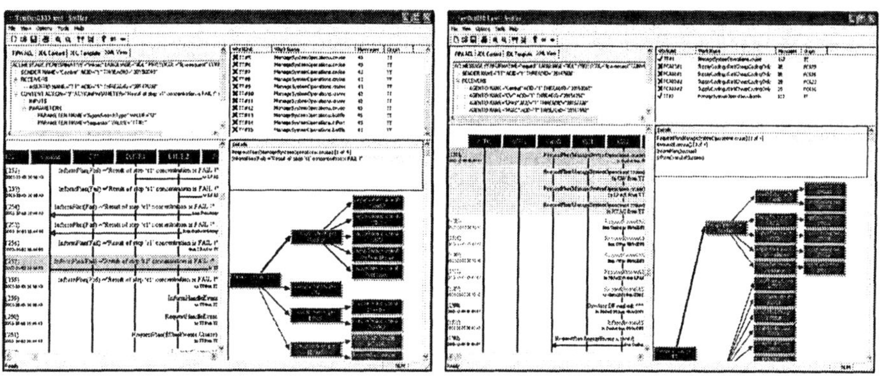

Figure 6 – Results of the first execution (left) and improved performance (right)

This experiment showed that the modification of the code eliminated the problem partially, since there were some failing conditions for the other missions. Without the tool to experiment with partial changes, this troubleshooting process would have been extremely hard and tedious using the real equipment. Progressively, as we continued debugging the system more errors appeared until the system was completely cleaned out to operate as expected.

We think that it is important to remark that the troubleshooting procedure described above does not replace the commissioning phase. On the contrary, it complements the final delivery of the solution by accelerating the process of eliminating errors from the system ahead of time and in arbitrary locations chosen by the designers of the system.

7. CONCLUSION

The above results give explanation of what could be done with the agent/control validation system. In this work, we presented a set of prototype tools and procedures well aligned with industrial automation. Other more complex interactions have been experimented with excellent results. One immediate observation is the reduction of the design time from the beginning of the modeling until obtaining a good working model. It has been also observed that the number of modification cycles increased but these were processed faster. This technique prevented our team from experimenting with real expensive equipment. The debugging and validation tasks were partitioned among multiple users. Each user pinpointed specific advantages and deficiencies of the system. Current research efforts are on the improvement of the new tools and on the establishment of a general methodology to create agents and agent validation environments.

7. REFERENCES

1. Brooks A.: "A Robust Layered Control System for a Mobile Robot", IEEE Journal of Robotics and Automation, 2(1), 14-23, 1986.
2. Chiu S., Provan G., Yi-Liang C., Maturana F., Balasubramanian S., Staron R., and Vasko D.: "Shipboard System Diagnostics and Reconfiguration using Model-based Autonomous Cooperative Agents", ASNE/NAVSEA Intelligent Ship Symposium IV, Philadelphia, PA, April 2001.
3. Christensen J.H.: "Holonic Manufacturing Systems: Initial architecture and standards direction", First European Conference on Holonic Manufacturing Systems, Hanover, Germany, 20pp, 1994.
4. IEC (International Electrotechnical Commission), TC65/WG6, 61131-3, 2nd Ed., Programmable Controllers - Programming Languages, April 16, 2001.
5. Mařík V., Pěchouček M., and Štěpánková O.: "Social Knowledge in Multi-Agent Systems". In: Multi-Agent Systems and Applications (Luck M., Mařík V., Štěpánková O., Trappl R. eds.) LNAI 2086, Springer-Verlag, Heidelberg, pp. 211-245, 2001.
6. Mařík V., Vrba P., and Fletcher M.: "Agent-based Simulation: MAST Case Study". Accepted by the 6th IFIP International Conference on Information Technology for Balanced Automation Systems in Manufacturing and Services (BASYS'04), Vienna, Austria, 2004.
7. Maturana F., Staron R., Tichý P., and Šlechta P.: "Autonomous Agent Architecture for Industrial Distributed Control". 56th Meeting of the Society for Machinery Failure Prevention Technology, Section 1A, Virginia Beach, April 15-19, 2002.
8. Maturana F.P., Tichý P., Šlechta P., and Staron R.: "Using Dynamically Created Decision-Making Organizations (Holarchies) to Plan, Commit, and Execute Control Tasks in a Chilled Water System". In Proceedings of the 13th International Workshop on Database and Expert Systems Applications DEXA 2002, HoloMAS 2002, Aix-en-Provence, France, pp. 613-622, 2002.
9. Shen W., Norrie D., and Barthès J.P.: "Multi-Agent Systems for Concurrent Intelligent Design and Manufacturing". Taylor & Francis, London, 2001.
10. Smith R. G.: "The Contract Net Protocol", High-level Communication and Control in a Distributed Problem Solver. In IEEE Transactions on Computers, C-29(12), pp. 1104-1113, 1980.
11. Tichý P., Šlechta P., Maturana F.P., and Balasubramanian S.: "Industrial MAS for Planning and Control". In (Mařík V., Štěpánková O., Krautwurmová H., Luck M., eds.) Proceedings of Multi-Agent Systems and Applications II: 9th ECCAI-ACAI/EASSS 2001, AEMAS 2001, HoloMAS 2001, LNAI 2322, Springer-Verlag, Berlin, pp. 280-295, 2002.
12. Wooldridge M. and Jennings N.: "Intelligent agents: theory and practice", Knowledge Engineering Review, 10(2), pp. 115-152, 1995.

10

MULTI-AGENT BASED FRAMEWORK FOR LARGE SCALE VISUAL PROGRAM REUSE

Mika Karaila
Energy & Process Automation, Research & Technology Department
Metso Automation Inc. FIN-33101 Tampere, FINLAND
+358-40-7612563 mika.karaila@metso.com
Ari Leppäniemi
Ari.leppaniemi@metso.com

Today's application engineers are committed to the reuse of programs for performance and economic reasons. Moreover, they increasingly have to complement application programs with less information and in shorter time. The reuse of already implemented programs is therefore fundamental. We have implemented a process automation specific framework that supports reuse of our domain specific visual language. The visual Function Block Language is used for power plant and paper machine controls.
The reuse framework discussed in this paper relies in identification and usage of templates, which are used for generating actual application software instances.
The framework automates data mining with software agents collecting metadata. The metadata is send ahead to the receiver agent that stores the data into the central database. Another agent analyzes stored data and performs template matching. Again another agent is called for more detailed template match comparison. Although the database is centralized, the agents can be distributed and run in intranet. The framework implementation is pure java based and runs on JADE-FIPA agent platform

1. INTRODUCTION

Normally computer programs are written using textual programming languages. The more sophisticated or domain specific environment programming can be done in visual way. CAD-like programming environment will support different kinds of symbols and connections describing methods or relationships between the actual objects or instances.

The process automation specific visual language is used for making customer specific process control software (mass customization). The application software is created with visual Function Block Language (Figure 1. An example program). Later on function block loops are compiled to byte-code that is executed on the control system.

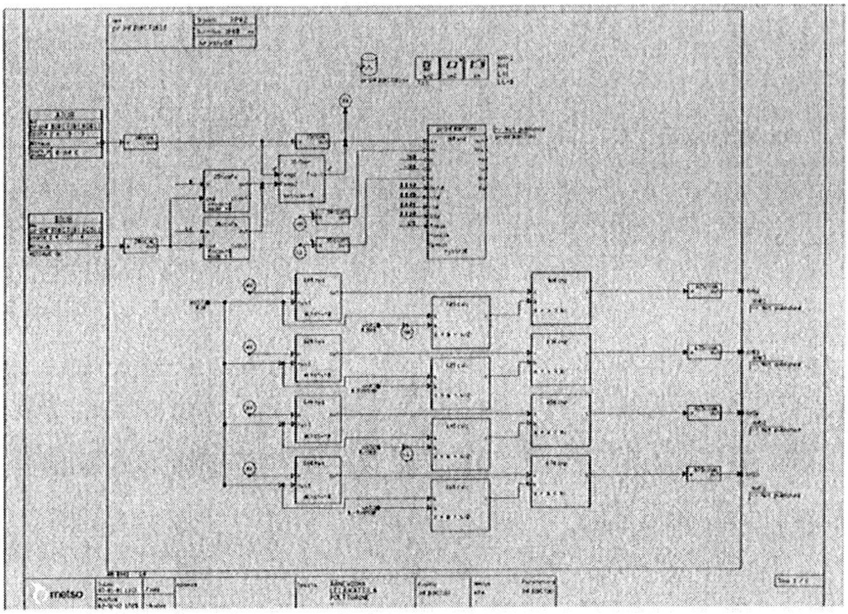

Figure 1 – An example program.

A function block is a capsulated subroutine. It will run functions according the given parameters and connections. Each parameter value reflects to component's functionality and connections are binding dynamic values to a function block. Each function block will allocate only the predefined amount of memory, because in process industry controls the real-time response and functionality must be predictable.

One function block diagram may represent actually many programs and document efficiently one program entity. If the program is larger, the program can be divided into multiple diagram pages with references.

In process industry each delivered project contains customer specific data and field devices. The program interface for the field devices is implemented with varying project specific addressing convention. An average project contains 5000-6000 loops / function block diagrams and over 20000 input/output-connections for the field devices.

When dealing with such a large amount of data, an efficient and successful project requires mass customization. Normally an engineer uses his/her own knowledge and earlier implemented programs in each project.

The basis of effective application implementation relies on usage of so called templates. Templates are application entities describing individual parts of process control software, without project specific definitions. Actual application instances are created when project specific data is combined to template. Our framework utilizes a practical way to identify and search these templates and implemented instances for project reuse.

2. FRAMEWORK ARCHITECTURE

The reuse framework is based on delivered project archives. These project library archives contain all implemented application solutions. Application instances and templates used are stored as DXF-files (Data eXchange Format) on directory structure corresponding to the projects process hierarchy. These archives are accessible for project engineers as mounted network disks.

The reuse framework developed binds these detached project libraries under single content management entity. The centralized content management solution stores only the essential application metadata from diagrams to content management server and allows the archived files remain in local project libraries. The stored metadata includes also links to actual application solution files.

The content management server contains search interface for finding appropriate application solutions for reuse. The stored metadata is used as search conditions and the desired solutions can be downloaded from local project libraries through the file links (Figure 2. Reuse framework, basic architecture).

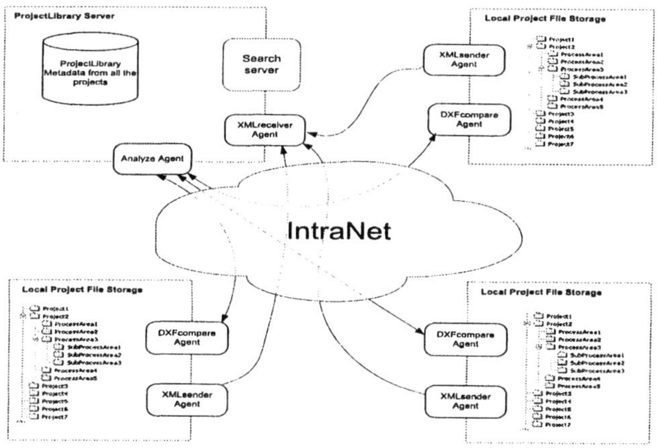

Figure 2 – Reuse framework, basic architecture

The metadata consists of the general part of data included to the diagram header: creator, modifier, creation time, modification time and other basic description fields. The general data is informative and practical as search conditions later on.

Moreover some data is read or calculated from the diagram objects: object count, primary function and statistical amounts of the following:

- Entities,
- Function blocks,
- Analog inputs / outputs,
- Digital inputs / outputs, and
- Connection type.

In addition, to a search criteria this kind of metadata is used for analyze and compare diagrams comprehensively in the database. Comparing actual diagrams files does the final and more detailed comparison. Since the actual file comparison is rather heavy process the preliminary comparison is essential for better performance.

Another performance related problem was solved by distributing tasks to agents running on local computers instead of centralized everything on content management server.

3. JADE-FIPA AGENT PLATFORM

The developed reuse framework is implemented on JADE-FIPA agent platform. JADE (Java Agent Development Framework) is a software development framework aimed at developing multi-agent systems and applications conforming FIPA standards for intelligent agents. It includes two main products: a FIPA compliant agent platform and a package to develop Java agents (JADE, 2004) (Bellifemine et al., 2004).

The agent platform can be split to several hosts, as has been done in developed reuse framework implementation. Only one Java application, and therefore only one Java Virtual Machine (JVM), is executed on each host. Each JVM is a basic container of agents that provides a complete run-time environment for agent execution and allows several agents to concurrently execute on the same host.

The JADE Agent class represents a common base class for user-defined agents. Therefore, from a programmer's point of view, a JADE agent is simply an instance of a user defined Java class that extends the base Agent class (Figure 3). This implies the inheritance of features to accomplish basic interactions with the agent platform (registration, configuration, remote management...) and a basic set of methods that can be called to implement the custom behavior of the agent (e.g. send/receive messages, use standard interaction protocols, register with several domains...).

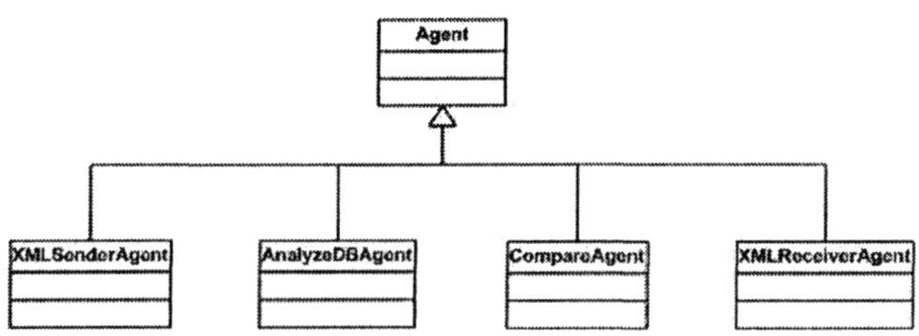

Figure 3 – Implemented application agents.

4. AGENTS & TASKS

The implementation of the developed multi-agent based reuse framework uses only simple JADE agent behavior classes. The agent communication is implemented as common JADE agent communication language (ACL) that is based on java remote method invocation (RMI) -communication.

The developed framework includes four types of agents. XMLsender-agents are executed on project library hosts. They agents detect new directories in the local project library disks. Agent will automatically process zip-compressed files searching essential application metadata. The XML-coded metadata is enveloped into an agent message and passed ahead to XMLreceiver-agent on content management server (Figure 4). The XMLreceiver-agent will receive metadata messages and store the data into the database.

Periodically executed Analyzer-agent performs analyzes in content management database. Analyses include project template summary counts, template identification and template matching to generated instances. When the Analyzer-agent identifies a matching template it will inform Compare-agent that will then compare the template with generated instance files locally on project library hosts. After the comparison Compare-agent replies to Analyzer-agent that updates comparing results into the database.

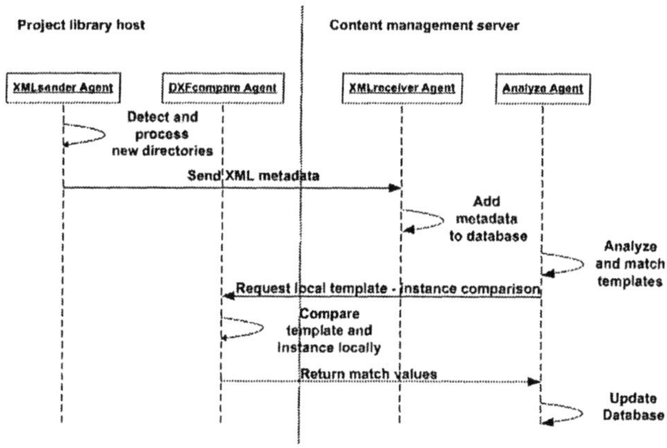

Figure 4 – Agent interaction.

4.1 XMLsender Agent

The XMLreport-agent detects new directories in the given environment according to last modification date of the file. Special JADE WakerBehavior is used to execute new directory search at regular intervals just after a given timeout is elapsed. New archived files are identified and processed. The searched metadata is enveloped into an agent message and passed ahead to XMLreceiver-agent over intranet. The essential data is also stored locally to XML-files for possible later use.

4.2 XMLreceiver Agent

XMLreceiver-agent receives XMLsender-agent's XML-coded metadata messages. Special JADE CyclicBehaviour class is used to control the message receiving process. The XML-messages are parsed to common attribute-value pairs and stored to content management database. XMLreceiver-agent is also able to update already existing metadata entries into the database.

4.3 Database Analyzer Agent

Periodically executed Analyzer-agent is used to process the database-stored metadata. Agent's main task is to identify and match templates used to generated instances. The identification process uses the similarities between primary function blocks, function block amounts, and certain function block attributes to match templates to instances. When Analyzer-agent identifies matching template-instance pairs it will request Compare-agent for more accurate template match comparison. Analyzer-agent will get comparison results from the Compare-agent and update the result value to the database.

Analyzer-agent is also used to perform certain project specific analyses. For example, the project summary analyses include etc. different loop type and IO connection counts and complexity numbers that can be used to support decision-making.

4.4 DXF Compare Agent

Compare-agent receives DBanalyzers matching requests. The agent compares the actual template and instance files and calculates match values. When no structural changes between the template and instance exist the match value equals 100. That is, only different parameter values may exist. Each structural chance diminishes match value with a certain amount. For example by deleting and adding one symbol the match value is decreased by two to 98.

Function blocks are compared first at element level: new and removed elements are identified. For the common existing elements, parameter values are compared. Most critical changes are structural changes that are actually viewable as added or removed elements. The comparison can also be visualized with different colors indicating added and removed elements and changed parameter values.

5. SEARCH CAPABILITIES

The versatile search tool is essential for engineers to find and download good application solutions for reuse. The developed framework contains Tomcat server based search tool enabling versatile search options. The search tool implementation takes advantage of java-bean, JSP-page and applet technologies and thus the users can access the search tool without any external program installations by using only a web browser.

The search interface (Figure 5) allows users to search application solutions according to collected metadata and agent performed analysis. The search can be focused to certain process areas and projects. The more detailed search criteria can include e.g. the main function of the program (function block like pid-controller or motor controller), the IO connection type used and the application creation time.

The search results include all the matching application solutions or templates. Each search match can be taken to more detailed inspection. The more detailed view represents all the relevant metadata of the current loop.

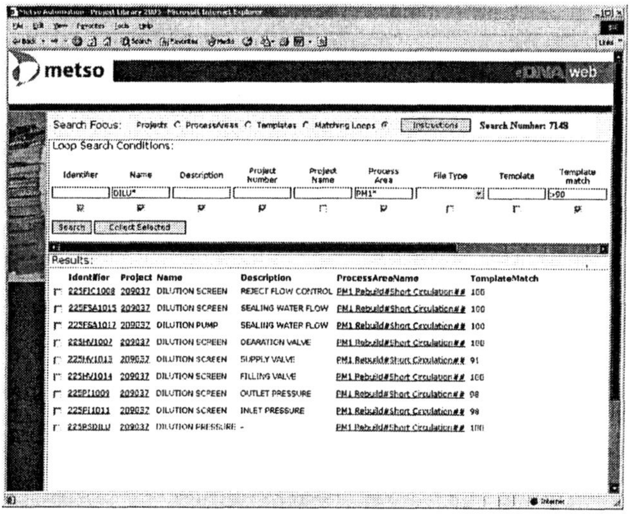

Figure 5 – Search interface.

The templates and instances used can also be graphically examined by using the DXF-viewer applet. DXF-viewer functionalities include also panning and zooming.

The search interface contains also possibilities to inspect complete project and process area analyses. Analyses include information about different kind of implemented IO connection and application loop amounts. These analyses serve also as the project summaries when complete projects are archived.

Analyses also contain essential key figures estimating project complexities and implementation methods. For example, the project summary analyses include complexity values that can be used to support decision-making concerning schedules and workloads for similar projects in future. Also marketing may use complexity figures as a support when pricing future projects.

6. CONCLUSIONS

The agent based reuse framework developed has enabled an efficient way for users to archive and share implemented solutions and knowledge. The automated agent-based application solution filing process together with search tool has proven to be an efficient and practical solution.

The current content management database size exceeds now 800 Mega bytes. Database contains over 200 projects and links together over 30 Giga bytes of compressed files. Metadata has been archived from approximately 600000 function block diagrams. The usage of search tool has become a part of application engineers working manners. Approximately 2000 searches are performed monthly.

Our experiences with JADE-FIPA agent platform have illustrated the flexibility and suitability of the multi-agent technologies to function in local and distributed environment. Moreover, the agent platform has been very stable. Although the current agent messaging has been tested only in local office network, the JADE supports also HTTP communication that also enables communication over Internet.

The analyses and template-matching processes implemented have allowed us to study more the real problem of finding a higher abstraction level for mass customization.

7. RELATED WORK

A similar agent framework is used also for traditional software reuse (Erdur et al, 2000). The framework is more advanced and contains more agents than this implementation. Another good study is related more to our visual language and the metadata handling. Younis and Frey have made a survey how existing PLC programs can be formalized (Younis et al., 2003). Even our template matching is on general level it binds instances and templates together for reuse and in future also for reverse engineering capabilities.

8. FUTURE DEVELOPMENT

The future development work of the reuse framework includes uploading new feature templates and instances to project library hosts. With this feature we are striving to get applications reused more quickly and efficiently.

The analyses methods will be further developed to use enhanced algorithms to match a template. Also, the differences between instance and templates can be already visualized in the CAD based engineering tools and it should be added also to search interface's applet window. This can be very good way to make first analysis from the variation similarities.

9. ACKNOWLEDGMENTS

We would like to thank professor Tarja Systä (Tampere University of Technology) for the support, Timo Kuusenoksa for the coding and many other people who provided helpful comments on previous versions of this document.

10. REFERENCES

1. Bellifemine Fabio, Caire Giovanni, Trucco Tiziana, Rimassa Giovanni, Jade Administrator's Guide http://sharon.cselt.it/projects/jade/ Accessed 23.1.2004.
2. Erdur Riza Cenk, Dikenelli Oguz, Agent Oriented Software Reuse (June 2000)
3. JADE-FIPA, http://sharon.cselt.it/projects/jade/ Accessed 23.1.2004.
4. Younis M. Bani, Frey G., Formalization of existing PLC Programs: A Survey (July 2003)

11

INTEGRATING MULTI-AGENT SYSTEMS: A CASE STUDY

Francisco Maturana[1] Raymond Staron[1] Fred Discenzo[1] Kenwood Hall[1]
Pavel Tichý[2] Petr Šlechta[2] Vladimír Mařík[2]
David Scheidt[3] Michael Pekala[3] John Bracy[3]

[1]Rockwell Automation
1 Allen-Bradley Drive, Mayfield Heights, OH 44124-6118, USA,
{fpmaturana, rjstaron, fmdiscenzo, khhall}@ra.rockwell.com
[2]Rockwell Automation Research Center
Americka 22, 120 00 Prague, CZECH REPUBLIC
{ptichy, pslechta, vmarik }@ra.rockwell.com
[3]The Johns Hopkins University Applied Physics Laboratory
11100 Johns Hopkins Road, Laurel Maryland 20723-6099
{David.Scheidt, Mike.Pekala, John.Bracy}@jhuapl.edu

Intelligent Agent technology provides an appropriate framework to integrate knowledge with efficient production actions in distributed organizations. Integration of knowledge depends on balanced information representation within and across heterogeneous organizations. Integrating information within a specific environment can be helped by the deployment of standards and common practices. However, it is harder to attempt such a smooth integration with the information of foreign organizations. It is the challenge of this paper to present an architecture that provides a first step towards successful integration of separate multi-agent systems in a real life control domain.

1. INTRODUCTION

Intelligent autonomous control provides the ability to address large complex systems. However, reliance upon a single intelligent controller presents a survivability problem, as damage to that centralized controller or to the communications infrastructure used to interact with actuators and sensors, can result in a loss in controllability. This is especially applicable when the system in question operates within a hazardous environment. In this paper, we describe two intelligent control systems that have been applied to provide intelligent control for auxiliary systems on capital ships. These platforms are exposed to potential threats throughout the course of normal operations, and survivability is therefore a key concern.

Multi-Agent Systems (MAS), where distributed autonomous intelligent agents collaborate to carry out control activities, provide a compelling means for achieving robust, survivable control. Recognizing this fact, Rockwell Automation (RA) and the Johns Hopkins University Applied Physics Laboratory (JHU/APL) have independently developed multi-agent architectures, as well as corresponding prototype implementations, to address the control of these auxiliary ship systems. These prototypes target a closed loop fluid distribution system used to regulate the temperature of devices aboard capital ships.

This paper provides some general background on agent-based control as it applies to this particular domain, discusses the architecture of these two MAS, provides some details related to their testing and validation, and concludes by describing a recent new effort to integrate these two agent frameworks in order to form a single collaborative control system.

2. AGENT-BASED CONTROL

In the ship domain, distributing control across multiple agents can be used to improve survivability and to reduce the complexity of the domain for individual controllers. Ideally, device controllers will be co-located with their subservient devices to decrease the likelihood that an intact, serviceable device will be unable to function due to loss of control. In order to satisfy ship-wide goals, distributed controllers must cooperate through either supervisory or peer-to-peer relationships. In this scenario, inter-agent collaboration can be achieved easily within the homogeneous MAS, but past experience indicates that multiple heterogeneous agent systems will need to coexist and collaborate. In our context, heterogeneous agent systems are more than just two different agents; it means different agent systems, syntax and semantics, programming language, and computing platforms and operating systems. Therefore, interoperability is an important dimension of MAS. One intention of this paper is to show how two independent infrastructures can be integrated to produce this type of heterogeneous collaborative control.

Each intelligent agent represents a physical process, machine or device and coordinates its operations with other agents, using standards such as FIPA (FIPA 2003). The MAS architecture is organized according to the following characteristics:

· Autonomy: Each intelligent agent makes its own decisions and is responsible for carrying out its decisions toward successful completion.

· Cooperation: Intelligent agents combine their capabilities into collaboration groups (clusters) to adapt and respond to diverse events and goals.

· Communication: Intelligent agents share a common language to express their beliefs, desires, and intentions.

· Fault tolerance: Intelligent agents possess the capability to detect and isolate equipment failures.

· Interoperability: Intelligent agents use a standard Agent Communication Language (ACL) to extend their collaboration into remote and heterogeneous agent systems.

3. INTELLIGENT AGENT ARCHITECTURES

The following sections describe the agent architectures that were developed by RA and JHU/APL. While these architectures were developed independently, they nevertheless exhibit many similar attributes, which are common to the domain, as well as intelligent agents in general. To date, implementations of both frameworks have been individually demonstrated on a hardware test bed. This includes demonstrations of standard ship operations under nominal conditions, as well as in the presence of faults. More specific details on the tests that were conducted can be found in (Alger et al. 2002) (Tichý et al. 2002) (Maturana et al. 2002).

3.1 RA Agent Architecture

Industrial automation environments are generally populated by multiple interconnected control devices. These devices include controllers and networks, all working in a synchronized manner to handle production actions and events. Although this technology has proven effective in the last 30 years, the current requirements imposed by larger and increasingly more diverse information volumes have redirected the effort towards more flexible systems. Next generation automation devices will be agent-based. These agents will need to be able to efficiently and effectively coordinate the activities of the controlled hardware. The following sections detail the agents developed by RA to satisfy these criteria.

3.1.1 Automation Architecture

In agent-based control, the controllers have an agent infrastructure for enabling component-level intelligence. With this, it is possible to distribute the intelligence among multiple controllers using different agent sizes and populations. Different network connectivity can be used to exploit the distributed intelligence dimension that is added by the intelligent agents.

3.1.2 Inter-Agent Architecture

The controllers have an agent infrastructure for enabling the component-level intelligence. With this, it is possible to distribute the intelligence among multiple controllers using different agent sizes and populations. The relationship among the agents is loosely coupled. The controller's firmware was modified with two additional infrastructures: (1) Distributed Control Agent (DCA) and (2) Intelligent Agent (IA).

The DCA infrastructure is a set of software interfaces added to the controller's firmware to enable and maintain agent tasks. It uses the base firmware to glue the components via a multithreaded system. Also, it uses the controller's interfaces to communicate with field devices (i.e., sensors and actuators). The IA extension co-exists with user level programs (i.e., ladder logic IEC 1131-3 (IEC 2001)). Here, the user downloads the application specific components, which are the rules and behaviors of the intelligent agents.

For inter- and intra-organization conversations, the intelligent agents emit messages outside their organization by wrapping Job Description Language (JDL) messages inside FIPA envelopes. Communication with remote agent systems (inter-organization) depends on having the listening nodes understanding the language and transport protocol of the sender nodes. FIPA compliance is one requisite but it is insufficient if lower layers of the communication stack do not understand each other.

Hence, our work is important because it also identifies the requirements to build an interoperable agent communication stack.

In JDL, information is encoded as a sequence of hierarchical actions with precedence constraints. These are requests and information messages. When an intelligent agent accepts a request, an instance of a plan template is created to record values that emerge during the planning process. Requests are propagated throughout the organization using the Contract Net protocol (Smith, 1980). The requests visit multiple agents and multiple negotiation clusters are formed.

The FIPA standard uses a Matchmaking mechanism to connect agents. In this implementation, we are FIPA compliant and the architecture includes the functionality of Directory Facilitators (DFs). A DF performs capability registration and matchmaking. For each capability request, a DF provides a list of agents that are able to provide the requested capability. The DF functionality is part of the DCA extension.

3.1.3 Intra-Agent Architecture

The intelligent agents are goal-oriented entities. They organize the system capabilities around system missions. Each mission is intended to satisfy a given set of goals, which are commonly expressed as a single cost or performance metric value.

Group goals emerge dynamically and are agreed upon by the intelligent agents through negotiation. For instance, an intelligent agent that detects leakage in a pipe of the cooling system establishes a goal to isolate the leakage. The intelligent agent then informs adjacent intelligent agents to evaluate the problem according to their local views and knowledge. This is the origin of a group-based goal, which is to isolate the leaking pipes. An intelligent agent has four components:

· Planner: This component is the brain of the intelligent agent. It reasons about plans and events emerging from the physical domain.

· Equipment Model: This component is a decision-making support system. Models of the physical domain are placed here to help the Planner evaluate different configurations. The Equipment Model provides metrics for proposed configurations.

· Execution Control: This component acts as control proxy, which translates committed plans into execution control actions. These actions are synchronized with the control programs. It also monitors events from the control logic and translates them into response-context events to be processed by the Planner component.·

Diagnostics: This component monitors the health of the physical device. It is programmed with a model of the physical device, against which a set of input parameters (e.g., bearing vibration of a pump) is evaluated.

The physical device (e.g., a valve) is operated according to a sequence of actions. These actions are encapsulated inside device drivers and classified according to the type of OEM. The device driver becomes a template library for the physical device. When an intelligent agent is created for a device, an instance of its device driver is copied in the controller's memory. The sensors and actuators associated with the physical device are connected to the automation controller using an industrial network. State variables are contained in the controller's data table.

The intelligent agents perform resource allocation via priority ranking and negotiation. Each intelligent agent provides a set of capabilities to the overall organization. These capabilities are combined to carry out distributed planning in three main phases: Creation, Commitment, and Execution. During creation, an intelligent agent initiates a collaborative decision making process (e.g., a load that will soon overheat will request cold water from the cooling service). In this process, multiple agents are involved in deciding the local setups. The intelligent agents offer a solution for a specific part of the request. Then, the intelligent agents commit their resources to achieve the task in the future. Finally, the plan is executed.

3.2 JHU/APL Agent Architecture

The original agent framework for autonomous distributed control that JHU/APL developed is called the Open Autonomy Kernel (OAK), and is described more thoroughly in (Alger et. al, 2002). OAK defines flexible inter- and intra- agent architectures, and is targeted towards "hard" control problems that involve complex, partially observable systems.

Internally within OAK agents, system knowledge is comprised of both directly observable knowledge (e.g., sensor readings and memory of past commands) and inferred or "hidden" knowledge (e.g., the actual state or current operational mode of each component). Control is addressed as a three-step process, consisting of *diagnosis*, *planning* and *execution*. See Figure 1. We refer to the process of deriving the inferred knowledge from the directly observable knowledge as diagnosis. Reconfiguration involves identifying a system state that will achieve some desired goal and identifying a sequence of actions that will successfully transition the plant into this state. Issuing and tracking the progress of these action sequences is the role of execution. To perform diagnosis, OAK agents utilize the L2 reasoning engine (Kurien 1999), which is one member of a family of model-based autonomy technologies (Williams, Nayak 1996).

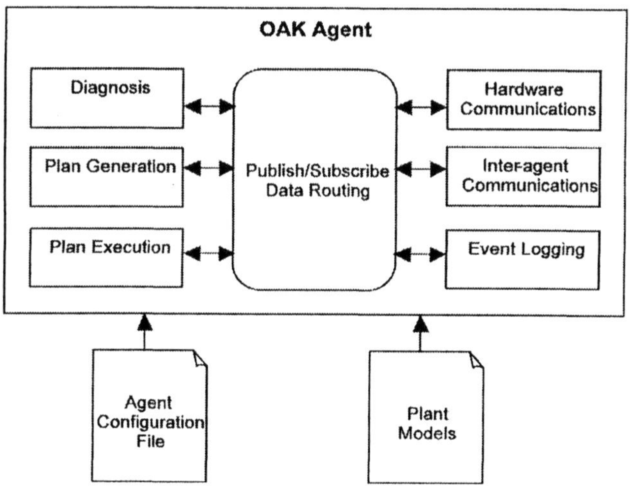

Figure 1– OAK agent architecture

OAK agents classified all external communications into two categories: communications with other OAK agents via the Agent Communication Framework (ACF) and communications with hardware devices. The types of information shared between agents consisted of: facts (assertions about the current state of the world), goals (statements of intent) and information related to agent coordination (e.g., establishing communications paths). The types of information shared between agents and hardware devices consisted of: commands (sent from agents to hardware) and observations (sent from hardware to agents). All inter-agent communications utilized an XML-based Agent Communications Language (ACL), while the individual hardware interfaces each determined the format of information passed between the agent and the hardware.

4. INTEGRATING THE AGENT FRAMEWORKS

The effort to integrate these two agent frameworks stems largely from a desire to leverage the relative strengths of each, while at the same time, demonstrating the effectiveness of common standards. In particular, our goal is to highlight a synthesis of the distributed reconfiguration capabilities of the RA agents and the powerful model-based diagnostic capabilities provided by OAK agents. A fluid control system was selected as the target domain as both organizations have extensive experience in employing agent based control in this capacity.

The first challenge faced by the team was determining the roles and responsibilities for the respective agent systems. One option involved partitioning the fluid control system along functional or spatial boundaries, and assigning a subset of the overall system to each agent network. In this case, however, the highly reconfigurable nature of the system under consideration makes it difficult to identify spatial regions that will tend to remain independent, especially in fault scenarios.

A functional decomposition poses certain difficulties as well, as effective control requires a high level of coordination among the various component types that comprise the fluid system. While both of these approaches remain options for future work, the initial approach selected involves assigning each framework a different portion of the overall control cycle. In this paradigm, RA's agents are responsible for performing reconfiguration, while APL's agents will perform diagnosis.

Adapting these agent frameworks to accommodate situations where a given agent is no longer responsible for the entire control cycle within its domain became one of the challenges of this effort. Previous deployments of these frameworks admitted both hierarchical and peer-to-peer relationship among agents, but assumed that the control cycle itself was largely encapsulated. In this new configuration there is increased motivation to have agents share additional pieces of information. For example, previously APL agents externalized only a succinct summary of their internal state for consumption by peer and parent agents. Now that the corresponding planning capability resides in a different agent, there is increased value in having agents share with the community more detailed diagnostic information, such as multiple hypotheses. In addition to modifying the types of information passed between agents, this task required standardization of the inter-agent communication mechanisms themselves.

4.1 Agent Capabilities

Once agent roles were defined, the next step involved defining the types of information that would be exchanged and the mechanism for exchanging them. The team decided upon a service-based architecture, where both white pages and yellow pages discovery techniques are used. In order to perform yellow pages discovery, a standard set of *agent capabilities* need to be defined. The overall information flow between the two agent systems can be succinctly summarized in terms of these capabilities.

4.1.1 RA Agent Capabilities

The set of services provided by RA agents include:

Register for Data – This capability allows an agent to register for the asynchronous notifications that are produced whenever a command is issued or a new sensor value is observed. Properly diagnosing the current state of the system requires both knowledge of past commands and knowledge of current sensor readings. From the perspective of APL agents, who are consumers of this data, this represents a change from the previous deployment where agents already possessed local knowledge of hardware commands (which were a product of the local planning process).

Control Load – This service provides the mechanism by which external agents can request that a particular load be activated or deactivated. Activating or deactivating a load implicitly invokes the distributed planning engine, which reconfigures the fluid system in order to meet this objective.

Set Priority – This service allows external agents to set the relative importance of loads in the fluid control system. In situations where the full set of desired loads cannot concurrently be activated, this priority is used to determine which loads will be supplied. Thus, external agents that have additional or higher-level information (e.g., knowledge of pending damage events) can use this capability to productively influence the results of the distributed planning algorithm.

4.1.2 APL Agent Capabilities

The set of services provided by JHU agents include:

Get Diagnosis – This capability allows agents to request diagnostic information from the agent. Diagnostic information consists of a set of ranked hypotheses that include each component within the agent's sphere of influence.

Get Diagnostic Model – This capability allows agents to request detailed information about the diagnostic model from the agent.

4.2 Standardization

Successfully implementing these agent capabilities requires a unifying communication language, syntax and semantics, and communication transport stack based on a Common Industrial Protocol (CIP). To this end, we use the FIPA/JDL specification to implement discovery services and to interpret application domain knowledge. Unifying the communication transport stack is a very important activity towards the standardization of the agent interoperability. In the intended system, we have different agent platforms, programming languages, and radically different agent platform containers, which include Windows XP workstations (C++ domain), Linux workstations (Java domain), and automation controllers (C++ domain), as shown in Figure 2. Our intention is to create a common stack in the form of a library, which will be written in Java and C++ languages to support the workstations and controllers.

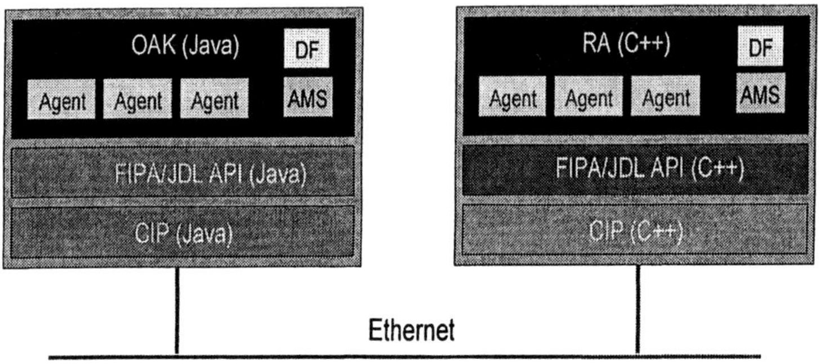

Figure 2– Integrated heterogeneous agent systems

For agent discovery, we use the FIPA specification for Directory Facilitators (DF), which provides matchmaking capabilities and Agent Management Services (AMS). The architecture offers the ability to create multiple global or local DF agents. Redundancy, in conjunction with a heart-beating mechanism, provides fault tolerance. Multiple knowledge propagation and retrieval techniques have been implemented, providing the system designer with the flexibility to trade off efficiency for redundancy.

Communications is defined using Job Description Language (JDL) messages. A common JDL messaging library has been developed and integrated into both MAS. Any non-primitive message content encapsulated within a JDL message is first encoded in XML, using schemas to define the valid syntax. Whenever possible, existing standards were leveraged in order to define this syntax. For example, the syntax adopted for diagnostic model descriptions is the XML Model-based Programming Language (XMPL) (Livingstone2, 2004). Although this specification was developed for the L2 reasoning engine, it has proven sufficiently general to be used with other model-based reasoning engines as well.

4.3 Implementation

In addition to implementing the libraries and specifications described above, this integration effort has also engendered changes to the agent frameworks themselves. Although a detailed description falls outside the scope of this paper, a successor to the OAK agent framework has been developed that strives to provide additional flexibility in defining agents and their capabilities. This new framework greatly simplifies tasks such as developing agents whose capabilities are not prototypical or change at runtime.

4.4 Use Case: Dynamic Modeling

As mentioned previously, fault tolerance is an important aspect of multi-agent systems that are exposed to harsh or dangerous environments. In these situations, the agent network must be robust against the loss of one or more individuals. Redundancy in the form of standbys is one traditional approach often used to address this issue. When the loss of an agent is detected, a new agent is brought online to replace it. As a variation on this concept, JHU/APL has begun exploring agents whose control domains are dynamic and responsive to failures. For example, when a peer agent is lost, neighboring agents can negotiate to determine who will assume responsibility for that agent's resources.

Part of assuming responsibility for an agent's resources involves the ability to perform diagnosis on that portion of the system that was under control of the deceased agent. For model-based agents, this implies obtaining or generating a model for this portion of the system. Thus, the ability to exchange model fragments, as described previously, represents an important prerequisite for many types of dynamic modeling approaches. In order to further the dual objectives of exercising our inter-agent communications standards and developing a framework that can be used to test dynamic modeling algorithms, we have developed a dynamic model management component for APL's next-generation intelligent agents. This component has the ability to maintain multiple model fragments, and to merge these model fragments into the agent's local model at runtime as appropriate.

This feature was exercised in a small scenario involving three child agents and a single parent agent. Each child agent has a local model consisting of a strict subset of the overall fluid control system. At runtime, these models are serialized and sent to the parent agent, using the standards described previously. The parent agent, using the dynamic model management capability, then assembles the overall system model. Subsequently, the parent agent used observations in conjunction with this model to diagnose faults in the system. Although only a small part of the overall solution, this demonstrates some of the utility in providing this type of inter-agent communication.

5. CONCLUSION

Integrating heterogeneous agent systems is not an easy task. Nevertheless, the fundamental pieces necessary for a first integration based on standards such as FIPA, Ethernet/IP, and CIP have been created. Progress made to date indicates that leveraging these standards, in conjunction with a common knowledge representation, will enable a seamless interaction between the two systems.

6. ACKNOWLEDGEMENTS

This work has been conducted as part of the Office of Naval Research "Machinery Systems and Automation to Reduce Manning" program under contracts N00014-00-C-0050 and N00014-02-C-0526.

7. REFERENCES

1. Alger D., McCubbin C., Pekala M., Scheidt D., Vick S. Intelligent Control of Auxiliary Ship Systems. Proc. of Innovative Applications in Artificial Intelligence, AAAI, Edmonton, CA, 2002.
2. Kurien J., Model-based Monitoring, Diagnosis and Control, PhD Thesis Proposal, Brown University Department of Computer Science, 1999.
3. Maturana F., Tichý P., Šlechta P., and Staron R., "Using Dynamically Created Decision-Making Organizations (Holarchies) to Plan, Commit, and Execute Control Tasks in a Chilled Water System". In Proceedings of the 13th International Workshop on Database and Expert Systems Applications DEXA 2002, HoloMAS 2002, Aix-en-Provence, France, pp. 613-622, 2002.
4. Tichý P., Šlechta P., Maturana F., and Balasubramanian S., "Industrial MAS for Planning and Control". In (Mařík V., Štěpánková O., Krautwurmová H., Luck M., eds.) Proceedings of Multi-Agent Systems and Applications II: 9th ECCAI-ACAI/EASSS 2001, AEMAS 2001, HoloMAS 2001, LNAI 2322, Springer-Verlag, Berlin, pp. 280-295, 2002.
5. Williams B., and Nayak P. A model-based approach to reactive self-configuring systems. In Proceedings of AAAI-1996.
6. Shen W., Norrie D., and Barthès J.P.: Multi-Agent Systems for Concurrent Intelligent Design and Manufacturing. Taylor & Francis, London, 2001.
7. Smith R.G.: The Contract Net Protocol. High-level Communication and Control in a Distributed Problem Solver. In IEEE Transactions on Computers, C-29(12), pp. 1104-- 1113, 1980.
8. International Electrotechnical Commission (IEC) TC65/WG6, 61131-3, 2nd Ed., Programmable Controllers - Programming Languages, April 16, 2001.
9. The Foundation for Intelligent Physical Agents (FIPA): http://www.fipa.org, 2003.
10. Livingstone2 Documentation, http://ic.arc.nasa/gov/projects/L2/doc, 2004.

12

ALARM ROOT CAUSE DETECTION SYSTEM

Milan Rollo, Petr Novák, Jiří Kubalík, Michal Pěchouček
Gerstner Laboratory,
Department of Cybernetics, Faculty of Electrical Engineering,
Czech Technical University in Prague,
Technická 2, 166 27, Prague 6, CZECH REPUBLIC
{rollo|novakpe|kubalik|pechouc}@labe.felk.cvut.cz

Production process control becomes complicated as the complexity of the controlled process grows. To simplify the operator's role many computer based control systems with integrated visualization clients have been developed. In many practical circumstances malfunction of one or more process components results in other related components entering the alarm states. Several alarms appear on the operator's display in a short time making it difficult for the operator to diagnose the root cause quickly. Within this paper we describe a solution of this problem based on a multi-agent system that processes all incoming alarms, identifies the root cause alarms vs. alarms arising in consequence of the roots and presents the diagnostic results to the operator visually.

1. INTRODUCTION

Nowadays, as the complexity of the controlled processes grows, more stress is laid down on the operator. Number of manufacturers offers software for automatic process control with built-in visualization clients or even complex solutions including controllers, sensors, communication buses and visualization devices (e.g. control panels, touch screens). All these components are dedicated to support the operator's role. They allow the operator to select the controlled component and assign limit values of variables to it. When the variable exceeds this value (or comes down under it) the operator is informed about it visually. This helps him to quickly recognize the origin of the problems and fix it up.

But the situation becomes more complicated for the operator when malfunction of single process component results in other related components entering the alarm states. In such a case several alarms may appear on operator's display simultaneously or in a short time and in a random order (which depends on the nature of the variables and speed of the sensors). Some of the alarms may not appear on the screen at all, because the control process is displayed schematically only and contains just principal components (due to the clarity reasons). This all makes it difficult for the operator to recognize the origin of the problem. In this paper we describe a diagnostic system that supports the operator's decision making when such situation occurs. Main reason to develop this system was, that in some kinds of production processes is necessary to fix up the problems as soon as possible to avoid economical losses or eventual exposure to danger.

Solution is based on the multi-agent system that models the production process, processes all incoming alarms and determines the root cause. New determination algorithms, based either on the topology of the production process or physical nature of individual components, were developed to this purpose. Results of the determination process may be presented visually to the several operators. This diagnostic system nicely illustrates capabilities of the multi-agent system based solutions like a modularity or adaptability.

1.1 State of the Art

Use of multi-agent systems for modeling and controlling the production processes grows rapidly and seems to be very promising. In general the agents technologies are suitable for domains that posses either of the following properties: (i) where highly complex problems need to be solved or highly complex systems to be controlled or (ii) solving problems or controlling systems, where the information is distributed and is not available centrally. In manufacturing agent technologies have been applied mainly in planning highly complex production, control of dynamic, unpredictable and unstable processes, diagnostics, repair, reconfiguration and replanning. Important application domains of agent-based applications can be also found in the field of virtual enterprises (e.g. forming business alliances, forming long-term/short-term deals, managing supply chains) and logistics (e.g. transportation and material handling, optimal planning and scheduling, especially in cargo transportation, public transport but also peace-keeping missions, military maneuvers, etc.)

There are several companies that have adopted the agent-based solutions in production already, e.g. Daimler-Chrysler car manufacturing (McFarlane, 2000), Rockwell Automation developed agent based solution for BHP Melbourne, Australia, or ExPlanTech/ExtraPLANT agent-based production system running in Modelarna Liaz pattern factory, CZ and Hatzapoulos, packaging company, Greece (Pěchouček, 2002). The most relevant is the application of agent technology in a Reconfigurable Shipboard automation system developed by Rockwell automation and Rockwell Scientific Company (Maturana, 2003). In this application the agents are brought down to the level of physical components of the shipboard chilling system and they monitor and control stability of the whole of the chilling machinery in distributed manner. The practical applications are in the focus of attention of the HMS consortium (Brennan et al., 2003)

2. SYSTEM DESCRIPTION

Infrastructure of the proposed diagnostic system is shown schematically in Figure 1. Within this infrastructure, a software agent models each process component. Relationships of agents represent the topology of the process components (theirs input/output links). Each agent receives alarms activated by the component it represents and collaborates with related agents to identify the potential root causes.

In order to do so effectively new detection algorithms, which utilize the process topology, component type and alarm type were developed. These algorithms complement the expert system (provided by the control software) and bring us a couple of advantages. Expert system based solutions require for each process control case separate knowledge base describing the relations and dependencies among the components. Such a knowledge base may not be available for the particular process or may contain incomplete information. Once an expert system is build up for a concrete process it provides an operator with most reliable results, however it doesn't cover the cases that occurred in the process never before. In contrast, the proposed agent based solution is general enough and is expected to work for an arbitrary manufacturing case.

The agent-based algorithms developed to date include rule-based and topology-based search methods, which appear to have promise for this purpose (see section 4).

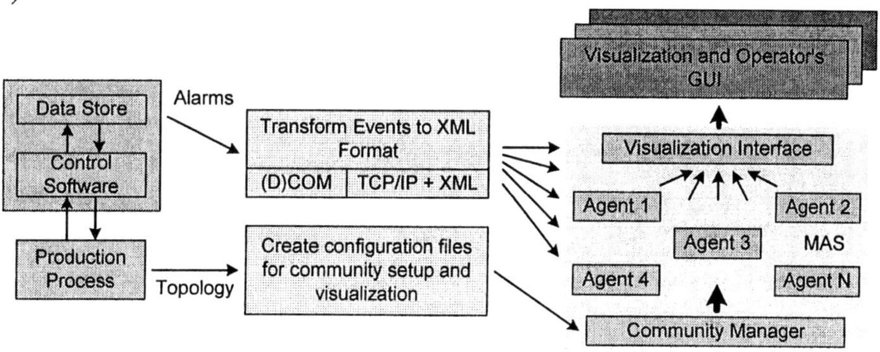

Figure 1 - Configuration of the multi-agent alarm root cause diagnosis system

2.1 Production Process Interface

In the past particular production processes were isolated and controlled by the software that used different forms of communication and data management. Technical progress in the area, especially the distributed control, brought couple of problems with interconnection among different production processes and control software. Industry leading companies thus formed a foundation dedicated to develop standard specifications. Most widely used solution and de facto standard in this area is currently the OPC (OLE for Process Control) (OPC, 2004).

OPC is based on the OLE/COM standard designed by Microsoft. For process automation systems that generate alarms and events the OPC Alarms and Events specification was developed. The OPC Alarms and Events server captures the alarms and events rose in the system, and makes them available to any client application that is interested in this information. It does not create the alarms and events; it only reports the alarms and events previously defined in the system using a standard communication interface (alarms and events in the system are automatically generated, based on the operating conditions and actions performed in the process plant).

OPC in this case serves as a kind of database (Data Store) that all clients can utilize. Information is in this database filled by the control software that gets it from the sensors (e.g. through the communication bus). This database is shared among all clients.

In this project we use the OPC Alarms and Events server to receive the alarms invoked in the process (either real process or its simulation). These alarms (each containing information about its source, type, date of rise, variable, etc) are sent to the multi-agent system to be diagnosed.

2.2 Multi-Agent System

Alarm root cause determination process itself is carried out by the multi-agent system. Input of the determination process is an alarm record from the OPC server (converted to the XML format).

Multi-agent system brings us a couple of advantages compared with using an expert system only:

- **Prediction ability** - system can predict the problems (determine root causes) in process only from the general rules or process topology and thus discover problems that appeared never before and will not be covered by the expert system rules.
- **Adaptability** - system can be easily fit on different process (unlike expert system that is tied to the concrete process).
- **Modularity** - process can be simply extended with a new type of the component. Adding some new rules based on the physical characteristics of this component we can predict the behavior of the process (even when we have no knowledge about the entire process). Using the expert system, user is enforced to add a concrete rules based on the observation of the process's behavior.

2.3 Operator's Visualization Interface

Result of the root cause detection algorithm is represented by the change of agents' output state. This should be displayed on the operator's screen in a comprehensive way. When a complex system is to be controlled a number of output windows providing different views and details of the process may be required. For this purpose a versatile visualization system (VISIO) has been developed. It enables among others to define multiple snapshots of different parts of the process; they can be displayed to the operator either all at the same time or just some of them.

Developed visualization system contains also multiple-operator support. Each operator can be informed only about events that happened in the part of the production process he is responsible for.

3. MULTI-AGENT SYSTEM IMPLEMENTATION

Agent community consists of four different types of agents (see Figure 2). Beside the three static agents (Agent Factory, Root Cause Analysis Agent and Visualization

Agent), each of them appears in the system only once, there are several Block Agents. Their number depends on the particular production process.

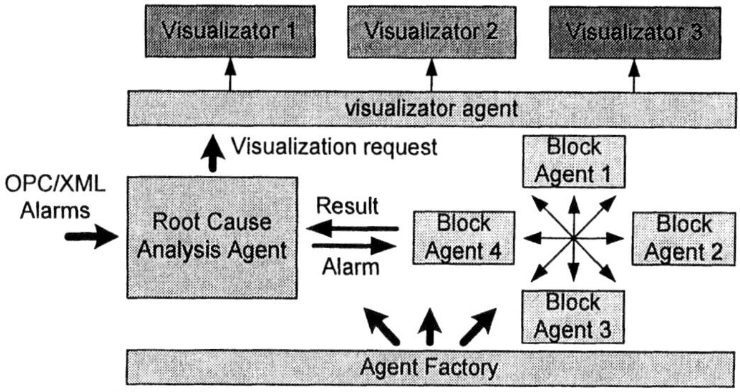

Figure 2 - Design of the multi-agent system

3.1 Graph Representation

Every component (pump, valve, tank, controller, etc.) is represented by one agent (Block Agent). Agents are connected by oriented links representing input/output relationships among components. Whole process thus forms an oriented graph. Detection algorithms utilize this feature during the communication among agents (prevention of circular communication inside the control loops).
Community manager (Agent Factory) generates agents according to the configuration file that contains record for each component: its name, agents linked to its input/output, information whether it generates alarms or not, type of component it represents (e.g. sensor, valve), the variable it monitors, etc. Note that the links connecting agents represent both material flow as well as control links present in the process (e.g. feed back control signals). This is important to make sure that no causal relation among components would be omitted when looking for the root cause alarm.

3.2 Agent Types

- **Agent Factory** - community manager used to startup the agent's community. It starts the agent platform (providing the necessary infrastructure), Root Cause Analysis Agent, Visualization Agent and certain number of Block Agents.
- **Root Cause Analysis Agent** (RCAA) - coordinator of the alarm determination process. It is connected to the OPC-XML interface and receives from the OPC Alarms and Events server all alarms that arose in the process. Alarms are then stored in the internal queue and processed on the FIFO basis. Each alarm is compared with the rules in the internal expert system first. In case that no matching rule is found, appropriate agent (corresponding to the source of alarm in the process) is informed about the event. Agents process the alarm using one of the inference modes and RCAA receives the result. Request to display this result on the operators' screens is then send to the Visualization Agent.
- **Visualization Agent** - is responsible for displaying the information about alarm states of all function blocks on the screens of all operators that are involved in.

Agent connects to the VISIO program (see section 2.3) that is running on the operator's side. Once the information about change of function block's alarm state arrives from the RCAA, Visualization Agent transforms it into format acceptable by the VISIO and sends it to all operators that has appropriate function block on their screen.

- **Block Agent** - every Block Agent represents a real physical entity in the process. Block Agent has a record about its inputs, outputs and its alarm state. It has also assigned functionality in the process (e.g. pump, valve, tank, etc.). When agent receives information about new alarm it invokes an inference process to determine the alarm type.

4. INFERENCE MODES

Multi-agent system includes two inference modes: topology-based search and rule-based search. Topology-based search is based only on the topology of the process, rule-based exploits some background (prior) knowledge about the process. This knowledge is stored in form of rules in the rules database.

4.1 Topology-Based Search

Topology-based search takes advantage of converting the process model into its graph representation. This algorithm is based on the input/output relationship only and doesn't take into account features of the particular process. This algorithm operates in two separate phases:

Backward (Upstream) Search:

- Starting from the currently processed function block, it goes back through all its inputs (as long as the block has some or until the loop is closed).
- Looks for the possible cause of the alarms.
- Algorithm runs until the first block with alarm in the chain is found (this block will be marked as root).

Forward (Downstream) Search:

- Informs all successors (all agents in the output direction) that they are not root alarms any more.
- Updates alarm types for visualization.

Finally the results of the both search phases are merged together. In this result each agent (function block) has its own record that contains its name, type of the determined alarm type and information whether the agent can generate alarm. These records are stored in a tree form.

All alarms are stored in a buffer as they come and are processed sequentially one by the other. Final result of the algorithm does not depend on the order in which the alarms are processed.

4.2 Rule-Based Search

The topology in itself represents only the lowest-level information about the monitored process. This makes the topology-based algorithm independent on the physical nature of the process. On the other hand, the topology itself cannot provide sufficient background for correct root cause detection in many cases. It is obvious that some rules, which take into account the knowledge about the process principles, may help to make the root cause detection more reliable.

A rule-based search algorithm utilizes the fact that individual agents model a specific component type (e.g. pump, valve, tank, heat exchanger). Each rule consists of two parts (condition and action) and is based upon the type of alarming component, the alarm type and the component topology. For example, in the case the agent receives a high-pressure alarm from a tank it searches for possible root cause on the input (e.g. malfunctioned pump) and output of the tank (e.g. closed valve).

Each agent has assigned a type of the component it represents (sensor, pump, valve, etc.), only agents that represent sensors can signal alarm. Each sensor has assigned a variable it monitors (pressure, temperature, etc.). When new alarm arises, rule-based search is invoked (provided that expert system fails to find a solution). This means that the corresponding agent searches its rule database for any rule whose condition part can be satisfied. If agent finds such a rule, it carries out actions specified in the action part of the rule. In case the agent cannot find any rule that could be applied the topology-based detection algorithm is initiated. Thus the topology-based search is considered to be a complementary option to the rule-based one. It is invoked only when the rule-based search fails to run.

5. EXPERIMENTS

The proposed diagnostic system was originally developed to improve the control process in hydrogen production plant. Because this process is very complex, simplified case study based on the distillation column was chosen to carry out the experiments (both processes have similar physical characteristics).

Instead of connection to the real physical process, mathematical model of the distillation column running in the Hysys (Hysys, 2004) simulation environment was used. This allowed us to better simulate the malfunctions and alarm explosions in the process. The DeltaV Automation System (DeltaV, 2004) was used as control software. Simulation environment and control software were connected via the OPC Server. Developed multi-agent diagnostic system was also connected to this server in order to receive the alarm records.

Distillation column used in this case study consists of more then fifty components (function blocks). We also carried out some additional scalability tests (measuring the number of messages and operational memory consumption), with several hundreds of components.

Multi-agent system was implemented in the Java using the JADE multi-agent platform (Jade, 2004). Other parts (Production Process Interface and Operator's Visualization Interface) were implemented in C/C++.

6. CONCLUSIONS AND FUTURE WORK

System described in this paper illustrates the capabilities of the multi-agent systems to solve problems in highly complex distributed environment. Some of the important features of multi-agents systems we utilize in this case are modularity and adaptability. System can be easily reconfigured or extended with new elements according to the changes in the real production process.

Obviously, root cause alarm diagnostics based purely on the topology knowledge about the process cannot reliably return the optimal solutions because it does not take into account the physical principle of the process. As the preliminary experiments showed the utilization of rule based search algorithm might considerably improve the performance of the diagnostic system.

There are still some open issues that need to be solved to make the diagnostic system more robust and reliable:

- Synchronization and cooperation between both inference modes (topology-based and rule-based search).
- Improve the connection with OPC Alarms & Events server - suppression of alarms identified as a consequence of the root alarm in the DeltaV.
- Increase of alarm priorities (operator's notification level) if alarm isn't suppressed within a certain period of time (e.g. when high pressure alarm on the distillation column will arise, alarm priority will be increased each minute until the alarm will be suppressed).

7. ACKNOWLEDGEMENT

The project work has been in part co-funded by NASA Hydrogen Research Effort - Software Agents and Knowledge Discovery and Data Mining Research for Complex System Safety, Health, and Process Monitoring project and by the Grant No. LN00B096 of the Ministry of Education, Youth and Sports of the Czech Republic.

8. REFERENCES

1. Brennan R., Hall K., Mařík V., Maturana F., and Norrie D. A real-time interface for holonic control devices. In Mařík, McFarlane, and Valckenaers, editors, Holonic and Multi-Agent Systems for Manufacturing, number 2744 in LNAI, pages 25–34. Springer-Verlag, Heidelberg, June 2003.
2. DeltaV - the Digital Automation System for Process Control. http://www.easydeltav.com, 2004.
3. HYSYS - Integrated Simulation Environment. http://www.hyprotech.com/hysys, 2004.
4. JADE - Java Agent Development Framework. http://jade.tilab.com, 2004.
5. Maturana F., Tichý P., Šlechta P., and Staron R. A highly distributed intelligent multiagent architecture for industrial automation. In Mařík, Muller, and Pěchouček, editors, Multi-Agent Systems and Applications III, number 2691 in LNAI, pages 522–532. Springer-Verlag, Heidelberg, June 2003.
6. McFarlane D. and Bussmann S. Developments in holonic production planning and control. International Journal of Production Planning and Control, 11(6):552–536, 2000.
7. OPC Foundation. http://www.opcfoundation.org, 2004.
8. Pěchouček M., Říha A., Vokřínek J., Mařík V., and Pražma V.. Explantech: applying multi-agent systems in production planning. International Journal of Production Research, 40(15):3681–3692, 2002.

13

A METHODOLOGY FOR SHOP FLOOR REENGINEERING BASED ON MULTIAGENTS

José Barata , L. M. Camarinha-Matos
New University of Lisbon
Quinta da Torre – 2825 114 Caparica – PORTUGAL
jab@uninova.pt -- cam@uninova.pt

Achieving shop floor agility is a major challenge for manufacturing companies. A multiagent based approach for shop floor reengineering where agility is achieved through configurations of contracts is briefly introduced. A methodology to agentify manufacturing components in order to participate in the multiagent community is presented. An experimental validation scenario is finally described.

1. INTRODUCTION

Shop floor agility is a central problem for manufacturing companies. Internal and external constraints, such as growing number of product variants and turbulent markets, are changing the way these companies operate and impose continuous adaptations or reconfigurations of their shop floors. This need for continuous shop floor changes is so important that finding a solution to this problem would offer a competitive advantage to contemporary manufacturing companies.

The central issue is, therefore, to design and develop techniques, methods, and tools are appropriate to address shop floors whose life cycles are no more static but show high level of dynamics. In other words, how to make the process of changing and adapting the shop floor faster, cost effective, and easy. The long history of industrial systems automation shows that the problem of developing and maintaining agile shop floors cannot be solved without an integrated view, accommodating the different perspectives and actors involved in the various phases of the life cycle of these systems. Moreover, supporting methods and tools should be designed and developed to accommodate the continuous evolution of the manufacturing systems along their life cycle phases – a problem of shop floor reengineering. The design and development of a methodology to address shop floor reengineering is thus an important research issue aiming to improve shop floor agility, and therefore, increasing the global competitiveness of companies.

A particularly critical element in a shop floor reengineering process is the control system. Current control/supervision systems are not agile because any shop floor change requires programming modifications, which imply the need for qualified programmers, usually not available in manufacturing SMEs. To worsen the situation, the changes (even small changes) might affect the global system architecture, which inevitably increases the programming effort and the potential for side-effect errors. It is therefore vital to develop approaches, and new methods and tools that eliminate or reduce these problems, making the process of change (re-engineering) faster and easier, focusing on *configuration* instead of *codification*. In this context, a multiagent based system, called **Co**alition **B**ased **A**pproach for **S**hop Floor **A**gility – **CoBASA**, was created to support the reengineering process of shop floor control/supervision architectures. CoBASA uses contracts to govern the relationships between coalitions' members (manufacturing agents). The main foundations of the system architecture was described in (Barata & Camarinha-Matos, 2003). This paper focuses on the reengineering methodology to be used with CoBASA. First, the main agents that compose the CoBASA architecture are described in section 2, while section 4 describes the steps required to operate CoBASA, which involves creating Manufacturing Resource Agents - MRAs that can be used as candidates in future manufacturing coalitions. Then, section 4 briefly describes the experimental setup. Finally, section 5 presents the conclusions.

2. THE CoBASA ARCHITECTURE

Although the the main aspects (components and basic interactions) of the CoBASA architecture have been described in (Barata & Camarinha-Matos, 2002, 2003; Camarinha-Matos & Barata, 2001) a brief overview is presented here for the sake of better understanding of this paper.

The basic components of the proposed architecture are Manufacturing Components, Manufacturing Resource Agents, Coordinating Agents, Clusters, Coalitions/consortia, Broker, and Contracts.

Definition 1 - Manufacturing Components
A manufacturing component is a physical equipment that can perform a specific function in the shop floor. It is able to execute one or more basic production actions, e.g. moving, transforming, fixing or grabbing.

Definition 2 - Manufacturing Resource Agents (MRA)
The MRA is an agentified manufacturing component, i.e. a manufacturing component extended with agent skills like negotiation, contracting, and servicing, able to participate in coalitions/consortia.

As it could be expected there are several types of MRAs, one for each manufacturing component type. Therefore it is expected to find robot MRAs, gripper MRAs, tool warehouse MRAs, etc. Each MRA is individualised by its basic skills and attributes. In the CoBASA society the basic members are not the physical manufacturing components but the MRAs. Each manufacturing component thus needs to be agentified (transformed into an MRA) before it can participate in the CoBASA society. Since skills represent a very important characteristic of a manufacturing component, and since these skills are implemented by manufacturing controllers, a MRA represents in fact the agentification of a manufacturing controller.

Every MRA should be able to: 1) adhere to a cluster, 2) participate in consortia/coalitions, and 3) perform the manufacturing operations associated to the skills it represents.

Definition 3 – Coalition/Consortium
A coalition/consortium is an aggregated group of agentified manufacturing components (MRAs), whose cooperation is regulated by a coalition contract, interacting in order to generate aggregated functionalities that in some cases are more complex than the simple addition of their individual capabilities.

A coalition is usually regarded as an organisational structure that gathers groups of agents cooperating to satisfy a common goal (Shehory & Kraus, 1995). On the other hand, the term consortium comes from the business area where it is defined as an association of companies for some definite purpose. Comparing both definitions it can be seen that they are quite similar because in both definitions there is the notion of a group of entities cooperating towards a common goal. Therefore, this common definition is adapted to the CoBASA context.

A basic coalition/consortium besides being composed of MRAs includes an agent that leads the coalition – Coordinating Agent (CA). In addition it can include as members other coalitions/consortia. The coordinator of a consortium is able to execute complex operations that are composed of simpler operations offered by the consortium members.

Definition 4 – Coordinating Agent (CA)
A CA is a pure software agent (not directly connected to any manufacturing component) specialised in coordinating the activities of a coalition, i.e. that represents a coalition.

As members of coalitions/consortia, MRAs can only play the member role while CAs can play both the coordinator role and member role. A simple manufacturing coalition/consortium is composed of some MRAs and one CA. However, a coalition/consortium can be composed of other consortia creating in this way a hierarchy of coalitions/consortia. Therefore a CA can simultaneously coordinate MRAs and others CAs.

It is worthwhile to emphasise an intuitive aspect: the fact that the set of skills offered by a coalition is composed of not only the basic skills brought in by its members but also more high level skills that result from a composition of those simpler skills. Therefore, some kind of skill composition is needed to generate new skills.

When forming a coalition/consortium there are no limitations on the type of agents that can be involved in but it is mandatory to know what are the available and willing to participate agents. It would be important that these agents could be grouped by their spatial relationships (or any other relevant relationship e.g. technological compatibility), i.e., manufacturing agents that could establish consortia should be grouped together because they share something when they are candidates to consortia. Therefore, there is a need for a structure (cluster) that group the agentified manufacturing components (MRAs) willing/able to cooperate and from which the agents share some concepts.

Definition 5 – Shop Floor Cluster
A cluster is a group of agentified manufacturing components (MRAs) commited to participate in coalitions/consortia and sharing some relationships, like belonging to the same manufacturing structure and possessing technological compatibility.

A shop floor cluster includes a kind of directory where the agents willing to participate in coalitions/consortia can register. This directory acts as the place where those agents become known and where they publish their skills. A special agent – the cluster manager (CMgA) – is responsible for keeping the directory and supporting the adhesion/withdrawal of agents.

The formation of a coalition/consortium, however, is not done by the cluster manager but rather by a specialised agent called broker.

Definition 6 – Broker
A broker is an agent that is responsible for the creation of coalitions/consortia in interaction with an external user. The broker agent gathers information from the cluster and based on the user preferences supervises/assists the process of creating the coalition/consortium.

The broker therefore interacts with the human, the cluster, and the candidate members to the consortium. Coalitions/consortia can be created either automatically or manually. At the current stage only the manual option is considered.

Contracts are the next important CoBASA mechanism, which is used to regulate the agent's interaction with a cluster as well as its behaviour within coalitions/consortia.

Definition 7 – Contract, according to the law (FindLaw, 2002)
"An agreement between two or more parties that creates in each party a duty to do or not do something and a right to performance of the other's duty or a remedy for the breach of the other's duty."

In the CoBASA architecture two types of contracts are considered: **cluster adhesion contract (CAC)**, and **multilateral consortium contract (MCC).**

Definition 8 – Cluster Adhesion Contract (CAC)
This contract regulates the behaviour of the MRA when interacting with a cluster. Since the terms imposed by the cluster cannot be negotiable by the MRA the contract type is "adhesion". The CMgA offers cluster services in exchange for services (abilities or skills) from the MRA.

The CAC includes terms such as the ontologies that must be used by the candidate, the duration of the membership, the consideration (a law term that describes what the candidate should give in turn of joining the cluster, usually the skills that the candidate is bringing to the cluster).

Definition 8 – Multilateral Coalition/consortium Contract (MCC)
This contract regulates the behaviour of the coalition by imposing rights and duties to the coalition members. The contract identifies all members and must be signed by them to be effective. The coalition leader (CA) is identified as well as its members. The members are entitled to a kind of award (credit) in exchange for their skills.

The important terms of this type of contract, other than the usual ones like duration, names of the members, penalties, etc., are the consideration and the individual skills that each member brings to the contract. The importance of contracts as a mechanism to create/change flexible and agile control structures (consortia) lays on the fact that the generic behaviours exhibited by generic agents are constrained by the contract that each agent has signed. This calls forth that different consortium behaviours can be achieved by just changing the terms of the consortium contract, namely the skills brought to the consortium.

The MRA was defined as a manufacturing component extended with agent skills, which corresponds to its agentification in order to be able to participate in the CoBASA society. The agentification could have been achieved by developing an agent that could simultaneously interact with the manufacturing equipment controller, and manage its CoBASA social activities (cluster joining and coalition participation) that includes, among others, the contract negotiation and composition of skills tasks. In addition, the requirements imposed by the need for interaction to be maintained with the controller (short response time, interaction protocol specificities, ...) suggest the use of a dedicated agent to interact with the controller. Therefore, as the two activities are both very demanding to be accomplished by one agent only, the adopted approach was to separate the functionalities and to have one dedicated agent to interact with the controller – Agent Machine Interface (AMI) and a generic agent specialised in the CoBASA social activities, namely, contract negotiation and skills composition – Generic Agent (GA).

3. THE STEPS OF THE METHODOLOGY

This section describes the main steps required to operate CoBASA. This involves creating Manufacturing Resource Agents - MRAs that can be used as candidates in future manufacturing coalitions, and creating, changing and deleting consortia. In this paper only the part referring to the join cluster is analysed since the other parts have been already discussed in (Barata & Camarinha-Matos, 2002, 2003)

Figure 1 shows these steps by clearly indicating the most important functionalities under CoBASA. While the steps for creating, changing, and deleting consortia must be executed any time the system is changed, the join cluster functionality needs only to be executed whenever a manufacturing component needs to be agentified. Therefore, the steps required to create a manufacturing agent are done only once in the life of a manufacturing component. Considering that in most situations the manufacturing components are reused the time spent with the agentification is affordable in comparison with its lifetime. This is an important fact since the agentification process is the most complex and time-consuming activity and, in addition, most of the used functionalities are creating, changing, and deleting consortia, which are simple and fast processes.

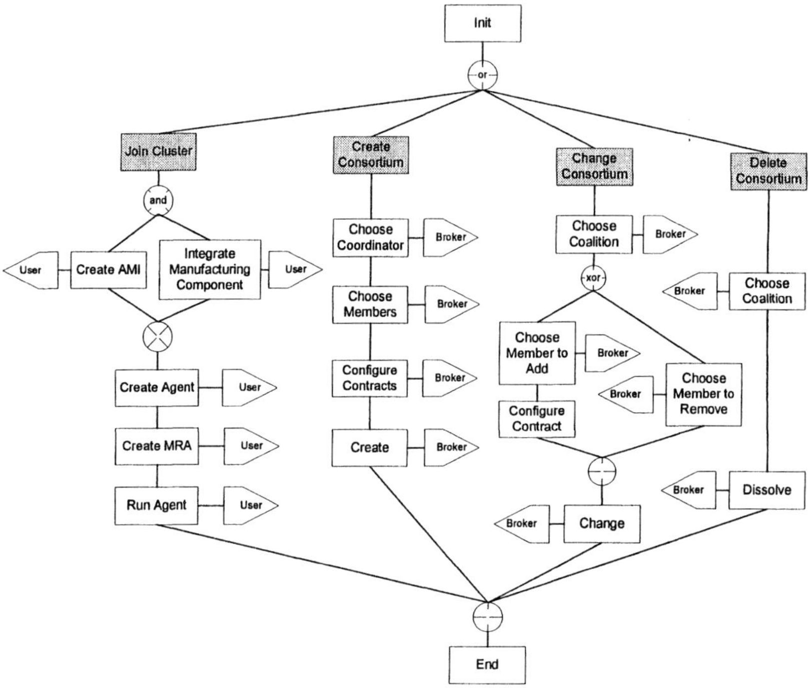

Figure 1 - Steps of the methodology

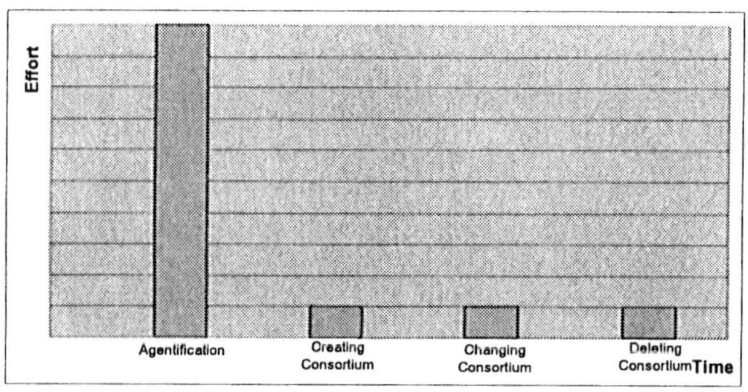

Figure 2 – Variation of the configuration effort for a manufacturing component

Figure 2 illustrates how the effort required to configure a manufacturing component evolves along its lifetime. The objective is not to indicate exact values about how much effort is required but rather to give an idea that the agentification process that is made initially, and only once, requires a bigger effort than the other phases of the component's lifetime. In the specific case of the manufacturing component considered in Figure 2, the manufacturing agent after being agentified, participated in a consortium, which was later on changed, and finally removed.

2.1 Join the Cluster

The branch *Join Cluster* in Figure 1 shows the steps that must be performed in order to have a manufacturing agent participating in a cluster. Therefore, before joining the cluster the manufacturing equipment must be transformed into a manufacturing agent (MRA). Joining the cluster corresponds to the practical situation in which a manufacturing component is added to a given manufacturing cell. It is supposed that the broker agent and the cluster manager agent are already running. It must be noticed that a MRA is more than an agentified manufacturing component and thus the following steps have been identified whenever a new manufacturing equipment needs to be transformed into a MRA: 1) ***Create the AMI, 2) Integrate Manufacturing Component, 3) Create Manufacturing Agent – Agentification of the Manufacturing Component, 4) Create the MRA, and 5) Run the agent***.

Steps 1 to 3 correspond to the agentification of the manufacturing component. At the end of step 4 a manufacturing resource agent exists composed of the configured generic agent plus the appropriate AMI for the manufacturing component. At the end of step 5 the agent has been registered in the cluster.

Create the AMI. In this step the user configures the generic AMI according to the functionalities and requirements of the manufacturing component to which the AMI is going to be connected. The AMI establishes the link between the generic agent (GA) and the manufacturing component. This agent is specific to the type of the manufacturing component to which it is connected in terms of the functionalities it offers.

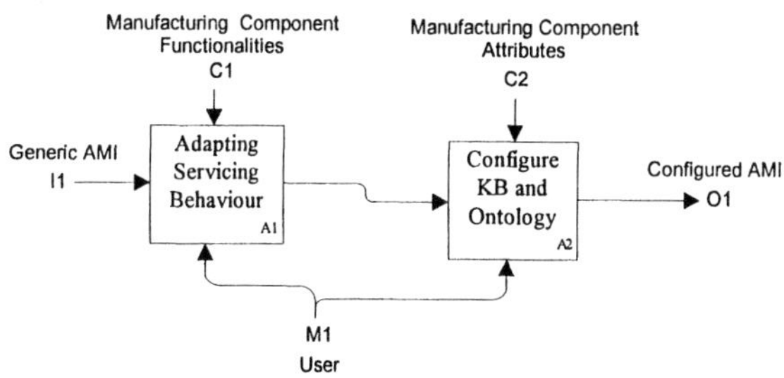

Figure 3 - Activities when configuring the AMI

The generic AMI includes roughly the behaviours required for user interface, and the generic behaviour to attend requests from the GA to which it will be connected. Configuring it corresponds to implement the specific functionalities of the agent to which it will be connected. This is represented in Figure 3 by the activity *Adapting Servicing Behaviour*. Furthermore, it is necessary to configure the individual KB and ontology of the agent to the individual attributes of the manufacturing component, which is represented by the activity *Configure KB and Ontology* in Figure 3.

Integrate Manufacturing Component. In this phase the functionalities of the manufacturing component controller are modelled using computer based abstraction mechanisms such as Remote Procedure Calls (RPCs) or distributed objects. When using RPCs, the functionalities of the controller are represented by the methods included in the RPC that models the controller. When modelling manufacturing controllers using objects or distributed objects, their attributes model the static characteristics of the controller being modelled while the methods model the controller's functionalities. This modelling is fundamental since many of today's manufacturing controllers do not provide an abstraction at this level and, hence, to be controlled, they need special commands usually sent via a RS232 protocol. This is the process of connecting the physical controller to the agent. This could be an easy task if every physical component was controlled directly by its own agent. However, outdated legacy controllers with closed architectures control most of the existing physical components.

Independently of how the commands are sent to the specific manufacturing controller the important thing to keep in mind is that it is necessary to adapt the computational abstraction of the manufacturing component to an abstraction level that can be used by agents. This is so because, in the end, the commands that the physical manufacturing controllers receive are originated in the agents that compose CoBASA (MRA agents).

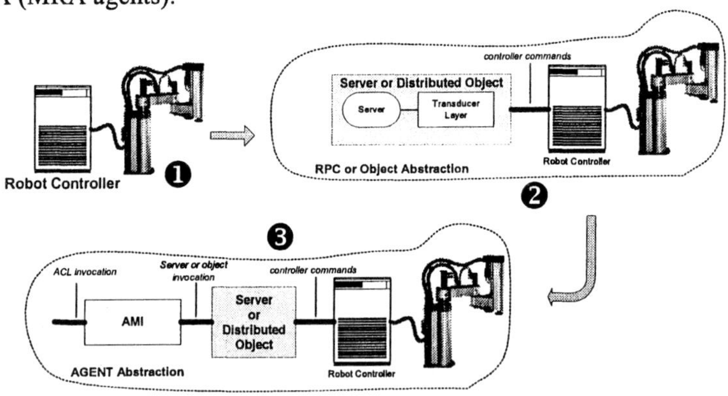

Figure 4 - Transformation of a manufacturing component into an agent

The objective of this phase is thus to build a server, using RPCs or a distributed object, using CORBA or DCOM, whose methods can be invoked from any computer system connected to a network. This approach decouples the physical manufacturing controller from the agent environment, which facilitates the integration. To integrate these legacy components in the agents' framework it is thus necessary to develop a software wrapper to hide the details of each component (Barata & Camarinha-Matos, 1995; Camarinha-Matos, Seabra Lopes, & Barata, 1996). The wrapper acts as an abstract machine to the agent supplying primitives that represent the functionality of the physical component and its local controller. The steps of this phase are indicated in Figure 4 with the numbers ❶ and ❷. At the end of this phase the manufacturing component can be commanded through the activation of a RPC or the invocation of one of the methods of the object that mimics it.

Agentification of the Manufacturing Component. The physical manufacturing component is effectively transformed into a manufacturing agent at the end of this phase, hence, the name agentification. This is represented in the number ❸ of Figure 4.

Steps 1 to 3 occur only when the physical manufacturing component is first integrated in the community of agents. All the other situations happen after the component has been already agentified, like for instance changes on the consortium.

The activity that must be done in this phase is connecting the AMI, which has been already configured in step 1, to the object or server that represents the physical manufacturing controller. The AMI accesses the wrapper services using a local software interface (proxy), where all services implemented by the wrapper/legacy controller are defined. Figure 6 illustrates how the AMI connects to the wrapper by showing some details of it that were not presented in Figure 4.

The generic AMI agent is a simple agent with a simple behaviour to accept requests from other agents. When a REQUEST is received, the AMI calls the wrapper to execute the requested service and when the command is executed, it sends back a DONE message to the enquirer agent. Each AMI implements the services supported by the physical component by configuring what the services of the proxy to which it is connected are and the name/address of the component. Because issuing Agent Communication Language REQUEST commands to the AMI does the activation of the manufacturing component, it can be stated, then, that this software layer provides agent abstraction (Figure 4, number ❸).

Create the MRA. At the end of this phase a Manufacturing Resource Agent (MRA) is created. MRAs are composed of a Generic Agent connected to the agentified manufacturing component produced in the last phase. In the CoBASA framework only manufacturing components represented in this way (MRAs) are able to join the cluster and, hence, participate in coalitions.

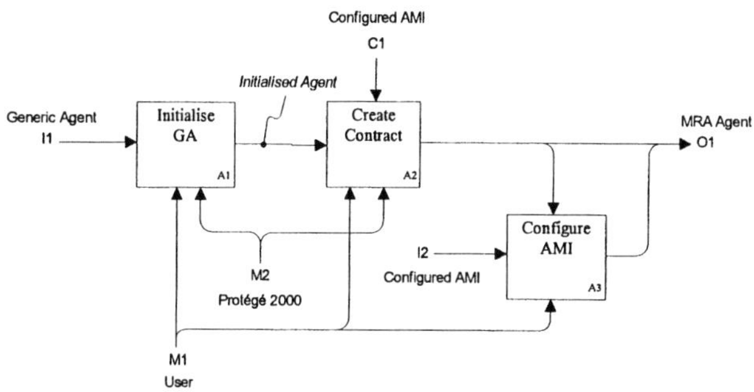

Figure 5 - Activities in the Create MRA phase

In Figure 5 the activities that must be executed to create a MRA are shown. In the activity *Initialise GA* the user initialises the Generic Agent, which, in this case, is used to supply negotiation skills to the MRA agent. It is then necessary to configure the initialisation and the ontology file to be used by the agent. At the end of the activity, the GA is an initialised instance of a generic agent.

The connection to the AMI is guaranteed by creating a consortium contract between the generic agent and the AMI, establishing in this way a consortium. The member promise part (AMI) of the contract contains the services supplied by the AMI. This is supported by the activity *Create Contract*, which is done by the user using the Protégé 2000 ontology management (Protégé-2000, 2000) environment to create the contract. The information existing in the *configured AMI*, which was configured before, constrains the output of this activity because the behaviour of the MRA agent depends on the skills brought in by the AMI to whom it is connected. The contract being created is attached to the variable *coordinated contracts* that contains all the contracts that this GA is coordinating. In this situation, the GA only coordinates, in fact, the AMI to which it is connected.

At the end of the *Create Contract* activity the MRA agent is complete and composed of the generic agent GA attached through *coordinated contracts* to its AMI. However, the AMI is not yet configured in the sense that it should only accept requests from the GA that is attached to it. This is done in the activity *Configure AMI* in which the AMI must also be configured to include the name of the generic agent to which it is connected. This guarantees that an AMI refuses any requests from unknown agents.

Figure 6 shows the various entities that compose the MRA. It must be remembered again that a MRA needs only to be created when the manufacturing component is used for the first time. Future uses of the manufacturing component through its representative (the MRA), in coalitions and possible different clusters, do not involve any changes to it.

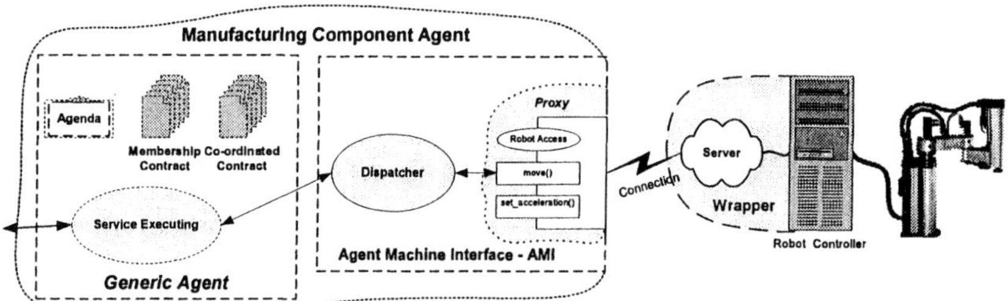

Figure 6 - The Manufacturing Resource Agent – MRA

Run the agent. This step is just to guarantee that the agent joins the cluster.

4. EXPERIMENTAL SETUP

The described methodology was validated against a scenario that could represent a real manufacturing environment, which is the case with the flexible manufacturing system (Novaflex), established at the UNINOVA (Figure 7).

The cluster is the Novaflex which aggregates all the manufacturing equipment. Different coalitions can be created out of this equipment. These coalitions represent no more than different ways of exploring the cell. Therefore, there are two ways of regarding the cell: the physical and the abstract. In the physical way the NovaFlex is a cell composed of several manufacturing equipments that are related by physical relations. It is possible to imagine that parts of the entire cell can be operated independently as well as that equipment can be added or removed. In the abstracted way the NovaFlex is a cluster composed of several manufacturing agents (agentified manufacturing equipment) whose entire set of skills represents the potential of this cluster to solve problems. Whenever a problem requiring a specific set of skills available in the cluster is needed a coalition can be created fitted with that specific set of skills. Furthermore, several problems can be answered simultaneously as long as the cluster includes members able to answer the various problem requirements. The interesting point about this vision is the dynamics of the coalitions.

Figure 7 – Partial view of Novaflex

It is important to remember that any manufacturing equipment that might need to be added to the NovaFlex physical infrastructure (physical view) must join the cluster NovaFlex (abstract view). Hence, adding equipment corresponds to joining the cluster while removing equipment corresponds to leaving the cluster.

The scenario used to test the prototype included a BOSCH SCARA based assembly system as well as an ABB based assembly system.

Figure 8 - The SCARA assembly system

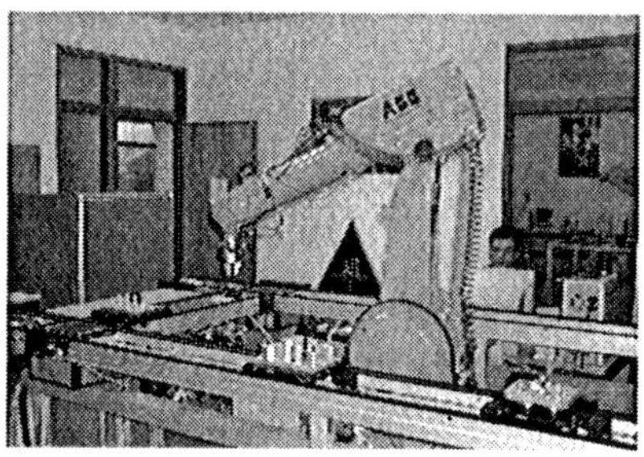

Figure 9 - The ABB assembly system

The SCARA assembly system (Figure 8) is composed of one robot BOSCH SCARA SR-80 equipped with a tool exchange mechanism, a tool warehouse composed of 6 individual slots, 4 different grippers, a feeder, and a fixture.

The ABB assembly system (Figure 9) is composed of one robot ABB IRB 2000 equipped with a tool exchange mechanism, a tool warehouse composed of 4 individual slots, 4 different grippers, a feeder, and a fixture.

Two coalitions were successfully formed out of each of these two subsystems.

5. CONCLUSIONS

This paper described a methodology to support the agility and reconfiguration of shop floor control systems. The CoBASA multiagent architecture that supports the shop floor reengineering process was briefly described and the methodology used to create the agentified manufacturing components was detailed as well as the experimental setup where the prototype was applied.

The first results of using this reengineering methodology proved the validity of the concept in terms of agility. Changes made to the manufacturing cell, either by adding or removing manufacturing components, were simple and involved only changes in the contracts that regulate the created coalitions. In addition the creation of different ways of exploiting the manufacturing cell (NovaFlex) was simple since it involved only the creation of new coalitions.

6. ACKNOWLEDGEMENT

This work was funded in part by the European Commission through the ASSEMBLY-NET Growth Network.

7. REFERENCES

1. Barata, J., & Camarinha-Matos, L. M. (1995). Dynamic Behaviour Objects in Modelling Manufacturing Processes. In Q. Sun & Z. Tang & Y. Zhang (Eds.), *Computer Applications in Production Engineering* (Vol. 1, pp. 499-508). London: Chapman & Hall.
2. Barata, J., & Camarinha-Matos, L. M. (2002). *Implementing a Contract-based Multi-Agent Approach for Shop Floor Agility*. Paper presented at the Holomas 2002, Aix-en-Provence.
3. Barata, J., & Camarinha-Matos, L. M. (2003). Coalitions of Manufacturing Components for Shop Floor Agility - The CoBaSA Architecture. *International Journal of Networking and Virtual Organisations, 2*(1), 50-77.
4. Camarinha-Matos, L. M., & Barata, J. (2001). Contract-Based Approach for Shop-Floor Re-engineering. In R. Bernhardt & H. H. Erbe (Eds.), *Cost Oriented Automation* (First ed., Vol. 1, pp. 141-148). Oxford - UK: Pergamon - Elsevier.
5. Camarinha-Matos, L. M., Seabra Lopes, L., & Barata, J. (1996). Integration and Learning in Supervision of Flexible Assembly Systems. *IEEE Transactions on Robotics and Automation (Special Issue on Assembly and Task Planning), 12*(2), 202-219.
6. FindLaw. (2002). *Legal Dictionary* [Web Site]. Retrieved November 2002, 2002, from the World Wide Web: http://dictionary.lp.findlaw.com/
7. Protégé-2000. (2000). *http://protege.stanford.edu* [web site]. Retrieved Jan 2002, 2002, from the World Wide Web:
8. Shehory, O., & Kraus, S. (1995). Coalition Formation among Autonomous Agents: Strategies and Complexity. In C. Castelfranchi & J. P. Muller (Eds.), *From Reaction to Cognition* (pp. 57-72). Heidelberg: Springer-Verlag.

14

AGENT-BASED DISTRIBUTED COLLABORATIVE MONITORING AND MAINTENANCE IN MANUFACTURING

Chun Wang[1], Hamada Ghenniwa[1], Weiming Shen[1, 2], Yue Zhang[1]

[1] *Department of Electrical and Computer Engineering*
The University of Western Ontario, London, Ontario, CANADA
[2] *Integrated Manufacturing Technologies Institute*
National Research Council Canada, London, Ontario, CANADA
cwang28@engga.uwo.ca; hghenniwa@eng.uwo.ca;
weiming.shen@nrc.gc.ca; yzhan88@uwo.ca

This paper proposes an agent-based approach for real-time collaborative monitoring and maintenance of equipments and machines. The proposed approach automates the maintenance management process by utilizing intelligent software agents. The maintenance schedules are achieved by automated real time negotiation among software agents representing geographically distributed field engineers and a broker. The paper presents a detailed system design and a prototype implementation of the collaborative monitoring and maintenance system which can be used by maintenance managers in manufacturing industry as a tool to improve the efficiency and quality of maintenance scheduling. Test results show that the proposed approach has potential for automating the maintenance management process in manufacturing environments.

1. INTRODUCTION

In today's modern manufacturing organization, the management of equipments and machines maintenance is one of the most important activities. Maintenance planning, scheduling and coordination focus on and deal with the preparatory tasks that lead to effective utilization and application of maintenance resources (Nyman, 2002). Traditionally this complex process is conducted by maintenance managers. After receiving reports of malfunctions or maintenance requirements, they allocate engineers for maintenance jobs at specific times and locations. Such allocation must take a number of factors into consideration, such as priority of and constraints between jobs, capability and availability of engineers, and costs. Although this manually controlled process is still widely adopted by manufacturers, it has a number of shortcomings in the context of today's global competitive manufacturing environments. Manufacturers are now under a tremendous pressure to improve their productivity and profitability. As an integrated part of production, maintenance management is required to be conducted in a more efficient way. In large scale or virtual enterprise manufacturing environments, plant floors are usually geographically distributed. In addition, environments in plant floors may change dynamically over time. Human maintenance managers may not be able to effectively handle this kind of large amount and dynamic changing information.

This paper proposes an agent-based approach for collaborative monitoring and maintenance of equipments and machines. The proposed approach automates the maintenance management process by utilizing intelligent software agents. The maintenance schedules are achieved by automated negotiation among software agents representing geographically distributed field engineers and a broker rather than a centralized scheduler.

The rest of this paper is organized as follows: Section 2 briefly discusses some background information regarding equipments and machines maintenance and agent based manufacturing; Section 3 proposes an agent based collaborative monitoring and maintenance for equipments and machines; Section 4 describes the system architecture design and system components; Section 5 presents an implemented prototype environment; Section 6 provides a brief discussion and discusses the future work.

2. RESEARCH BACKGROUND AND LITERATURE

Effective maintenance of machines and equipments is very important for companies to sustain their manufacturing productivity and customer satisfaction. Traditional approaches, such as Total Productive Maintenance (TPM) (Venkatesh, 2003) and Reliability Centered Maintenance (RCM) (Moubray, 2001), mainly tackle this issue from the perspective of management science. TPM is a maintenance program concept evolved from Total Quality Management (TQM). Philosophically, it resembles Total Quality Management in several aspects. RCM is a highly structured framework that overturns many widely held beliefs about preventative maintenance. It was originally developed by the civil aviation industry and is now finding applications in various kinds of industrial and service organizations.

In light of the advances of Internet/Web technologies and e-business systems, e-Maintenance has been considered as an integral part of e-Manufacturing. Koc and Lee proposed a system framework for the next generation e-maintenance systems, called Intelligent Maintenance System (IMS) (Koc and Lee, 2002). In this framework, the maintenance system is an Internet-based and Web-enabled predictive maintenance technology which consists of three levels. Firstly, at the product/machine/process level, the focus is on predictive intelligence. Secondly, at the system level the working equipments and machines are compared with symptoms under different conditions. Thirdly, at the enterprise level the focus is on the Web-enabled agent to achieve near-zero-downtime performance through smart asset optimization.

In today's highly distributed and dynamic manufacturing environments, agent based technologies have been proposed to overcome the limitations that traditional hierarchical and centralized control systems show. In this paradigm, the central controller is decomposed functionally or physically into several controllers, usually encapsulated as autonomous agents, each one devoted to a small portion of the overall system. The overall coordination is achieved by communication and negotiation among agents in the system. Once manufacturing systems are appropriately modeled as multi-agent systems and suitable negotiation protocols are adopted, such agent-based systems may yield a global performance which is flexible, robust, adaptive, and fault tolerant. A comprehensive survey of agent based manufacturing systems can be found in (Shen et al., 1999).

3. COLLABORATIVE MAINTENANCE OF EQUIPMENTS AND MACHINES

Agent based system technology has been proposed as a promising approach to address various issues in modern manufacturing industry, such as manufacturing scheduling and shop floor control (Shen et al., 2000; Shen, 2002), and supply chain management (Fox et al., 1993). In this paper we propose an agent based approach for collaborative maintenance equipments and machines with a focus on distributed, dynamic and open manufacturing environments.

In this approach, the maintenance management of equipments and machines is modeled as a multi-agent system. The system consists of five kinds of agents: diagnostic agent, broker agent, field engineer agent, directory facilitator agent, and help desk agent. Diagnostic agent is used to provide monitoring and diagnostic services to the equipments. It collects current failure and degradation data, analyzes the collect data, generates maintenance requirements and sends them to a broker agent. The broker agent plays a role similar to maintenance managers in the traditional maintenance process. After receiving the maintenance requirements, the broker agent checks with one of the directory facilitators to find capable field engineer agents currently registered within the system. The field engineer agent is a personal assistant of a field engineer. Information regarding the engineer's capabilities (types of services he can provide), availabilities (time table), and cost, is part of the knowledge of field engineer agent. A field engineer may dynamically change its profile and preferences. The broker agent negotiates with the field engineer agents to work out a suitable schedule for them to achieve the maintenance goals. In cases that the field engineers have technical difficulties during the process of equipment maintenance, they may locate a suitable remote help desk service by utilizing the negotiation mechanisms built in field engineer agents. The help desk services are provided by in-house engineers. Field engineers can communicate with in-house engineers through digital data, audio, and images which are encapsulated in a collaborative discussion mechanism.

At the system level, all agents need to register with one of the directory facilitator agents regarding their services provided. Director facilitator agents register with each other. In this way, any agent in the system can virtually find an agent with the services it needs as long as it is alive in the system scope. In this approach, the equipment maintenance management is conducted as an automated procedure. Maintenance task planning, scheduling, and coordination are achieved by automated negotiation among agents. This mechanism provides an automation infrastructure for maintenance management process. At the operational level it can dramatically increase the process efficiency. In addition, the efficiency of automated negotiation mechanism makes use of real time distributed scheduling mechanisms, especially in large scale and dynamic environments. This will increase the quality of the maintenance schedules.

The robust, flexible, and distributed natures of the agent based collaborative maintenance system make it easy to be deployed in various computer networks, even in global computing network environments. Although self-diagnostic agents can theoretically provide monitoring and diagnostic services to any equipment on the network, it is preferred that they sit close to the equipments or are connected directly to the equipments being monitored through a local network to provide reliable monitoring, diagnostic services with fast responses.

4. SYSTEM ARCHITECTURE AND COMPONENTS

4.1 System Architecture

In this work, we propose a six-layer system architecture (Figure 1) for the agent based distributed collaborative equipment maintenance system.

As shown in Figure 1, the Help Desk Services layer provides field engineers with remote help desk services. It contains help desk agents which can interact with field engineer agents to set up a conference session for in-house experts and a field engineer. The Maintenance Services layer provides maintenance services for the equipments. It contains field engineer agents representing a batch of field engineers and other necessary resources, such as tools and spare parts. Brokering Service layer consists of several agents which are responsible of finding suitable field engineers for a specific maintenance task and scheduling a time slot for it.

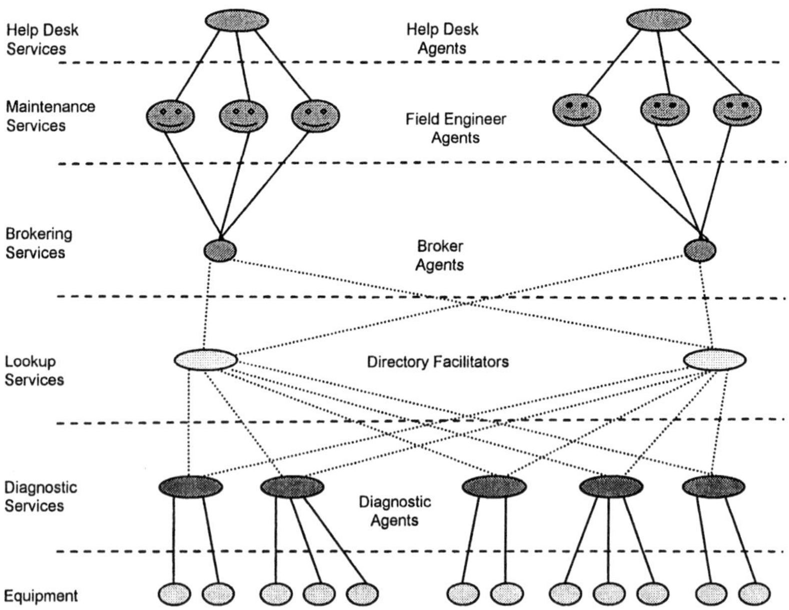

Figure 1- System structure for agent based collaborative maintenance

The Lookup Services layer contains Directory Facilitators agents. All agents in various service layers need to register with these Directory Facilitators agents. The Diagnostic Services layer includes diagnostic agents. They provide monitoring and diagnostic services to equipments. The Equipment layer consists of various equipments that need to be monitored and maintained.

4.2 Agent Architecture

The agent architecture adopted for the agents in our collaborative maintenance system is Coordinated, Intelligent Rational Agent (CIR-Agent) architecture (Ghenniwa, 2000). In the CIR-Agent model, an agent is an individual collection of primitive components. Each component is associated with a particular functionality that supports a specific agent's mental state as related to its goal. The agent's mental state regarding the reasoning about achieving a goal, in the CIR model, can be in one of the following:

1. Problem solving: determines the possible solutions for achieving a goal.
2. Pre-interaction: determines the number and the type of interdependencies as well as the next appropriate domain action.
3. Interaction: resolves the problems associated with the corresponding type of interdependencies. The mechanisms used in the interaction are called interactive devices.
4. Execution: affects the world.

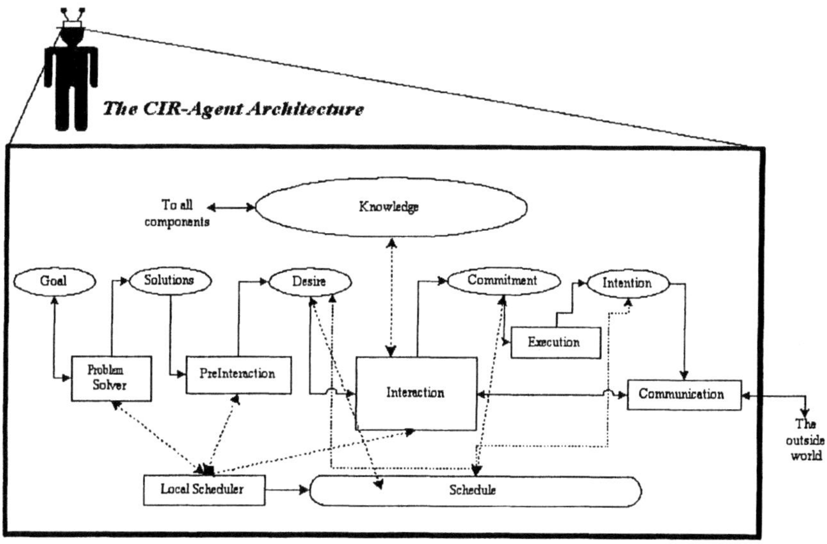

(a) Detailed Architecture of CIR-Agent

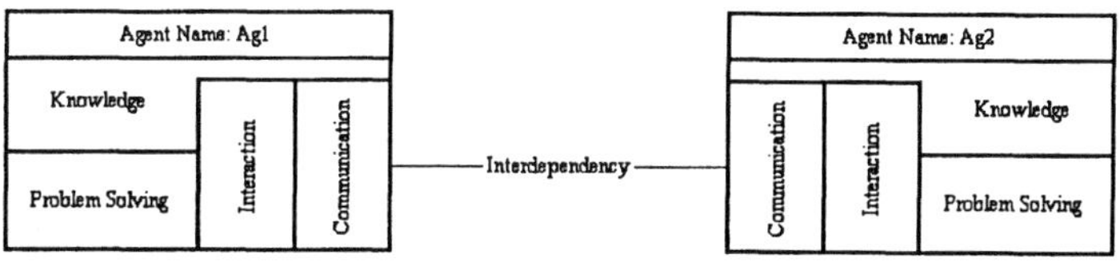

(b) Logical Architecture of CIR-Agent

Figure 2-Detailed and logical CIR-Agent architectures

Based on these mental states, the CIR-Agent's architecture can be considered as a composition of four components: problem solver, pre-interaction, interaction, and execution. In the context of our agent based collaborative monitoring and maintenance system framework, we mainly focus on the problem solvers and interaction devices of different agents in the system.

4.3 Diagnostic Agent

The main functionalities of the diagnostic agent include monitoring, diagnostic and requirement generation services. In terms of CIR-Agent architecture, these should be encapsulated in the problem solver of diagnostic agents. In addition to monitoring current failures, diagnostic agents should monitor the degradation states of equipments as well. Based on the information, it generates different maintenance requirements with different priority in terms of the degrees of urgency.

The Interaction devices that diagnostic agents use to request maintenance services are FIPA Query protocols. Certainly, it needs to be connected to the equipments. The implementation of the connection between diagnostic agents and the equipments may depend on the data ports that the equipments can provide.

4.4 Broker Agent

Broker agents are used to schedule maintenance tasks received from diagnostic agents on capable field engineers. The problem solver of the broker agents is basically a scheduling engine. It assigns tasks to engineers by automated negotiation with field engineer agents. During the scheduling, several constraints, such as degree of urgency, field engineers' capability, availability, and costs, the availability of necessary tools need to be taken into consideration. At the same time, the scheduling engine selects the best possible solution that satisfies diagnostic agents' requirements according to their various objectives.

Broker agents use FIPA Query protocols to receive maintenance requirements from diagnostic agents and FIPA Contract Net protocol to assign tasks to field engineers through field engineer agents.

4.5 Field Engineer Agent

Field Engineer agents are basically the personal assistants of field engineers. Field engineers may put their profiles, preferences, current situations, and any dynamic changes into the field engineer agents. The problem solver of these agents is a scheduler dedicated to the field engineers. It is aware of all the user input information. It maintains an up-to-date schedule and historical data for a field engineer. It can automatically negotiate the task allocations with broker agents based on the user information it has. If they want, field engineers are able to involve into the negotiation process through a user interface. In addition, the problem solver of field engineer agents can locate a help desk agent and start a collaborative discussion with in-house engineers behind the help desk agent as well if a field engineer needs help during the maintenance process.

Interaction devices used by field engineer agents include FIPA Query protocols and FIPA Contract Net protocol. FIPA Contract Net protocol is used to negotiate task allocation with broker agents. Both FIPA Query protocols and FIPA Contract Net protocol can be used to locate a help desk agent. If a field engineer agent knows which help desk agent it wants, it will go directly to the agent though FIPA Query protocols. Otherwise it will find a suitable one though FIPA Contract Net protocol.

4.6 Help Desk Agent

The problem solver of help desk agents is a scheduler as well. It schedules a set of discussion sessions with field engineers based on the requirements from field engineer agents and their priority. In addition, it needs to consider the current availability and capability of in-house engineers. Help desk agents use FIPA Query protocols and FIPA Contract Net protocol to interact with field engineer agents. Real time discussion and collaborative design are also supported by help desk agent.

4.7 Directory Facilitator agent

Directory facilitator agents provide agent registration and look up service. They use standard FIPA Directory Facilitator interfaces to provide services.

4.8 Collaboration between agents

In this system architecture, the collaboration mechanisms adopted between agents are economically inspired negotiation protocols, e.g. FIPA Contract Net and FIPA Query. Agents exchange information in the framework of these protocols. The preference models and negotiation strategies inside agents are not modeled in this system architecture. We leave them to domain specific implementations.

5. PROTOTYPE IMPLEMENTATION

To validate the proposed approach, a prototype environment has been implemented in a wireless network of mobile robots, desktop PCs, Pocket PCs, and a Smart Board. Figure 3 shows the actual prototype environment.

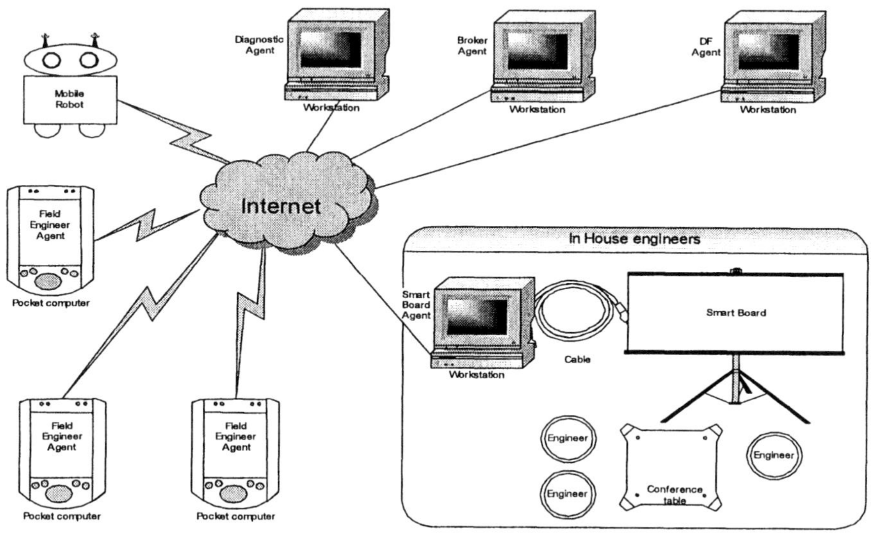

Figure 3-Agent based collaborative maintenance system prototype implementation

In Figure 3, the equipment is a Magellan mobile robot which is connected to the diagnostic agent through a wireless network. iPAQ pocket PCs are used as the platforms of field engineer agents. The major collaborative discussion tool in this implementation is a smart board connected to a help desk agent. In this implementation we call the help desk agent as smart board agent, as it mainly provides the smart board service through which field engineers and in-house engineers discuss the maintenance tasks.

The diagnostic agent, the broker agent, the directory facilitator agent and the smart board agent are running on Pentium 4 PCs connected to the network.

Table 1 shows the hardware and software implementation details of the prototype environment.

Table 1-Hardware and software implementation details of the prototype

Subsystem	HardWare Platform	SoftWare Platform	Application Package	NetWork Interface
Mobile Robot	Magellan Robot	Linux OS, OmniORB	Diagnostic Information Collection	Wireless LAN
Diagnostic Agent	Pentium 4 PC	Windows 2000, JadeLeap	Diagnostic Agent Package	100M Ethernet
Broker Agent	Pentium 4 PC	Windows 2000, JadeLeap	Broker Agent Package	100M Ethernet
Field Engineer Agent	iPAQ Pocket PC	Windows CE, JadeLeap	Field Engineer Agent Package	Wireless LAN
Smart Board Agent	Pentium 4 PC	Windows 2000, JadeLeap	Smart Board Agent Package	100M Ethernet
Smart Board			iBid	Serial Port
DF Agent	Pentium 4 PC	Windows 2000, JadeLeap	JADE DF Agent	100M Ethernet

Figure 4 shows the graphical user interface of the broker agent. Four windows appear simultaneously in the interface for displaying Maintenance requirements, available field engineers, system status and the current schedule. The current schedule is presented in the form of a Gantt chart. The Gantt chart is the usual horizontal bar chart with the x-axis representing the time and the y-axis, the various field engineers (machines in terms of classical scheduling). The bars inside the Gantt chart represent operations of maintenance jobs. The operations belong to the same job have identical color. The coordinates of operations determined by identities of field engineers and times reflect the allocation of the operations to field engineers over time.

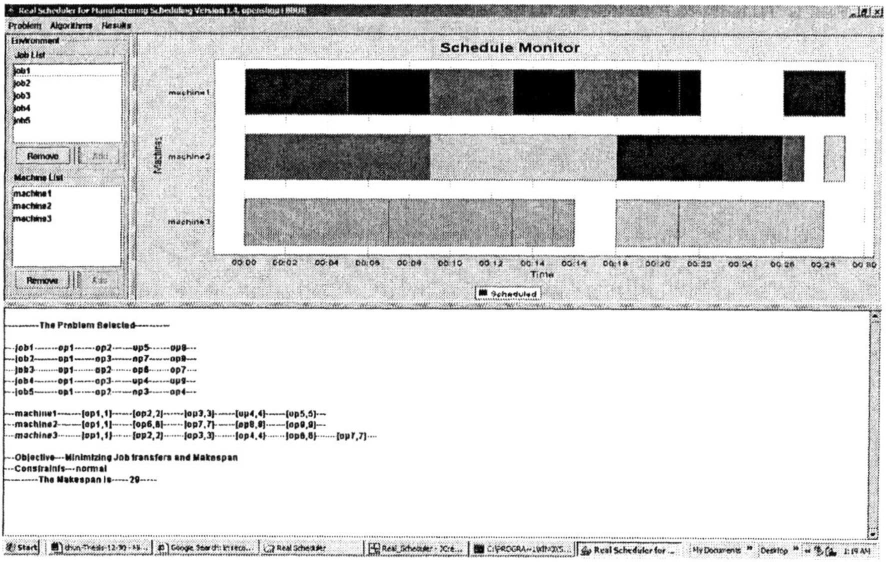

Figure 4 -User interface of broker agents

6. DISCUSSION AND FUTURE WORK

In this paper, we proposed, designed, and implemented an agent based system to support distributed collaborative equipment monitoring and maintenance. The process is fully automated by utilizing intelligent agents. The system also supports the collaboration between field engineers and in-house engineers. Based on this agent based framework, efficient maintenance schedule can be generated by utilizing sophisticated distributed scheduling mechanisms to allocate maintenance tasks to field engineers over time. In addition, this highly open and flexible system framework can be easily integrated into e-Manufacturing and e-Business environments to provide maintenance services as part of the entire e-Manufacturing or e-Business framework.

The implemented system prototype environment has been tested in several scenarios of the maintenance management process. Results show that the proposed approach has potentials in terms of automating the maintenance management process in manufacturing environments.

For future extension, other negotiation mechanisms, such as various auctions, will be integrated into the system to provide further adaptability and efficiency in the dynamic, distributed environments. More effective distributed scheduling mechanisms, equipment monitoring and diagnostic techniques need to be integrated into the system as well.

7. ACKNOWLEDGEMENT

We would like to acknowledge the financial support provided by Timelog International Inc., Barrie, Ontario, Canada (http://www.timeloginternational.com/) and Materials and Manufacturing Ontario (http://www.mmo.on.ca/). Also we would like to thank Wafa Ghoniam, Bashar Duheidel and Samer Hammoud who participated in the prototype implementation of the project.

8. REFERENCES

1. Fox, M.S., Chionglo, J.F., and Barbuceanu, M., The Integrated Supply Chain Management System Internal Report, Dept. of Industrial Engineering, University of Toronto, 1993, http://www.eil.utoronto.ca/public/iscm-intro.ps
2. Ghenniwa, H., and Kamel, M., Interaction Devices for Coordinating Cooperative Distributed System, *Automation and Soft Computing*, Vol.6, No.2, 2000, pp.173-184
3. Koc, M and Lee, J. A System Framework for Next-Generation E-Maintenance Systems, Intelligent Maintenance System Center, http://www.uwm.edu/Dept/IME/IMS, 2002.
4. Moubray, J. Reliability-Centered Maintenance, Industrial Process Inc. 2001.
5. Nyman, D., and Levitt, J., Maintenance Planning, Scheduling and Coordination, Industrial Process Inc. 2002.
6. Shen, W and Norrie, DH Agent-Based Systems for Intelligent Manufacturing: AState-of-the-Art Survey. Knowledge and Information Systems, an International Journal, 1(2), 129-156, 1999.
7. Shen, W., Lang, S., Korba, L., Wang, L., and Wong, B., Reference architecture for internet-based intelligent shop floor, *Proceedings of the SPIE International Conference on Network Intelligence: Internet Based Manufacturing,* Boston, MA, Nov. 8, 2000. Vol. 4208, PP. 63-71.
8. Shen, W., "Distributed Manufacturing Scheduling Using Intelligent Agents", *IEEE Intelligent Systems*, Jan./Feb., 2002, pp. 88-94
9. Smith, R.G. "The Contract Net Protocol: High-Level Communication and Control in a Distributed Problem Solver," *IEEE Transactions on Computers*, C-29(12):1104-1113, 1980.
10. Venkatesh, J. An Introduction to Total Productive Maintenance, http://www.plant-maintenance.com/articles/tpm_intro.shtml, 2003.

15

MOBILE ACCESS TO PROCESS KNOWLEDGE: AN AGENT-BASED APPROACH

Leendert W. M. Wienhofen
CognIT a.s,Meltzersgt. 4, N-0257 Oslo, NORWAY
{Leendert.Wienhofen@cognit.no}

This paper discusses a methodology and its enabling technology using mobile devices and process workflow roles to bring work relevant information directly to users' fingertips. We present one complete solution for knowledge management in the process industry. The proposed solution is accessed from mobile devices and combines a state-of-the-art enterprise knowledge server with a multi-agent system. After a short introduction of the technologies as well as an architecture sketch, three different scenarios in an aluminium production setting are presented. The goal of the presented system is to contribute to knowledge sharing on all levels of the organization.

1 INTRODUCTION

A large number of process industry enterprises such as Shell, Statoil, Hewlett Packard and Norsk Hydro, have already taken the step to implement knowledge management systems in one way or another 141617. Other large corporations as well as articles in magazines such as Harvard Business Review and the Knowledge Management Magazine also indicate this 311121618. Having control over the knowledge in an enterprise makes it easier to find groups of experts within the organisation which reduces the chances for 're-inventing the wheel' and therefore makes the enterprise more efficient 8.

1.1 Caveat

Technology for carrying out the scenarios described in chapter 3, 4 and 5 is already available, however the knowledge management system (see paragraph 2.1) and the agent technology (paragraph 2.2) have not yet been combined into one program. Tools for knowledge management from the CORPORUM® suite by CognIT a.s (in particular Knowledge Server and Knowledge Factory 57) have successfully been incorporated at a number of process industries, among others Hydro Aluminium 14and TMG 19. The JADE Multi-Agent System 4has been implemented and tested in the AmbieSense system 12(which includes proximity detection by means of Bluetooth).

2 BACKGROUND AND REQUIREMENTS

Many different ways of describing a knowledge management system exist. Some software vendors claim their groupware solution is a full-fletched knowledge management system. However, software alone will not suffice, it is merely a method to serve the purpose. A full-scale knowledge management approach must embrace the whole organisation including the workers that do not have a desk. Everybody must be a part of the knowledge life cycle.

This paper addresses knowledge management in an industrial context, in particular the situation from the industrial workers point of view. The aluminium production process, which is referred to in the scenarios, takes place in an approximately 900 meter long production line where the bauxite ore is the input at one end and the products or half-fabricates are the output at the other end. Many processes need to be executed in order to manufacture a product from bauxite ore [9].

Industrial workers, although carriers of vital competence for the industrial plant that employs them, have in general not been recognized as knowledge workers, and therefore have not been included in the knowledge sharing network in a systematic fashion. Their voices are only heard in meetings where they are often confronted with a situation involving rhetoric that may seem foreign or even hostile to them. We should acknowledge that knowledge workers of this kind have little practice in expressing their ideas and therefore a high percentage of these workers keep their knowledge to themselves and often see this tacit knowledge as inferior to ideas and concepts voiced by trained academics. We should also acknowledge that, as a knowledge worker, the operator (the worker) is a consumer of information; a fact that is often disregarded both by the management and by the worker himself. The result typically leads to aloofness and organizational detachment. Hence it is a basic objective to bring them in and involve them in ways coined in Nonaka's SECI model 15(see also the end of paragraph 2.1). The majority of the workers in industrial plants have must access to the corporate memory, both as donors as well as receivers. This model describes the knowledge creation of firms as conversion of tacit knowledge into explicit knowledge and vice-versa. The interaction builds to a continuous spiral reaching from individual to organizational level. The SECI model contains the following four elements:

- Socialization, where tacit knowledge is shared through shared experiences.
- Externalization, where tacit knowledge is converted into explicit knowledge with the help of metaphors and analogies.
- Combination, where explicit knowledge is systemized and refined e.g. by utilizing information and communication technologies and existing databases.
- Internalization, where explicit knowledge is converted into tacit knowledge, e.g. by learning by doing.

2.1 Knowledge Management System

In order to support the sharing of knowledge, an IT system should be put in place. All of the features mentioned in this paragraph are to be made available to the user using platform independent intuitive access (for example by web-interface). Features to be included in this system are:

1) A database for storing work processes (workflow), descriptions and definitions (also called best practice documents). This content is represented in a virtual pyramid shape, value chains being at the top (see Figure 1):
    1.1) Value chains, to show the overall workflow.
    1.2) Processes, to show a higher level of detail
    1.3) Flow charts, to show the workflow in a process. In case the process is carried out by employees with different roles in the organisation, the flow chart is split up horizontally to represent each role in the co-operative work. (examples of a role are "operator" or "procurement manager")
    1.4) Process descriptions, to give a crisp textual definition of the process. It defines the input, output and roles associated to the described process.
    1.5) Role definitions, a textual description of each role explaining the field of work and the certificates/diplomas needed in order to carry out this role.
    1.6) Activity matrices show a step-wise approach to each of the activities in a flow chart. Activity matrices can include links to 'one point lessons', see paragraph 3.2 for a scenario that describes such lessons.
2) Methods for approving and verifying processes in order to ensure content integrity.
3) A database and/or web space for storing procedures. Procedures are documents that for example describe legislation, or steering documents.
4) A feedback database, which is a tool for continuous improvement. See paragraph 3.3 for a scenario describing the use of the feedback database.
5) A process network, which links the work processes and other content to the feedback database. When the feedback function is used, the process network will find the right person to notify.
6) An agent enabled search engine, which crawls both the Internet and the Intranet in order to continuously look for information that is relevant to either processes or procedures.

Workers are gathered from different plants to share their experiences for a particular process (SECI: Socialization and Externalization), which usually are defined as a procedure for each specific plant (SECI: Combination). Together they define the best practice for this process and enter this in the database.

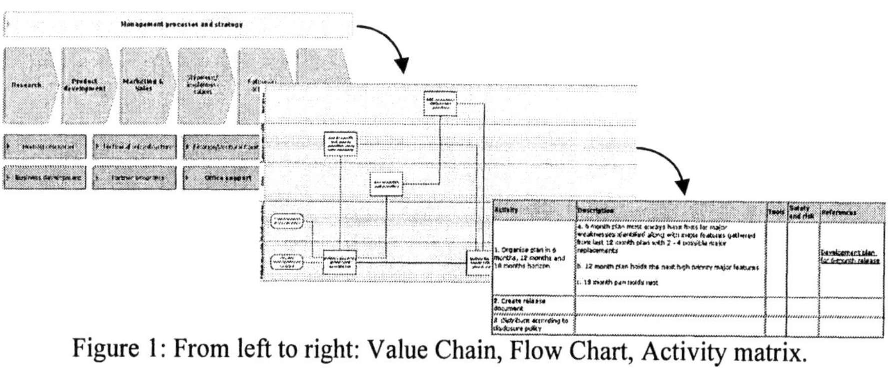

Figure 1: From left to right: Value Chain, Flow Chart, Activity matrix.

2.2 Multi-Agent Technology

Typically vendors supply (parts of the) above-mentioned solution as a portal, where the users must click their way through the content in order to get to the part relevant for them. Yet, the use of a multi-agent system (MAS) enables the user to gain direct access to relevant processes without the need of navigating through the whole value chain and processes.

Ferber 10writes that an agent can be a physical or virtual entity that can act, perceive its environment (in a partial way) and communicate with others. Further it is autonomous and has skills to achieve its goals and tendencies. A MAS supplies an environment for the agents, defines relations between all the entities, a set of operations that can be performed by the entities and the changes of the universe in time and due to these actions.

In the solution presented (see Figure 2 for a graphical overview), agents work in the background to retrieve knowledge that is relevant to the role and location of the user. This behaviour can be compared to the search agent in the AmbieSense architecture, which uses advanced content retrieval mechanisms as well as case-based reasoning systems in order to retrieve information which is relevant based on the users' context, hereunder location and personal preferences 12. In case the agent finds more than one knowledge element (such as a flow chart or a one-point lesson describing how to operate a certain machine the operator just passed), it will give the user the option to choose what he/she wants to get more detailed information about. Other than that, the user can also still browse the whole process hierarchy, starting at the value chain level and clicking to the desired level of detail, thereby bypassing the findings of the agent. This way, the user still has full control.

A variety of agents roam about on this system, each of which has a specific task, yet the agents showed in the scenarios are limited to:

- A notification agent, running on the knowledge management platform. This agent sends notifications to employees who can make use of this information because their role is somehow connected to the information. Notifications can be about changes in the process flow, updates in procedures, new feedback, emergency situations, etc. The agent also forwards news about procedures (for example new legislation), found by the search engine. The use will be explained in more detail in the scenarios in chapter 3.
- A mobile device agent, running on the mobile device. The mobile device agent is aware of the role of its user, and has different types of behaviours (in order to achieve the 2nd and 3rd behaviour the agent has a direct link into the best practice part of the knowledge management system):
  - It reacts on notifications sent by the notification agent and makes sure the receiving system handles the message in an appropriate way (an emergency message will have absolute highest priority and will be displayed on the mobile device directly upon reception. See scenario 2 for a case on handling emergencies).
  - Handling of proximity detection of Bluetooth beacons (location awareness), in order to supply the user with relevant process information.

- o Feedback handling

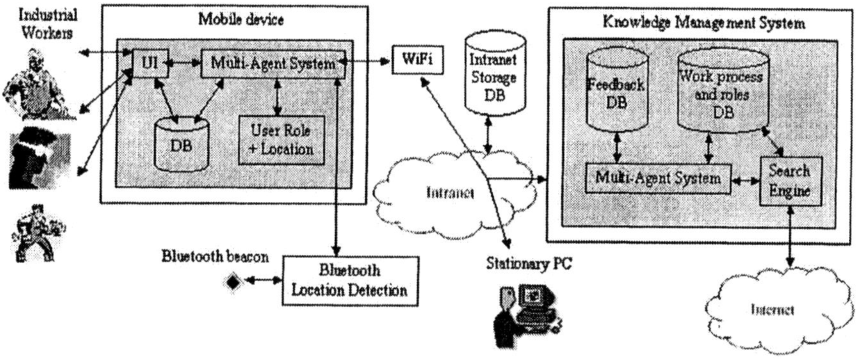

Figure 2 System Architecture

2.3 Mobile Devices

Bluetooth 6and WiFi (wireless network) 20enabled mobile devices running agents that are linked to the knowledge management solution, in particular to the role of the user and the process he/she is responsible for in the best practice environment, are to be made available to the industrial workers. An example of a mobile device that can be used is the HP iPAQ Pocket PC h4350.

By placing Bluetooth beacons around in the production hall the mobile device can identify its current location with a 5-10 meter radius. These beacons can be programmed with its exact coordinates and the associated processes in its vicinity. The user of the mobile device no longer has to click his/her way through the value chain in order to reach the relevant process. The mobile device agent automatically finds the processes that are relevant based on location and role of the user, and presents these processes on the mobile device as a list from which the user can pick the one that is most pertinent to the situation. Wireless network access gives the user the ability to access all content of the knowledge management tool and also allows the agents on the mobile device to be able to stay continuously updated and communicate with other agents. A similar approach is used in the AmbieSense project 12.

Stationary computers should be placed at strategic locations such as the lunchroom and control room. The computers that are placed in 'public' areas can be used by all workers to view the content of the knowledge management system (which is relevant for this plant) on a screen that is easier to read than the small screens the mobile devices are usually equipped with.

3 SCENARIO 1: NORMAL USE

As mentioned in the introduction, the mobile device gives access to all content stored in the knowledge management tool. The user can browse the value chain and processes via the wireless network connection. When the user is near a Bluetooth beacon, a list over relevant information is displayed; this can be the flowchart for the process or a one-point lesson for this particular part of the production line. Since the agent on the mobile device is aware of the role of the user, it will only show content that is relevant for this role. An operator will for example not be presented with information on routines or procedures that are not relevant to his/her current function.

Since the screens of mobile devices are not suitable for displaying large amounts of content and large value chains and flowcharts, the user has the possibility to click a button to have the presented information sent as an e-mail to him/herself or to bookmark it. This way the user can access the same information later from a stationary PC.

3.1 Process News

The agents on the search engine are always on the lookout for news that is relevant for a process and will automatically publish hits such as a change in legislation (SECI: combination). Process leaders also have the possibility to add news themselves, for example about an upcoming internal seminar on this process' topic. In either case, the news is pushed to the mobile device if relevant for the user.

3.2 One-point Lessons

The knowledge management system offers the ability to distribute so called one-point lessons, which are either one page with very clear instructions or a small video about a particular part in the process, always very focused on just this part. An example of this is how to dispose chemical waste. Since one-point lessons are very focused, they are generally only relevant when the user actually is in the direct vicinity of this process. Since Bluetooth enables proximity-detection, users can automatically be alerted that there is a one-point lesson available. If the user wants to know more about this particular part of the process, he/she can click a button and either read the document or watch the movie clip showing how this task is done (SECI: Internalization). This is especially useful for apprentices who can get a good overview over how processes are carried out.

Another possibility is to equip the mobile device with a camera in order to give the user the possibility to create a one-point lesson video (SECI: Externalization). Before the video is made available to all workers, it needs to be approved by the process responsible. An agent can automatically file the video in the approval queue for this process and will be made available from the point where the video was taken after approval.

3.3 Feedback

The feedback database is a very important part of the knowledge management system as it contributes to continuous improvement of the represented processes. Professor Mintzberg from the McGill University has stated 13: "If companies depended on dramatic, top-down change, few would survive. Instead, most organizations succeed because of the small change efforts that begin at the middle or bottom of the company...". This pinpoints the essence of the feedback database, which in fact is mostly fed by the people at the bottom of the company, the industrial workers.

When a user finds a mistake in any of the elements of the best practice part of the knowledge management system, he/she has the ability to send feedback about this issue (SECI: Combination). The user does not need to know who is responsible for the part of the system he/she is supplying feedback to, as this is defined in the process network, meaning the feedback will automatically reach the right person. Other than sending feedback about best practice issues, the user can also submit feedback on the status of machinery, or about very general issues such as the quality of the food in the cantina, etc.

4 SCENARIO 2: EMERGENCY SITUATION

An emergency situation occurs; for example triggered when someone pushes an alarm button or a monitoring system has detected that a certain level is over its threshold: Emergency procedures need to be followed. Regardless of which method is used to trigger alarm procedures, the location where the alarm is triggered is always known. Both alarm buttons and sensors have fixed locations and therefore it is possible to know where the nearest Bluetooth beacon is. The alarm will be pushed to the notification agent on the mobile devices that are in the vicinity of the location where the emergency procedure was triggered. An alarm will be broadcasted to the devices (regardless of the role of the user of the device), accompanied by the emergency procedure itself. In addition to the local broadcast, the alarm will also be sent to users that have a role associated with this emergency, regardless of their location. Examples of associated roles are the process responsible, fire fighters and a paramedic team. The superior can either rush to the emergency site, or give orders to the worker via a camera station.

5 SCENARIO 3: SCHEDULING

When the agent system is integrated with the groupware solution for e-mail and calendar applications, automated scheduling is a possibility. A notification on an internal seminar (SECI: socialization), as mentioned in the process news scenario, can be accompanied by a button or link to book this event directly in the user's agenda.

6 CONSIDERATIONS

Some considerations that need to be addressed when implementing such a system are:

- Possible interference of industrial equipment caused by the use of mobile devices using Bluetooth and WiFi must be investigated.
- Securing access rights, for example: the finance process should not be available other places than in the finance department
- Corporate espionage: How to prevent the competitor from laying hands on

sensitive information in a wireless network situation?
- Users may get an increased feeling of surveillance, since the location of the mobile device is known.

7 CONCLUSIONS

To make sure that every employee in a company contributes to and benefits from the knowledge sharing cycle, everyone must be able to access the corporate memory. By equipping employees with mobile devices linked to a knowledge management solution by means of agent technology the need for the user to search for relevant information can be eliminated, as the relevant information presents itself based on location and the role of the user. This enables the employees to learn much faster about the processes, and by means of the feedback database, give them influence to how to carry out this process in the most effective way. This interaction contributes to knowledge sharing on all levels of the organization, effectively using the ideas behind the SECI model.

8 REFERENCES

1. AmbieSense Consortium (2003). *Deliverable 8, The AmbieSense Multi-Agent* System Architecture Report. AmbieSense project, IST 2001- 34244
2. AmbieSense Consortium (2004). *Deliverable 9, Intelligent, Personalised Agents for Mobile Use* Report. AmbieSense project, IST 2001- 34244
3. Argyris, C. (1991). *Teaching Smart People How To Learn*. Harvard Business Rev.. May-Jun 91.
4. Bellifemine, F. Poggi, A. and Rimassa, G. (2000). *Developing multi-agent systems with a FIPA-compliant agent framework.* In: Software - Practice And Experience, 2001 no. 31, p. 103-128
5. Bremdal, B., Johansen, F., Spaggiari, C., Engels, R., Jones, R. (1999). *Creating a Learning Organisation Through Content Based Document Management.* In: Proceedings of Halden Program Group Meeting (HPG-352/12). OECD.
6. The official Bluetooth website, url: http://www.bluetooth.com/
7. CognIT web-site, url: http://www.cognit.no
8. Davenport, T. H., Prusak, L. (1998). *Working Knowledge: How Organizations Manage What They Know.* Harvard Business School Press.
9. European Aluminium Association, url: http://www.eaa.net/
10. Ferber, J. (1999). *Multi-Agent System: An Introduction to Distributed Artificial Intelligence.* Harlow: Addison Wesley Longman.
11. Gannon, A. (1998). *Knowledge Management at Hewlett-Packard.* Knowledge Management Vol 1, #3 Dec/Jan 1998.
12. Garvin, D. (1998) A. *Building a Learning Organisation.* Harvard Business Review. Jul-Aug 98.
13. Hesselbein, F., Johnston, R. (2002). *On Mission and Leadership, a Leader to Leader Guide.* Jossey-Bass, ISBN: 0787960683
14. Hydro Aluminium Metall, ULA Magasin Nr 1 – Juni 2001 (pp. 11-19)
15. Nonaka, I. and Takeuchi, H. (1995). The Knowledge-Creating Company: How Japanese Companies Create the Dynamics of Innovation. New York: Oxford University Press.
16. Stata, R. (89). *Organisational Learning – The Key to Management Innovation.* Sloan Management Review, Spring 1989.
17. Stokke, P. S. and Bremdal, B. A. (1994). *Information Refineries and the Manufacturing of Industrial and Corporate Knowledge.* Proceedings International Seminar on the Management of Industrial and Corporate Knowledge, ISMICK 94, Compiégne, France.
18. Speh, M. (1998). *Kmunity Building at Shell.* Knowledge Management Vol 1, #3 Dec/Jan 1998.
19. TMG International AB, url: http://www.tmg.no
20. WiFi, IEEE standard 802.11x, url: http://www.ieee.org

16 RELIABLE COMMUNICATIONS FOR MOBILE AGENTS – THE TELECARE SOLUTION

Octavio Castolo, Luis M. Camarinha-Matos
New University of Lisbon / Uninova
Monte Caparica, 2829-516 Caparica, PORTUGAL
{lgc, cam}@uninova.pt

The mobile agents area represents an emerging technology that extends the distributed computing mechanisms and shows a high potential of applicability. However, the problem of reliable communications among mobile agents persists. In this paper a practical solution for reliable communication among mobile agents is described and its characteristics and limitations are discussed. The suggested approach is implemented in the TeleCARE platform for elderly care.

1. INTRODUCTION

Progress in agent development platforms is making this technology a serious approach for developments in complex distributed systems. While an agent can be considered as an independent software entity that shows several degrees of autonomy, running on behalf of a network user (Hendler, 1999), the mobile agent is often seen as an executing program that can migrate autonomously, from machine to machine in a heterogeneous network (Gray et al., 2000); that is, an agent with mobility property. The mobile agent concept has been applied to a variety of application domains including electronic commerce, network management, information dissemination, spacecraft, and network games (Kotz et al., 2002). Applications to remote operation (Vieira et al., 2001), including remotely operated robots, manufacturing systems, remote surveillance systems and remote elderly care, have been suggested. Another interesting area is the use of mobile agents in virtual laboratories (Camarinha-Matos et al., 2002), to share expensive equipment, or to operate machines in hazardous environments, among others.

A fundamental issue in the development of mobile agent systems is the reliability of agent-based applications. The need for reliable inter-agent communications is one of the key requirements. Since mobile agents roam on different hosts (or machines), communication among them is not a trivial issue.

In this paper a practical solution for reliable inter-agent communication, which was developed in TeleCARE project, is presented. The paper is structured as follow: Section 2 presents the motivation for this work, where some approaches for (reliable) communication are described; in Section 3 a general overview of the TeleCARE system is given; Section 4 describes the architecture and strategy to provide reliable communication for mobile agents; and, finally, Section 5 presents the conclusions and suggests further work.

2. MOTIVATION AND RELATED WORK

A classical approach for communication in mobile agents is to offer mechanisms for local messaging, where agents talk with others only if they are living at the same host. This approach is enough for many typical applications of the mobile agent paradigm, since the agents roam among several remote hosts in order to make use of local resources in each one (Fuggeta et al., 1998). This approach can include event notification for group communication (Lange and Oshima, 1998), tuple spaces (Picco et al., 1999), among other features. There are however scenarios that require communications among remote agents. Some of these scenarios are related to mobile agent management and monitoring, in the "master-slave" case, or to accessing resources offered by other agents, in the "client-server" case. Other examples can arise within the context of a distributed application, a mixture of mobile agents and message exchange can be used to achieve different functionalities (Murphy and Picco, 2002).

Two typical approaches to message delivery are broadcasting and forwarding, both using server/hosts capabilities in order to deliver messages. A simple broadcast scheme (see Figure 1(a), based on (Murphy and Picco, 2002)) assumes a spanning tree of the network hosts that any host can use to send a message. The source host (sender) broadcasts a copy of the message to each of its neighbors, which on their turn broadcast the message to their neighbors, and so on until the leaf hosts are eventually reached. However, this does not guarantee the delivery of the message; in the process of broadcasting it might occur that when a message is being broadcasted to a host, at the same time, the destination agent is migrating in the opposite direction and destination delivery will not occur. Some approaches use a simple forwarding scheme (see Figure 1(b)) that keeps a pointer to the mobile agent at a well-known location. Upon migration, the mobile agent notifies the last place of its new location in order to make possible a future communication, leaving references (forwarding pointers) to the location where the agent currently is; or, in other variations of the scheme, the mobile agent must inform the home place in order to enable further communication. However some messages sent during the "migration and update process" might get lost.

In general, practical approaches to communications in mobile agents are based on the aforesaid. Some agent systems, such as Aglets (Lange and Oshima, 1998) and Voyager (Glass, 1999), employ a forwarding schema by associating to each mobile component a proxy object that "points" to the agent's home. Others, e.g. Mole (Baumann et al., 1997), assume that an agent never moves while engaged in communication; if migration of any of the parties involved in a communication takes place, the communication is implicitly terminated. Mole also exploits a different forwarding scheme that does not keep a single agent's home; rather it maintains a trail of pointers (forwarding pointers) from source to destination for faster contact (Baumann and Rothermel, 1998). Finally, some systems, e.g. D'Agents (Gray et al.,

2002), provide mechanisms that are based on remote procedure calls, and transfer to the application developer the task of handling a missed delivery.

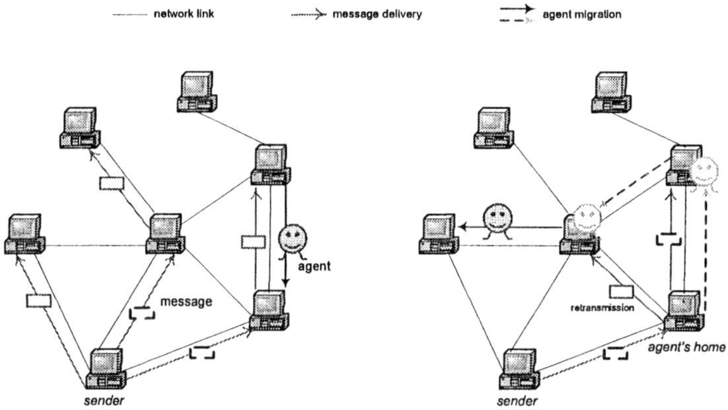

(a) Spanning tree broadcasting (b) Forwarding

Figure 1 – Missing message delivery in simple broadcast and forwarding schemes

Advanced approaches to reliable message delivery in mobile agents have been proposed in (Murphy and Picco, 2002), (Assis-Silva and Macêdo, 2001; Liu and Chen, 2003), (Ranganathan et al., 2000), and (Roth and Peters, 2001). In (Murphy and Picco, 2002) the approach supports uni- and multicast, but failures are not tolerated. In (Ranganathan et al., 2000) failures are considered, but the approach only supports unicast (peer-to-peer communication). In (Assis-Silva and Macêdo, 2001; Liu and Chen, 2003) an approach based on mobile groups and group communication is presented; this approach requires synchronism among mobile agents for well-functioning, considering only environments with synchronized events. In (Roth and Peters, 2001) the establishment of a public global tracking service for mobile agents, using dedicated tracking servers to storage the global name of each agent in the system, is proposed.

3. THE TELECARE SYSTEM

Due to the growing numbers of elderly population there is an urgent need to develop new approaches to care provision. Tele-assistance and provision of remote care to elderly living alone at home represents a very demanding case of a distributed system. Developments in this area have to cope with some important requirements, namely (Camarinha-Matos et al., 2004):

- Openness, in order to accommodate a growing number of new services and supporting devices.
- Support for heterogeneity, as different users have different needs and might possess a diversity of legacy systems (e.g. computers, home appliances, domotic infrastructures).

- Scalability, in order to allow the integration of a variable number of users in a tele-care community.
- Reliability of the system in terms of continuity of the service.

The mobile agents paradigm offers interesting characteristics that address some of these requirements. In fact, moving the code to the place where actions are required enables timely response, autonomy and continuity of service provision with reduced dependency on network availability and delays. Since new mobile agents can be built and deployed for remote execution whenever needed, higher levels of flexibility and scalability are achieved. By investing on the level of autonomy / decision-making capability of mobile agents, it is possible to conceive solutions that smoothly adapt to different user environments.

The convergence of a number of technologies such as multi-agent systems, federated information management, safe communications and security over Internet (Pozo et al., 2004), hypermedia interfaces, rich sensorial environments, increase of intelligence of home appliances, and collaborative virtual environments, represents an important enabling factor for the design and development of virtual elderly support community environments.

In this context, the IST TeleCARE project (Camarinha-Matos and Afsarmanesh, 2002) aimed at designing and developing a configurable mobile agents' framework focused on virtual communities for elderly support. Figure 2 shows a blocks diagram of the proposed architecture for a layer to be installed in each node of the TeleCARE organization (Camarinha-Matos et al., 2003).

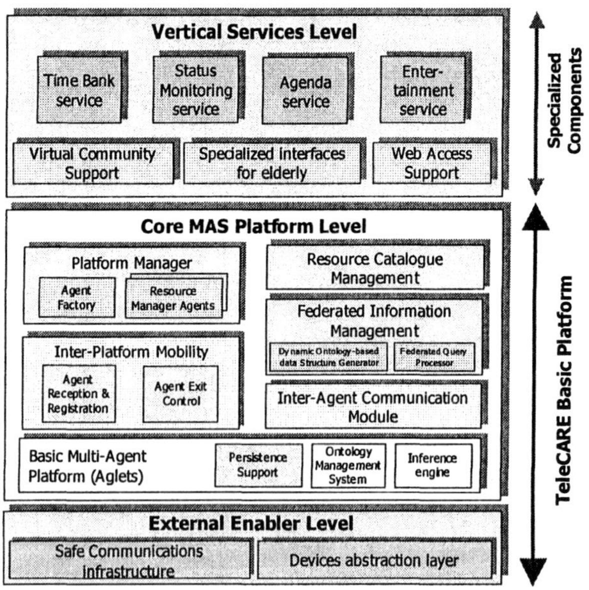

Figure 2 – The TeleCARE system architecture

The three-tier TeleCARE platform comprises:

❒ The **External Enabler Level**, which supports the communication and interfacing to the external devices, and other nodes.
❒ The **Core MAS Platform Level**, which is the core layer of the platform architecture. It supports the creation, launching, reception, and execution of stationary and mobile agents as well as their interactions / coordination.
❒ The **Services Level**, that consists in a variety of application services that can be added to the basic platform in order to assist the elderly, care providers, elderly relatives, etc.

3.1 Multi-Agent Infrastructure Design

The design and construction of the Multi-Agent Infrastructure of the TeleCARE project (the TeleCARE extended MAS Platform) comprises several modules of the **Core MAS Platform Level** of the TeleCARE system architecture, which is shown in Figure 3. From the multi-agents' perspective, the main modules of the TeleCARE Core MAS Platform are the following:

- Basic Multi-Agent Platform,
- Inter-Platform Mobility,
- Inter-Agent Communication Module, and
- Platform Manager.

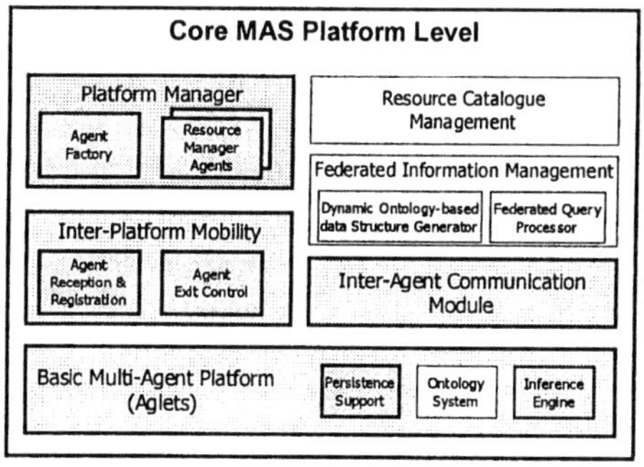

Figure 3 – Components of the TeleCARE Basic Platform

3.1.1 Basic Multi-Agent Platform

The *Basic Multi-Agent Platform* is the basic engine of the **Core M AS Platform Level**. The Aglets (Lange and Oshima, 1998) open-source system has been chosen as the multi-agent development tool mainly because it provides a strong support for agent mobility. The Aglets system is complemented with three additional modules that extend its functionality: (*i*) an *Ontology System* for knowledge modeling, using Protégé (Protégé-200); (*ii*) an *Inference Engine*, which uses Jinni (Jinni2004), a Prolog-like inference machine; and (*iii*) a *Persistence Support Service* (developed in TeleCARE) that allows system recovery in the case of a break-down.

3.1.2 Inter-Platform Mobility

The Aglets provides basic inter-platform mobility mechanisms. Nevertheless, in order to implement a security level for accessing TeleCARE resources, which is a critical issue in the TeleCARE domain, such basic mechanisms need to be extended. For this purpose two additional components are included:

i) The **Agent Reception & Registration** is responsible for accepting or refusing incoming agents. It implements the following main functions:
   - Accept / refuse incoming mobile agents,
   - Register agent, and
   - Notify sender (and origin) of accepted incoming agent.

ii) The **Agent Exit Control** is responsible for the "logistics" of sending out an agent. The main function implemented is:
   - The control of outgoing mobile agent.

Complementing these two main components, the *Inter-Platform Mobility* module also comprises a function to:
- Log information registration regarding agent's migration.

3.1.3 Inter-Agent Communication

The Aglets system provides a simple mechanism for inter-agent communication. However this mechanism is not sufficient for reliable communication between mobile agents. Therefore, this module implements additional communication services, namely:
- Extended message exchange mechanism, and
- The use of FIPA ACL.

3.1.4 Platform Manager

The *Platform Manager* is responsible for the configuration and specification of the operating conditions of the TeleCARE Platform in each site, in order to ensure the platform is working adequately. This module comprises functionalities for (*i*) system configuration, (*ii*) system supervision, (*iii*) definition of users and categories, and (*iv*) GUI for both programmers' interaction and users' interaction.

Additionally, this module has two main sub-modules:
- The *Agent Factory* is the module that can help service developers in the implementation of Vertical Services.
- The *Resource Manager Agents* module provides a common and abstract way of dealing with devices and home appliances in TeleCARE.

3.2 The TeleCARE Platform

The TeleCARE Platform is the environment where the TeleCARE agents (named as Agents in the following) will live. These Agents can be stationary or mobile.

The TeleCARE Platform supports two types of Agents: (*i*) the System Agents that are responsible for the good-functioning and management of the TeleCARE Platform; and (*ii*) the Application Agents, which are all the other Agents defined by the user in order to perform any task or service, or to build the TeleCARE applications. The former are stationary Agents, unique for each host considered as TeleCARE Platform; the latter can be stationary or mobile Agents, depending on the application. Several modules of the TeleCARE Basic Platform (at the Core MAS Platform Level) are defined as System Agents.

For purposes of simplification, in this paper only the following System Agents will be considered: (*i*) the *Agent Registry*, (*ii*) the *Agent Reception Control*, and (*iii*) the *Agent Exit Control*. These Agents form the *Inter-Platform Mobility* module, and their characteristics will be described below. Other System Agents of the platform are part of the modules *Federated Information Management*, *Resources Catalogue Management*, and *Platform Manager*, but they are not fundamental in the process of reliable communication support.

On the left side of Figure 4 the two types of Agents in a TeleCARE Platform are shown. On the right side a stylized representation of a TeleCARE Platform, where the System Agents are represented as blocks into the platform, is depicted (please, notice that the Agent Systems can be considered as a part of the agents' platform and not like agents themselves).

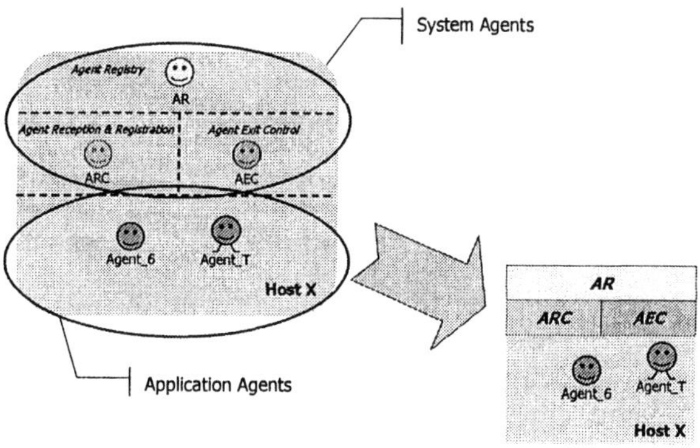

Figure 4 – Types of agents in the TeleCARE Platform

3.3 The TCAgent Class

The *TCAgent* class, illustrated in Figure 5, represents the base class for all System and Application Agents in TeleCARE.

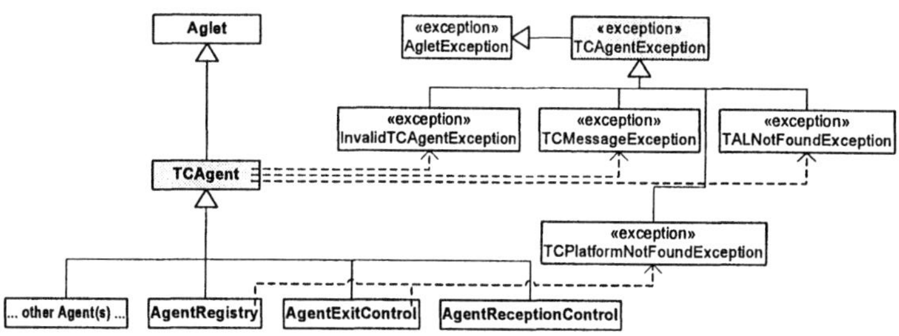

Figure 5 – TeleCARE classes' hierarchy

The *TCAgent* is the key class in the TeleCARE API. It is the abstract class that the developers must use as the base class to create customized Agents. Every class that inherits from it can be instantiated as an Agent. Some classes of exceptions are defined in order to deal with the possibility of unusual or unexpected system behavior, such as, e.g., the non-existence of the remote TeleCARE Platform to where the Agent is going to travel to.

The *TCAgent* defines methods for controlling its own life cycle, which are: (*i*) methods for dispatching, deactivating and disposing the Agent; (*ii*) methods for communication; and (*iii*) methods for implementing persistency support mechanisms.

Some of these features are already provided by the basic Aglets framework. The *TCAgent* uses them within the context of the TeleCARE extended functionality. The extended functionalities of the *TCAgent* are: (*i*) Agent registration and localization, (*ii*) communication through structured content messages, and (*iii*) fail-safe Agent execution (mainly using persistency mechanisms).

Furthermore, some of these functionalities, mainly the first one, are executed using the support offered by System Agents. As mentioned above, System Agents are in charge of managing all stationary and/or mobile Agents inside the platform. Thus, the *TCAgent* internally communicates with System Agents in order to achieve the mentioned features and functionalities. The communication processes are totally transparent to the developers.

3.4 The Passport

The Agent's passport is a mechanism that allows for some levels of security to protect TeleCARE communities, i.e., it is a "gate" for accessing and using the TeleCARE resources. The passport is also used for migration control (in a similar way as described in (Guan et al., 2003), but without using visas) and locating Agents, and it is encapsulated in messages sent by Agents as well. The principal characteristics of the passport are: (*i*) the passport is unique for every Agent, (*ii*) the passport is part of every Agent, and (*iii*) the passport can be partially assigned by the developer, but cannot be modified by him/her.

After the creation of an Agent, its passport constitutes a proof of its identity. It is the official "travel document" recognized by any TeleCARE site of the network. Any mobile agent that intends to migrate to another platform must have a valid passport. The passport structure is shown in Figure 6, and it is composed of the following fields:

- ***TAL*** – The *TeleCARE Agent Locator*, which is an identifier used by the system for locating an Agent. With information provided by TAL, the system can find the proxy of any agent, no matter where it is (for instance, to send it a message), in almost all cases. It contains data of the Aglets' identification of the agent (a string of 16 hexadecimal characters), the host where the agent has born, and the host where the agent is currently living.
- ***TLAID*** – The *TeleCARE Logical Agent Identification*, which is used to validate an agent at any platform, and to locate an agent (using human understandable data) as well. The developers can identify any TeleCARE agent with the information provided by the TLAID, given any parameter of the two substructures the compose it:
  - ***TLAD*** – The *TeleCARE Agent Data* that contains specific human readable identification of the Agent, namely its name and type; and

- ***TLUD*** – The *TeleCARE User Data* that encloses human readable identification of the user who created the agent, namely the role and ID of the user, and the domain node of the TeleCARE Virtual Organization that the origin host (or platform) of the agent belongs to.

- ***agentVal*** – It is used for assigning the duration time of the Agent's passport.
- ***itineraryDone*** – It indicates the itinerary traveled by the Agent, and stores a list of the last visited hosts.

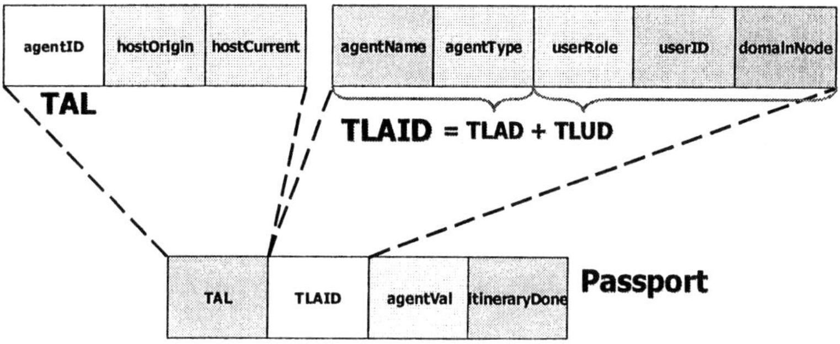

Figure 6 – The TeleCARE passport structure

3.5 The Inter-Platform Mobility Module

The module is composed of a set of stationary Agents at each TeleCARE Platform that provide its main functionalities. These Agents, which are the System Agents described in Section 3.2, are the following:

❐ The **Agent Registry**, that keeps a record / register of every Agent currently living and/or that was created in the platform. This register consists of a copy of the passport of each Agent. For instance, whenever an Agent needs to send a message to another Agent, it first gets the Receiver's TAL from the local (or remote) *Agent Registry*. A timer to refresh the register is included as well.

❐ The **Agent Reception Control**, that is responsible for the reception of the incoming mobile Agents. Depending on their passports, these Agents can be accepted or refused. Whether an arriving Agent is accepted in the local platform or not, the *Agent Exit Control* of the remote platform is notified.

❐ The **Agent Exit Control**, that controls the outgoing of mobile Agents. Every time an Agent is going to leave the platform, its passport and destination (as an available and valid TeleCARE Platform) are first checked.

4. RELIABLE COMMUNICATION IN TELECARE

4.1 Communication Mechanism

As abovementioned, TeleCARE MAS is built on the top of Aglets. Some extensions of Aglets messaging were developed in order to allow that:

- An Agent can communicate with other Agents if it knows some information about them. With the knowledge of, for instance, an Agent's location and some other parameters of its TLAID, an Agent can be easily reached (by another Agent) without further effort, in order to establish a contact (see Figure 7).

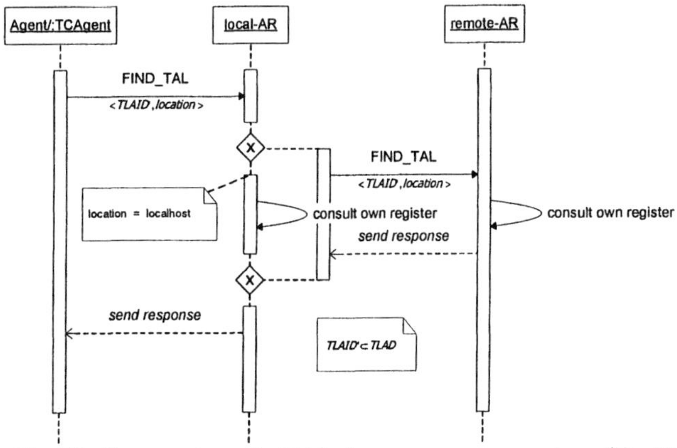

Figure 7 – Finding an Agent's TAL from some parameter of its TLAID

- An Agent that receives a message knows whom the Sender is. A typical situation is when an Agent sends a message to a resource manager Agent (asking it to perform some action on a resource); before granting access to the resource, the Sender's permission to use the resource must be first verified.
- Messages from non-Agents are adequately managed. There might be applications that require message exchanging with non-Agents; the Receiver must have a way of identifying such type of messages.
- Messages are certified / verified. Before being processed by the Receiver, the outgoing message has to be certified by encapsulating the passport of the Sender into the message to be sent. On the Receiver's side, the incoming message may also be verified, obtaining the passport of the Sender from the message. The verification process would succeed if the message carries a known/valid passport, otherwise the message should not be handled (see in Figure 8).

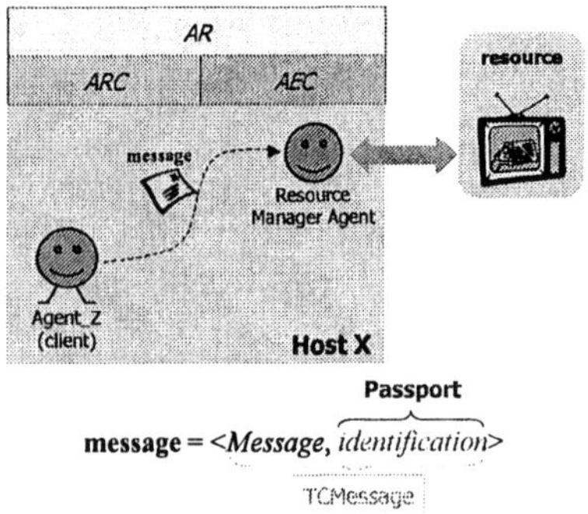

Figure 8 – Accessing to the TeleCARE resources

For dealing with the certification / verification processes a class named *TCMessage* was developed. This class acts like a wrapper of Aglets' *Message*, but enriched with the identification of the Sender (see Figure 8), and adding a method for handling message exceptions between Agents in different (remote) platforms.

4.2 Protocol of the Reliable Communication

In (Vieira, 2001) a mobile agent is defined as a 6-tuple $< id, code, state, ho, hc, t >$, where id is the unique identification of the agent, $code$ represents the code of the agent, $state$ is the vector of its static state, ho is the platform/host where the agent was created, hc is the host where the agent is currently living, and t is a particular instant in the life of the agent. For TeleCARE we redefined this definition into:

Definition 1. *The representation of an Agent at any instant is a 4-tuple* $< passport, code, state, t >$, *where passport is the unique identity of the Agent, code represents the code of the agent, state is the vector of its static state, and t is a particular instant in the life of the Agent.*

Definition 2. *The passport is a 4-tuple* $< tal, tlaid, agentVal, itineraryDone >$. *The TAL is a 3-tuple* $< id, ho, hc >$. *The TLAID is a 2-tuple* $< tlad, tlud >$. *The TLAD is a 2-tuple* $< agentName, agentType >$. *The TLUD is a 3-tuple* $< userRole, userID, domainNode >$.

All parameters have been defined above (Section 3.4). All passport parameters are constants in the life of an Agent, except the parameters hc and $itineraryDone$, which have their value changed in each migration event of the Agent.

An example [hypothetical] scenario of a mobile agent system is depicted in Figure 9. In this case all hosts are TeleCARE platforms. Every Agent that is created at any host can autonomously roam among all platforms, when necessary, in order to carry out its tasks. Communication between Agents is achieved as it is explained afterwards.

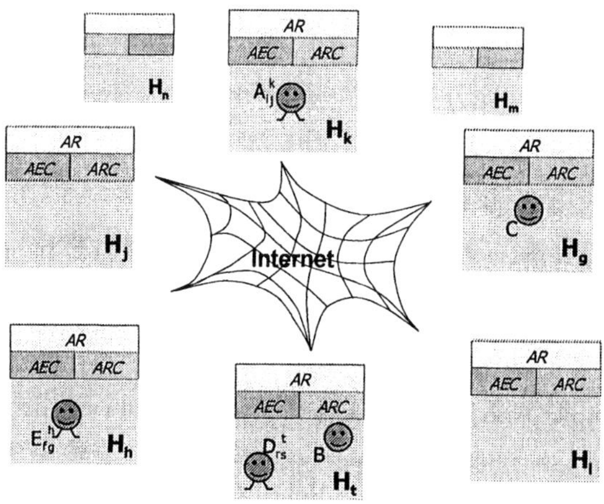

Figure 9 – Schematic scenario of a TeleCARE system

Let $A_{i_j}^k$ be the i-th Agent that born in the j-th host (H_j) and is currently living in the k-th host (H_k), where j is not necessary different from k. Let B be a stationary Agent that shall send a message to $A_{i_j}^k$. For the first contact between both Agents, B must know, at least, either H_j or H_k and, if needed, some parameters of $TLAD'$ of the agent to which it shall communicate, $A_{i_j}^k$ (where $TLAD' \subset TLAD$). Depending on the searching data, B will receive a null set, a set of several Agents $A_{x_j}^k$ (and in this case a refinement to obtain $A_{i_j}^k$ is necessary), or the specific Agent $A_{i_j}^k$. Because the Receiver always knows who the Sender is, a reliable communication can be established between both Agents.

Let's consider the following cases:

- If $A_{i_j}^k$ migrates to another host (H_l), changing now to $A_{i_j}^l$, the communication between both Agents will be assured, in the new first contact, if H_j is reachable.
- If for any reason H_j is not reachable (the communication channel fails or the host H_j crashed), the communication between both Agents, B and $A_{i_j}^l$, can be reestablished if $A_{i_j}^l$ sends a message to B.
- If a new Agent C wishes to start a communication with $A_{i_j}^l$, H_j is unreachable, and C does not know H_l, the communication cannot be done.
  - A special case occurs when $l = j$ and H_j is unreachable, which has the result that communication between C (or B) and $A_{i_j}^l$ cannot succeed.
- If $A_{i_j}^l$ migrates to a new location H_m, changing to $A_{i_j}^m$, and H_j is unreachable, H_j will not have any updated reference to where $A_{i_j}^m$ is currently living.
- If H_j is now reachable and C tries again to establish a communication with $A_{i_j}^m$, the communication cannot be done because H_j does not know the current value hc of $A_{i_j}^m$, erasing the $A_{i_j}^l$ record (the last record of the Agent before H_j was unreachable) of it own register; considering that $A_{i_j}^x$ is probably dead.
  - A special case occurs when $A_{i_j}^x$ dies. If H_j is reachable, it receives the notification of the death of $A_{i_j}^x$. If H_j is unreachable, when it seeks for $A_{i_j}^x$ at the last known location H_x and does not find it, H_j erases the record of $A_{i_j}^x$ on its register, and return a null set to all Agents that want to establish a communication with the missing $A_{i_j}^x$.
- If H_j is reachable and $A_{i_j}^m$ moves to location H_n, changing to $A_{i_j}^n$, H_j will be notified and it will store the record of $A_{i_j}^n$ on its own register. Now if C tries to start a communication with $A_{i_j}^n$, it will be succeeded.

All these cases can be generalized to a situation where two mobile Agents $A_{i_j}^k$ and $D_{r_s}^t$ want to establish a reliable communication, within the restrictions aforesaid, between them. Also, if the Agent $E_{f_g}^h$ (or B, in the case of considering a stationary agent) needs to send multicast messages to $A_{i_j}^k$ and $D_{r_s}^t$, it can get a reasonable success.

The proposed solution fails when the following three situations appear:

a) The agent's home, H_j, of the Receiver, $A_{x\,j}^{\,x}$, is unreachable, and
b) The Receiver migrates to another host, and
c) The Sender, B, initiates the communication process and does not know what the itinerary of the Receiver is.

5. CONCLUSIONS AND FURTHER WORK

Mobile agents technology can be used in several industrial environments, such as remote operation and monitoring of machines and equipment, tele-robotics, or even in services such as tele-health, tele-assistance and remote elderly care. Some of these domains require maximum reliability of agent-based applications, where communications among agents is one of the key requirements.

In this paper a practical solution for reliable communication in mobile agents systems, as developed in the TeleCARE platform, is described. This solution extends the approach of forwarding pointers in order to achieve reliable communication among mobile agents in almost all cases—some applications developed using the TeleCARE platform, and, in consequence, the proposed solution for reliable communication, can be found in (Camarinha-Matos, 2004).

It is also shown what the involved components on Agent communication of the TeleCARE platform are. This explanation is needed in order to discuss the characteristics and limitations of the solution. Since mobile agents systems have asynchronous characteristics, the solution focuses on guaranteeing, within the discussed limitations, reliable fault-tolerant communication for mobile agents without the restrictions of synchronous schemes. Nevertheless more effort is necessary in order to diminish exceptions as far as possible.

6. ACKNOWLEDGEMENTS

This work was funded in part by the IST program of the European Commission. The authors thank the contribution of the TeleCARE consortium members, and specially thank the participation of João Sarraipa, Ana Inês Oliveira, Filipa Ferrada and João Rosas, in the accomplishment of the reported results.

7. REFERENCES

1. Assis-Silva FM and Macêdo RJA. Reliable communication for mobile agents with mobile groups. IEEE/ACM ICSE 2001 Proc. Workshop on Soft. Engineering and Mobility; Toronto, Canada, 2001.
2. Baumann J, Hohl F, Radouniklis N, Rothermel K, Straßer M. Communication concepts for mobile agent systems. In Rothermel K and Popescu-Zeletin R (eds.) Mobile Agents: 1st Int. Workshop MA'97; LCNS 1219, Springer: Berlin, pp. 123–135, 1997.
3. Baumann J and Rothermel K. The shadow approach: An orphan detection protocol for mobile agents. In Rothermel K and Hohl F (eds.) Mobile Agents: 2nd Int. Workshop MA'98; LCNS 1477, Springer: Berlin, pp. 2–13, 1998.

4. Camarinha-Matos, Luis (ed.) Tele-Care and Collaborative Virtual Communities in Elderly Care. Proc. 1st Int. Workshop on Tele-Care and Collaborative Virtual Communities in Elderly Care (TELECARE 2004), in conjunction with ICEIS 2004, INSTICC Press, ISBN: 972-8865-10-4; Porto, Portugal, 13 April 2004.
5. Camarinha-Matos LM and Afsarmanesh H. Design of a virtual community infrastructure for elderly care. PRO-VE 2002, 3rd IFIP Working Conf. on Infrastructures for Virtual Enterprises; Sesimbra, Portugal, May 1–3, pp. 439–450, 2002.
6. Camarinha-Matos LM, Viera W, Castolo O. A mobile agents approach to virtual laboratories and remote supervision. Journal of Intelligent and Robotic Systems, KAP, pp. 1–22, 2002.
7. Camarinha-Matos LM, Castolo O, Rosas J. A multi-agent based platform for virtual communities in elderly care. ETFA-2003, 9th IEEE Int. Conf. on Emerging Technologies and Factory Automation; Lisbon, Portugal, 16–19 September 2003.
8. Camarinha-Matos LM, Rosas J, Oliveira AI. A mobile agents platform for telecare and teleassistance. In (Camarinha-Matos, 2004), pp. 37–48, 13 April 2004.
9. Fuggeta A, Picco GP, Vigna G. Understanding code mobility. IEEE Trans. on Soft. Engineering, 24(5):342–361, 1998.
10. Glass G. ObjectSpace Voyager Core Package Technical Overview. ObjectSpace, Inc., White Paper, 1999.
11. Gray R, Kotz D, Cybenko G, Rus D. Handbook of Agent Technology; chapter Mobile Agents: Motivations and State-of-the-Art Systems. AAAI/MIT Press, 2000.
12. Gray RS, Cybenko G, Kotz D, Peterson RA, Rus D. (2002). D'Agents: applications and performance of a mobile-agent system. Software: Practice and Experience, 32(6):543–573, 2002.
13. Guan S-U, Wang T, Ong S-H. Migration control for mobile agents based on passport and visa. Future Generation Computer Systems, no. 19, Elsevier Science B. V., pp. 173–186, 2003.
14. Hendler, James. Is there an intelligent agent in your future? Nature, Web Matters, 11 March 1999. From http://www.nature.com/nature/webmatters/agents/agents.html
15. Jinni2004: Java and Prolog Software for Internet Programming. From http://www.binnetcorp.com/
16. Kotz D, Gray R, Rus D. Future Directions for Mobile Agent Research. Technical Report TR2002-415, Department of Computer Science, Dartmouth College, 2002.
17. Lange D and Oshima M. Programming and Deploying Mobile Agents with Aglets. Addison-Wesley Longman, Inc., Reading, MA, 1998.
18. Liu C-F and Chen C-N. A sliding-agent-group communication model for constructing a robust roaming environment over Internet. Mobile Networks and Applications, no. 8, KAP, pp. 61–74, 2003.
19. Murphy A and Picco GP. Reliable communication for highly mobile agents. Autonomous Agents and Multi-Agent Systems, KAP, pp. 81–100, 2002.
20. Picco GP, Murphy AL, Roman G-C. LIME: Linda meets mobility. In Garlan D (ed.) Proc. 21st Int. Conf. on Soft. Engineering; pp. 368–377, 1999.
21. Pozo S, Gasca R, Gómez MT. Securing mobile agent based tele-assistance systems. In (Camarinha-Matos, 2004), pp. 63–72, 13 April 2004.
22. Protégé-2000: The Protégé Ontology Editor and Knowledge Acquisition System. From http://protege.stanford.edu/
23. Ranganathan M, Bednarek M, Montgomery D. A reliable message delivery protocol for mobile agents. Proc. Joint Symposium ASA/MA2000; LCNS 1882, pp. 206–220, 2000.
24. Roth V and Peters J. A scalable and secure global tracking service for mobile agents. 5th Int. Conf. MA 2001; LCNS 2240, Springer: Heidelberg, pp. 169–181, 2001
25. Vieira, Walter. Adaptive Mobile Agents for Remote Operation. PhD thesis, New University of Lisbon, 2001 (in Portuguese).
26. Vieira W, Camarinha-Matos LM, Castolo LO (2001). Fitting autonomy and mobile agents. ETFA-2001, 8th IEEE Int. Conf. on Emerging Technologies and Factory Automation; Antibes – Juan les Pins, France, vol. 2, pp. 471–480, 15–18 October 2001.

17

AN EMPIRICAL RESEARCH IN INTELLIGENT MANUFACTURING: A FRAME BASED REPRESENTATION OF AI USAGES IN MANUFACTURING ASPECTS

Mohammad R. Gholamian, Seyyed M. T. Fatemi Ghomi
Department of Industrial Engineering
Amirkabir University of Technology, Tehran, IRAN
{Gholamian, Fatemi}@aut.ac.ir

This paper tries to stimulate empirical research into the overall impacts of intelligent system implementations in manufacturing aspects. To reach this goal, a schema of intelligent applications is provided for each aspect as frame base structure, meaning the knowledge of intelligent applications in that specific aspect. Then, a semantic network is developed for intelligent manufacturing based on hierarchical structure of manufacturing systems to provide Meta knowledge of intelligent manufacturing applications. The paper is concluded with discussions of application performance.

1. INTRODUCTION

Over the past several years, there has been an increasing trend in use and development of artificial intelligence (AI) in various application areas such as machine learning, planning and robotics, modeling human performance, expert systems, automated reasoning and even in philosophy [17]. In practical fashion, advances in artificial intelligence coupled with reduction in cost of computer hardware and software, have made possible, the introduction of AI at different industrial sectors [15]. But, there are few sectors that have experienced as rapid a push towards this technology as manufacturing. In recent years, the intelligent systems have been widely used in manufacturing aspects. Many of these systems, such as advance manufacturing systems (AMS), computer integrated manufacturing (CIM), flexible manufacturing systems (FMS), manufacturing resource planning (MRPII), CAD/CAM, NC/CNC numerical control machines are being developed for production and operation management and present a cross fertilization of ideas from manufacturing and AI that is named Intelligent Manufacturing (Int.Man).

Intelligent manufacturing can be broken down in two major areas based on its level of application [1]:

1) Strategic intelligent manufacturing (Str.Int.Man) dealing with what, how and where subjects of production activities.
2) Tactical intelligent manufacturing (Tac.Int.Man) dealing with timing and quality of production activities.

But unfortunately, the impacts that intelligent systems are having in these environments have not been investigated for the most parts. Most studies on intelligent manufacturing focused on either technical aspects or validation issues. No one has taken a systematic view to this subject and address in implementing these systems.

In this study, several independently basic and important aspects in each area will be discussed systematically. So, the frame based representation has been developed for each aspect such that each frame explains applications of AI implementations in its aspect. In fact, the frames are explanations the knowledge of intelligent applications in their specific aspects. These capsules of knowledge are integrated as Meta knowledge which is the knowledge about the use and control of domain knowledge. The integration is performed using semantic network followed by hierarchical structure of intelligent manufacturing concepts. In fact, the Meta knowledge is explanation of AI implementations in intelligent manufacturing.

The frame of intelligent manufacturing slots with labels, describing usage of intelligent systems as attributes (or properties) and possible values for each attribute. Although a wide range of AI applications can be suggested but specially following ones are selected [3]:

1) rule based reasoning systems (RBR)
2) model based reasoning systems (MBR)
3) case based reasoning systems (CBR)
4) frame based reasoning systems (FBR)
5) probabilistic reasoning (PBR)
6) fuzzy logic
7) neural networks (NN)
8) Meta-heuristics

Then the frame of Int.Man can be developed as follows [28]:

Int.Man Frame	
Super Class	Intelligent Systems, Manufacturing
Sub Class	Str.Int.Man, Tac.Int.Man
RBR	Rule Base, Hybrid Systems
MBR	Model Base, Hybrid Systems
CBR	Case Base, Hybrid Systems
FBR	Frame Base, Hybrid Systems
PBR	Bayesian Reasoning, Dempster-Shafer Evidence Logic, Hybrid Systems
FUZZY	Fuzzy Rule Base, Fuzzy Case Base, Fuzzy Frame Base, Fuzzy Operation Research (FOR) , Fuzzy Clustering, Fuzzy Numbers, Hybrid Systems
NN	ART family networks, Hopfield networks, Boltzmann Machine, Kohonen networks, Feedforward networks, Time Delay Neural Networks (TDNN), Maximum Neural Networks (MNN), Fuzzy Neural Networks (Neuro-Fuzzy)
Heuristic	Genetic Algorithm (GA), Simulated Annealing (SA), Tabu Search (TS), Ant-Colony (ACO),

In The next sections, the frame base of each manufacturing aspect will be described. The paper is concluded to Meta knowledge of intelligent manufacturing and descriptions about the range of applications

2. STRATEGIC INTELLIGNT MANUFACTURING FRAME (Str.Int.Man)

In this section three aspects, which are directly related to strategic manufacturing operations, will be described as follows:

2.1 Aggregate Planning (AP) Frame

Aggregate planning (AP) is an OR model of production planning. The major aim of AP is to determine aggregate quantity of product, for each time period in a future interval of time (called planning horizon), such that minimum total cost is obtained.

Intelligent systems are generally used to generate decisional rules. Rules are utilized to establish production rate, workforce level required, overtime requirements, inventory level, capacity and costs as a rule base aside with mathematical model. In fact, the rule base is used to be auxiliary of mathematical model. HMMS (Holt, 1960) is a sample of such systems.

In addition, rules can be defined fuzzily using linguistic variables and values. Rinus developed fuzzy rules for production and workforce level in HMMMS such as follows [30]:

IF D_t is VH **AND** I_{t-1} is SL **AND** W_{t-1} is RH **THEN** P_t is SH
IF D_t is RH **AND** I_{t-1} is VL **AND** W_{t-1} is SL **THEN** ΔW_t is Positive

In Summarize of above explanations, the frame base of "aggregate planning" will be illustrated as follows:

Object Name: Aggregate Planning (AP)

Class: Str.Int.Man

Properties:

RBR	HMMS (Holt, 60)
MBR	∅
CBR	∅
FBR	∅
PBR	∅
FUZZY	Fuzzy HMMS (Rinus, 82)
NN	∅
Heuristics	∅

2.2 Facility Location (FL) Frame

Facility location is the subject of locating one or more new facilities with respect to existing facilities; such that minimum transportation cost is provided [10]. There are various applications of intelligent systems in facility location problems:

- Fuzzy logic is used in definition of discrete location problems as fuzzy integer programming (FIP) models.
- Meta-heuristics are widely used in quadratic assignment problems (QAP) and facility layout problems. Since the QAP is familiar with TSP, and is graded as NP-Complete problems, various GA, SA, TS and ACO methods are developed

in this context. In addition, an application of neural networks (i.e. MNN) is also developed in this context.

- Finally, rule based systems are used in layout and material handling problems. In former case, the rules are defined based on the frames of material handling devices such as follows [10]:

  IF (material is of unit load type) **AND** (truck.load/unload level is between 30 and 45 feet) **AND** (truck.load is less than 2500 lb) **THEN** (side-loading outrigger truck is desired)

In Summarize of above explanations, the frame base of "facility location" will be illustrated as follows:

Object Name: Facility Location (FL)

Class: Str.Int.Man

Properties:

RBR	FADES (Fisher & Nof, 84), EXIT (Malmborg, 89), KBML (Heragu & Kusiak, 90)
MBR	∅
CBR	∅
FBR	∅
PBR	∅
FUZZY	FIP (Darzentas, 87)
NN	MNN (Tsuchiya et. al, 96)
Heuristics	GA, SA, TS, ACO

2.3 Forecasting (FC) Frame

Forecasting is the prediction, projection or estimation of the occurrences of uncertain future events or levels of activity. In manufacturing, forecasting is used to predict changeable circumstances such as revenues, costs, profits, prices, technological changes and (in most cases) demand [27].

The model bases are the most eminent systems developed in forecasting aspect. The system includes certain numerous forecasting models and specific models in the scope of brands to provide the analysis capability.

But often no computer-based model can easily incorporate all that is needed to make a sound business decision. In such cases, rule bases can be used to capture the basic judgments that are necessary in forecasting systems. Express is a sample of such RBR systems.

Unfortunately, these systems are very data intensive and data processing is very difficult. Instead, fuzzy rule base can be used with linguistic interpretation. The fuzzy knowledge can be acquired either form experts linguistically or with a set of historical data using Sugeno rule based system. In addition a hybrid of fuzzy rule base and MBR can be used to support both specifications.

Finally some applications of neural networks are developed in forecasting problems as Neuro-identification of time series [13]. Temporal processing networks specially TDNN [9] and simple feedforward networks are sample of such applications.

In Summarize of above explanations, the frame base of "forecasting" will be illustrated as follows:

Object Name:	Forecasting (FC)	
Class:	Str.Int.Man	
Properties:	**RBR**	Express (Manzano, 90)
	MBR	Model Base, Model Base & Fuzzy Rule Base
	CBR	∅
	FBR	∅
	PBR	∅
	FUZZY	Fuzzy Rule Base, Fuzzy Rule Base & Model Base
	NN	TDNN(Lang & Hinton, 88), Feedforward (Billings, 92)
	Heuristics	∅

Now, using above frame bases and based on inheritance rule, the parent frame of strategic intelligent manufacturing can be illustrated as follows:

Str.Int.Man Frame	
RBR	Rule Base
MBR	Model Base, Model Base & Fuzzy Rule Base
CBR	∅
FBR	∅
PBR	∅
FUZZY	Fuzzy Rule Base, FOR, Fuzzy Rule Base & Model Base
NN	MNN, TDNN, Feedforward
Heuristics	GA, SA, TS, ACO

3. TACTICAL INTELLIGNT MANUFACTURING FRAME (Tac.Int.Man)

In this section, six important aspects are selected to be discussed as efficient aspects of tactical intelligent manufacturing:

3.1 Scheduling (SCH) Frame

The main aim of scheduling is allocation of resource overtime to perform a collection of tasks. Scheduling itself includes a set of various subjects such as single machine problem, parallel machine problems, flow shop scheduling, job shop scheduling, project scheduling, FMS scheduling. Most of the papers published in AI usages in manufacturing aspects, are commonly related to this aspect.

In job shop scheduling, there are a wide range of heuristic rules developed in various areas such as Lisp, Prolog, Itp, OPS5, and Smalltalk [12]. In addition various RBR, FBR and fuzzy RBR systems are developed for various job shop scheduling problems [19-24]. Following is a sample of FBR rules developed on object-oriented fashion:

IF job[i].time < job[j].time **AND** job[i].duedate > job[j].duedate **THEN** job[i] precedes job[j]

Fuzzy rule bases are also used in other subjects. In FMS, fuzzy rules are applied in release and machine scheduling; similar to following rule [30]:

IF waiting time is long **AND** slack time is short **THEN** date criterion is urgent (0.5).

In addition fuzzy numbers are used in project scheduling instead of PERT networks.

But since, most of scheduling problems are NP-Complete, a wide range of Meta heuristic development methods and neural network optimization methods [5], are used in this context.

In Summarize of above explanations, the frame base of "scheduling" will be illustrated as follows:

Object Name:	Scheduling (SCH)	
Class:	Tac.Int.Man	
Properties:	**RBR**	OPT (Jacobs, 83), ISIS (Fox, 83), PATRIARCH (Morton et al. 84), MARS (Marsh, 85), PEPS (Robbins, 85), RPMS (Lipiatt & Waterman, 85), OPIS (Dw & Smith, 86), PLANEX (Zozaya & Gorostiza, 89), SURE (Thalman & Sparn, 90), HESS (Deal et al, 92), ESRA (Solotorevsky, 94)
	MBR	∅
	CBR	∅
	FBR	Enterprise (Marlone, 83), Yams (Parunall, 86), CORTES (Fox & Sycora, 89), KBMS (Cholawsky, 90), PARR (McLean, 91)
	PBR	∅
	FUZZY	OPAL (Bensana, 88), Fuzzy PERT (Prade, 79), Fuzzy FMS (Hintz & Zimmermann, 89), FLES (Turksen, 93)
	NN	Hopfield network , Kohonen network
	Heuristics	GA, SA, TS, ACO (McMullen, 2001)

3.2. Inventory Control (INV) Frame

The inventory models are developed to response two important questions:

1) How much (quantity) to order [Q].
2) When to order [LT].

There are various inventory models developed based on marketing problems. In addition some inventory systems are developed which are the complex of various marketing subsystems. MRP (material requirement planning), MRP II (material resource planning) and ERP (enterprise resource planning) are samples of these integrated systems [27].

Since the inventory models are widely developed in various marketing subjects, various intelligent systems are developed for approximately all reasoning systems. Specially FBR systems are successfully used in MRP II and ERP systems [12], [22] and CBR systems are successfully used in inventory planning [23-24]. Following is sample of inventory fuzzy rules:

IF the current inventory level is much higher than the preferred level **AND** the direction is decreasing at medium rate **THEN** production rate to be moderately slowed down.

In addition, fuzzy mathematical models are developed for various inventory models under uncertainty conditions. As an example it can be mentioned to fuzzy aggregate inventory planning with fuzzy numbers which is solved based on Bellman-Zadeh's rule of conjunction [30].

In Summarize of above explanations, the frame base of "inventory control" will be illustrated as follows:

Object Name:	Inventory Control (INV)	
Class:	Tac.Int.Man	
Properties:	**RBR**	INTELLECT (AICORP), NCR, ADS
	MBR	∅
	CBR	Case Base
	FBR	MAPLEX (Walls & Gilbert, 89) PAREX-CO (Martins & Wedel, 90)
	PBR	Bayesian Reasoning
	FUZZY	FDP (Sommer, 81), FMIP (Kaprzyk & Staniewski, 82) FLP/FIP (Zimmermann & Pollatschek, 84), FNLP
	NN	∅
	Heuristics	∅

3. 3 Quality Control (QC) Frame

Quality control is application of some statistical techniques to control the production process and improve quality of products with minimum cost. Generally, control charts and acceptance sampling plans are the well-known techniques used in quality control. Recently new concepts such as QFD, TQM, quality assurance (QA), six sigma and ISO standards are successfully used as quality concepts.

Since the structure of quality control is essentially statistical, the Bayesian reasoning can be used in decisional levels [26]. In addition, acceptance sampling plan can be preformed using CBR systems; the historical rejections are saved as cases and then the retrieval process determines acceptance or rejection of inspection. Quality control can be defined fuzzily using fuzzy numbers and fuzzy rules in the structure of quality techniques; the control limits in fuzzy control charts and acceptance/rejection rules in fuzzy sampling plan may be defined fuzzily instead of using crisp values.

Finally as marginal application, neuro-fuzzy and feedforward neural networks are used to train monitoring sensors. The method is successfully used in CNC machines to reach real time machining control [2] and machine condition monitoring [11].

In Summarize of above explanations, the frame base of "quality control" will be illustrated as follows:

Object Name:	Quality Control (QC)	
Class:	Tac.Int.Man	
Properties:	**RBR**	∅
	MBR	∅
	CBR	Case Base
	FBR	∅
	PBR	Bayesian Reasoning
	FUZZY	Fuzzy Rule Base, Fuzzy Numbers
	NN	Feedforward (Jan, 92), Neuro-fuzzy (Javadpour & Knapp, 03)
	Heuristics	∅

3.4 Maintenance (MTC) Frame

Maintenance is a branch of quality control with the aim of maintaining the currently available machinery and equipment to avoid failures and to improve applicability and reliability of facilities. The maintenance models can be classified as follows:
1) Decision models in facility replacement.
2) Inspection models
3) Decision models in partial and fundamental maintenance.

Generally the models are statistical structure and very complex. So, probabilistic reasoning methods can be used in decisional levels successfully. On the other hand, the various models can be saved in model base and then the reasoning is performed using online information of machine situation [19]. In contrast, RBR systems are generally used in operational levels [22]. Following, is sample of such rules:

IF main spindle does not turn after switching on, **THEN** failure will be located on

Similarly case bases are used in operational levels as diagnosis systems. Case bases are very powerful in diagnosis processes [25] specially when the system is developed using fuzzy neural networks [16].

Finally maintenance models can be defined fuzzily in possibility conditions instead of probability conditions. So, fuzzy dynamic programming models (FDP) or maintenance models with fuzzy numbers can be developed and then is solved using fuzzy arithmetic.

In Summarize of above explanations, the frame base of "maintenance" will be illustrated as follows:

Object Name: Maintenance (MTC)

Class: Tac.Int.Man

Properties:

RBR	EXMAS (Milacic, 88), XPS
MBR	Model Base
CBR	Case Base, Case Base & Neuro-Fuzzy
FBR	∅
PBR	Bayesian Reasoning, Evidence Logic
FUZZY	FDP, Fuzzy Numbers
NN	Neuro-Fuzzy & Case Base
Heuristics	∅

3.5 Group Technology (GT) Frame

Group technology (GT) is a management philosophy that attempts to group products with similar design (shape oriented) or manufacturing characteristics (process oriented) or both. One of the most important applications of GT is Cellular Manufacturing (CM). The main objective of CM is to identify machine cells and part families concurrently and to allocate part families to machine cells in a way that minimizes the intercellular movement of parts [10]. In fact, the main problem of GT (and CM) is clustering and classification. Neural networks are very powerful in classification; specially ART family networks (ART1, ART2, ART MAP, fuzzy ART and so on) are used in clustering excellently [4]. Similarly, other self organizing neural networks such as Kohonen network can be used in classification process. As an alternative, fuzzy clustering [23] can also be used to cluster the parts.

GT can be performed using rule bases. Rules are defined such that map physical features to external shape features [22]. Meanwhile, Products can be represented as frames; then FBR may be widely used in GT process [12], such as following rules:

IF 40 < item.length ≤ 80 **AND** 15 < item.diameter ≤ 30 **AND** item.material is Alloy steels **THEN** item.calss = Hole family group.

Similarly, products may be made up as a case and then CBR can be used in classification process alone or with rule base [19].

In Summarize of above explanations, the frame base of "group technology" will be illustrated as follows:

Object Name: Group Technology (GT)

Class: Tac.Int.Man

Properties:

RBR	SAPT (Milacic, 87), Rule Base & Case Base
MBR	∅
CBR	Case Base, Case Base & Rule Base
FBR	Frame Base
PBR	∅
FUZZY	Fuzzy Clustering (Bezdek, 81)
NN	ART, Kohonen, Feedforward
Heuristics	∅

3.6 Process & Product Design (PPD) Frame

Process and product design are of main duties of manufacturing, related to designing complex products and also designing production process. Globally some advanced manufacturing systems such as CAD/CAM, CAPP, CACE and CAPM are categorized in this aspect.

There are a wide range of rule bases and frame bases which are developed to decide exactly how to design a part and how to manufacture it [6-7], [12], [19-24]. Followings are sample of rules (from ARL) and sample of frames (from CPMAPII) used in process and product design:

IF part = hood outer **AND** material = Aluminum Alloy **AND** application = forming **AND** forming = stretch drawing **THEN** Guidelines = test val: left 5:right 8:line 2″.

Object: Gas Fuse

Properties:

Classification	Electronic.Components
Identification	Through.hole
Shape	Axial
Type	Two.leads

The rules and frames can be defined fuzzily [8]; such as following fuzzy object-oriented rule [14]:

IF bareboard.width is high **AND** bareboard.height is less **AND** bareboard.length is low **THEN** exterior cover # is "35-256"

Similarly, the process and product cases can be defined and then designing is performed using CBR [19]. Cases may be defined fuzzily. For example Main et. al developed fuzzy case based system along with feedforward neural network for fashion shoe design [18]. Finally, as marginal application, neural networks can be used in training of system designers such as NC, CNC and DNC machines [29].

In Summarize of above explanations, the frame base of "process and product design" will be illustrated as follows:

Object Name:	Process & Product Design (PPD)	
Class:	Tac.Int.Man	
Properties:	**RBR**	GARI (Descote & Lathom, 83), Proplan (Philiphs, 84), CABPRO (VanDyna, 85), XCUT (Brooks, 87), ARL (Demeri, 90), KDPAG (Chen & Qin, 98)
	MBR	∅
	CBR	PDA (Bhrvani, 90) , Fuzzy Case Base
	FBR	TIES (Ford Company), Himapp (Berenji & khoshnevis, 86), DLMS (Johnson, 89) and ISPA (Bozenhardt, 90) CPMAPII (Dagnin & Council, 90), Fuzzy Frame Base
	PBR	∅
	FUZZY	Fuzzy Rule Base, Fuzzy Frame Base, Fuzzy Case Base
	NN	Feedforward
	Heuristics	∅

Now, using above frame bases and based on inheritance rule the parent frame of strategic intelligent manufacturing can be illustrated as follows:

Tac.Int.Man Frame	
RBR	Rule Base, Rule Base & Case Base
MBR	Model Base
CBR	Case Base, Case Base & Rule Base, Case Base & Neuro-fuzzy
FBR	Frame Base
PBR	Bayesian Reasoning, Evidence Logic
FUZZY	Fuzzy Rule Base, Fuzzy Clustering, Fuzzy OR, Fuzzy Number, Fuzzy Case Base, Fuzzy Frame Base
NN	ART, Hopfield, Kohonen, Neuro-fuzzy, Feedforward, Neuro-fuzzy & Case Base
Heuristics	GA, SA, TS, ACO

4. CONCLUSIONS

In previous sections, various frame bases are introduced based on various aspects of manufacturing. The frame bases are representative of AI applications in these aspects. Now, let suppose the frame as capsules of knowledge; then based on hieratical structure of intelligent manufacturing, following semantic networks can be developed. The arcs illustrate the attributes which are inherited. The network includes knowledge about the operation of knowledge-based systems in intelligent manufacturing; which is the same Meta knowledge. This Meta Knowledge enhances the efficiency of AI applications by directing to the most promising aspects.

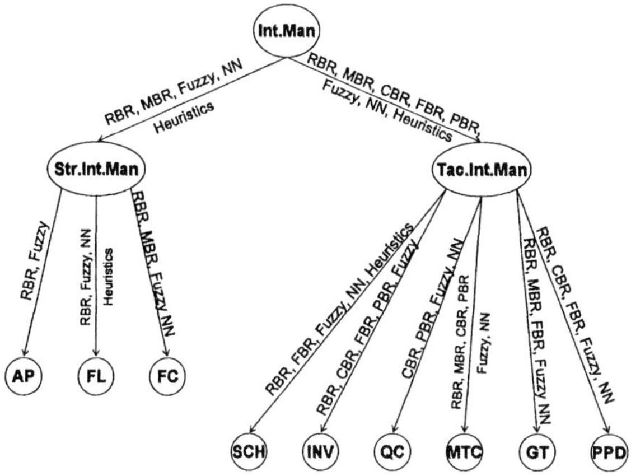

Figure 1 – Semantic network of intelligent manufacturing

Such as shown, rule based systems are approximately used in all aspects; perhaps because the rule bases are the earliest intelligent systems and naturally have found various applications. But the rule bases are symbolic knowledge bases; whilst manufacturing aspects are generally numerical. Such as shown, RBR is mostly used in SCH and PPD and also in some aspects such as MTC and AP is used marginally and auxiliary.

Unlike, fuzzy rule bases have more comprehensive structure. While the system uses linguistic variables and values, the inference is performed on numerical processing and crisp solutions can be generated. Although application of fuzzy systems is not much more than RBR, but because of data driven property and uncertainty suggestion, the growth acceleration is incrementally increasing.

Fuzzy logic provides more robust capability for vague and uncertain information. Hence, various fuzzy OR models are developed in manufacturing aspects and then solved using fuzzy arithmetic and fuzzy relations.

Other reasoning systems seem to be more causal:

- FBR is used in the aspects with hierarchical structure, such as group technology, inventory systems (i.e. MRP II and ERP) and scheduling.
- CBR on the other hand is used in the most tactical aspect; since these aspects accompany with decisional operations such as purchasing or non-purchasing, reject or accept and replacement or continue. In addition CBR have found success in the aspects related to design of products. (I.e. GT and PPD).
- The application of PBR and MBR are restricted to special conditions of related aspects.
- Similarly, Meta heuristics are generally developed in NP-Complete and NP-Hard problems. The numbers of paper which are published in this context is very much and the applications are expanding to the other aspects.
- Finally, the applications of NN are very different. Some optimization networks such as MNN, Hopfield and Kohonen networks are used in NP-Complete problems. Others have used in specific applications such as clustering (ART networks), forecasting (TDNN) and training the other reasoning systems (Neuro-fuzzy).

In general speaking, while the acceleration of rule base applications is decreasing, fuzzy systems and hybrid applications have found attainable growth in various manufacturing aspects.

5. REFERENCES

1. Byrd TA, Hauser RD. Expert systems in production and operations management: research directions in assessing overall impact. International Journal of Production Research 1991; 29 (12): 2471-2482.
2. Dagli CH. Artificial neural networks for intelligent manufacturing. London: Chapman & Hall, 1994.
3. Durkin J, Durkin J. Expert systems: design and development. New Jersey: Prentice Hall, 1998.
4. Frank T, Kraiss KF, Kuhlen T. Comparative analysis of Fuzzy ART and ART-2A network clustering performance. IEEE Transactions on Neural Networks May 1998; 9 (3): 544-559.
5. Fu L. Neural networks in computer intelligence. New York: McGraw Hill, 1994.
6. Gero JS. Artificial intelligence in engineering: robotics and processes. New York: Elsevier, 1988.
7. Goslar MD. Expert systems in production and operation management. Book of proceedings of the 4th international conference, South Carolina: University of South Carolina, 1990.
8. Grabot B, Geneste L. Management of imprecision and uncertainty for production activity control. Journal of Intelligent Manufacturing 1998; 9: 431-446.
9. Haykin S. Neural network: a comprehensive foundation, Second Edition. New Jersey: Prentice Hall, 1999.
10. Heragu S. Facility design. Boston: PWS Publishing, 1997.
11. Javadpour R, Knapp GM. A fuzzy neural network approach to machine condition monitoring. Computer & Industrial Engineering 2003; 45: 323-330.
12. Kusiak A. Expert systems: strategies and solutions in manufacturing design. Dearborn: SME Publication, 1988.
13. Leondes CT. Fuzzy theory systems, techniques and applications, Vol. 3, 4. New York: Academic Press, 1999.
14. Lee SC, Lee JY, Kuo YH. A framework for object-oriented fuzzy expert systems. IEEE Proceedings of NAFIPS/IFIS/NASA Dec 1994: 214-222.
15. Liebowitz J. The handbook of applied expert systems. New York: CRC Press, 1998.
16. Liu ZQ, Yan F. Fuzzy neural network in case based diagnostic system. IEEE Transactions on Fuzzy Systems May 1997, 5 (2): 209-222.
17. Luger GF. Artificial intelligence: structures and strategies for complex problem solving, Forth Edition. New York: Addison Wesley, 2002.
18. Main S, Dillon TS, Khosla R. Use of fuzzy feature vectors and neural networks for case retrieval in case based systems. Biennial conference of NAFIPS June 1996: 438-443.
19. Maus R, Keyes J. Handbook of expert systems in manufacturing. New York: McGraw Hill, 1991.
20. Meyer W. Expert systems in factory management: knowledge based CIM. New York: Ellis Harwood, 1990.
21. Metaxiotis KS, Askounis D, Psarras J. Expert systems in production planning and scheduling: A state-of-the-art survey, Journal of Intelligent Manufacturing 2002; 13: 253-260.
22. Milacic VR. Intelligent manufacturing systems, Vol. I, II, III. New York: Elsevier, 1991.
23. Oliff MD. Expert systems and intelligent manufacturing. New York: North-Holland, 1988.
24. Rzevski G. Artificial intelligence in manufacturing. New York: Springer, 1989.
25. Schenker DF, Khoshgoftaar TM. The application of fuzzy enhanced case-based reasoning for identifying fault-prone modules. Third IEEE International High Assurance Systems Engineering Symposium November 1998; 90-97.
26. Sriram RD. Intelligent systems for engineering: a knowledge based approach. London: Springer, 1997.
27. Tersine RJ. Principles of inventory and materials management, Forth Edition. New Jersey: Prentice Hall, 1994.
28. Turban E, Aronson JE, Liang TP. Decision support systems and intelligent systems, Seventh Edition. New Jersey: Prentice Hall, 2004.
29. Want J, Takefuji Y. Neural networks in design and manufacturing. New Jersey: World Scientific, 1993.
30. Zimmmermann HJ. Fuzzy set theory and its applications, Third Edition. Boston: Kluwer, 1996.

18

PREFERENCE BASED SCHEDULING FOR AN HMS ENVIRONMENT

S. Misbah Deen[1] and Rashid Jayousi[2]

[1]*DAKE Group, Computer Science Department, University of Keele, Keele, Staffs. ST5 5BG, ENGLAND.*
deen@cs.keele.ac.uk

[2] *Computer Science Department,AlQuds University, Jerusalem, ISRAEL*
rjayousi@science.alquds.edu

This paper presents a model for preference-based multi-agent scheduling suitable for Holonic Manufacturing Systems in which holons can cooperate in producing a satisfactory global schedule. The goodness of the scheduling model has been verified by a theoretical behaviour model and confirmed by simulation, using a number of Assembler holons as the scheduler agents of manufacturing tasks. The result of this study, which we found to be satisfactory, has been presented in the paper.

1. INTRODUCTION

In Holonic Manufacturing Systems (HMS) holons (interpreted here as agents) cooperate together to manufacture products. We may assume a coordinator holon has a global task (joint task) to create a product with the help of a set of Assembler holons as cohorts, each cohort scheduling its local component of the global task in an environment where some of the scheduling slots (e.g. time-slots) of each cohort will have been already occupied by previously allocated tasks from other coordinators. From the perspective of the coordinators, the ideal situation is when all the required Assemblers are idle, so that the waiting time in between the Assemblers can be reduced to zero, while from the perspective of an Assembler there should not have any idle time at all, so that it does not lose say financially. Thus contentions are inevitable, there is no global optimum, only negotiated compromises.

Recognising this reality, we have developed a model based on user-defined preferences, which are expressed on resources used in task scheduling, such as machines, time or labour. It is not possible to meet all the preferences due to contention, and therefore we can define the *best solution* as the solution that meets as many preferences as theoretically possible, while satisfying all the constraints. However, since a the theoretical best is not really practicable, we opt for a *good* solution that lies within the upper and lower bounds of theoretical predictions.

In this paper we address the problem of how to derive a preference-based solution in presence of contention, where awarding preferences to the solution of one task can only be done by depriving/removing preferences from that of another. Deprivation and particularly removal of preferences creates a high non-linearity leading to non-convergence. We propose a general model that produces a good solution in finite time (section 2), backed by a theoretical performance model (section 3) and a simulation study that verifies the correctness of the theoretical model (section 4). This work extends that published in [2].

Preferences are being used in many applications including document ordering, learning and storage in an Web-based environment [1], electronic commerce [4], product design [5], agent-based routing [6], distributed meeting scheduling [7, 10], advanced information retrieval [8], fuzzy ranking [11] and cooperative decision-making [12]. Preferences are used in most of these papers as a simple ranking system, and most of the applications occasionally require human intervention at some point. Also most applications do not guarantee the convergence of distributed computation. None of these papers mentions explicitly the cascading effect or a mechanism for dealing with rescheduling. Numerous researchers use agent technology to resolve the manufacturing scheduling problem. Shen and Norrie [9] give an overview of recent projects, and how they deal with the scheduling.

Our multi-agent approach is based on what we call Cooperating Knowledge-Based Systems (CKBS), in which holons cooperate together in solving a global task through a Cooperation Block (CB) where one holon acts as the coordinator and the others as cohorts. This is an engineering paradigm as opposed to the mentallistic paradigm of distributed AI/Multi-agent systems (DAI/MAS) [3], but it blends ideas from both distributed databases and DAI/MAS.

2. THE SCHEDULING MODEL

We assume a (global) task T, subdivided to lower-level tasks, to be referred to as subtasks (T_1 ... T_n), each subtask T_i having preferences on the resources (strictly speaking resource instances) used for their scheduling. Dependencies among subtasks, including precedents constraints can lead to a heterarchical structure. The task T is executed by the coordination, while each subtask T_i is allocated to an agent (Assembler) A_i. It is often impossible to satisfy all the preferences due the following reasons: (i) contention with the preferences of other agents, (ii) processing cost and (iii) intractability leading to non-convergence.

We assume each subtask to specify preference values on the resources required for the allocation of each subtask. A coordinator can specify very high preference values for its subtasks greedily. To control this greed, we use a market based cost model. The coordinator must state how much it is prepared to pay to achieve its preference. A task then is expressed as follows:

$$T{::}\ [(G_1 \ldots G_m), (P_1 \ldots P_n), (V_1 \ldots V_n)\ (O_1 \ldots O_n)]$$

where ($G_1 \ldots G_m$) are the set of the task constraints, typically task dependencies, ($P_1 \ldots P_n$) are a set of (preferred) resource instances, such as end-times, ($V_1 \ldots V_n$) are the corresponding preference values on these resource instances, and ($O_1 \ldots O_n$) are the corresponding offer prices – the prices the coordinator is prepared to pay to get these preferred resource instances. In other words, the coordinator is prepared to pay price O_i for resource (instance) P_i with preference value V_i. The offer price has to be checked against the actual cost (cost price) of the requested allocation as discussed below.

Associated with the allocation of each task T_i is a cost C_T, which is meant to be covered by the offer price. The cost C_T is used to terminate a branch if the agreed cost is exceeded. Each coordinator can accumulate the payments it has received from other coordinators, and use this to buy preference values in the future. C_T can be less than the offer price, depending on the negotiation but no offer is accepted if it is less than C_T. Cost C_T has two components: initial (or basic) cost C_I, and a refinement cost C_R. Component C_I is the cost of finding an allocation for the task ignoring its preferences (on resources). Such an allocation might fortuitously satisfy some or even all preferences. If not, then further processing (called refinement on allocation) over many iterations may be needed to gain more preference values. In any iteration, this task may be reallocated to a preferred resource instance, removing another task from that resource instance. potentially recursively. This cost C_R is the estimated relocation cost of those dislocated tasks, and can be a cascaded cost due to task dependencies. It is expressed as follows:

$$C_T = C_I + \sum_{i=1}^{i=m} C_R$$

where m is the number of refinement needed to find a solution. The basic algorithm is outlined in the following pseudo code, in which an initial total preference value V_t is obtained at first, which is subsequently improved by refinement iterations. At each iteration, multiple candidate allocations are evaluated, from which the one that provides the maximum preference value, say V_m, is selected within the allowed cost (i.e $O_I \leq C_T$). However, this improvement is accepted only if $(V_m - V_t) \geq \Phi$ where Φ is a preference cut-off value (see section 4). If this improvement is accepted, then V_m becomes the new V_t, and a new iteration begins. But if this gain does not exceed Φ, the execution is terminated. In practice, iterations are continued for several more times before termination. The actual process is more elaborate as can be gleamed from section 4.

```
For each coordinator agent:
Get an initial solution, say total preference value Vt
If (initial solution satisfies all the preference values)
        Accept the solution
else{
       Begin an iteration to maximise the preference values
       Seek new allocation by negotiating with other tasks.
       Find the new preference values and costs of all possible new solutions.
       Select from these solutions the one with Vm for (OI ≤ CT)
       If  (Vm – Vt) ≥ Φ,
              Accept this solution, and set Vt to Vm
              Proceed to the next iteration.
     else
              Accept the earlier value of  Vt and Terminate
   }
```

3. THEORETICAL DISTRIBUTION MODEL

Because the scheduling model described above is highly nonlinear, verification using the available mathematical techniques is difficult, as we cannot relate the arbitrary user values, such as the preference values, cost values and preference cut-off values, effectively by a mathematical formula However, we have been able to produce a behavioural model by examining the iterative allocation process and the associated incremental preference gain, as presented below. After each iteration I, the total remaining preference value that is yet to be satisfied is given by R_I. During the allocation process, described in section 2, the final global preference gain, G, gradually decreases per iteration as the iteration number I increases, eventually converging onto a minimal value for the total remaining preference value not achieved. We can express R_I for iteration I in an exponential form as:

$$R_I = A + Be^{-\lambda I}$$

where the constant A is the height of the plateau when $e^{-\lambda I}$ tends to zero. Hence A = R_{min}, which is the remaining minimum preference value at $e^{-\lambda I} = 0$, to be referred to as the residue R_{min}. Constant B is R_{max}, which is equal to ($R_1 - R_{min}$), and constant λ gives the curvature of the distribution related to the rate of preference gain. Thus the equation can be written as follows:

$$R_I = R_{min} + R_{max}\, e^{-\lambda I}$$

The initial allocation is made ignoring preferences. This is the zeroth iteration (I = 0) when $R_0 = R_{min} + R_{max}$,. At the next iteration (I = 1), $R_I = R_{min} + R_{max}\, e^{-\lambda I}$. As I increases, $e^{-\lambda I} \rightarrow 0$ and $R_I \rightarrow R_{min}$.

Our extensive investigation to determine the factors that affect the value shows (not presented here) that predominant factor is the clustering of preferences on resource instances. For example if only one subtask can be allocated to a time-slot say 11 am on a machine. but three subtasks jostle for it, then only one of these can be satisfied even in theory, the other two contributing to preference losses.. We have have captured the clustering effect in our theoretical estimation of R_{min}. In our formulation, we use two parameters, the distribution density d, and the *Preference Reduction Function* ρ, as explained below for preferences on a single resource type, which can be imagined as the end-times for subtasks. If we have t subtasks, all having preferences over the same m (<t) resource instances (eg end-time slots), then the density d = t/m. This is the effect of clustering, which can only be resolved by allocating some of these t subtasks away from m, but still as close to their preferred resource instance as possible, so that the preference loss is minimised. The function ρ is used to determine this preference loss.

Given a subtask, we may assume that its preferred resource instances are placed in a convenient preference-value order. Thus, if the end times are the resources, then the time-slots are the resource instances, which would be in time order. Hence, if a preference value V has been attached to the instance k and if ρ is the percentage decrease for each slot away from k on either sides of k, then V will change to V (1 - jρ) for both the instances (k-j) and (k+j)). For example if a preference V is expressed for the hourly slot 10 am, then it will be V(1 - ρ) for both 9 am and 11 am slots. Note (1 - jρ) = 0 if jρ > 1.

We assume we have t=2h subtasks with preference over m=2n resource instances, where t > n (see Figure 1). Each resource instance as a time-slot, preference values decreases on either sides of the most preferred time-slot, this decrease is given by ρ, as discussed above. In order to find the preference loss formula we need to evaluate the average movement from slot 1 at the right side of the midpoint m_l for the right half slots and hence half the subtasks (t/2).

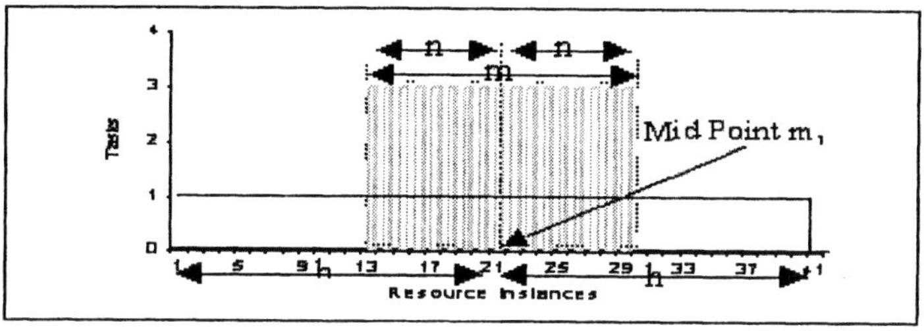

Figure 1- Preference Distribution

In Figure 1 above, more than three subtasks have been given for each slot in the central region, which must be declustered with one subtask per slot (indicated by the lower horizontal line above the X-axis). To do this we move subtasks away from the midpoint, we evaluate the shifts and then calculate the total preference loss, which will give us R_{min}..

The first d subtasks at slot 1 moves to slots 1,2,3,...d, with displacements [0, 1, 2, .., d-1], then the displaced d subtasks moves to slots d+1, d+2, .. 2d, with displacements [d-1, d, d+1, .. 2(d-1)], then the displaced d subtasks moves to slots 2d+1, 2d+2, .. 3d, with displacements [2(d-1), .. 3(d-1)], and so on. Generally speaking the displaced d subtasks moves to

slots (n-1)d+1, (n-1)d+2, .. nd, with displacements
[(n-1)(d-1), (n-1)d, (n-1)(d+1), .. n(d-1)].

The summation of the displacement of each sets of d subtasks:

$$[0, 1, 2, .., d-1] = (d-1)d/2$$
$$[d-1, d, d+1, .. 2(d-1)] = [(d-1)+2(d-1)]d/2= 3(d-1)d/2$$
$$[2(d-1), .. 3(d-1)] = [2(d-1)+3(d-1)]d/2= 5(d-1)d/2$$

......

.....

$$[(n-1)(d-1), .. n(d-1)] = [(n-1)(d-1)+n(d-1)]d/2= (2n-1)(d-1)d/2$$

If we sum the right hand side we get

$$(1 + 3 + 5 + 7 + ... + (2n-1))(d-1)d/2$$
$$= [1 + 2n -1](n/2)(d-1)d/2 = nn(d-1)d/2$$

This also applies for the subtasks on the left half, hence the total displacement for the whole distribution is n n(d-1)d. To get the average movement per subtask we divide by t = 2nd, the average movement per subtask becomes:

$$nn(d-1)d/(2nd) = n(d-1)/2$$

If the preference loss per unit shift is ρ, then the preference loss per subtask is:

$$\rho n (d-1)/2 \qquad \text{Eqn 1 (Preference Loss Formulae)}$$

Upper and Lower Bounds of Preference Loss

First assume that the subtasks do not have any precedent constraints and hence can be allocated freely. In that case we can evaluate an weighted average of the total preference loss as follows:

$$[P_1*L_1 + P_2*L_2 + \dots\ \dots\ P_q*L_q]/q$$

where L_i is the preference loss (in percentage) of Assembler A_i (calculated from the preference loss formula given above), and then P_i is the total preference of A_i. This is the minimal loss based on density d_i in each Assembler A_i and hence is the predicted lowest limit of R_{min}.

Assume that the t subtasks were distributed over s resource instances in iteration 0 due dependencies (including precedent constraints), where $s \geq t$. Therefore, each subtask will occupy on average s/t (≥ 1) slots rather than t/t (= 1) slot. In that case the density $d_i = t/m$ will change into a revised density $D_i = s/m$ for Assembler A_i. Its effect is the same as replacing d by D in Eqn 1, which will then yield a revised estimate of the preference loss that includes the effect of dependencies. This will give the upper limit of R_{min}. Therefore our simulation should yield a value for R_{min} that lies between these two limits.

4. SIMULATION STUDY

In this section we aim to present the results obtained from a simulation study for scheduling in a distributed manufacturing environment using the algorithm described earlier in section 2. We have implemented the scheduling model in a demonstrator using a Java platform.

For the simulation study we have used a set of coordinators CA_1, CA_2, .. CA_n, each CA_I with a (global) task T_i , each task T_i being further subdivided into subtasks: $T_{ij} \mid \{j = 1, 2, ..\ m\}$, one or more subtasks being allocated to a target agent (say an Assembler holon) A_k. In our implementation we have used up to n = 14, that is, up to 14 coordinators, up to six subtasks (m= 6) in each task and three target agents, each target agent sharing many subtasks of different coordinators. The subtasks have precedent subtasks and their allocation to the target agents can be conjectured as the allocation to Assembler agents in manufacturing. The preferred resource type used in our simulation is end-time slot of a subtask. Each subtask T_i has a preference value V_i and an offer price O_i that the task is willing to pay for preference satisfaction. Also the costs C_i and C_r, are set by the coordinator, and indicate the price that the task should pay for a preferred allocation (see section 2). Our objective is to find a schedule that satisfies as much preferences as possible using our proposed scheduling model.

According to our algorithm initially each subtask is allocated the earliest possible slot that satisfies precedence constraints. In the following iterations the coordinator will accept a negotiation for an exchange if its offer price $O \geq$ the cost C for the expected preference gain *g*. After negotiations, an actual exchange with preference gain *g'*, where $g \geq g'$ is proposed. Note *g'* is $(V_m - V_t)$, in the algorithm in section 2. However we use a special preference cut-off value Φ such that if an iteration does not improve the preference gained by at least this Φ, then this gain is rejected, in order to prevent too many insignificant gains and iterations. Thus the coordinator will accept it if $g' \geq \Phi$, but pay pro-rata to C^*g'/g. If the exchange is unsuccessful, the negotiation will continue for another possible exchange. If no improvement on preferences can be made, allocation is made according to the previous allocation. In our implementation the preference values and offer prices are assigned randomly using the random method available in Java Math package. We have conducted several hundreds of experiments with many permutations and combinations in order to verify the basic properties of our model and to verify the mathematical model described in section 3.

4.1 Verification of Basic Properties

In order to show that the solution converges to the same final value independent of the initial order of subtask processing, we carried out an experiment with 24 subtasks of all mixes but with non-conflicting end times slots (preferred resource). If these subtasks are allocated in the arrival order (which was the end time order) over three target agents without paying any attention to their preferences, 100% preference values will be automatically achieved. We then allocated these tasks in the reverse order without taking any preference into account. This yielded 30% preference gain. On this distribution we applied our model and re-allocated the subtasks, this time (iteration 1) taking preferences into account. This first iteration achieved 100% gain. We repeated this experiment with different initial order, and each case 100% gain was achieved at the first iteration. In these experiments the value of preference cut-off Φ was kept fixed at 5%.

These experiments confirmed that our model behaves, as we expected, and that it does lead to convergence. *A significant point is that this model produces results which are independent of initial allocations (i.e the order of subtask processing), so difficult to achieve in machine scheduling.*

In the next set of experiments we investigated if higher offer prices by some coordinators can distort the results significantly. We have used six coordinators, each task T_i of the coordinator CA_i having *m* number of subtasks, *m* varying from 3 to 6. Initially all subtasks are allocated on the first available (time) slots in the (global) task order $T_1, T_2, \ldots, T_6$, at a given offer price, but without considering the

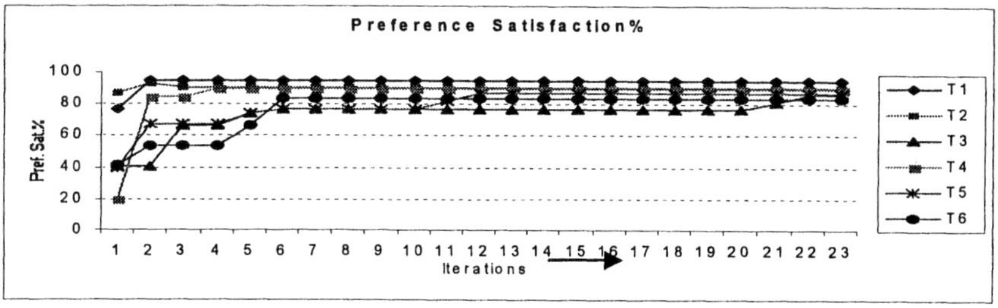

Figure 2 - Preference satisfaction of tasks

preferences. Then a series of iterations was carried out with the same Φ value of 5%. At each iteration, the offer price of one of the coordinators was raised by 5% and its subtasks reallocated, taking only its preferences into consideration.

The results presented in Figure 3 which shows the changes in the preferences satisfied during the allocation of each task (T_1, T_2, ..., T_6) in which only preferences of that task was taken into account. So our conclusion is that higher offer prices do not make any significant change, and therefore our model produces stable preference gain. Next we have examined if the preference gain and the cost of one task is affected by increasing offer prices of the other coordinators. We have conducted this experiment for each task T_i, but selected arbitrarily to show it for T_3 in Figure 3, which was typical for all other tasks.

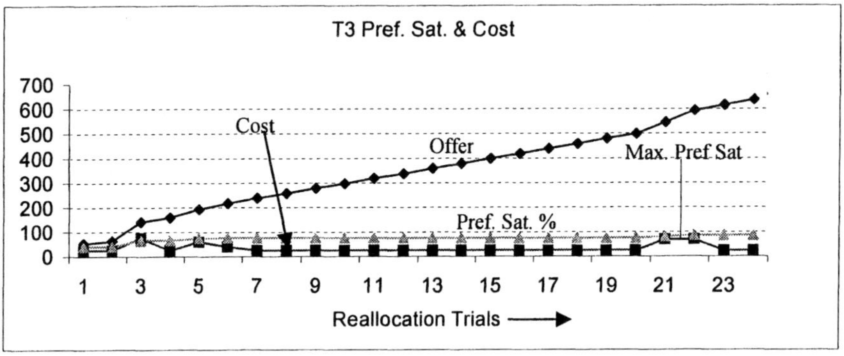

Figure 3 - T_3 preference satisfaction offer & cost variation

We can observe from Figure 3 that the preference gain and cost for T_3, against the accumulated offer prices of the other coordinators. Evidently the increase in their offer prices did not affect the preference gain of T_3 in any significant way. This result is typical of all T_i's.

4.2 Verification of Our Model

These second sets of experiments were carried out to verify that the preference gain over iterations obeys the theoretical model. These results are presented below for two categories:

Observance of the exponential rule at a variable Φ value

Observance of the upper and lower bounds

4.2.1 Variable Φ Value.

In this study we used 14 tasks (coordinators) and 54 subtasks distributed over 3 target agents, and carried out three experiments for different Φ (2%, 5%, 10%), the same set of preference values, and the same preferred end-time slots of each subtask. As shown in Figure 4, the fits on the results of the iterations for preference gain confirm the exponential pattern predicted by the theory.

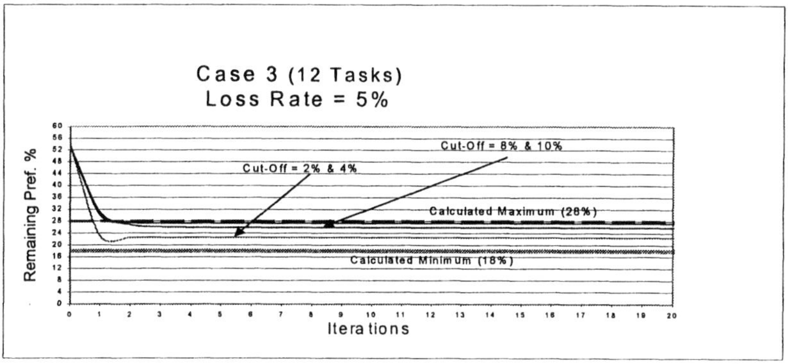

Figure 4 - T_3 preference satisfaction offer & cost variation

4.2.2 Boundaries

The objective of this experiment is to compare the predicted results from the theoretical model with the obtained results from the implementation. In this experiment we used the results from the previous experiments as well as new cases. In some experiments we changed the value of the preference loss rate (explained in the previous section). We used different number of tasks with different number of coordinators, 12, 24, 29, 54 tasks and 4, 6, 7 and 14 coordinators, respectively. For the purpose of this paper we show a summery of part of the is shown in Table 1.

Table 1 - Results Summary

Case	Number Of Tasks	Pref. Loss Rate	Calculated Results		Experimental Results
			Minimum	Maximum	
1	12	5	4.9	11.6	8.5
2	12	10	9.8	23.3	16.7
3	28	5	9.3	27.5	22.8
4	54	5	29.2	58.6	50.8
5	54	5	18.9	28	25.6

Clearly the experimental results on preference loss lie within the upper and the lower bounds of the theoretical predictions.

5. CONCLUSION

In this paper we have presented a cooperative scheduling approach based on user-defined preferences that can be applied in HMS applications. In this context, we have described a distributed allocation technique and a theoretical model to assess its correctness, which we have verified by conducting a simulation study. We have used a cost-based negotiation approach to ensure that the system can converge to a good solution within the upper and lower bounds of our theoretical prediction. The allocation is independent of initial allocation, it converges, and furthermore the convergence can be achieved faster for crude allocation, should it be desired. We see the potential of applying this approach in HMS and other agent-based cooperative scheduling applications, including distributed project managements. Finally although not part of the HMS project, this work is a spin-off from it. We are indebted to our HMS partners for the various ideas and discussions from which this work has indirectly benefited.

Acknowledgements

The authors are indebted to the partners of the Holonic Manufacturing Systems (IMS/HMS) project for providing inspiration behind this work. One of the authors (RJ) is indebted to Al-Quds University (Jerusalem) for funding this research.

6. REFERENCES

1. K. Burke and K. Aytes: Preference for Procedural Ordering in Distributed Groups…, Proc. of the 34th Hawaii Int. Conference on System Sciences (HICSS-34), IEEE Computer Press (2001)
2. S.M. Deen and R Jayousi: Preference Based Task Allocation in Holonic Manufacturing, the 13th International Conference on Database and Expert Systems Applications (DEXA'02), published by the IEEE Computer Society (2002) 573-577
3. S. M. Deen and C. A. Johnson: Formalizing an Engineering Approach to Cooperating Knowledge-Based Systems. TKDE 15(1) (2003) 103-117
4. K.H. Joo, T. Kinoshita and N. Shiratori: Agent-based Grocery Shopping System Based on User's Preference, Proc of the 7th Int. Conf. on Parallel and Distributed Systems (ICPADS'00 Workshop: Flexible Networking and Cooperative Distributed Agents, IEEE (2000) 499-505
5. T. Keinonen: Expected Usability and Product Preference, Proceedings of DIS'97 Conference Designing Interactive Systems, August, Amsterdam. ACM (1997) 197-204
6. S. Rogers, C. Fiechter, and P. Langley: An Adaptive Interactive Agent For Route Advice, Proceedings of the 3rd Int. Conference on Autonomous Agents, Seattle: ACM Press (1999) 198-205
7. S. Sen, T. Haynes and N.Arora: Satisfying user preferences while negotiating meetings, International Journal of Human-Computer Studies, Academic Press, London (1997)
8. Y. W. Seo and B. T. Zhang: A Reinforcement Learning Agent for Personalized Information Filtering, Proceedings. of International Conference on Intelligent User Interface(2000) 248-251
9. W. Shen, D.H. Norrie,: Agent-Based Systems for Intelligent Manufacturing: A State-of-the-Art Survey, An extended HTML version of the paper published in Knowledge and Information Systems (KAIS), an International Journal, 1(2), (1999), 129-156.
10. T. Shintani, T. Ito, and K Sycara: "Multiple Negotiations among Agents for a Distributed Meeting Scheduler", Proc. of the 4th Int. Conference on Multi Agent Systems (ICMAS'2000), poster
11. J. Wang: Ranking Engineering Design Concepts Using a Fuzzy Outranking Preference Model", Fuzzy Sets and Systems 119 (2001) 161-170
12. S. T. C. WONG: "Preference-Based Decision Making for Cooperative Knowledge-Based Systems", ACM Transactions on Information Systems, Vol. 12, No. 4 (1994) 407-435

19 OPTIMIZATION ALGORITHM FOR DYNAMIC MULTI-AGENT JOB ROUTING

Leonid Sheremetov[1], Luis Rocha[1], Juan Guerra[2], Jorge Martinez[2]
[1] *Mexican Petroleum Institute,*
Av. Lazaro Cardenas, 152, Col.San Bartolo Atepehuacan, México, D.F., CP 07730,
[2] *Computer Science Research Centre of National Technical University (CIC-IPN),*
Av. Juan de Dios Batiz esq. Othon de Mendizabal s/n Col. Nueva Industrial Vallejo,
México, D.F., C.P. 07738, MEXICO
e-mail: sher@cic.ipn.mx

Current research in multi-agent heterarchical control for holonic systems is usually focused in real-time scheduling algorithms, where agents explore the routing or process sequencing flexibility in real-time. In this paper we investigate the impact of the dynamic job routing and job sequencing decisions on the overall optimization of the system's performance. An approach to the optimization of local decisions to assure global optimization is developed within the framework of a Neural Collective Intelligence (NECOIN). Reinforcement learning (RL) algorithms are used at the local level, while generalization of Q-neural algorithm is used to optimize the global behaviour. A simulation test bed for the evaluation of such types of multi-agent control architectures for holonic manufacturing systems integrating discrete-event simulation facilities is implemented over JADE agent platform. Performance results of the simulation experiments are presented and discussed.

1. INTRODUCTION

The worldwide competition and the highly specified customers' requirements towards product quality, delivery time, and services force the industry to a permanent optimization of the production. As a consequence, logistics gets a new focus on optimization of the production process in a very dynamic environment. Current research in multi-agent heterarchical control for holonic systems is usually focused in real-time scheduling algorithms, where agents explore the routing or process sequencing flexibility in real-time (Denkena et al., 2002, Heragu et al., 2002, Sheremetov et al., 2003, Usher, 2001). Though there are a lot of results on scheduling heuristics and dispatching rules, few researchers have studied the influence of these approaches on the overall optimization of the production system performance. Since most of these solutions and techniques are based on local optimization criteria, these decisions do not assure the overall business optimization at the global level because of the conflicts between the local goals (Julka et al., 2002). Traditional centralized techniques usually cannot assure global optimization either due to the inherent complexity of the problem.

In this paper, the problem of job routing (JR) is addressed within the context of the NEural COllective INtelligence (NECOIN) theory (Wolpert & Kagan, 1999) and the adaptation of the Q-neural algorithm (Rocha-Mier, 2002). According to our approach, agents construct previously unknown model of the environment through learning and interaction between them in a distributed fashion. The approach looks for balancing the agents' efforts to achieve the short-term or local goal (shortest path selection) and a long-term or global goal – overall production optimization (Shimbo & Ishida, 2003). According to our definitions, a production system within the NECOIN framework is a large multi-agent system where:

- Its objective is the decentralization of control and communication.
- Each entity of the system is represented as an agent with autonomous behaviour and a local utility function.
- The learning process consists of adapting the local behaviour of each entity (agent) with the aim of optimizing a given global behaviour.
- The agents execute Reinforcement Learning algorithms at the local level while generalization of Q-neural algorithm is used to optimize the global behaviour.

In this paper we investigate the impact of the dynamic job routing on the overall optimization of the system's performance. A simulation test-bed for the evaluation of such types of multi-agent control architectures for holonic manufacturing systems integrating discrete-event simulation facilities and implemented over JADE agent platform (AP) is described. This test-bed can be also used to compare different approaches to job routing on a common basis (Brennan & O, 2000). The case study deals with production of hypothetic products on the shop-floor level. Performance results of the simulation experiments are presented and discussed.

2. NECOIN framework for the JR problem

This work proposes a model of the production system (PS) within the framework of the NECOIN theory. In our approach, an agent can represent any entity of the PS. In contrary to the model described in (Sheremetov et al., 2003), the materials in the PS are represented as objects forming part of the environment. Therefore, every agent can change or influence these environment objects. The details of the objects are stored as attributes. We define the following elements within the NECOIN framework for the JR problem:

- Order-agent that has the knowledge on final products orders: PO
- Set of n machine-agents (MA): $M = \{M_1, M_2, ..., M_n\}$
- Set of s operations executed by machine $i: OP_i = \{O_1, ..., O_S\}$
- Vector of non-negative values of r features for each operation $O_i : \overrightarrow{V_i} =< v_1^i, ..., v_r^i >$, $e.g.\ v_1^i =$ average time. These features vary from one machine to another.
- Set of n storage-agents (SA) denoting raw material providers: $S = \{S_1, S_2, ..., S_n\}$
- Set of s objects corresponding to a type of raw material: $MP = \{MP_1, ..., MP_S\}$
- Set of n final product storage agents (FPSA): $FP = \{FP_1, ..., FP_n\}$

- Set of n objects corresponding to a type of final product: $P = \{P_1,..., P_n\}$
- Vector of non-negative values of r features for each product $P_i : \overrightarrow{PV_i} =< pv_1^i,..., pv_r^i >$, e.g. pv_1^i - product priority.

In this work, each agent has the following features:

- The set of environment states $X = \{x_1, x_2, x_3,...\}$. Knowledge (usually incomplete) about other agents is considered to be part of the environment state. For example, in some cases a MA might make decisions without knowledge that a supplier has frequently failed on due dates.
- The capacity of agent to act is represented as a set of actions: $A_i = \{a_1, a_2,..., a_k\}$.
- The relationships between the agents in the PS are defined by: $R = \{r_1, r_2, r_3,...\}$. For each neighbour agent, the following parameters are considered: a) its relationship to the current agent (customer, supplier), b) the nature of the agreement that governs the interaction (production guarantees), and c) the inter-agent information access rights (the agent's local state to be considered during the decision-making process).
- The priorities of every agent are represented by: $Q = \{q_1, q_2, q_3,...\}$ These priorities can help in sequencing incoming messages for processing.
- The local utility function (LUF) is represented as follows:
  $$Q_{(x(t),a_{x(t)})}(t+1) = Q_{(x(t),a_{x(t)})}(t) + \alpha\left[r(t+1) + \gamma \min_{a_{x(t+1)}} Q_{(x(t+1),a_{x(t+1)})}(t+1) - Q_{(x(t),a_{x(t)})}(t)\right],$$
  where: α is learning rate, γ is reduction rate.
  This equation represents the Q-learning (Sutton et al., 1998) equation used in RL. The Q-values $Q_{(x(t),a_{x(t)})}$ give an estimation of the PS. The way in which the Q-values are updated can be considered as one of the most important problems to solve in our framework. The reinforcement for the performed action is represented by $r(t+1)$. This function of reinforcement represents the partial time of product production and is composed of: a) transition time, b) waiting time, and c) operation time.
- The set of control elements: $C = \{c_1, c_2, c_3,...\}$. A control element is invoked when there is a decision to be made while processing a message. For example, in order to determine each destination of materials, a routing-control algorithm would be utilized.
- Every agent has a message handler responsible for communication.

3. Q-Neural Algorithm for Job Routing Task

To address the JR problem, the adaptation of the Q-neural algorithm (Rocha-Mier, 2002) is proposed and described. The behaviour of the Q-neural was inspired by the Q-routing algorithm (Littman & Boyan, 1993), the theory of NECOIN and the algorithms based on the behaviour of the colonies of ants. Learning is done at two levels: initially, at the agent's level locally updating the Q-values by using a RL rule, then, globally at system level by the utility function's adjustment. The control messages allow updating knowledge of the PS by updating the Q-values, which are approximated by using a function approximator (look-up table, neural network, etc.). In Q-neural, there are 5 types of control messages:

- *An 'environment-message' (flag_ret=1)* generated by an intermediate MA after the reception of a raw material if the interval of time ω has already passed.
- *An 'ant-message' (flag_ret=2)* generated by the FPSA according to the interval of time *w_ants* when a final product arrives at the final product storage.
- *An 'update-message' (flag_ret=3)* generated in the planning phase every *ε_update* seconds to ask the neighbouring MA for their estimates about the operations of the FP.
- *An 'update-back-message' (flag_ret=4)* generated after the reception of an update-message in order to accelerate learning of the environment.
- *A 'punishment-message' (flag_ret=5)* used to punish a MA using a congested resource.

The Q-neural algorithm includes 3 parts: planning, ant-message and punishment algorithms working as follows. Exploration can involve significant loss of time. In Q-neural, a mechanism of planning (within the meaning of this term in RL) was developed at the local level of each agent. This mechanism consists of sending an update-message every ε_update seconds. This update-message will ask for the Q-values estimates of all the products, which are known at that moment by the neighbours.

When an environment-message arrives at the FPSA, an ant is sent in return if the period of time *ω_ants* has already passed. This ant exchanges the statistics obtained on its way. When it arrives to the SA, it dies. The ant updates the Q-value of each MA through which the raw material passed before arriving at the FPSA.

In some cases, different MAs from the same tier can have the same best estimate (prefer the same route). If they act in a greedy way, congestion occurs in the queue. To avoid congestions, a MA must sacrifice its individual utility and use another route. In order to address this problem a punishment algorithm is developed forcing an agent who receives a punishment message to calculate the second best estimate.

Finally, the Q-neural algorithm is defined as follows:

Initialize at $t = 0$: All the Q-values $Q_{x(t),a_{x(t)}}$ with high values, the RL parameters:

$\alpha, \gamma, exploration, w, w\_ants$

REPEAT

Update the instant t

***if** a raw material is received by machine M_i*

Read the input vector X from the raw material header and environment variables

Send the message to the agent M_x where the raw material arrives with the value of the reinforcement function $r(t+1)$ and the estimation $Q^{\mu}_{(x(t),a_{x(t)})}(t)$

Execute the operation O' and choose the action $a_{\bar{x}(t)} = M'$ in function of the input vector X by using the strategy ε_greedy derived from $Q_{(x(t),a_{x(t)})}(t)$

Send the raw material $a_{\bar{x}(t)} = M'$

At the next time step, receive the message from the machine M' with the value of the reinforcement function $r(t+1)$ and the estimation $Q_{x(t+1),a_{x(t+1)}}(t+1)$

Apply the Q-learning update rule:

$$Q_{(x(t),a_{x(t)})}(t+1) = Q_{(x(t),a_{x(t)})}(t) + \alpha\left[r(t+1) + \gamma \min_{a_{x(t+1)}} Q_{(x(t+1),a_{x(t+1)})}(t+1) - Q_{(x(t),a_{x(t)})}(t)\right]$$

REPEAT

Algorithm of planning ε _update

Algorithm of punishment

Algorithm of ants

Planning, ant-message and punishment algorithms are described in more details in (Rocha-Mier et al., 2004).

4. Implementation of the Q-neural Algorithm in the Multi-agent Framework

The above-described model has been implemented using JADE AP in order to test the performance of the developed algorithms. A generic layout of the simulated PS is shown in figure 1. The first tier consists of suppliers of raw materials and is represented by the central storage. Raw materials are distributed among the machines organized into several different tiers. For simplicity, we consider that the operation lists corresponding to each machine of the tier are identical. Finally, all the processed parts are stored at the storage of the final products.

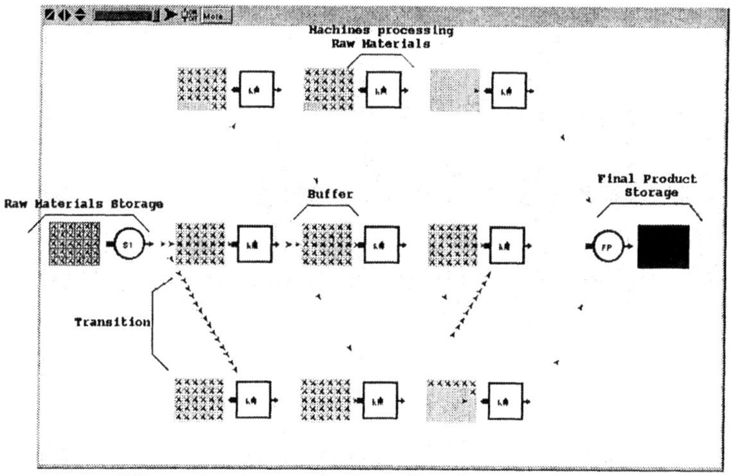

Figure 1 – A generic shop-floor layout scheme

An agent represents each entity in the model; there are *n* MAs at each layer, one SA and one FPSA. Also, there is a Scheduler Agent (SCHA) whose role is that of a synchronization ticker for the discrete event simulator organizing events during the simulation. Nevertheless, the event queue behaviour (currently owned only by this agent) can be transferred to the bulk of agents for an actual implementation in a physical environment. A User Interface Agent (UIA) is used to define experiments, start, resume, and finish system's operation.

The simulation begins when the SCHA broadcasts a *startUp* message to the whole group. This way, agents can perform internal initialization tasks mainly related to the tables (known as Q-tables) and variables to be used by the Q algorithms. In the case of the SA, it will load the *Technological Processes* and *Orders* lists that will be processed. Also, it will reply to the SCHA with the initial *Raw Materials* list to be released.

Raw materials, intermediate products and final products are not physically present. The information concerning each of them is passed via message interchange between the SA and MA. A raw material becomes a final product after having travelled along the machine network. Initially, it is created with an operation vector that decreases at each step, until it gets empty and an FP is located at the corresponding storage.

Fig. 2 shows a Collaboration Diagram illustrating the planning phase of the Q-neural algorithm, which involves the SCHA and three MAs. Also, a sample raw-material transfer and the corresponding arrival to the FPSA are shown.

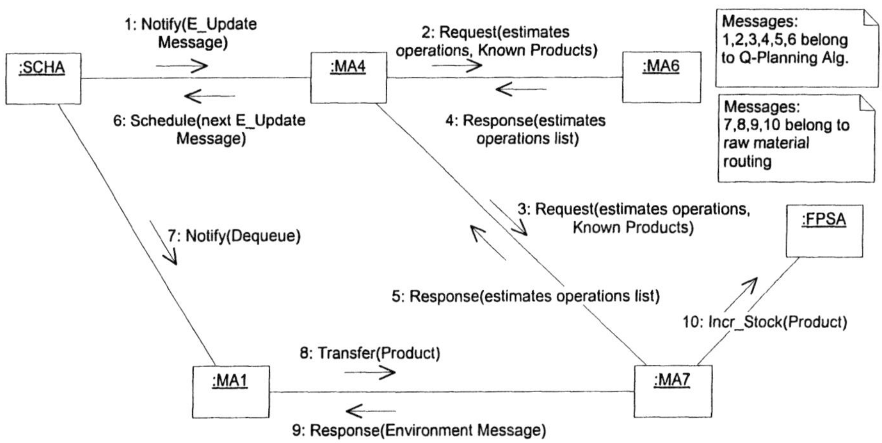

Figure 2 – Collaboration Diagram: planning phase of the Q-neural algorithm

As it can be noticed, both events are first dispatched by the SCHA. The SCHA warns MA each time they must take some product from their internal queue and get it processed. Before forwarding the modified product, an MA will consult the Q-tables in order to decide the best route to follow. Also, there is a ping-like operation that ensures the selected machine is still alive. Otherwise, the second best alternative will be chosen and the corresponding MA will be notified to queue the product for further processing. In consequence, the Q-tables are changed since a new product has been set in the link between the current machine and its neighbour.

Data back-propagation to the previous stage is achieved after a MA commits to queue a product. This mechanism helps ensuring that each machine has information to optimize product routing. Routing algorithms are embedded in MA's body. This results in information updating aimed in optimizing the decision making process on the best route selection for each stage.

It must be noticed that the SA works partially as a MA for the following reason: the SA also keeps a Q-table for making decisions on where to send the raw material after this is released. If the MA detects that there are no pending operations to accomplish for the processed product, it will notify the FPSA. In turn, this agent will modify the stock numbers for such product.

Under a centralized approach, a single entity evaluates continuously the best alternative for every agent's strategy within the system. However, computing the numbers for a large group could be unfeasible under time and resources constraints. Under our distributed approach, each agent shares its local estimates with the neighbours, thereby laying the ground for a collective solution with fewer resources than those required by a centralized entity.

The optimization is reached by informing the neighbourhood status; this information is used by reinforcement rules and conjoined with the collective algorithms. This way, the Q-tables are continuously updated to reflect the agent's perception of the world. They improve their decision-making as more raw materials flow through the system. Agents only have to activate the corresponding behaviour according to a schedule that gets adjusted during the system execution.

5. Case study description and performance results

In the previous section, we have shown how the different agents can interact with each other to carry out the control of the production processes. We distribute agents and product objects over the network, and have them interact with each other to simulate the communication and cooperation of the actual controllers distributed in a production plant. In this section, we will present an example to demonstrate the interaction model of the agents and the resulting optimization of the production processes. We present an example of some of the performance analysis that has resulted from the model described in this paper to investigate the impact of the dynamic JR decisions on the overall systems performance.

For the experiment configuration, we used one SA, 3-tier production scheme and a FPSA. Tier A is composed of MA_1 and MA_2 performing operations O1 and O2 taking 18 and 34 time units and 19 and 20 time units at different machines respectively. Tier B is composed of MA_3 performing operations O3 (36) and O5 (19), MA_4 with O4 (30) and MA_5 with O5 (14) and O6 (29). Finally, tier C is composed of MA_6 with O7 (30) and O8 (18) and MA_7 with O8 (19). During the initialization stage, MAs search their local Directory Facilitator. They receive a list of partners from the next tier of the PS. Also, MA asks next-tier neighbours about their capabilities. This information is used to initialize the Q-tables.

There are three different products over which, three operations must be accomplished: P1, operations O1, O4 and O5; P2, operations O2, O4 and O7; P3, operations O2, O5 and O7. The corresponding demand for each of them is: between time units 1 - 50, P1-type raw materials for 30 products must be released from the SA. Between 21 - 60, P2-type raw material for 40 products and between 31 - 70, P3-type raw material for 60 products must be released from the SA. All of them travel through the PS until they reach the FPSA at the other side of the network.

Starting simulation, the SA sends message indicating the MA_1 to add a new product to the buffer's queue. However, this is only a notification; the actual processing will not take place until the SCHA informs the MA_1 to do so. After the operation is completed, the agent is responsible for routing the product to the next tier where two events are triggered: a neighbour's request to add the intermediate product and a SCHA's request to add a new event for the corresponding machine to check its buffer for pending jobs.

For communication purposes, domain ontology is used. This encodes different types of events that agents must be aware of. The ontology consists of a set of numerical constants for events such as: "raw material release" (RELEASE_MP), "final product report" (INCR_STOCK) and "product leaves machine buffer" (DEQUEUE). Message structure between MA and SA is slightly different from conversations with the SCHA. This latter manages only events notifications from and to the group of agents. MA and SA, on the other hand, implement action requests for operations. As shown in fig. 3, MA_1 requests MA_2 to add a product to MA_2's queue. That is the way products travel from tier to tier. In the last conversation, MA_3 requests MA_4 and MA_5 to inform their capabilities in order to update MA_3's Q-tables. Each MA responds to the query with the list of operations it is able to perform and associated processing time using FIPA-request protocol. These operations are specified as "operations concept" using domain ontology.

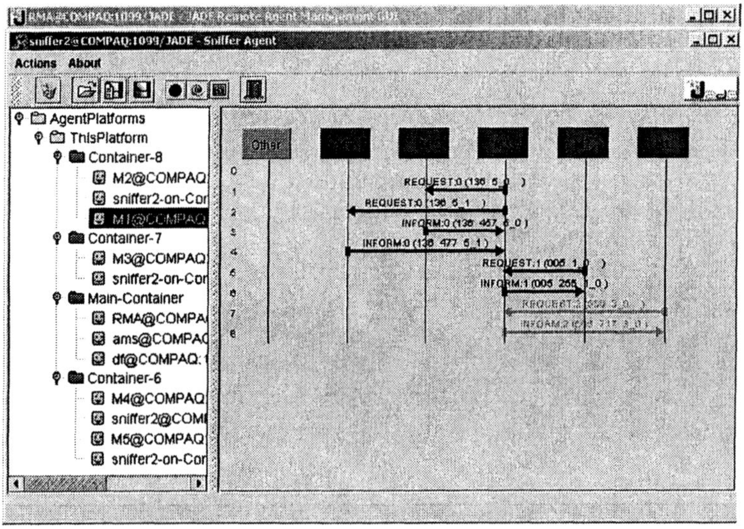

Figure 3 - Message interchange among a set of MAs (Sniffer Agent screenshot).

As shown in fig. 3, MAs are distributed along the JADE containers according to the production tier they belong to. Also, container facilities work as a bridge for experimenting with distributed locations. There is an option to connect physically distant machines that resembles the common layout of a real PS (as shown in Sheremetov et al., 2003). In other words, presented implementation is two-folded. It can be used as a test-bed for trying different configurations, and if required, it is intended to function as an implementation in a real scenario.

Planning and punishment sub algorithms were applied each second of simulation time and ant sub-algorithm was applied each 0.5 sec. The general parameters were: learning rate = 0.8 and exploration rate = 0.08.

At the second stage of the experiments, an adaptation of the Q-routing algorithm (Littman et al., 1993) within the framework of the JR problem was compared with the Q-neural algorithm, described in this paper. The comparison of these two algorithms can be found in Fig. 4. This figure shows the number of products produced (arrived at the FPSA) vs. the average production time.

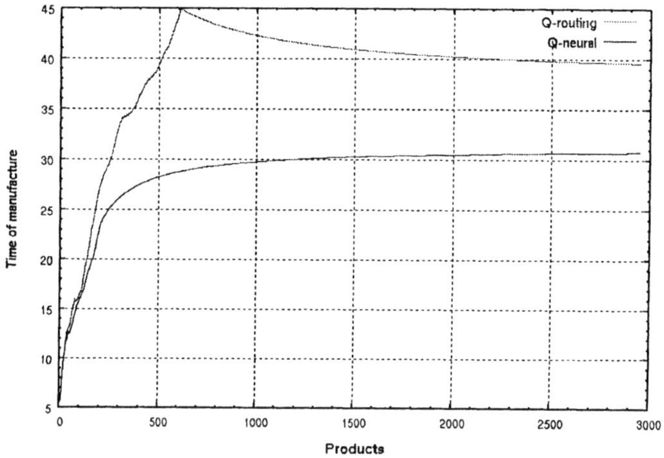

Figure 4 - Comparison of the results of the Q-routing and Q-neural algorithms

Fig. 4 shows the adaptability and better performance of the Q-neural algorithm due to the following. MAs based in Q-routing, make their decisions in a greedy way to optimize their local utility functions. This conduces to a buffer's saturation and a decrement of the global utility function as a result of this greedy behaviour. However, the MAs based in Q-neural, make their decisions taking into account both local utility and the global PS performance. As a result, the performance of the scenario is improved thanks to the adaptation to the changes of the PS environment.

6. DISCUSSION AND CONCLUSIONS

Today's challenge is to optimize the overall business performance of the modern enterprise. In general, the limitations of traditional approaches to solving the JR problem are due to the fact that these models do not correspond to the reality because of incomplete information, complex dynamic interactions between the elements, or the need for centralization of control and information. Most of heuristic techniques on the other hand, do not guarantee the overall system optimization.

In this paper, JR problem is addressed within the framework of NECOIN theory. In order to optimize the global behaviour of PS on the shop-floor level, learning process using RL algorithms to adapt agent's local behaviour is used. This model is implemented in the agent-based parallel modelling and simulation environment over the JADE platform. Being the agglutinating centre of the enterprise information infrastructure, an AP also serves as an experimental test-bed for the implementation of the models developed in this paper. By means of this, we can easily implement the algorithms tested in the simulated environment into the real-world applications.

The experiments on the comparison of this approach with that reported in (Sheremetov et al., 2003) on a common platform is under development. The tested control systems will have varying production volumes (to model the production system with looser/tighter schedules) and disturbance frequencies, so that the impact of the JR and sequencing decisions in various manufacturing environments can be evaluated. The communication protocol's behaviour between agents is under investigation using communication network simulation tools like Network Simulator (NS-2). We also pretend to compare our algorithms with other classic optimization methods using the developed multiagent test-bed.

Though we do not have yet the results of these experiments, we conclude that the JR problem is well situated for the application of the NECOIN theory. In addition, the adaptive Q-neural algorithm provides better throughput and reliability than other algorithms. In future work, an adapted model of the CMAC Neural Network will be used for the Q-values approximation. More complicated punishment algorithm will be developed to adjust the local utility functions.

7. ACKNOWLEDGMENTS

Partial support for this research work has been provided by the IMP, within the project D.00006.

8. REFERENCES

1. Brennan RW, O W. A simulation test-bed to evaluate multi-agent control of manufacturing systems, In Proc. of the 32nd Winter simulation conference, December 10-13, Orlando, Florida., 2000.
2. Denkena B, Tonshoff HK, Zwick M, Woelk PO. Process Planning and Scheduling with Multiagent Systems. In V. Marík, L. Camarinha-Matos, H. Afsarmanesh (Eds.) Knowledge and Technology in Product and Services. Selected papers of the IFIP, IEEE International Conference BASYS'02, Kluwer Academic Publishers, 2002; 339-348.
3. Heragu SS, Graves RJ, Byung-In K, St-Onge, A. Intelligent agent based framework for manufacturing systems control. IEEE Transactions on Systems, Man and Cybernetics, 2002; 32(5): 560–573.
4. Julka N, Srinivasan R, Karimi I. Agent-based supply chain management-1: framework. Computers and Chemical Engineering, 2002: 26.
5. Littman M, Boyan J. A Distributed Reinforcement Learning Scheme for Network Routing. School of Computer Science, Carnegie Mellon University, 1993.
6. Rocha-Mier, Luis. Apprentissage dans une Intelligence Collective Neuronale: application au routage de paquets sur Internet. PhD thesis, Institut National Polytechnique de Grenoble, 2002.
7. Rocha-Mier L, Sheremetov L, Contreras M, Osuna C, Romero M, Villa L, Hernandez A L. Global Supply Chain Management based on Collective Intelligence. In Proc. of the 2nd World POM Conference and 15th annual POM Conference. Cancun, Mexico, April 30 - May 3, 2004.
8. Sheremetov LB, Martínez J, Guerra J. Agent Architecture for Dynamic Job Routing in Holonic Environment Based on the Theory of Constraints. Honolic and Multi-Agent Systems for Manufacturing (HoloMas 2003), V. Marik, D. McFarlane, P. Valckenaers, eds.: Springer Verlag, 2003; LNAI 2744: 124-133.
9. Shimbo M., Ishida, T. Controlling the learning process of real-time heuristic search. Artificial Intelligence, 2003; 146: 1–41.
10. Sutton R, Barto A. Reinforcement Learning: An Introduction. The MIT Press, 1998.
11. Usher, John M. Negotiation-based in job shops via collaborative agents, Third International ICSC Congress on World Manufacturing WMC'2001, Rochester, N.Y., Sept. 24-27, 2001.
12. Wolpert D, Kagan T. An Introduction to Collective Intelligence. Technical Report NASA-ARCIC-99-63, NASA Ames Research Center, 1999.

20

AGENT SYSTEM APPLICATION IN HIGH-VOLUME PRODUCTION MANAGEMENT

Martin Rehák[1], Petr Charvát[2] and Michal Pěchouček[1]

[1] *Gerstner Laboratory*
Czech Technical University in Prague
Technická 2, Prague 6, 166 27 CZECH REPUBLIC
{pechouc,rehakm1}@labe.felk.cvut.cz
[2] *CertiCon, a.s.*
Václavská 12, Prague 2, 120 00 CZECH REPUBLIC
charvat@certicon.cz

Based on actual industrial project on which the Gerstner laboratory has collaborated, we present a multiple-level scheduling approach as a mean to efficiently apply agent-based planning systems in high-volume production environment. Brief description of efficient and reconfigurable high-level scheduler based on linear programming, as well as design elements of low-level agent-based planning are included

1. INTRODUCTION

Agent systems are currently predominantly used in project-oriented production management (Pěchouček et al., 2002) where they offer a significant competitive advantage by easily adopting to naturally very dynamic environment. In this work, we study an application of multi-agent systems in highly specialized and high volume manufacturing plant. Underlying research is an extension of real industrial application project executed for major automotive production plant in Eastern Europe.

First, we shall specify the client requirements on system function and behavior. Then, we shall introduce a concept of multi-level partially distributed planning realized in cooperation between dedicated high-level planning agent and agents representing real-world physical entities. High-level planning agent, its model, algorithm and current implementation is presented in section 4 and low-level planning and production management in the section 5. Conclusions are drawn in the section 6, together with future work directions.

2. CLIENT REQUIREMENTS AND PROBLEM STATEMENT

Automotive industry operates in high volumes and on very low margins, thus it focuses a lot of attention on process optimization. Such optimization can be specified by following generic requirements, derived from the project specific requirements drawn by our client. For through motivation of these criteria, see (Goldratt, 1990).

- Minimize the stock through the production chain, thus decreasing the financial and storage costs.
- Maximize the production uniformity, to be able to use the industrial means in an efficient manner and to avoid overtime cost.
- Minimize the unnecessary handling of products between successive steps of the production process to further reduce human resources and other manipulation related costs.
- Allow the integration with production surveillance and management tools.
- Allow real-time or almost real-time re-planning in case of demand changes or production anomalies.
- Allow easy and straightforward process reconfiguration in the future (strengthening the bottlenecks of the production process)

It is interesting to note that the quantitative (first three) criteria listed above are completely contradictory and that a satisfactory optimum is their weighted, context dependent combination.

Problem statement

Factory in question contains three serially organized production lines, as shown on the figure.

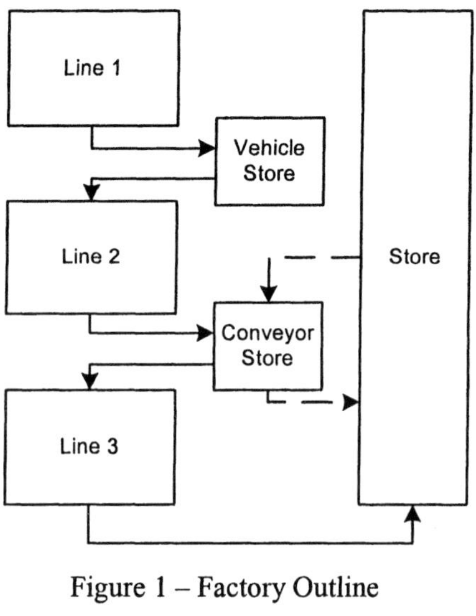

Figure 1 – Factory Outline

The factory production means can be described as three lines in a series, with two buffer stores and one main store used for final product storage before delivery and for the intermediary product storage as well. The intermediary product can be bought from outside or shipped directly to clients, as it may be a part of deliveries described below. Material store is not represented on drawing, as the material is delivered to different positions on production lines when required. Demand is formalized as a matrix defining how many products of a given type shall be delivered on a given day.

3. MULTI-LEVEL PLANNING

In order to successfully plan in this environment, we shall note that parallelism is limited to using pre-completed parts instead of completing them locally, or to producing partly assembled components for the completion in other plants. Such production is represented by dashed paths in the figure.

Therefore, the use of classic negotiation techniques is highly constrained due to the fact that there are almost no alternative paths and the whole planning problem is reduced to pure scheduling in a static environment, featuring an enormous solution space. In this case, a single, dedicated planning agent has a considerable advantage of global knowledge that allows it to use a heavyweight but efficient scheduling methods. On the other hand, a today's user will typically require a rapid process reconfiguration ability to follow process or manufacturing equipment evolutions and a close integration with on-line production surveillance & analysis tools. This is the area where agent based distributed and adaptive systems have significant advantages (Bussmann et al., 2001).

The solution we propose is based on a compromise between these two approaches. We have divided the scheduling process to two distinctive phases and distributed different scheduling tasks between these phases. *High level scheduling* agent is responsible for the fulfillment of the first two requirements, minimum stock and maximum production uniformity, together with ensuring that the deliveries are feasible. Output of this planning stage is the size of lots to produce on different lines during the given day.

Low level scheduling agents process the output of high level scheduling and organize the work on their lines during the appropriate time period by distributing the lots of products into appropriate timeslots. In this phase, we handle the production continuity, ensuring that the line 2 immediately consumes the product produced on line 1 and that the same applies for Lines 2 and 3. Production surveillance may be connected to this process for dynamic rescheduling.

In the scope of our case, two levels of planning are completely sufficient. With increasing problem scope, or while extending it to extra-enterprise environment (Mařík et al., 2002) we will probably need to add extra layers of planning above the current high-level scheduling. This will allow an effective choice of production site for each task. As a negotiation or auction methods are probably best suited for this layer, an additional future requirement on high-level scheduling is its speed, allowing us to answer the bids for production instantly.

The planning agents may be directly integrated with the embedded holonic systems in production equipment, provided by manufacturers. This will empower both the high-level and low-level scheduling by giving it more information about the equipment and allowing it not only to react to problems when they happen but also to proactively predict problems and include contingency in the plan.

4. HIGH-LEVEL SCHEDULING

This section describes the model and algorithm used by high-level scheduler to determine how many products to produce in a given time interval. First, we will formally restate the problem (simplified) solved by high-level scheduler, explicitly specifying the limitations to respect and the parameters to optimize. We will also briefly discuss some interesting aspects of the solution and implementation.

Formal Problem Statement

Let's have J products groups ($j = 1, 2, \ldots, J$). The time horizon is divided into T time intervals ($t = 1, 2, \ldots, T$). The production can be scheduled on P processors ($p = 1, 2, \ldots, P$). Let's also have the expedition demand d_{jt}, for each product group j and time interval t. Further, we denote q_{pjt} the production volume of the product group j in the time interval t on the processor p and I_{jt} as stock of the product group j at the end of the time interval t. Mark I_{j0} the initial stock of the product group j. We intuitively request all elements of Q to be positive and integer.
Denote $z_{jp} \in \{0, 1\}$ the ability to produce the product group j on the processor p_3. The production volume of the product group j in the time interval t on all processors is then

$$Q_{jt} = \sum_{p=1}^{P} z_{jp} q_{pjt}$$

The production relationships is described by square matrix $S_{J \times J}$. For the production of one item of the product group x we need $s_{x,y}$ items of the product group y. The production demand is then

$$v_{jt} = \sum_{x=1}^{J} s_{xj} Q_{xt} \qquad \forall j,t$$

We can describe the relation among the production, stock and demand as

$$I_{j,t-1} + Q_{jt} - I_{jt} = d_{jt} + v_{jt} \qquad \forall j,t$$

Let the C_{pt} is the capacity of the processor p in the time interval t. For production of one item of the product type j $r_{jp} > 0$ units of processor p are necessary. Then we can describe the constraint on the production induced by the capacity of the processor with following relation

$$\sum_{j=1}^{J} r_{jp} z_{jp} q_{pjt} \le C_{pt} \qquad \forall p,t$$

The L^{min}_{pjt} is the minimal lot-size. We also request the lot-size (number of products of type j produced in day t on processor p) to be either zero or more then this minimal lot size value due to material handling efficiency issues and non-zero switching times.
Limitations do apply also on the stock of the products produced and ready for shipment. Typically, a certain amount of pieces of each product (c_{jt}^{min}) is kept in reserve in order to be able to replace the non-produced or incorrectly produced pieces of this type. On the other side, the capacity of the store is physically limited

to certain amount of pieces of all products, weighted by their respective space consumption. This capacity is denoted M_t.
Our goal is to satisfy all orders, respect all the limitations described above and to minimize the stock and production variability. Stock minimization can be described as

$$\sum_{t=1}^{T}\sum_{j=1}^{J} I_{jt} \rightarrow \min$$

and production variability can be described using the following relations:

$$\sigma_p = \frac{1}{T}\sum_{t=1}^{T}\left|\omega_{pt} - \overline{\omega}_p\right| \rightarrow \min$$

with

$$\overline{\omega}_p = \frac{1}{T}\sum_{t=1}^{T}\omega_{pt} \qquad \textit{where} \qquad \omega_{pt} = \frac{\sum_{j=1}^{J} r_{jp} z_{jp} q_{pjt}}{C_{pt}} \in \langle 0,1 \rangle .$$

Problem Solution Elements

After careful deliberation and several experiments, linear programming was chosen for implementation of high-level scheduler agent. The main factors beyond this choice were its speed, robustness and an ability to detect the constraint preventing us from achieving our goal. However, the application of this method in our case is not straightforward, because the problem as specified is not entirely linear. We had to resolve following issues:

1. the lot-size has to be an integer,
2. the production uniformity relation is non-linear,
3. lot-size has to be either zero or more than the minimum value.

Integer values of lot sizes are an issue that is easy to resolve. Either we can use integer extension of standard LP algorithm, which is NP-hard and complicates the solution, or we can simply ignore this issue and round the results of LP algorithm. The rounding error caused by this approach is (in most cases, as well as in our case) insignificant compared to total number of products.
Non-linearity of the production uniformity relation is an issue that is much harder to resolve. We have opted for an alternative approach that modifies the conditions of the original model by requesting the production not to divert form the average required production value by more than certain percent. In practice, we replace the condition presented above by following inequalities.

$$\sum_{j=1}^{J} r_{jp} z_{jp} q_{pjt} \le C_{pt} \quad \Rightarrow \quad B_p^{low} C_{pt} \le \sum_{j=1}^{J} r_{jp} z_{jp} q_{pjt} \le B_p^{high} C_{pt}$$

Average load used in the relation to determine the boundary values can be calculated per processor, per processor group or for the whole plant, depending on client preferences.
Third problem, minimum lot size issue was solved by iterated runs of the algorithm. In the first run, we use the model as described above and we use the resulting

production matrix Q to modify the conditions for the second run according to the following condition.

$$q_{pjt} \leq \alpha L_{pjt}^{\min}$$

If the condition is satisfied, then q_{pjt} is fixed to 0. Otherwise, we require it to be bigger than the minimum lot size. The value of the parameter α can either be set manually by operator, or deduced by the system from its previous experience.

Implementation

For the implementation of our current high-level scheduler, a free third party LP solver was used, together with communication and data transformation wrapper. The whole scheduling takes less then 1 second on standard PC (with 28 days, 50 products and 3 processors), thus completely satisfying the performance requirements resulting from frequent re-planning and future possible requirements resulting from the integration with another negotiation-based planning layer.

In the case that the solution cannot be found, we can use the solver with modified problem to identify the critical limitation and to communicate it to operator or other appropriate system component.

Linear programming model is also rather easy to extend or modify in case of plant reconfiguration. We can simply add new production lines or products, together with their properties and modify the *Z* matrix, describing the ability of processor to produce different products.

In the next part, we will propose equally reconfigurable solution for low-level scheduling and production surveillance as an extension to today's traditionally less flexible solutions.

5. LOW-LEVEL SCHEDULING AND PRODUCTION CONTROL

Low level scheduling will process the required daily quantities determined by preceding high-level scheduling. Its task is to order the lots on the lines during the given day to minimize the manual manipulation and material handling related to product switches. In our model, we associate a cost to switching between two types of products on a line. This cost is relatively low for products that share a major part of their components, but grows with increasing differences between products.

Single processor problem statement

For a single line (processor), we may describe the succession of different products as an oriented graph, where the nodes represent different lots and the evaluations of edges connecting them represent the cost of succession of these particular lots in the order determined by the orientation of edges (See figure for simple example).

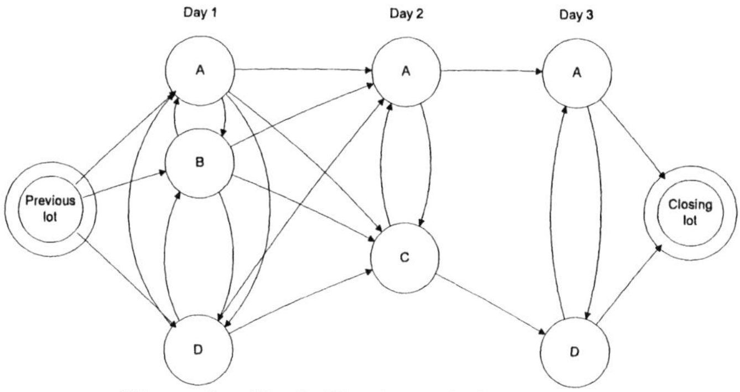

Figure 2 – Single line lot ordering problem

Note that the lots scheduled for one day are connected in both directions, as any of them may precede another, while the connections between successive days are only from past to the future. When we establish such graph for all the lots during the planning period, we may note that finding an optimum order is equivalent to finding a Hamiltonian path (a path passing through all nodes) through this graph, connecting the first and last lot. Such task is NP complete (see (Blazewicz et al., 2001) for alternative prove) and there is no trivial way of finding an optimal solution. However, the frequent changes of delivery orders, together with non-uniformities of the production process make the complete and uninterrupted execution of the plan highly unlikely. Therefore, we have opted for the use of greedy algorithm, which seeks the local optimum and selects the batch with locally cheapest transition, expecting that future gain from accepting locally suboptimal decision is highly unlikely to be collected anyway.

Extension to multiple processors

In reality, few industrial processes are executed by single processing unit and our case is no exception to this rule. This poses us in front of another obstacle, as we must ensure the continuity between successive processors. These processors may have incompatible preferences concerning the products to manufacture in a given moment. In this stage, the agent nature of low-level scheduler may prove to be advantageous, as the negotiations between different processor-representing agents would be able to efficiently organize the lot order in a given day.

In accordance with principles defined by the theory of constraints, we may see that within a single day scope, our bottleneck is actually defined by the high-level scheduling output. Therefore, throughout the negotiations, we shall prioritize the agents representing the components with load closest to nominal in a given day. This shall ensure the feasibility of fulfillment of the goals specified by the high-level scheduler. Other components switching costs may be discounted compared to the bottleneck switching. This approach uses the information prepared by the high-level scheduling to make the low-level optimization more efficient and relatively straightforward.

Production surveillance and dynamic re-planning

As already stated above, a smooth execution of any plan is rarely natural and often requires many local or even major global adjustments. Today, such adjustments are decided by human managers based on their observations of the production process and their experience, even if automatic solutions start to appear (Bussmann et al., 2001). We propose that the processor planning agents shall be also used for online process surveillance and that the information gathered by these agents may be used both to increase the experience of agents and to react immediately to current situation, using the agent experience (Mařík et al., 2002) and (Pechoucek et al., 2000). This approach would allow us to eliminate many false alarms, connected with slow production start or short-time material inaccessibility, but can sooner detect potential major problems, especially by looping back the information form quality assurance stations.

6. CONCLUSIONS AND FUTURE WORK

The work described in this article is based on an industrial application project to which the Gerstner laboratory has contributed by its experience in industrial process planning and control. Major features of the design presented in this article is the separation of high-level and low-level scheduling, enabling us to benefit both of the global view of the dedicated scheduling component and flexibility, reactivity and potential learning ability of agent based systems. We've successfully integrated a linear programming methods into the project solution and demonstrated the complementarity between this classical approach and dynamic agent systems.
Even if the final implementation of this particular project will not probably directly use agent framework (due to system integration and technology transfer issues), we are looking forward to integrate the planning component and other features from this project with current, more project oriented software tools to extend their reach to high-volume manufacturing.

7. REFERENCES

1. Blazewicz, J, et al. Scheduling Computer and Manufacturing Processes – 2. ed. Springer Verlag, Berlin 2001
2. S. Bussmann, K. Schild: An Agent-based Approach to the Control of Flexible Production Systems, in: Proc. of the 8th IEEE Int. Conf. on Emergent Technologies and Factory Automation (ETFA 2001), p.481-488 (Vol.2). Antibes Juan-les-pins, France, 2001.
3. Goldratt, Eliyahu M. The Theory of Constraints, Croton-on-Hudson, N.Y.: North River Press, 1990.
4. Mařík, V., Pěchouček, M., and Štěpánková, O.: Social Knowledge in Multi-Agent Systems, In: *Multi-Agent Systems and Applications (M. Luck et. al, eds.)*, LNAI 2086 Tutorial, Springer-Verlag, 2001, pp. 211-245.
5. Mařík, V. - Pěchouček, M. - Vokřínek, J. - Říha, A. Application of Agent Technologies in Extended Enterprise Production Planning. In: EurAsia-ICT 2002: Information and Communication Technology. Berlin : Springer, 2002, p. 998-1007. ISBN 3-540-00028-3.
6. Pěchouček, M. - Říha, A. - Vokřínek, J. - Mařík, V. - Pražma, V. ExPlanTech: applying multi-agent systems in production planning. In: International Journal of Production Research. 2002, vol. 40, no. 15, p. 3681-3692. ISSN 0020-7543
7. Pěchouček, M., Mařík, V., Štěpánková, O., 2000, Role of Acquaintance Models in Agent-Based Production Planning Systems. In M. Klusch and L. Kerschberg (eds), Cooperative Information Agents IV – LNAI 1860, (Heidelberg: Springer Verlag), p. 179-190

21

MULTI-AGENT BASED ROBUST SCHEDULING FOR AGILE MANUFACTURING

Toshiya Kaihara* and Susumu Fujii**

* *Graduate School of Science and Technology, Kobe University*

** *Department of Computer and Systems Engineering, Kobe University*

{kaihara, fujii}@cs.kobe-u.ac.jp, *JAPAN*

Scheduling is the problem of allocating resources to alternate possible uses over designated period of time. Contract mechanisms use prices derived through distributed bidding protocols to determine an allocation. A robust manufacturing scheduling protocol based on multi-agent paradigm is proposed in this paper. We define all the manufacturing units, such as machines and jobs, as economic agents, which conduct strong robustness against practical manufacturing conditions. A contract mechanism with bidding protocol corresponding to market structure is proposed. We study the dynamism of the proposed scheduling protocol, and confirm its validity by several simulation experiments.

1. INTRODUCTION

There is a growing recognition that current manufacturing enterprises must be agile, that is, capable of operating profitably in a competitive environment of continuously changing customer demands. It is important to realise the total productivity, efficiency and flexibility in factory management under such an environment. Scheduling problem is one of the major issues on the effective manufacturing management in the agile environment. Distributed autonomous manufacturing control is recently introduced, and several distributed scheduling methodologies are proposed by several researchers (Sugimura, 1994; Kaihara, et al., 1997; Kaihara, et al., 1998; Rabelo, et al., 1998).

Recently the utilisation of multi-agent system in manufacturing application increases, such as robot assembly planning, multiple mobile robot control and so forth (Deneubourg, 1991). Multi-agent paradigm has several characteristics to overcome the current scheduling problems in the agile manufacturing environment (Ishida, 1995). The capacity of a single scheduling rule to achieve efficiently for any length of time will be in doubt - only autonomous and coordinated paradigm will succeed (Walsh, 1998). By a social goal we mean a goal that is not achievable by any single agent alone but is achievable by a group of agents. The key element that distinguishes social goals from other goals is that they require cooperation; social goals are not, in general, decomposable into separate subgoals that are achievable independently of the other agent's activities. In other words, any agent cannot simply proceed to perform its action without considering what the other agents are doing. The attainment of social goals appears to require a coordination of agent actions (Kaihara, 1996).

Solving scheduling problems with and for distributed computing systems presents particular challenges attributable to the decentralised nature of the computation. System modules represent independent entities with conflicting and competing scheduling requirements, who may possess localised information relevant to their utilities in such an environment. To recognise this independence, we treat the modules as agents, ascribing each of them autonomy to decide how to deploy resources under their control in service of their interests. It is assumed that the agents can communicate with messages in which they may convey some of their private information.

Our goal is to propose a decentralised universal scheduling concept which is robust against several environmental changes despite its simple architecture. We present a new distributed scheduling concept based on the Contract Net Protocol (CNP) (Smith, 1980), which is one of the negotiation protocols taking the metaphor of market behaviour. The task allocation is realised by a negotiation process between agents called manager that has tasks to be executed and agents called contractor that may be able to execute those tasks. These agents negotiate each other by exchanging mutual messages. In the negotiation, decision-making criteria are necessary for agents to select a contracting partner to send a message. Therefore, to decide of appropriate criteria is very important because the criteria affect the system performance (Ishida, 1996).

In this paper, after a brief explanation of CNP, the criteria on basis of utility in each agent are formalised for the decentralised manufacturing scheduling. We demonstrate the applicability of the CNP based scheduling concept by simulation experiments. Finally it is proved the proposed concept can provide several advantages on decentralised manufacturing scheduling.

2. COOPERATIVE SCHEDULING CONCEPT

2.1 Contract Net Protocol (CNP)

The Contract Net Protocol is based on multi-agent paradigm, which explored a distributed approach to problem-solving using a "negotiated" mutual selection process for task allocation. A CNP based problem-solver is a collection of nodes in manager and contractor roles. A top level task is allocated to a manager node, which generates subtasks and issuing task announcements for them to some subset of the potential contractor nodes (a process called task announcement). Contractors bid on tasks they desire and are qualified for.

The manager selects the highest rated bid, and allocates the task to that contractor, possibly monitoring the contractor's progress toward solution. When several contractors supply final reports of individual subtask results, the manager is responsible for integrating and supplying a global solution. The manager-contractor relation is recursive, and nodes simultaneously may be managers for some tasks and contractors for others.

2.2 Scheduling problem

Generally scheduling problem involves several criteria, and a solution that minimises optimality of all the criteria does not exist. It is required for scheduling algorithm to search a Pareto optimal solution. We treat two types of general criteria about scheduling in this paper as a basic study, f: lead time and g: throughput.

<u>Notations</u>

Let J_i denote job i (i=1,...,N), M_j machine j (j=1,...,L), K_i the number of operations in job i, O^j_{ik} operation i (i=1,...,N), j (j=1,...,L), k (k=1,...,K_i), TO^j_{ik} process time and $STO^l_{ik}O^l_{i'k'}(O^l_{ik} \neq O^l_{i'k'})$ set-up time between O^l_{ik} and $O^l_{i'k'}$. We introduce the following assumptions in our scheduling model:

- Operational order in job J_i is given and fixed.
- Machine M_j deals with one product at the same time.
- Process time TO^j_{ik} varies and depends on machine M_j.

Then the objective function in our scheduling problem is described as

$$\min((\sum_{i=1,N} f_{Ji})/N) \cap \max(g_{Mj'}) \tag{1}$$

where

f_{Ji} : lead time of job j_i

$g_{Mj'}$: throughput of machine $M_{j'}$ $\quad g_{Mj'} = \min(\forall g_{Mj(j=1,L)})$

Generally, these criteria, lead time and throughput, are in trade-off relationship. Shortening lead time requires small WIP (Work In Process) size, that causes small throughput in the production. Conventional scheduling methodologies apply heuristic rule based approach, but they can't handle such a trade-off relationship appropriately.

2.3 Machine agent

Machine agents try to process as many products as possible so as to maximise the individual throughput. Their utility function is defined as follows:

$$U^l_{machine} = \max \sum_{i=1,N} \sum_{k=1,Ki} Ocomp^l_{ik} \tag{2}$$

where

$Ocomp^l$: the number of completed operations in machine l

Machine agents adopt the following scheduling policy to satisfy the utility function described in (2);

$$Select^l_{machine} = \exists i.(\min(TO^l_{ik} + STO^l_{i'k'}O^l_{ik})) \tag{3}$$

where the operation $O^l_{i'k'}$ is followed by O_{ik} consecutively in machine l.

2.4 Job agent

Job agents try to proceed as fast as possible so as to minimise the individual lead time. Their utility function is defined as follows:

$$U^n_{job} = \min \sum_{k=1,Ki} TO_{nk} \tag{4}$$

where

TO_{nk}: process time for O_{nk} in job n

Job agents adopt the following strategy to satisfy their utility function defined in (4):

$$Select^n_{job} = \exists l.(\min TO^l_{nk}) \tag{5}$$

The job agents can't acquire set-up time, because they have no idea which job agent comes next with their local scope. Only the machine agents can hold the set-up information.

3. SCHEDULING PROTOCOL

3.1 Scheduling model

In most of conventional server-client scheduling models, process machine and job are normally defined as server and client, respectively. However, they can behave bilaterally in the metaphor of general market. In this paper we assume two types of scheduling model in terms of agent role in CNP shown in table 1.

Table 1 Scheduling model

	Manager	Contractor
Case1: Contract_Mac	Machine	Job
Case2: Contract_Job	Job	Machine

Needless to say, Case 1 and Case 2 correspond to PULL logic and PUSH logic in factory management, respectively. Therefore this classification is quite natural in manufacturing scheduling.

3.2 Scheduling protocol

We propose a new distributed scheduling concept based on the CNP. The task allocation is realised by a negotiation process between agents called manager. A principle feature is mutual selection mechanism between manager and contractor. In this section, we describe the proposed scheduling protocol in the case 1 (Manager: machine), as an example.

Step 1: Task announcement

After completed a process, machine l constructs task announcements t_l for possible processing service and distributes it by broadcasting to all jobs with requesting information. Bidding time, when the task validity expires, is also included in the information.

Step 2:

After job n receives task announcements, evaluates its own eligibility. If the task satisfies the eligibility, go to *Step 3*. If not, ignore the task. If it receives multiple tasks at the same time, select the most favourite task measured by equation (5).

Step 3: Bid

Job n send a bid with the requested data to machine l.

Step 4: Task allocation

When the bidding time expires, Machine l selects the most appropriate returned bid measured by (3) and allocates the task to that bidder by award notification, and go to *Step* 5. Fail messages are sent to all the other bidders, then go to *Step* 6.

Step 5: Process execution and report

Job n performs the task allocated to it and reports results produced from the performing task to the machine l.

Step 6:
Job waits other appropriate task announcements sent by machines.

4. EXPERIMENTAL RESULTS

4.1 Simulation model

A virtual primitive factory, which is installed the proposed scheduling protocol, has been constructed as a simulation model in order to analyse the scheduling dynamism of the protocol. In this paper we assume the factory has no internal disturbances, such as machine faults or higher priority lot, as a basic study.
Experimental parameters are defined as follows:

- *L*: the number of machines
- *Kind*: the number of Job types
- *Kn*: the number of operations in job *n*
- *SimTime*: simulation period
- *ProcTime*(*Mc*, *Vc*): process time distribution
- *SetUpTime*(*Mp*, *Vp*): set up time distribution
- *ArrivalTime*(*Ma*, *Va*): arrival time distribution
- *Lot*(*Mr*, *Vr*): lot size distribution
- *Bidding period*: BiddingPeriod

where
(M*: average, V*: Standard Deviation) in regular distribution

We prepared the following 3 types of conventional heuristics rule-based scheduling algorithms for the comparison in this experiment:

FIFO: first in fist out
SPT-A: shortest processing time
SPT-B: shortest (processing + set up) time

Two kinds of typical manufacturing conditions, "high-volume & low-variety" and "low-volume & high-variety" are examined as the simulation scenario.

4.2 Large lot size manufacturing

The proposed scheduling protocol is evaluated and compared with the conventional heuristic rule-based scheduling in *Kind* = 3, as an example of "high-volume & low-variety" manufacturing. Simulation results are shown in Figure 1, 2, 3.

As described in 2.2, lead time and throughput are two major criteria. At first, Figure 1 shows the relationship between set up time and lead time, the first criterion. It is obvious that the proposed methods show better performances than conventional approaches. Additionally, if we focus only on the proposed approaches, Case-2 is better than Case-1. Figure 2 indicates the same tendency in terms of throughput (= yields). We analysed the relationship between lead time and throughput in Figure 3. Two new parameters are introduced in Figure 3 for a simple analysis as follows:

$$Lead\ Time\ Rate(i) = Lead\ Time_i/\max(Lead\ Time_{j(j \in P)}) \quad (6)$$

$$Throughput\ Rate(i) = \min(Yield_{j(j \in P)})/\ Yield_i \quad (7)$$

where

P: the set of all the examinations
i, *j*, *p*: an examination $i, j, p \in P$

Figure 3 shows the proposed methods have higher robustness compared with conventional scheduling algorithms in term of set up time influence. Especially Case-2, job plays manager, performs the best of all the methods.

In our agent definitions, machine agents try to increase their throughputs and job agents aim at shortening their lead time. Finally a scheduling solution is acquired as the result of their negotiations, and that means the scheduling dynamism is characterised by the mutual selection of the heterogeneous agents with different criteria.

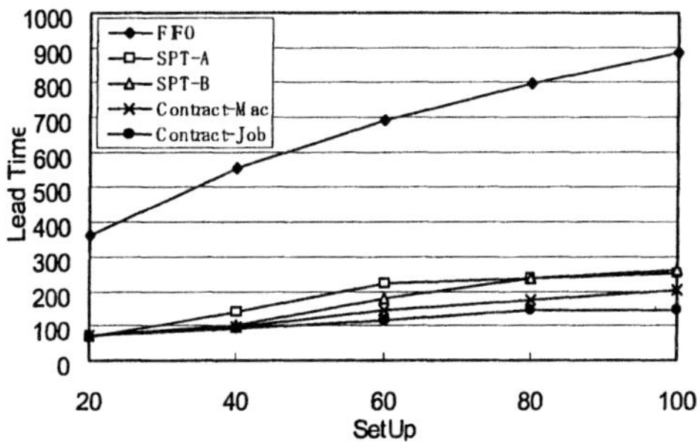

Figure 1 Large lot size manufacturing (Lead Time) *L*=3, *Kind*=3, *Kn*=1, *SimTime*=3600, *ProcTime*(50,10), *SetUpTime*(*Mp*,*Vp*): *Mp*={20,40,60,80,100}, *Vp*={4,8,12,16,20}, *ArrrivalTime*=50, *Lot*=3, *BiddingPeriod*=0

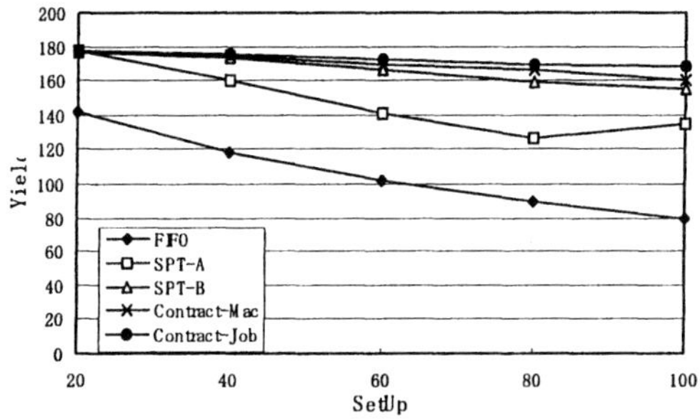

Figure 2 Large lot size manufacturing (Yield) *L*=3, *Kind*=3, *Kn*=1, *SimTime*=3600, *ProcTime*(50,10), *SetUpTime*(*Mp*,*Vp*): *Mp*={20,40,60,80,100}, *Vp*={4,8,12,16,20}, *ArrrivalTime*=50, *Lot*=3, *BiddingPeriod*=0

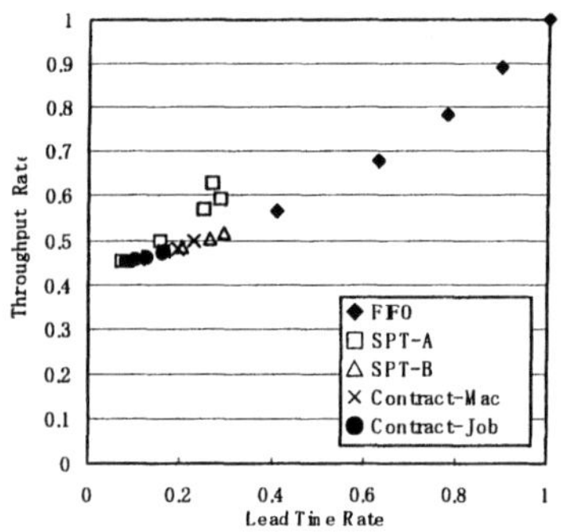

Figure 3 Large lot size manufacturing (Lead time-Throughput) *L*=3, *Kind*=3, *Kn*=1, *SimTime*=3600, *ProcTime*(50,10), *SetUpTime*(*Mp*,*Vp*): *Mp*={20,40,60,80,100}, *Vp*={4,8,12,16,20}, *ArrrivalTime*=50, *Lot*=3, *BiddingPeriod*=0

The results shown in Figure 1, 2, 3 indicate that the mutual selection mechanism amongst the heterogeneous agents plays an important role in the scheduling robustness against set up time. Conventional heuristic rule-based approach can't handle with multi-criteria scheduling demands. It is obvious that our approach is effective in terms of the flexibility against the multi-criteria.

By the comparison between Case-1 and Case-2, it is obvious that the careful construction of the decision process is also important even in the proposed approach as well as the conventional ones.

4.3 Small lot size manufacturing

The performance of the proposed scheduling protocol is compared with the conventional approach in *Kind* = 30, as an example of "low-volume & high-variety" manufacturing. Simulation results are shown in Figure 4.

It is clear that the general tendency of the results is almost equivalent to the large lot size manufacturing described in 4.2. It has been confirmed that our approach performs well with robustness in general case. One obvious difference is that Case-2 performs much better than Case-1, compared with the large lot size manufacturing. That points out an important characteristic of the proposed approach.

Our approach is based on the mutual selection amongst the heterogeneous agents. However, first selection is carried out by Job agents and Machine agents as contractors in Case-1 and Case-2, respectively. Machine agent behaviour, shown in equation (3), is to minimise (process + set up) time for the maximum throughput, that is required especially in small lot size manufacturing. These consideration lead the fact, that is the contractor's willingness influences the final scheduling solution more than manager's decision. As the result, the negotiation process is conducted by the contractors more strongly than the managers in the proposed approach.

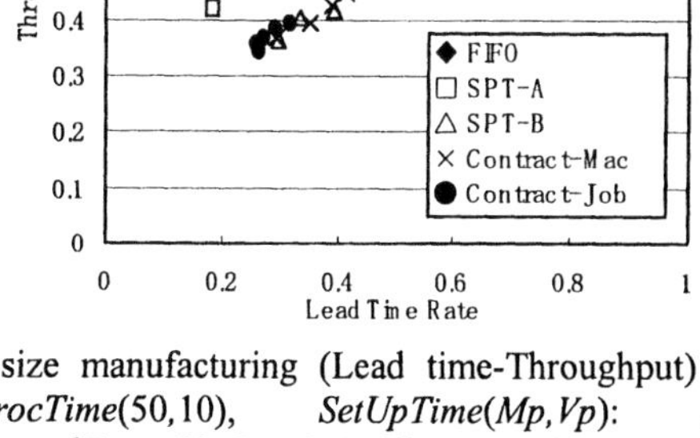

Figure 4 Small lot size manufacturing (Lead time-Throughput) L=3, *Kind*=30, *Kn*=1, *SimTime*=3600, *ProcTime*(50,10), *SetUpTime*(*Mp*, *Vp*): *Mp*={20,40,60,80,100}, *Vp*={4,8,12,16,20}, *ArrrivalTime*=50, *Lot*=3, *BiddingPeriod*=0

5. CONCLUSIONS

We have proposed a new scheduling concept that takes into consideration the special requirements of decentralised manufacturing environment. The highlights of the system are that it maintains the higher robustness under multi-utilities in trade-off relationship, such as lead time and throughput.

In this paper, we introduced multi-agent based negotiation protocol, CNP, into scheduling algorithm. After a brief explanation of CNP concept, the criteria on basis of utility in heterogeneous agents, named manager and contractor, were formalised for the decentralised manufacturing scheduling. We demonstrated the applicability of the CNP based scheduling concept by simulation experiments and clarified several important dynamism of the proposed scheduling protocol. Finally it has been proved the proposed concept can provide several advantages on decentralised & distributed manufacturing scheduling.

There are two obvious extensions. The first is to elaborate the negotiation protocol, possibly by exploiting some complexity with bidding period. The second extension is to analyse the robustness against dynamic disturbances in manufacturing system, such as machine failure.

6 REFERENCES

1. Deneubourg, J. (1991), The dynamics of collective sorting robot-like ants and ant-like robots, Proceedings of the First International Conference on Simulation of Adaptive Behavior, The MIT Press.
2. Ishida, T. (1995), Discussion on Agents, Journal of Japanese Society for Artificial Intelligence, Vol.10 No.5, pp.663-667.
3. Ishida, T., Y. Katagiri and K. Kuwabara (1996), Distributed Artificial Intelligence, Corona Publishing Co. Ltd.
4. Kaihara, T. (1996), A Study on Multi Agent Scheduling - An Analysis if the self-organization Phenomenon-, Journal of the University of Marketing and Distribution Sciences -Information, Economics & Management Science-, Vol.5, No.2, pp.113-120.
5. Kaihara, T. and S. Fujii (1997), A self-organization scheduling paradigm using coordinated autonomous agents, Rapid Product Development (N. Iwata, T. Kishinami and F. Kimura Eds.), Chapman & Hall, London, pp.489-498.
6. Kaihara, T. and S. Fujii (1998), An evolutionary scheduling paradigm using coordinated autonomous agents, Innovation, Globalization of Manufacturing in the Digital Communication Era of the 21st Century (G. Jacucci, G. J. Olling, K. Preiss and M. Wozny Eds.), Kluwer Academic Publishers, Boston, pp.553-563.
7. Rabelo, J. R., L. M. Camarinha-Matos and H. Afsarmanesh(1998), Mutigent perspectives to agile scheduling. BASYS 1998, pp. 51-66.
8. Sugimura, N. (1994), Autonomous distributed scheduling, Journal of the Society of Instrument and Control Engineers, Vol.33, No.7, pp.585-589.
9. Smith, R.G. (1980), The Contract Net Protocol: High-level Communication and Control in a Distributed Problem Solver, IEEE Trans. Comput., Vol.C-29, No.12, pp.1104 - 1113.
10. Walsh, W. E., M. P. Wellman, P. R. Wurman, J. K. Mackie-Mason (1998), Some economics of market-based distributed scheduling, Proceedings of the eighteenth international conference on distributed computing systems.

22

FUSION-BASED INTELLIGENT SUPPORT FOR LOGISTICS MANAGEMENT

Alexander Smirnov, Mikhail Pashkin, Nikolai Chilov,
Tatiana Levashova, and Andrew Krizhanovsky
St.Petersburg Institute for Informatics and Automation of the Russian Academy of Sciences, 39, 14-th Line, 199178, St.-Petersburg, RUSSIA
{smir, michael, nick, oleg, aka}@mail.iias.spb.su

Since the Internet has grown an easily accessible and popular place for business applications the problem of knowledge integration based on using Web tools and dealing with knowledge representation and processing has become actual. The paper presents a knowledge fusion agent as a part of multi-agent system addressing knowledge logistics. This agent is based on a constraint satisfaction technology. The task statement for knowledge fusion agent is presented in an ontology form. This flexible task representation via transparent and open ontology formalism enables creating routing plans for a transportation system of a virtual production network.

1. INTRODUCTION

Rapid development of the Internet has caused a huge amount of information about different problem areas to become available for users. Since the information is represented in various formats and by different tools, the problems of format compatibility, search tools implementation, recognition and fusion of knowledge from distributed sources/resources have become critical. The necessity of the knowledge fusion (KF) approach development for global understanding of going on processes and phenomena, dynamic planning and global knowledge exchange has developed. The KF methodology is a new direction of knowledge management in the part of knowledge logistics (Smirnov et al., 2003a).

In nowadays conditions knowledge is becoming an important resource. The main characteristics of it are the following: (i) knowledge is a critical source of long-term competitive advantage; (ii) knowledge is more powerful than natural resources; (iii) knowledge resource has cost, location, access time and life-time; (iv) knowledge worker is an owner of knowledge and a member of a team/group.

Related to this, along with development of computing machinery and information technologies, there arose a need of systems working with knowledge, i.e. dealing with knowledge creation, classification, synthesis, analysis, storage, search and mapping.

Intelligent agents are a very hot research topic that significantly changed the way distributed systems are working. Multi-agent system technology was chosen as the basis for KF systems (Smirnov et al., 2002).

Logistics systems play an important role in manufacturing companies, especially based on the concept of the virtual production network (Golm & Smirnov, 2000). An intelligent support, based on technologies of Web intelligence, intelligent agents, and open services, may significantly enhance the logistics system abilities (e.g., reduce costs and times of delivery). Therefore the paper proposes an application of knowledge logistics as an intelligent service for creation efficient routing plans (as one of the major logistics tasks in virtual production network management) under given constraints and preferences. This application is illustrated via a case study of delivering goods to customers.

Numerous logistics techniques are known, from the traveling salesman problem to complex dynamic problems. Vehicle rooting problem (VPR) as part of logistics has received a lot of attention in the literature because many real world transportation problems are related to it (Tarantilis et al., 2004; Ruiz et al., 2004). VPR can be briefly described as a set of *N* clients or customers with known demands d_i, $i \in 1,\dots,N$, that have to be served from a central depot with a fleet of *t* delivery trucks of known capacity *Q*. VPR is a problem of a high computational complexity (NP-hard), so the large number of approaches focusing on using heuristics and constrained satisfaction techniques. An integration of ontology management and constraint satisfaction solving VPR is presented in the paper.

The paper is organized as follows. Section 2 elucidates the knowledge logistics concerned with the ontology approach. Section 3 presents the KF agent design, implementation and its decisive role in the integration of ontology management and constraint satisfaction. Section 4 describes a case study. Main features of the system "KSNet" due to using the KF agent are presented in the conclusion.

2. ONTOLOGY-DRIVEN KNOWLEDGE LOGISTICS

Knowledge logistics addresses the problem of acquisition of the right knowledge from distributed sources, its integration and transfer to the right person within the right context, at the right time, for the right purpose. This problem in the approach is considered as a configuration of network that includes end-users, loosely coupled knowledge sources, and a set of tools and methods for knowledge processing located in e-business environment. Such network of loosely coupled sources was referred to as knowledge source network or "KSNet".

The application of intelligent agents representing their knowledge via ontologies (Weiss, 2000) was motivated by the need of knowledge logistics systems for flexibility, scalability, and customizability. The multiagent system architecture based on the FIPA Reference Model (FIPA, 2004) was chosen as a technological basis for the definition of agents' properties and functions since it provides standards for heterogeneous interacting agents and agent-based systems, and specifies ontologies and negotiation protocols.

The formalism of object-oriented constraint networks (Smirnov et al., 2003a) was chosen as the abstract model for ontology representation. According to the formalism knowledge can be described by classes, attributes, domains, constraints, and methods. This perspective of knowledge representation correlates well with the semantic metadata representation concept being developed under the Semantic Web project (Semantic Web, 2004).

The thorough comparison of multi-agent systems (KRAFT, InfoSleuth) designed for knowledge fusion operations vs. the system "KSNet" has been done in previous works (Smirnov et al., 2001). The system "KSNet" has a distributed multiagent architecture (Smirnov et al., 2002). Components of the system can be allocated at different hosts and connected via TCP/IP protocol. Users and experts can work with the system via a Web-based interface.

3. KNOWLEDGE FUSION AGENT

The KF agent is responsible for KF operations. It is shown below that KF agent has a distributed nature and uses a constraint satisfaction technology.

3.1 Solver Middleware Design

The system uses ontologies for user request processing. A user request in free form defines both the problem statement and what data has to be retrieved from ontology library (OL) and from knowledge sources (KSs). Thereby the problem statement is changed from one request to another. The "on-the-fly" compilation mechanism in combination with ILOG (ILOG, 2004) is proposed to solve these varying problems. In a rough outline this "on-the-fly" compilation mechanism is based on the following concepts (Figure 1):

- a pre-processed user request defines (1) which ontologies are to be extracted from an ontology library (OL), and (2) which KSs are to be used;
- C++ code is generated on the basis of information extracted from (1) the user request (goal, goal objects, etc.), (2) appropriate ontologies (classes, attributes, and constraints), and (3) suitable KSs;
- the compilation is performed in an environment of the prepared in advance C++ project;
- failed compilations/executions do not fail the system work in whole; an appropriate message for the user is generated.

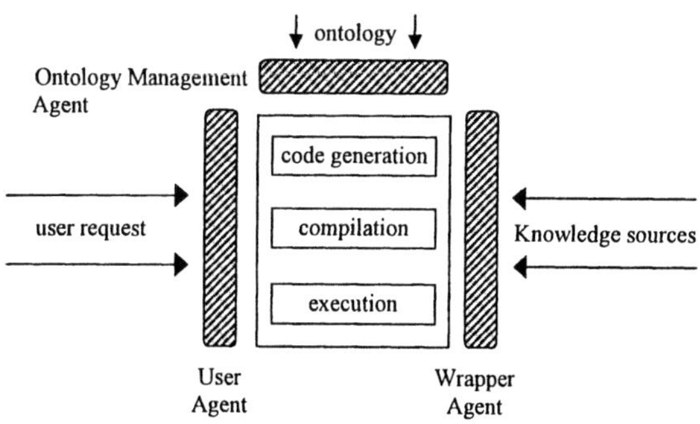

Figure 1 – Solver "on-the-fly" compilation mechanism

3.2 Solver Middleware Implementation

The KF agent is directly coupled with the solver. The described earlier mechanism of the "on-the-fly" compilation assumes that the KF agent gathers required data from other the system "KSNet" agents (translation agent, ontology management agent, configuration agent, and wrappers), generates a solver (performs the "on-the-fly" compilation), and launches the solver to generate a solution. In more detailed fashion this particular task performed by the KF agent requires: (i) a pre-processed user request prepared by the translation agent; (ii) an appropriate ontology demanded from the OL found and extracted by the ontology management agent; (iii) knowledge received from KSs by wrappers.

ILOG Solver has been chosen as a constraint solver because it has the following valuable features: (i) it is based around some C++ libraries; (ii) it has functionalities for managing constraints; (iii) it has solver; (iv) user defines solving process; (v) user defines constraints propagation mechanisms.

The KF agent and ILOG Solver interact via the XML-RPC (XML-RPC, 2004) protocol (remote procedure calling via HTTP as a transport and XML as an encoding). XML-RPC is designed to be as simple as possible, while allowing complex data structures to be transmitted, processed and returned.

XmlRpc++ library (XmlRpc++, 2004) is used as an implementation of the XmlRpc protocol written in C++. XmlRpc++ is designed to make it easy to incorporate XmlRpc client and server support into C++ applications.

The agent performs KF operations based on Application Ontology (AO) and knowledge acquired from KSs. The implementation of the KF agent uses such fundamental ideas of programming languages as object-oriented approach and logic programming. ILOG Configurator (ILOG, 2004) was chosen as a generic tool for object-oriented constraint programming. It provides a library of re-usable and maintainable C++ classes. These classes define objects in the application domain in a natural and intuitively way so that it is possible clearly distinguish the problem *representation* from the problem *resolution*. Therefore, if a problem statement changes then it is not necessary to rewrite the entire code as in case of "pure" C++. In the given case the problem statement is defined by data retrieved from OL, therefore the problem of minimal code modification, fast and error-free is the important task here.

The essence of the proposed "on-the-fly" compilation mechanism is to write AO elements (classes, attributes, constraints) to a C++ file directly. The KF agent creates a C++ file based on these data and inserts program source code into a program (Microsoft Visual Studio project) prepared in advance. The program is compiled in order to create an executable file in the form of dynamic-link library (DLL). After that the KF agent calls a function from DLL to solve the task. The UML sequence diagram (Figure 2) shows the KF agent's scenario at the stages of knowledge obtaining, solver compilation and execution.

The KF agent uses the mentioned technique of the dynamic code generation (with ILOG Configurator commands embedded) to produce a solution set satisfying requirements of the user request and AO elements (classes, attributes, constraints, etc.).

The generated code (C++ file) consists of several parts:

- the ontology management agent passes a part of the program based on data from the OL;
- the wrapper passes a part of the program using local/remote KSs;
- the KF agent generates a part of the program based on user request processing as well as user requirements;
- a predefined part of code (unchangeable): strategy definition and automatic answer generation.

Thus the C++ file is created on the basis of a special template. This template allows researchers and developers to comprehend and realize in more explicit and well-defined form: (i) what information is needed to solve a task; (ii) which KSs are required; (iii) which agent is responsible for delivering particular specified information block.

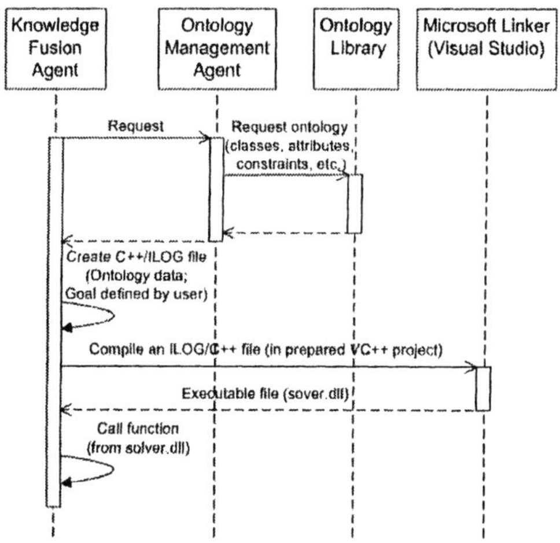

Figure 2 – The UML sequence diagram shows the KF agent scenario

4. CASE STUDY

A simplified version of the ontology describing a virtual production network used for the example is demonstrated in Figure 3. Based on this ontology the task ontology of the "configuration" task is built. The tasks "Staff definition" and "Cargo Allocation" are shown as dashed since they were out of the scope of the project. The other tasks were solved using ILOG (Solver, Configurator and Dispatcher) and specially written procedures (e.g., procedure that estimates impact of the weather on roads availability and transportation time).

The task of *allocation* is devoted to finding the most appropriate location for a distribution center considering such factors as location of the producers, nearby cities and towns, availability of facilities (e.g., communications), and decision maker's choice and priorities. This task was solved by ILOG Solver. The task of *capacity* and *bill-of-material* (BOM) definition was based on predefined information and specially created procedures taking into account such factors as demand forecast and others. The objective of the task of *resource allocation* is to define the most appropriate suppliers taking into account their capacities, prices, transportation costs, availability, etc. This task was solved using ILOG Configurator. The above tasks were described in detail in (Smirnov et al., 2003b; Smirnov et al., 2003c).

The task of *routing*, the paper concentrates on, is devoted to finding a pareto-optimal set of routes of delivery of products from selected suppliers considering such factors as communications facilities (e.g., locations of airports, roads, etc.), their conditions (e.g., good, damaged or destroyed roads), weather conditions (e.g., rains, storms, etc.) and decision maker's choice and priorities.

Presented example illustrates finding a routing plan for the same conditions but with different user preferences, namely: *minimize time; minimize time, then costs; minimize both time and costs; minimize costs, then time; minimize costs.* In (Figure 4) results for the case *minimize costs* are presented. For illustration of the results a map is generated that uses the following notations. Green dots are the cities of the region. The city with red edge (Aida) is the city where the dealer is located. The cities with blue edges are the cities where customers are located (Libar, Higgville, Ugwulu, Langford, Nedalla, Laki, Dado). Transportations routes are shown as lines. The grey lines are routes that are not used for transportation in the solution, the blue lines routes used for transportation, and the red lines are routes unavailable due to weather or for some other reasons. E.g., the routes through the city of Zaribe are not available because of the flooding. The colored trucks denote the routes of particular vehicles/vehicle groups.

Figure 5 represents a comparison of the routing plans created for different criteria. As it can be seen while importance of one of the parameters increases (e.g., importance for costs increases from left to right) the value of the parameter decreases (the red line with diamonds for the costs) and vice versa (the green line with squares for the time).

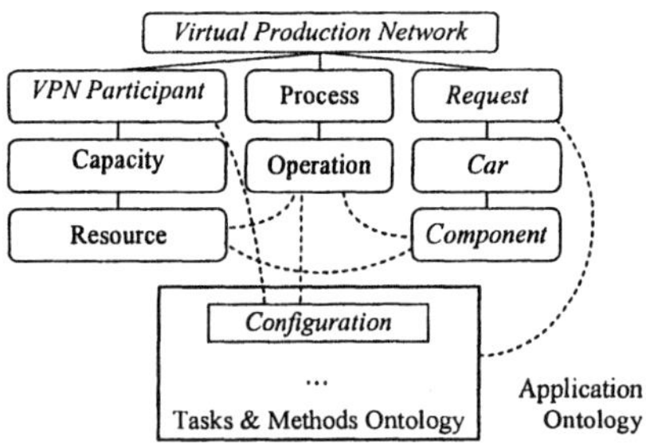

Figure 3 – Virtual Production Network ontology

5. CONCLUSION

The system "KSNet" where the presented KF agent plays one of the key roles has the following abilities: (i) ontology representation paradigm facilitates to process and understand natural language; (ii) ontology library based on the common vocabulary and notation can be considered as a dynamically created source of metaknowledge, (iii) user profiles and request ontologies support the personalization requirement; (iv) translation of ontologies from advanced formats)

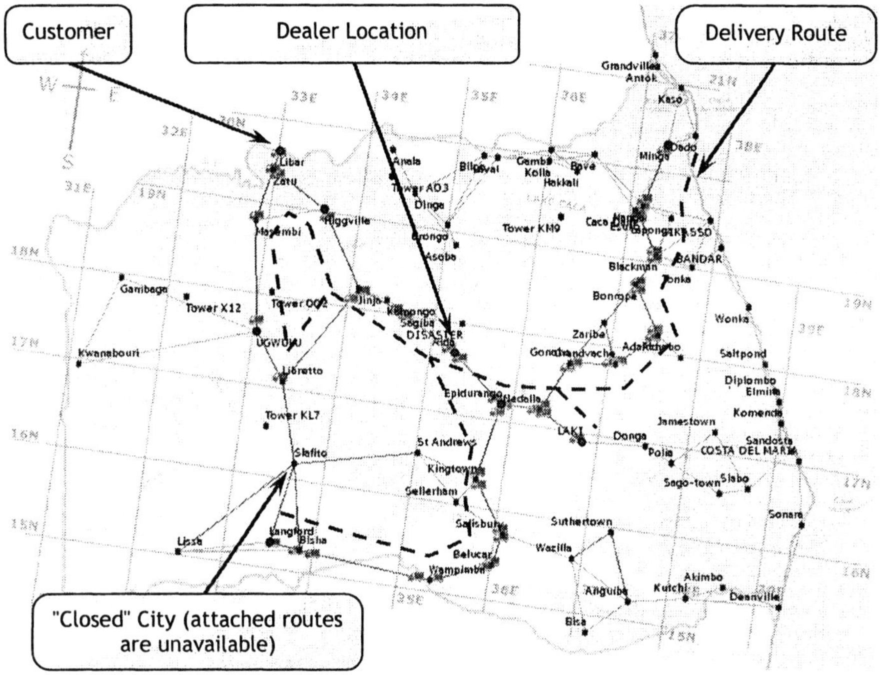

Figure 4 – Routing plan for the minimize costs preference (in this solution one vehicle/vehicle group is used to provide minimum of costs)

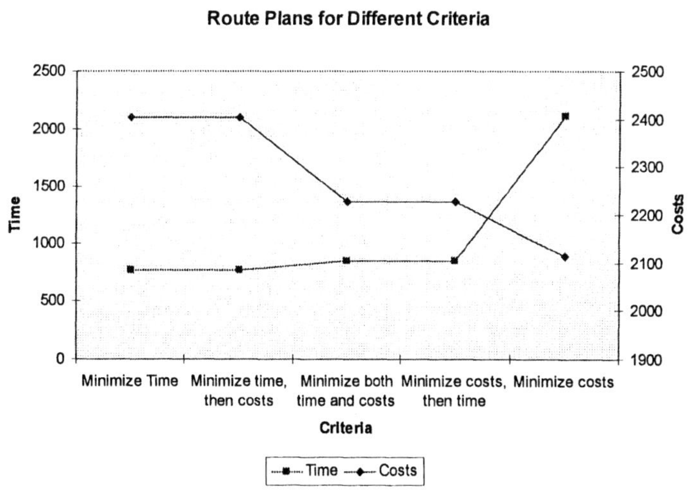

Figure 5 – Routing plans for different criteria (time and costs minimization preferences)

(e.g., DAML+OIL) into internal representation and out of it enables knowledge sharing and reuse; (v) the technology of constraint satisfaction & propagation enables to perform automatic reasoning on the Web.

Using the knowledge fusion agent based on the mechanism of the "on-the-fly" compilation allows generating new knowledge that is not available in existing knowledge sources independently on the application domain of the current problem. Here presented approach can be used for solving such real-world tasks as logisitics, configuration, management in various application domains of distributed nature including manufacturing, transportation, healthcare, etc.

4. REFERENCES

1. Bowman M., Tecuci G., and Ceruti M., Application of Disciple to Decision Making in Complex and Constrained Environments, Proceedings of the 2001 IEEE Systems, Man, and Cybernetics Conference 2001.
2. FIPA Foundation for Intelligent Physical Agents Documentation 2004. URL: http://www.fipa.org
3. Golm F., and Smirnov A. Virtual Production Network Configuration: ACS-approach and tools // Advances in Networked Enterprises: The Proceedings of the 4th IEEE/IFIP International Conference on Information Technology for Balanced Automation Systems in Production and Transportation (BASYS 2000). Berlin: Kluwer Academic Publishers; Bosten/Dordrecht/ London, 2000; 103 – 110.
4. ILOG Corporate Web-site, 2004. URL: http://www.ilog.com
5. Ruiz R., Maroto C. and Alcaraz J. A decision support system for a real vehicle routing problem, European Journal of Operational Research 2004; 153(3):593-606.
6. The Semantic Web Community Portal 2004; URL: http://www.semanticweb.org
7. Smirnov A., Pashkin M., Chilov N., & Levashova, T. Multi-Agent Architecture for Knowledge Fusion from Distributed Sources. Proceedings of 2nd International Workshop of Central and Eastern Europe on Multi-Agent Systems (CEEMAS'2001). Krakow, Poland, September 26-29, 2001; 403-412
8. Smirnov A., Pashkin M., Chilov N. & Levashova T. Agent-Based Knowledge Fusion in Scalable Information Environment: Major Principles and System Framework. Proceedings of the First International ICSC Congress on Autonomous Intelligent Systems (ICAIS'2002), ISBN: 3-906454-30-4, Geelong, Australia, 2002.
9. Smirnov A., Pashkin M., Chilov N., Levashova T. and Haritatos F. Knowledge Source Network Configuration Approach to Knowledge Logistics. International Journal of General Systems. Taylor & Francis Group 2003; 32(3): 251-269.
10. Smirnov, A. V., Pashkin, M. P., Chilov, N. G., Levashova, T. V. Agent-Based Support of Mass Customization for Corporate Knowledge Management. In: Engineering Applications on Artificial Intelligence 2003; 16(4): 349-364.
11. Smirnov, A. V., Pashkin, M. P., Chilov, N. G., Levashova, T. V. Knowledge Logistics in Information Grid Environment. The special issue "Semantic Grid and Knowledge Grid: The Next-Generation Web" (H. Zhuge, ed.) of International Journal on Future Generation Computer Systems 2003, 20 (1):61—79.
12. Tarantilis C.D., Diakoulaki D. and Kiranoudis C.T. Combination of geographical information system and efficient routing algorithms for real life distribution operations, European Journal of Operational Research 2004; 152(2) 437-453.
13. XML-RPC Web-site, 2004. URL: http://www.xmlrpc.org
14. XmlRpc++ Library Web-site, 2004. URL: http://xmlrpcpp.sourceforge.net
15. Weiss, G. (ed.): Multiagent Systems: a Modern Approach to Distributed Artificial Intelligence. The MIT Press, Cambridge, Massachusetts, London, 2000.

PART B

NETWORKED ENTERPRISES

23

INTELLIGENT AND DYNAMIC PLUGGING OF COMPONENTS – AN EXAMPLE FOR NETWORKED ENTERPRISES APPLICATIONS

Moisés L. Dutra; Ricardo J. Rabelo
Federal University of Santa Catarina, BRAZIL
moises@floripa.com.br; rabelo@das.ufsc.br

This paper presents an approach to minimize the problem of reduced functional flexibility in the complex industrial systems, where they are bought as a whole package or module, quite expensive, even though they are not used at all or do not fit the enterprise's needs completely. The approach is based on the idea of a dynamic and intelligent plugging of software components. This plugging will occur only when the components functionalities are effectively needed, adapted to the current computing environment in use. The plugging is made on demand, applying a new perspective to the Application Service Providers, under the form of a Federation of Application Providers.

Keywords: Components, Application Service Providers, Plugging on Demand, Functional Flexibility.

1. INTRODUCTION

Substantial investment on financial, technological and computing resources has been required from the companies to deal with the problem of increasing complexity of enterprise systems. This is even more problematic if it is taken into account that more than 90% of the companies are small ones, implying that most of the systems solutions that might leverage their competitiveness cannot be acquired.

This paper presents an approach to that problem, providing a model where the system to be used by a company is dynamically and intelligently built up and adapted at execution time according to the current user needs, having the system's *kernel* as the basis. The envisaged scenario is based on the systems paradigm where the user should work only with the necessary software functionalities, only when (s)he needs, at the necessary environment (Dutra et al., 2003).

In spite of some results achieved by a number of international initiatives / research projects towards larger systems functional flexibility, the situation can still be considered primary. The best that software vendors have been doing nowadays is to offer smaller, less cost and easier installable versions of their software (e.g. ERP systems) in the form of "*My system*", more adapted to the needs of a given company.

Actually, this business model does not solve the essential problem. Software modules remain with a high degree of granularity and they are made available just as instances of a wider system, i.e. companies have to buy them with their full subset of functionalities no matter what their effective needs are. This is a very important aspect as practice shows that most of the software functionalities are not used by the end-users. Therefore, if they do not use them, why should companies pay for them and waste computing resources to host them?

The proposed approach increases very much the functional flexibility of a system, where companies can use only the functionalities they need, when they need and at the required computing environment (PC, palm, etc.). The system is no longer developed as a monolithic system but rather as a set of small software *components*, independent and logically integrated, providing the full set of the system functionalities when put together. In this approach a new vision of Application Service Providers is given, transforming them into distributed *Application Providers* of components.

The model has been validated in a scenario of virtual enterprises, where the supporting tools used by its members to manage businesses can be adapted to members' current needs.

This paper is organized as follows: Section 2 stresses the main technologies used. Section 3 addresses the dynamic plugging technique. Section 4 presents the proposed model. Section 5 depicts the implemented prototype and preliminary results. Section 6 provides the main conclusions.

2. INVOLVED TECHNOLOGIES

This section depicts the three main technologies used as the basis for the proposed approach. These technologies have been chosen as they allow the construction of open, interoperable, adaptive systems, providing a larger system life cycle.

2.1 Components

A software component is a unit of composition with contractually specified interfaces and explicit context dependencies, which can be deployed independently and is subject to composition by third parties (Szyperski, 1997). A component-based development provides a more flexible approach than the traditional software development method, in which the system is designed globally by deploying and integrating small modules inside the same application.

Components can be designed to execute simple or complex tasks, with variable granularity, i.e., they can be implemented as simple functions or even as larger and complex modules. (Beneken et. al, 2003) see several advantages of using components: they can run in adaptive environments; can be exchanged or partly deployed partly without failure of the whole; can be written in different languages, using different technologies, in different operating systems; explicit interfaces allow to connect and decouple cooperating parts of the overall system.

2.2 Application Service Providers

According to (Dewire, 2002), an *Application Service Provider* (ASP) provides a contractual software-based service for hosting, managing, and providing access to an application from a centrally managed facility. For a certain periodically fee, the ASP provides content and other services for users connected through the Internet or any other network platform, and the users do not need to be concerned with software versions and upgrades. ASP provides access to applications that are located outside the client work environment. Several specialists believe that, with the appearance of the Internet, it would make more sense to provide software as a service than to sell it as a product "closed in a box" (Stardock Corporation, 2000).

Despite being a good model, it presents some relevant limitations when observed under the envisaged functional flexibility scenario: its processing is logically and physically centralized (the component is executed in the ASP); the granularity of its modules is very large; and they usually are not adapted at all to the client needs.

2.3 Peer-to-Peer

Peer-to-Peer (P2P) is an architecture where the resources and service sharing are made directly among the involved system peers, without the intervention of a central server (Parameswaran et al., 2001). The term "peer-to-peer" refers to a class of systems and applications that employ distributed resources to perform a function in a decentralized manner (Milojicic et al., 2002). Therefore, a P2P-based system is suitable to support large scale and geographically distributed / decentralized systems.

3. DYNAMIC PLUGGING OF COMPONENTS

In traditional approaches of component-based systems, the final system is "fully integrated" during design time. Each component is a "mini subsystem", which can be developed, deployed and tested separately. Its replacement by another component does not affect the global system operation, thus supporting some level of functional flexibility.

However, this flexibility is not as large as it could be. Firstly, because the traditional plugging of components is static and manually done. Secondly, because once plugged, the component remains in the same system even if it is no longer needed.

Some authors have made contributions in that direction, such as (Lauder, 1999) (Seiter et al., 1999), applying patterns for dynamic plugging. These patterns were based in generic frameworks to support the plugging (usually of an inherited class) in runtime. However, in the approach proposed in this paper, the dynamic plugging of components occurs transparently to the user, without any framework and on demand. The plugging is intelligent as it should adapt itself to the current computing environment (hardware and software), to the sources of download and to the type of components.

3.1 Designing Dynamic Components

Dynamic components need to be firstly adaptable to several types of hardware and operating systems, including PCs and mobile devices. As said before, the components granularity can vary substantially. The focus in this work is on components of small granularity in order to better fit the needs of the client system.

The two main component models are the *Enterprise Java Beans* (EJB) (Sun Microsystems, 2002) and the *CORBA Component Model* (CCM) (OMG, 2002). In both models a structure called *Container* is required to support the plugging and the components execution. A container provides supporting services for the component life cycle, transactions management, communication security, and events notification. In the dynamic model proposed in this paper, the container is no longer required. This provides a more agile transfer and component plugging, and it creates the basis for a solution independent of technologic, thus enlarging the system life cycle. The same direction has been followed by some international efforts (Agedis, 2003) (Adapt, 2004).

The communication between the application and the components (dynamically plugged in) can be carried out in several ways, e.g. by changing registered messages in the operating system, by *Application Programming Interfaces* (APIs) (Coach 2004), by *Dynamically Linked Libraries* (DLLs), by local components management (like COM) and distributed models (like CORBA) (Calim 2001) (Combine 2002), and simply by direct access (Liang et al., 1998), where the component functionalities are used directly, without the need of an integration middleware.

An application that enables dynamic plugging is composed of its core functionalities and "pluggable" areas where the components can be plugged in by means of their interfaces, which enables the communication between the component and the external world. The components' interfaces must be extremely well defined (parameters, generality and communication) so that plugging can be accomplished successfully.

4. PROPOSED APPROACH

In general, the proposed approach works as follows (Figure 1). The company has the software kernel, comprising its essential functionalities. When the client (user or groups) calls for a system option / functionality whose code is not presented in the kernel, a *requisition* for the associated code component is dispatched to the representative (*Coordinator*) of a central of components (*Federation of Application Providers - FAP*) that will search, over the Internet, for the most suitable repository (*Application Provider - AP*) that can supply that particular need / component. Once it has been found, a *peer-to-peer* communication is established between the application and the repository, and the component is sent out to the client application to be plugged in, in a transparent way, according to the requisition's specifications.

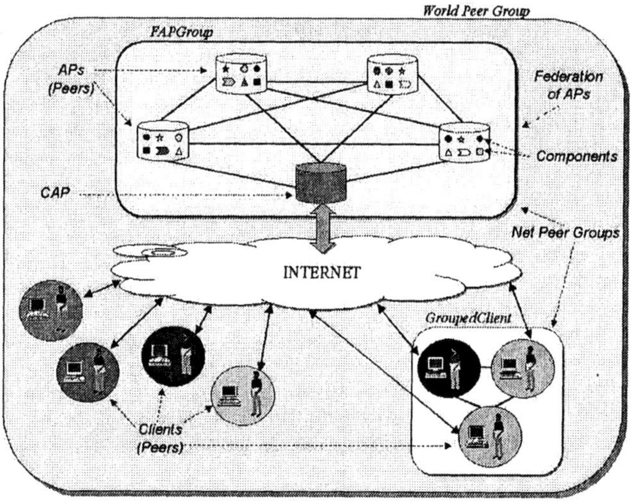

Figure 1 – Global Architecture

4.1 Application Provider (AP)

An AP is a repository of components. Unlike the traditional Application Service Providers (ASP), here the components dynamically plugged in run at the client's host, meaning that APs do not provide services but rather applications / components. Moreover, the proposed model is totally distributed, with the components coming from several APs located at different places, selected based on some decision criteria (e.g. geographic distance, bandwidth). Different versions / implementation models of the same component can be available in the APs, but all of them must follow the pre-designed component's interface.

4.2 Federation of Application Providers (FAP)

The FAP was introduced in the model to give scalability to the global architecture. It represents a cluster of APs from which the components should come. The FAP has a coordinator (*Coordinator of Application Provider - CAP*), which is the visible external entity to the FAP clients. The CAP is in charge of: i) seeking the AP which better matches the component advertisement; ii) creating a log file of all the received requests and the respective APs that were selected to supply the components; and iii) managing the components *contract.*

A contract is directly related to the business model involving the APs and the clients. For instance, a company can pay for the components a fixed monthly fee, or based on the number of components plugged in.

AP Structure

An AP has five cooperative modules (Figure 2). The first module is the *CAP Listener*, which receives component requests. Each request is checked by the *Specification Valuator*, which analyzes the request, validates it and searches for it in the *Component Repository*. The *Component Sender* makes the component transfer to the client, and the *Unsuccessful Message Sender* informs the client about the lack of the component.

CAP Structure

The CAP structure comprises six modules (Figure 3). The *Client Listener* receives the component requests from the clients. These requests are validated by the *Request Valuator*, which analyzes the received specification and verifies the client's contract terms. The *Component Advertisement Researcher* seeks the component's advertisement inside the FAP. The *FAP Listener* waits for the answer of the advertisement search, and the *Request Forwarder* redirects the request to the AP which has posted the advertisement. The *Unsuccessful Message Sender* will notify the client either if the FAP does not have the requested component advertisement or if the client's request was not approved by the request valuator.

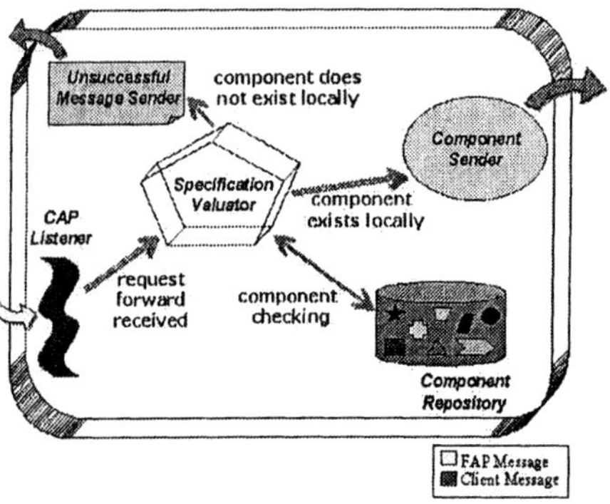

Figure 2 – AP Structure

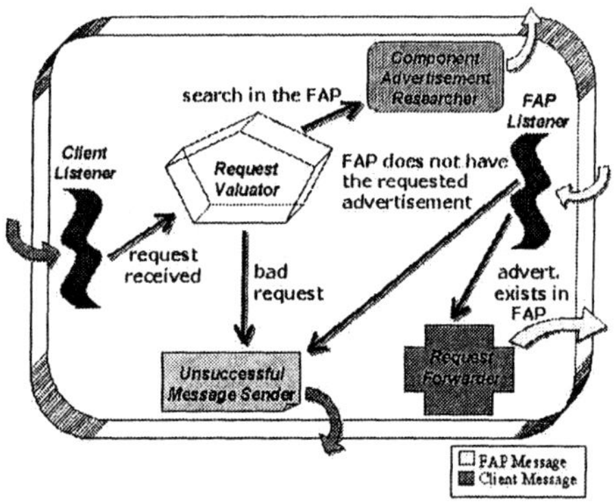

Figure 3 – CAP Structure

4.3 FAP Client

A FAP client is represented by a computer or a group of computers in a local network which hosts applications that request components from the FAP "server" (Figure 4). These applications in turn can be i) stand-alone; ii) distributed applications running either in a single computer or in several computers; and iii) the same application running its copies in several computers. For the cases *ii* and *iii* it is called *grouped client* (Figure 5) This allows computers to run more than one FAP-Client-application, developed in different languages and in different platforms, i.e. heterogeneous applications can request heterogeneous components no matter their languages and operating systems are. Grouped clients have just one contract with the FAP.

A FAP client has a module called *Component Management Module*, which manages the entire plugging process and that is composed of four sub-modules:

- *Component Fault Treater*: It acts whenever the system recognizes that the needed component is currently not present in the client. The treater then looks for the component in the local repository / cache. If it is found, a notification is sent to the *Component Plugger*. If not, the *Request Dispatcher* is called.
- *Request Dispatcher*: It builds the component request specification based on the client needs and environment characteristics, and sends the request to the FAP.

- *FAP Listener*: It waits for an answer from the FAP concerning the component that was requested. In the case of a positive answer, the *Listener* receives the message that encapsulates the component itself, stores it in the local cache, and calls the *Component Plugger*; otherwise a failure notification is sent to the application.
- *Component Plugger*: It performs the dynamic plugging itself. It loads the component from the cache so that the application can use it thereafter.

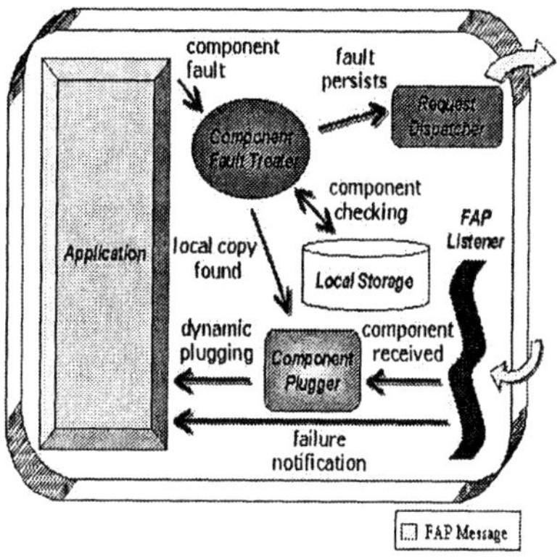

Figure 4 – FAP Client Structure

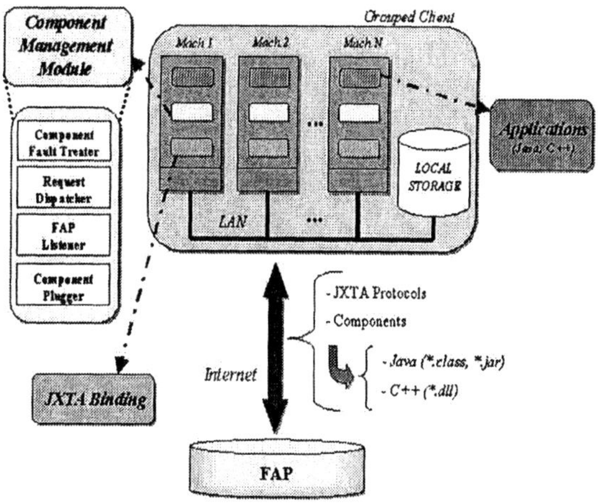

Figure 5 – Grouped Client

Each FAP client has a central local repository to cache the components already transferred. Once the components are used they can be later discarded from the application kernel (depending on the business model in use), but they are kept stored in the cache for future use, i.e. the client does not need to request it again to the FAP. This results in lower network latency and provides more agility to the global plugging process. The components can be transparently added to or removed from the application without interfering in the other components already plugged in. A component can be integrated with other (heterogeneous or not) components as well as with legacy ("re-engineered") systems. For the automatic component updating, the FAP client has a process that frequently asks the CAP about new versions. If a new version is found close to a certain AP, the same global plugging process is triggered again.

The approach proposed in this paper has some general similarities with the model proposed in (Camarinha-Matos et al., 2001) for Virtual Organizations, which is called *service federation*. It represents an approach to support the interoperation among heterogeneous, autonomous and geographically distributed entities. The services are available at service providers that publish their services in a catalogue that can be consulted by the users whenever they need and wherever they are. The service providers form a cluster, composing a federation. Each of those entities interested to provide services should announces them in a *catalogue* that will serve as the central source of information for clients. Once a given client selects a given service, the catalogue sends the service's interface to the client application so that a direct / remote service invocation can be carried out between the application and the service provider.

Both approaches, *service federation* and the one being proposed here (*federation of application providers - FAP*), involve distributed and autonomous clusters of providers and allow transparent access to what the user / client application requires, no matter where it is. Yet, both make use of a kind of central broker. However, the FAP approach presents some differences, namely:

i. FAP does not provide services, but system components.
ii. FAP does not require that providers are previously registered in the "broker" as the supporting platform that is used (JXTA – see section 5.1) is able to look for the components in the APs automatically.
iii. FAP does not provide the services' interfaces to the client application. It finds the most suitable AP for the required component, a transparent P2P connection is established between the AP and the application, the plugging process is carried out, and the "service" is executed locally.
iv. The FAP client application is the one which sees what is missing, therefore there is no human intervention.

5. PROTOTYPE

In order to test and to preliminary validate the proposed model, a prototype has been developed taking a virtual enterprise (VE) application into account. This application consists of a VE management system that provides several functionalities to the end-user (Rabelo et al., 2002). Applying the proposed FAP approach on that system meant to rethink it with the objective of defining what would be the system's kernel (i.e. the FAP client) and hence its "optional" functionalities (i.e. the pluggable components to be put available in APs).

The prototype was based on only one of the macro-functionalities of the system, called *Ad-hoc Reports*, which provides a number of managerial reports about a given VE (Figure 6). The user has several report options, such as the list of the VE members, the parts being produced, and the involved sales and shipping orders. The display of these options is executed by the ad-hoc's kernel. When the user selects, for instance, the report option *sales orders*, the system detects that this function is not there and requests the respective component to the CAP. After the whole plugging process is accomplished (see Section 4), the component is executed and then other graphical interfaces are shown, listing all the sales orders related to that given VE. In this simple case, the purpose of the component is to have access to the local database and to get those orders using SQL queries. It also has the purpose to provide the user with detailed data about each of these sales orders (from the database too), shown in the interface in the bottom of the figure 6.

It has to be noted that the way the component's graphical interfaces were shown (i.e. in HTML in that case) was specified (besides some other basic parameters) in the requisition for the component sent out by the FAP client regarding its computing environment and needs. For instance, another component with the *same* functionality but built up to run over another operating system (e.g. Linux) and non-web environment can exist in FAP. Therefore, the system does not need to have all possibilities to show the ad-hoc reports embedded in its kernel. Only the required possibility is (dynamically) linked to the kernel and exactly when it is needed, providing an effective functional flexibility.

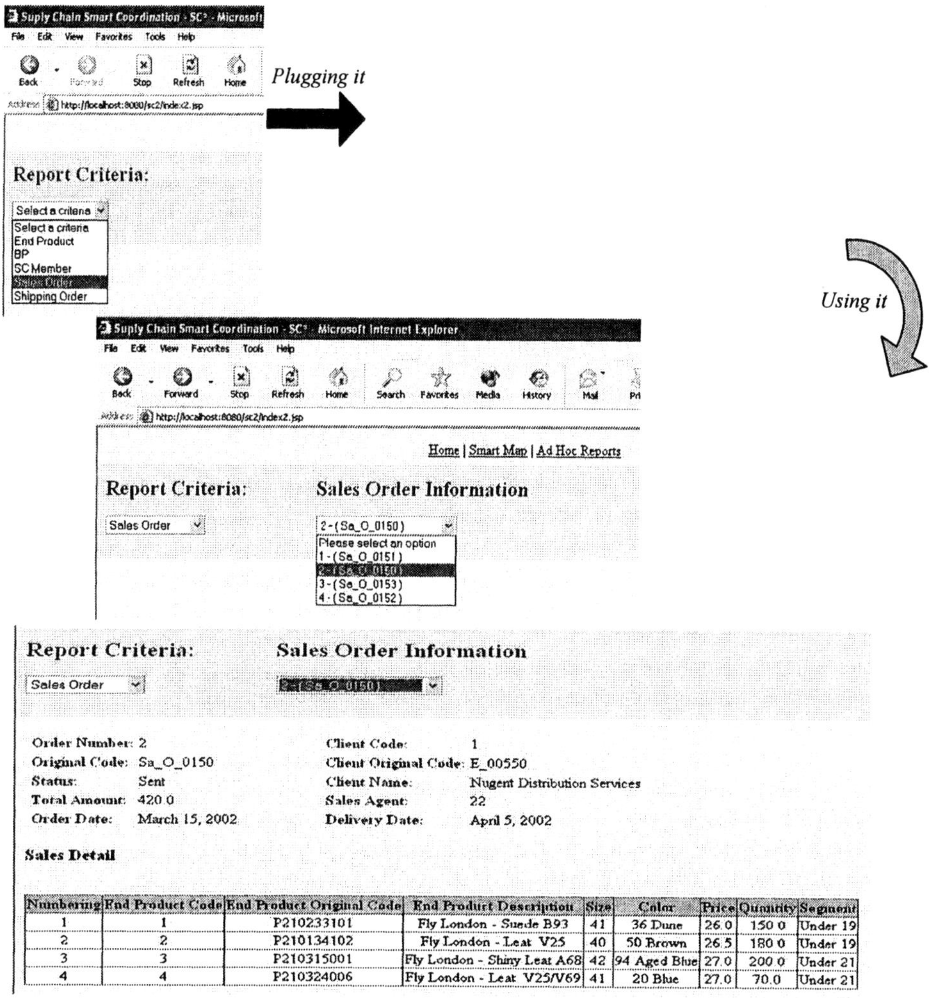

Figure 6 – Prototype Interface

The JXTA Model

In order to implement the envisaged decentralized / distributed model the JXTA peer-to-peer platform (JXTA Project) was chosen. JXTA is a very recent technology and it is constituted by a networked programming platform to cope with distributed computing and interoperable platforms, independent of operating systems and programming languages (Gong, 2002). JXTA has a set of communication protocols, each one containing one or more messages. This platform supports the several issues required in the proposed FAP model, namely location protocols, content search and transfer of data among peers, also including confidentiality, integrity, authentication, access control, auditing, cryptography and data security.

Every FAP client is seen as a JXTA peer, possessing a local JXTA Binding to enable the communication with the other peers. All peers belong to a group called *World Peer Group*, which is the logical reference of the JXTA structure. The peers can also belong to subgroups (*Net Peer Groups*). In the case of FAP a peer group was created, called *FAPGroup*. It means that every AP is part of this group. Yet, grouped clients (see section 4.3) also are represented as groups, belonging to the *World Peer Group*.

The APs' structures are implemented using the JXTA protocols. The communication process (messages and data transfer) has used the *Pipe Binding* and the *Peer Discovery* protocols. *Pipes* work as communication channels, and they can be of type *Pipe In* and *Pipe Out*. A *pipe in* is created to establish a communication and to wait for external connections, whereas a *pipe out* is used to locate some pipe and to connect with it. This localization is done using advertisements, which is the basic structure to announce the components specifications to the federation of APs. In the JXTA platform the advertisements are expressed in the XML standard.

This prototype has been developed using the following tools: Java SDK 1.4.2.02, JXTA platform version 2.1.1, Borland database Interbase 6.0, webserver TomCat 5.0.16 and JDBC driver FireBirdSQL 1.0.0., in the Windows XP environment.

6. CONCLUSIONS AND NEXT STEPS

This paper presented an approach called Federation of Application Providers (FAP) as a more flexible alternative to the Application Service Providers (ASP) existing in the market. Using FAP an application can be seen as a building block system, where its functionalities – implemented as software components – are dynamically plugged into the system in the exact moment they are required, adapted to the current computing environment. This approach creates rooms for new business models as the components come from a group of distributed, interoperable and autonomous application providers, which supply components, not services. Moreover, this approach makes it possible for small companies to acquire modern software (e.g. ERPs), usually too costly and with many unnecessary functionalities.

The development of dynamic and multi-platform components, with rigorous specifications, has been seen as a prominent approach to maximize code reusability. More comprehensive scenarios can arise from when large repositories of generic components were built up, especially if they were based on reference models for processes, information and ontologies. The preliminary results reached with the developed prototype showed that the FAP model seems very promising in supporting systems functional flexibility.

The model is strongly based on the JXTA platform, which supports most of the system requirements in a generic peer-to-peer scenario.

Further research should involve a deeper reflection on business models and the contracts between the FAP and their clients, and on how these aspects should be connected with the dynamic plugging and unplugging processes (specially in Java-based components), without human intervention. Yet, a more complex application should be developed to comprise multi-language components and hence to evaluate their interoperability.

Acknowledgements

This work was partially supported by CNPq – The Brazilian Council for Research and Scientific Development and in the scope of the IFM project (www.ifm.org.br).

7. REFERENCES

1. Adapt, *Middleware Technologies for Adaptive and Composable Distributed Components* – http://adapt.ls.fi.upm.es/adapt.htm, in March 2004.
2. Agedis, *Automated Generation and Execution of Tests Suites for Distributed Component-based Software* – http://www.agedis.de/, in March 2004.
3. Beneken, G., Hammerschall, U., Broy, M., Cengarle, M. V., Jürjens, J., Rumpe, B., Schoenmakers, M, *Componentware – State of the Art 2003.* In Understanding Components Workshop of the CUE Initiative at the Univerità Ca' Foscari di Venezia, Venice – Italy, October 7th-9th 2003.
4. Calim (2001) – *Corba Architecture for Legacy Integration and Migration* – http://dbs.cordis.lu/fep-cgi/srchidadb?ACTION=D&SESSION=118262004-3-5&DOC=2&TBL=EN_PROJ&RCN=EP_RCN:53593&CALLER=IST_UNIFIEDSRCH, in March 2004.
5. Camarinha-Matos, L. M., Afsarmanesh, H., Kaletas, E., Cardoso, T. F. (2001), *Service Federation in Virtual Organizations, in Proceedings of IFIP TC5 / WG5.2 & WG5.3 Eleventh Int. PROLAMAT Conf. On Digital Enterprise – New Challenges*, Kluwer Academic Publishers, pp 305-324, Hungary, 2001.
6. Coach (2004) – *Component Based Open Source Architecture for distributed Telecom Applications* – http://dbs.cordis.lu/fep-cgi/srchidadb?ACTION=D&SESSION=121472004-3-5&DOC=11&TBL=EN_PROJ&RCN=EP_RCN:61829&CALLER=IST_UNIFIEDSRCH, March 2004.
7. Combine Project (2002) – http://www.opengroup.org/combine/overview.htm , March 2004.
8. Dewire, D. T., *Application Service Providers - Enterprise Systems Integration,* 2nd Edition, pag.449-457. Auerbach Publications, 2002.
9. Dutra, M. L., Rabelo, R. J., *Dynamic Functional Instantiation of Industry Systems* [in Portuguese], in proceedings of VI Brazilian Symposium of Intelligent Automation, Brazil, September 2003.
10. Gong, L., *Project JXTA: A Technology Overview* – Sun Microsystems, Inc., October 29, 2002.
11. JXTA Project - http://www.jxta.org/ – in February 2004.
12. Lauder, A., C++ Report Magazine, *Pluggable factory in Pratice*, pp. 27-32, v. 11, n. 9, Oct 1999.
13. Liang, S., Bracha, G. (1998), *Dynamic Class Loading in the Java Virtual Machine. Proceedings OOPSLA '98*, Vancouver, Canada, October, 1998.
14. Milojicic, D., Kalogeraki, V., Lukose, R., Nagaraja, K., Pruyne, J., Richard, B., Rollins, S., Xu, Z., *Peer-to-Peer Computing* , Technical Report HPL-2002-57, HP Labs. 2002
15. Object Management Group, CORBA Componentes – formal/02-06-65.2002.
16. Parameswaran, M., Susarla, A., Whinston, A. B. (2001), *P2P Networking: An Information-Sharing Alternative* – IEEE Computer Society's – Computing Practices, pag. 31, July 2001.
17. Rabelo, R. J.; Klen, A. P.; Klen, E. R., A Multi-agent System for Smart Coordination of Dynamic Supply Chains, Proceedings PRO-VE'2002, pp. xx-yy, 2002.

18. Seiter, L., Mezini, M., Lieberherr, K., *Dynamic component gluing. In Ulrich Eisenegger, editor, First International Symposium on Generative and ComponentBased Software Engineering*, Springer , 1999.
19. Stardock Corporation (2000), *ASPs – A Primer – April/2000 –* http://www.stardock.net/media/asp_primer.html – in February 2004.
20. Sun Microsystems, *Enterprise JavaBeans Spec. version 2.1*, 2002.
21. Szyperski, C., *Component Software: Beyond Object-oriented Programming*, Addison-Wesley, 1997.

24

WEB SERVICES / AGENT-BASED MODEL FOR INTER-ENTERPRISE COLLABORATION

Akbar Siami Namin and Weiming Shen
National Research Council of Canada
{akbar.siami, weiming.shen}@nrc-cnrc.gc.ca

Hamada Ghenniwa
The University of Western Ontario
hghenniwa@eng.uwo.ca
CANADA

Web Services technology is a promising computing paradigm for applications integration over the Internet. Use of Web services and related technologies facilitates the implementation of virtual enterprises across heterogeneous hardware and software platforms. This paper proposes a Web services / agent-based model for inter-enterprise collaboration. It presents a multi-agent model in different levels of the enterprise's system architecture to accomplish a suitable selection of a registered service, to check the status of a process, to realize users' requests, and to react to them in a collaborative way with other agent-based Web services. Moreover, the paper proposes a multi-agent model to define a dynamic workflow capable of coordinating and monitoring the workflow processes.

1. INTRODUCTION

Global competition has forced manufacturing enterprises, particularly SMEs (Small and Medium-sized Enterprises), to increase their productivity and profitability through optimal resource utilization. On the other hand, changing customer demands and manufacturing environments make resource utilization more and more unpredictable and unstable. Conventional congregations of enterprises operating together try to solve the above problem by production outsourcing to achieve maximal group benefits. The advantages of Virtual Enterprises enabled by information and communication technology provide new ways to facilitate inter-enterprise manufacturing resource sharing, and therefore improve the profitability of SMEs (Camarinha-Matos and Afsarmanesh, 1999). However, implementation of virtual enterprises is not an easy job, since it is usually related the integration of hardware and software environments as well as serious privacy and security concerns.

Web services technology is a promising computing paradigm for integrating legacy applications over the Internet. Using Web services and the related standards facilitates the implementation of virtual enterprises in heterogeneous hardware and software platforms. It can also address well the privacy and security issues. In particular, it has following important advantages:

- Platform Independency: Web services technology provides communication among participants at the application level. Enterprises will be able to maintain their heterogonous legacy systems while seamlessly sharing these resources in a virtual homogeneous environment;
- Loosely Coupled Components: Web services can be modified, replaced, and removed with minimum or without affecting on the collaboration;
- Service Registry: Enterprises are able to advertise their services effectively through "*Universal Description, Discovery, and Integration* (UDDI)" registry. It also enhances both the creation and evolution levels of a virtual enterprise's life cycle;
- System Modularity/Reusability: Methods and data in a Web service can be reused in several applications regardless of their platforms.

However, Web services do not provide a complete solution for virtual enterprises or enterprise collaborations. In particular, it does not support fully automatic and dynamic collaborations among enterprises.

Software agents have emerged as a promising technology for dealing with cooperation and decision-making in distributed applications. Agent based manufacturing has become a new paradigm for next generation manufacturing systems, together with other manufacturing paradigms such as Holonic Manufacturing Systems, Agile Manufacturing, Reconfigurable Manufacturing, etc. (Shen et al., 2001).

Software agents have been developed with sophisticated interaction patterns. They are efficient in enforcing automatic and dynamic collaborations. Agent-orientation is an appropriate design paradigm for e-Business systems with complex and distributed transactions, especially for Web services. In services realization, software agents are very instrumental to provide a focused and cohesive set of active service capabilities (Li et al., 2004)

Software agents can be considered as an appropriate paradigm to overcome some shortcomings of Web services for enterprise collaborations:

- A Web service is just a self-describing software component such that it does not have enough knowledge about its environment, users, software components, and outside world in general. In contrast, software agents are capable of reasoning, and interacting with other entities;
- Web services are discoverable by XML-based UDDI standard. Current standard of the UDDI is only able to recognize terms "syntactically". The main challenge in service discovery is how to find services, which are "semantically" the same as clients' desires. Software agents operate at the knowledge level, at which they are able to reason semantically on the service requesters.

This paper proposes a Web services / agent based model for collaborative virtual enterprises. We discuss the integration of software agents and Web services as a suitable solution for setting up a virtual enterprise. The rest of the paper is structured as follows. Section 2 reviews the related work. Sections 3, 4, and 5 present a Web services / agent-based architecture, enterprise model, and UDDI model for inter-enterprise collaboration, respectively. Section 6 discusses some implementation issues of developed prototype. Section 7 provides some conclusions and discusses the future work.

2. RELATED WORK

Virtual enterprises based on software agents and Web services have a growing appeal. There have been significant research efforts to integrate software agents in a VE (Rabelo et al., 2001; Marik and Pechoucek, 2003).

The use of agents in Workflow Management Systems (WfMS), has been discussed in several papers (Yan et al., 2001). In an approach presented by Chang and Scott (1996), each workflow has been represented by multiple agents as personal, actor, and authorization agents. These agents perform actions on behalf of the workflow participants and facilitate interaction with other participants or organizations.

In ADEPT (Jennings et al., 1996), the multi-agent architecture consists of a number of autonomous agencies. A single agency consists of a set of subsidiary agencies, which are controlled by one responsible agent. Each agent is able to perform one or more services. None of these attempts adopts agent technology to compose workflow execution engine dynamically. The logic for workflow processing is hard-coded and thus it is hard to reuse the workflow execution engine for other business process.

The "Shadow Board Agent" architecture by orchestrating multiple Web services in an agent based transaction model was proposed in (Jin and Goschnick, 2003). Each participant is wrapped as a Web service and uses an agent-oriented approach to engineer each Web service as a software agent. Despite using agent-based architecture, the model lacks initial defining of workflows, monitoring mechanism, and appropriate rating mechanism.

Interleaving Web services composition and execution, using software agents and delegation have been discussed in (Maamar et al., 2003). Although the approach was based on software agent for Web services composition, however the model mostly focuses on the selection of services involved in the composition rather than composition itself.

A conceptual model for Web service reputation has been proposed in (Maximilein and Singh, 2001). In this model, There has been defined a "*Web Service Agent Proxy* (WSAP)" to access each service. A WSAP is an agent that acts as a proxy for clients of Web services.

3. A WEB SERVICES / AGENT-BASED MODEL

In this section, we propose a service oriented / multi-agent based model for virtual enterprises. We focus on integrating Web services and software agents inside the internal structure of a typical enterprise as well as adopting software agents inside the UDDI registry. On the enterprise side, we propose a goal-based model to define dynamic workflows. Also we introduce other agent-based components for coordination and monitoring purposes. On the UDDI side, we introduce some agent-based components to assist service requesters towards choosing the most suitable service provider. The proposed architecture is depicted in Figure 1.

On the enterprise side, A "*Proxy Agents Layer (PAL)*" is defined, which is sited on the top of other software units. PAL acts as an interface between the enterprise and outside world. A corresponding interface is defined for the UDDI server as well. The interface is named as "*Discovery Agent Layer (DAL)*", which is sited on the top of the UDDI server.

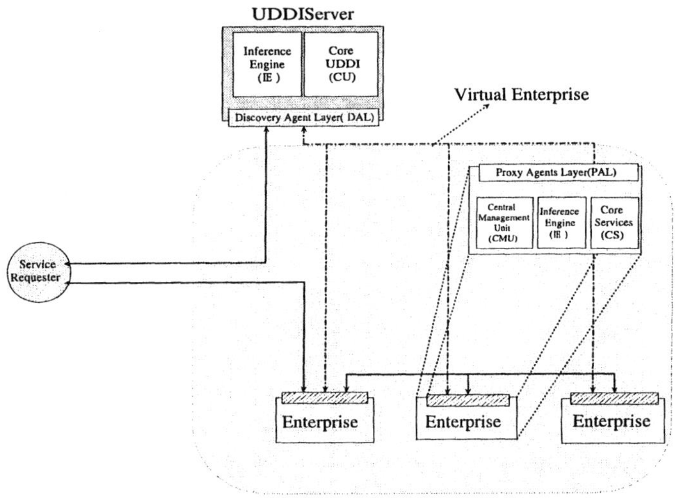

Figure 1 – A Web services/multi-agents based model for enterprise collaboration

As a simple scenario, a service requester (as a client), asks the UDDI server through the "Discovery Agent Layer (DAL)" to assist in locating a "suitable" and "reliable" service provider. Consequently DAL, in collaboration with some other components that have been defined in the UDDI server, presents the "recommended" service providers to the requester. Then, the requester, by choosing a suitable service provider, interacts with the provider (enterprise) through its "Proxy Agents Layer (PAL)".

4. ENTERPRISE MODEL

The detailed model of a virtual enterprise is depicted in Figure 2. All interactions between an enterprise and outside world are carried out by the "Proxy Agent Layer (PAL)". The "*Core Services (CS)*" unit consists of two components: Web services and corresponding database. The definition and functionalities of the Core Services (CS) are the same as current paradigm of Web services including the enterprise database. The other two units, namely, "*Central Management Unit (CMU)*" and "*Inference Engine (IE)*" are responsible for defining, managing workflows and knowledge representation. The agentified Web services will run on the service provider's side and do not affect the traffic of the network. The defined local agents communicate with the "Proxy Agents Layer (PAL)" to find out the existing Web services. Hence, among defined local agents, only PAL and Ontology Agent (OA) have knowledge about the Web services. The knowledge is acquired through communicating with the Web services.

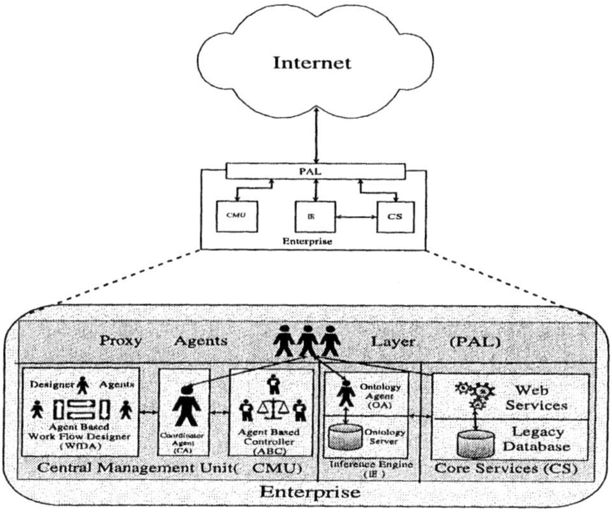

Figure 2 – A Web services / agent-oriented architecture for an enterprise

4.1 Central Management Unit (CMU)

Central Management Unit (CMU) is an "organizer" component with direct contact with the PAL. The CMU is responsible for:

- Accepting requests from PAL to design workflows;
- Informing PAL to locate suitable service providers and delegate the tasks of workflows to them;
- Assigning the tasks to the qualified enterprises;
- Coordinating the workflow amongst enterprises;
- Monitoring the progress and status of the delegated tasks.

CMU is responsible for most of the activities required at the creation and evaluation stage of a typical virtual enterprise's life cycle. These activities include designing workflows, partner searching, agreement, and monitoring the status of the delegated tasks.

4.1.1 Workflow Designer Agents (WfDA)

A workflow can be considered as a goal or set of goals. In other words, a service consumer is concerned about achieving the service; rather than "how" to gain it. To this extent, we let the defined agent-based units design the appropriate sub goals in runtime.

The most important activity of WfDA is to design a "cross-enterprise workflow specification" at runtime, by which it is possible to create a dynamic, not only in specification but also in assigning tasks to enterprises. The start point in WfDA is to achieve a request, or goal that has been asked by a service consumer interested in sharing resources as services and creating a virtual enterprise.

4.1.2 Coordination Agent (CA)
The Coordination Agent (CA) carries out all interactions between CMU and PAL. CA is responsible for coordinating and managing the processes of a workflow, which have been designed at WfDA. The main duties of CA are as follows:
- Communicating with PAL;
- Interacting with WfDA;
- Coordinating the involved enterprises;
- Communicating with Agent-Based Controller (ABC) by (1) informing ABC to create a corresponding controller agent in order to compare the plan and progress; and (2) receiving information from ABC, regarding the progress of the delegated tasks.

4.1.3 Agent-Based Controller (ABC)
In our model, after assigning each task to an individual enterprise, CA informs ABC to create a corresponding "controller agent" and monitor the progress and status of the underlined process, which is carried out by the enterprise. By getting a feedback from the enterprise, the controller agent compares the results of the enterprise's activities with the plan and informs CA to make a decision.

4.2 Inference Engine (IE)

In the current technology of Web services, requesters, by acquiring some meta-data from WSDL, realize how to exchange business data with the underlined Web services. These kinds of syntactically "invoking" services cannot cover "requesting" based on semantics. In fact, an enterprise needs an intelligent component such as an "*Ontology Agent (OA)*" to realize and discover the exchanged messages and map them to existing services. The structure of IE is similar to defined inference engine in UDDI server and we describe it in detail in the following sections.

4.3 Proxy Agents Layer (PAL)

The Proxy Agents Layer (PAL) can be considered as a complementary component for Web services technology in order to change a passive enterprise to a proactive entity capable of involving in transactions proactively. All communications between an enterprise and its outside world will go through the PAL. Moreover, PAL is responsible for exchanging data among internal components (CMU, IE, and CS) of an enterprise. From another point of view, PAL can be considered as a wrapper, by which the functionalities and complexities of the internal structure of an enterprise are encapsulated from outside visions.

PAL must be able to realize the format of current set of Web services standards such as: SOAP, WSDL, and UDDI. The main responsibilities of PAL include:
- Routing all incoming or outgoing messages to suitable software components either inside or outside the enterprise. The interaction can be:
  1) Receiving a request for an available existing service;
  2) Communicating with the Inference Engine (IE) in order to realize any ambiguity about the meaning of the used terminology;

3) Communicating with the Central Management Unit (CMU) to decompose the requested service.

- Contacting with the UDDI server in order to either look for an enterprise or report the quality of used services.

5. UDDI SERVER MODEL

The proposed architecture for a UDDI server is depicted in Figure 3. All interactions between a UDDI server and its outside world are carried out by the Discovery Agent Layer (DAL). DAL acts as an interface that accepts requests from outside world and analyzes them by collaborating with other defined internal components. In the following subsections, we describe each internal component in detail.

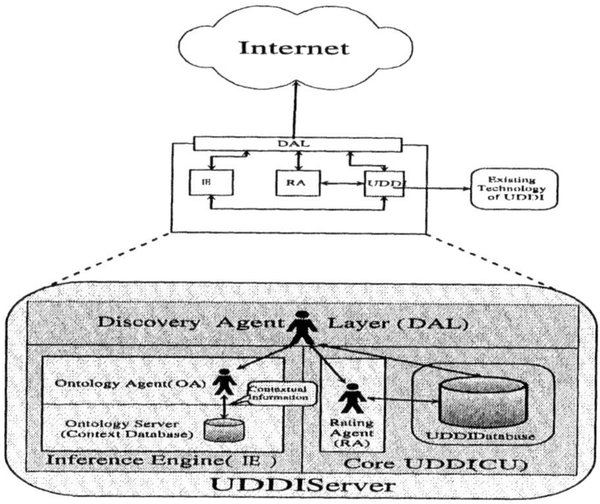

Figure 3 – An agent-oriented UDDI architecture

5.1 Discovery Agent Layer (DAL)

The Discovery Agent Layer (DAL) acts as an interface between the internal components of the UDDI server and the outside world. DAL can behave like a router to transfer incoming data to a suitable component located inside the UDDI server. DAL should be able to realize Web services standards such as UDDI, WSDL, and SOAP. The main responsibilities of DAL include:

- Routing incoming and outgoing messages to a suitable internal component or outside world;
- Exchanging information between the internal components of the UDDI server;
- Assisting the service requesters in finding suitable service providers by locating and suggesting the most qualified and matched services;
- Managing all received information regarding performances of an enterprise and updating the qualification of service providers' profiles.

5.2 Inference Engine (IE)

In our model, the ontology related components include an agent-based entity called "Ontology Agent (OA)" and a knowledge base component called "Ontology server", which is responsible for knowledge representation.

The Ontology Agent (OA), accepts the terminologies that have been used in the request, "interprets" them into a specific format, known as "context information", and submits the interpreted data to the ontology server to "process". The process on the context information is carried out by some "classifications" of some "context types".

Eventually, OA sends back to DAL the meaning of the terms with an understandable format to search at the Core UDDI.

5.3 Core UDDI (CU)

We consider two subunits inside the Core UD (CU): the UDDI-Database and the Rating Agent (RA). The UDDI database is the same as the current technology of the UDDI registry.

The RA provides some information in terms of qualification of services for assisting service requesters in choosing the most reliable enterprises. RA, by expanding the UDDI's knowledge of existing services, evaluates the quality of underlined services. RA updates the rating data by considering some measures such as availability, performance, reliability, and response time. The rating information is accessible for any service requester as public information.

6. PROTOTYPE IMPLEMENTATION

In order to prove the feasibility of the proposed model, a simple prototype has been implemented. The prototype is a simplified distributed system, which represents integration and resource sharing through a cooperative distributed system. It consists of following entities (Figure 4):

- Three enterprises as service providers capable of providing some services or resources;
- An agent based Web portal behaving as a gateway through which end users send their requests to the registered Web services.

The prototype is developed using popular Web programming tools and languages under the Windows NT/2000 environment. Java API for XML-based Remote Procedural Call known as JAX-RPC has been used to create and deploy all Web services.

As a simple scenario, there are three companies that possess some expensive machines offered as services to customers. A service requester, interested in using these machines through communicating with the Web portal, asks for the bids. Consequently, the Web portal, which has knowledge about the services and their providers, sends a "*Simple Object Access Protocol (SOAP)*" message ("*Call for Bids*") to enterprises. Thus, any service provider based on their facilities and availabilities proposes its bid to the Web portal. The Web portal, by analyzing the received bids, suggests the most suitable service to the customer.

Each service provider has its own databases, services, and configuration files, which protects the privacy of their internal resources. The underlined Web services are deployed on the Apache Tomcat Web container. The Web portal is created by Java Servlets, deployed on Apache Tomcat Web container as well. Figure 5 shows two snapshots of the implemented prototype.

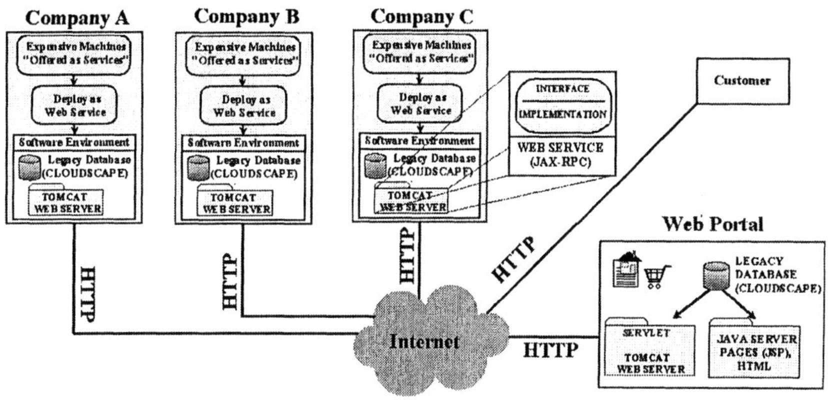

Figure 4 – The manufacturing resources sharing scenario

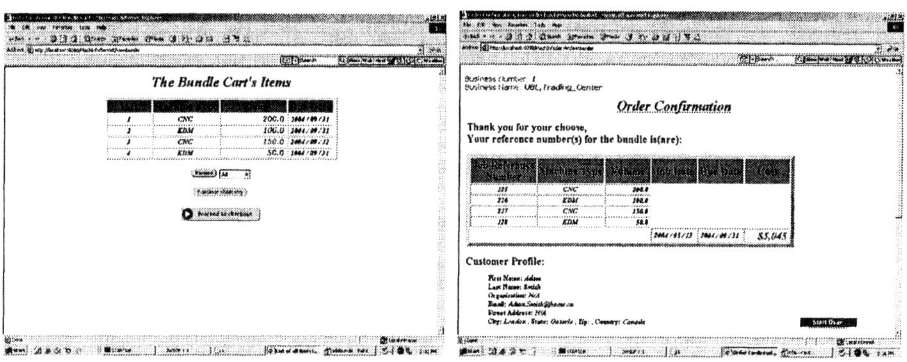

Figure 5 – Snapshots of the implemented prototype

7. CONCLUSION AND FUTURE CONTRIBUTION

In order to carry out customers' requests, enterprises need to collaborate with each other and share their resources. Enterprise collaboration, permanent or temporary, requests for a higher level of technology, which allows enterprises to integrate their applications regardless of platforms, data structures, or models. In this paper, we discuss the feasibility of using software agents and Web services technology for adopting into a collaborative environment. The paper proposes a Web services / agent-based approach towards setting up a distributed environment for inter-enterprise collaborations.

On the UDDI side, the paper defines some components such as Discovery Agent Layer (DAL) as an interface between the UDDI server and its outside world, Inference Engine (IE) responsible for realizing user's requests semantically, and

Core UDDI (CU), which consists of some components including the current technology of the UDDI.

On the enterprise side, some key components have been defined such as Central Management Unit (CMU) to define a dynamic workflow, Proxy Agents Layer (PAL) as an interface between the service and outside world, Inference Engine (IE) responsible for analyzing the incoming requests semantically, and Core Services (CS) including the current technology of Web services. From the implementation point of view regarding the complexity of workflows, the proposed model is believed to be feasible. The workflows are distributed among participants and consequently each participant takes its own responsibility of designing its delegated workflow and each workflow is viewed as a service that is offered by a provider.

As the future work the following challenges are to be addressed:

- Creating a Standard Ontology Server: In order to realize and offer services to end users, a Web service or the UDDI server should have a unified, standard, and complete ontology server, by which ontology agent interprets and reasons about the user requirements.
- Creating a Genetic Algorithm for Rating: The rating algorithm should be applicable for different domains and gives a standard and unified solution for different UDDI servers. Moreover, the algorithm should be dynamic enough to accept user-defined attributes for rating and taking them into account.
- Dealing with Security Issues: Without secure transactions, enterprises may not involve in any kind of business. The Proxy Agents Layer should be intelligent enough to prevent malicious requests to get inside the enterprise.

REFERENCES

1. Camarinha-Matos, L.M. and Afsarmanesh, H. Infrastructure for Virtual Enterprise, Kluwer Academic Publishers, Norwell, Massachusetts, 1999.
2. Chang, J. and Scott, C. "Agent-based workflow: TRP support Environment (TSE)", Computer Networks and ISDN Systems, pp 1501-1511, 1996.
3. Jennings, N. R., Faratin, P., Johnson, M.J., Norman, T.J., O'Brien, P. and Wiegand, M.E. "Agent-based business process management", International Journal of Cooperative Information Systems, pp. 105-130, 1996.
4. Jin, T. and Goschnick, S. "Utilizing Web Services in an Agent Based Transaction Model (ABT)", Workshop on Web Services and Agent-based Engineering, Melbourne, Australia, 2003.
5. Li, Y., Ghenniwa, H. and Shen, W. "Agent-Based Web Services Framework and Development Environment", to appear in Computational Intelligence, November 2004.
6. Maamar, A., Sheng, Q.Z. and Benatallah, B. "Interleaving Web Services Composition and Execution Using Software Agents and Delegation", AAMAS'03 Workshop on Web Services and Agent-based Engineering, 2003.
7. Marik, V. and Pechoucek, M., Agent Technology for Virtual Organization, in L. M. Camarinha-Matos, H. Afsarmanesh, (ed.), Processes and Foundations for Virtual Organizations, Kluwer Academic Publishers, pp. 243-252, 2003.
8. Maximilein, M. and Singh, M.P. "Reputation and Endorsement for Web Services", ACM SIGecom Exchange vol. 3, no. 1, pp. 24-31, 2001.
9. Rabelo, R.J., Afsarmanesh, H. and Camarinha-Matos, L.M. "Federated multi-agent scheduling in virtual enterprises", in Proceedings of PRO-VE'01, pp.145-156, 2001.
10. Shen, W., Norrie, D.H. and Barthes, J.P. Multi-Agent Systems for Concurrent Intelligent Design and Manufacturing, Taylor and Francis, London, UK, 2001.
11. Yan, Y., Maamar, Z. and Shen, W. "Integration of Workflow and Agent Technology for Business Process Management", Proc. of CSCWD 2001, London, Ontario, Canada, pp. 420-426, July 2001.

25

INTEROPERABILITY AMONG ITS SYSTEMS WITH ITS-IBUS FRAMEWORK

Osório, L., Barata M., Gonçalves C., Araújo P., Abrantes A., Jorge P.
ISEL, Instituto Superior de Engenharia de Lisboa, DEETC department, Lisboa, PORTUGAL
{lo, mmb, paraujo, aja, pmj}@isel.ipl.pt

Gomes, J. Sales, Jacquet G., Amador, A.
BRISA, Auto-estradas de Portugal, DID Innovation and Development Department, Carcavelos, PORTUGAL
{Jorge.Gomes, Gastao.Jacquet, António.Amador}@brisa.pt

This paper presents and discusses the extended Via-Verde business model from the point of view of the underlying requirements for the virtual business processes. The extension of Via-Verde concept to other services beyond the motorway toll collection has increased the number of independent companies involved. The complex networked scenario resulted on a proposal of an intelligent transport system – interoperability bus (ITS-IBus) (Gomes et. al. 2003) (Osorio, et. al. 2003-a) aiming to promote a generalized interoperability among heterogeneous (multi-vendor) technological subsystems. The ITS-IBus initiative has been developing since then, a reference implementation of a peer-to-peer service based framework with pluggable feature and a set of common agreed interfaces for coupling different technological systems. An important objective is to increase the quality of the offered services by establishing a flexible execution and coordination framework for the collaborative distributed business processes.

1. INTRODUCTION

There is a need for a continuous effort to promote interoperability among heterogeneous (multi-vendor) information and communication technology (ICT) systems. The fast evolution during the last three decades has contributed to a significant number of new products, some of them, unique solutions taking advantage of market opportunities. In several cases, costs reflect the low market scale if not "one-of-a-kind" solutions requiring extra-dedicated efforts. This leads to expensive management costs for a life-cycle solution justified by the lack of well-established technologies and methodologies (standards) and above all, a low reutilization level of components available in the enterprise.

This situation is common to scientific and technological areas in fast evolution processes, it happens until a generalized understanding of new concepts and technologies get somehow a consensus among the interested communities. In ICT area some important (ad hoc) normalization bodies like W3C, OMG, IETF and open source community like Linux, Apache, Java Community Process, has contributed with significant dynamics to the mentioned required consensus. Concepts like pervasiveness and ubiquity have their origins in the need for a generalized interoperability not only at technology level but also at process definition level. According to existing initiatives Brownsword (Brownsword, 2004) address interoperability from two perspectives: 1) establishment of enhanced software development engineering practices and 2) development of models like the North Atlantic Treaty Organization (NATO) C3 Technical Architecture (NC3TA) Reference Model for Interoperability (NMI). In our work, we are following the second perspective under a twofold approach. On the one hand individual systems supporting toll solutions are redesigned / (re)assembled based on open interfaces and on the other hand a service based model was adopted to offer flexible support to business distributed processes execution and coordination (Osório, 2003-a).

Two years ago, BRISA initiated this discussion when addressing new business challenges involving extended business models with the participation of different companies with complementary responsibilities. This inter company cooperation scenarios have established new requirements some from technological area and others from business process management domain. Considering a recurrent business level problem, associated with distributed information consistency, it was identified a need to move from file base information exchange to distributed business process integration. This distributed business process integration requires a different approach to the interoperability challenge among contributing technological subsystems as a consequence of the "disintegration" of monolithic systems into a group of specialized services "orchestrated" by some service specialized on business process execution based on its representation (Osório, 2003-a). Other strategies exist to redefine enterprise integration approaches considering reutilization of legacy systems associated to other more service or component oriented approaches like the programmatic integration servers as defined by Gartner (Pezzini, 2003).

Focusing on ITS domain, the interoperability among dedicated short-range communication systems (DSRC) have received several efforts, namely by European Committee for Standardization (CEN) through the technical committee TC-278 (Osório, 2003-b). Nevertheless, this area requires further investments considering the lack of interoperability among existing DSRC systems, situation that makes difficult to create a pan-European electronic fee collection system (EFC). Beyond different communication rates between the on board unit (OBU) and the road side equipment (RSE), low data rate (LDR) and a medium data rate (MDR) systems, the interfaces at fee collection application level presents minor differences considering message structure and the underlying technologies adopted by systems from different suppliers.

The ITS-IBus initiative aims to contribute to these interoperability obstacles by following an open initiative approach involving technological system suppliers and other end-user companies like Brisa. This paper discusses not only the underlying motivations for ITS-IBus but also the adopted strategy. The discussion, while centered on interoperability strategy and process modeling aspects aims also to contribute to a balanced approach between technology and process expertise domains. There is a need to develop technological frameworks able to offer the process domain experts a set of flexible tools proper to help them to develop solutions answering to the (collaborative) business needs. The workflow model and more recently works on web services choreography address coordination of long-running interactions between distributed components (Muehlen, 2004).

2. DISCUSSING ITS-IBUS MISSION

There is an open discussion about the right strategy to address the new emergent complex challenges created by the crescent cooperation among companies when sharing common business objectives. This challenge is grounded on another unsolved but rather old set of problems related with the intra-company integration. From eighties, a growing investment has been done to understand the company as a holistic system based on processes and underlying technologies including social and organizational aspects (Vernadat, 1996). Several company models have been developed to abstract the complex systems, people and relations, some guided by researchers in management and others originated in different technological areas from manufacturing, engineering and computer science. These efforts have contributed with different formalizations to the innovation processes. In a broad sense, these were consequence of a continuous innovation in computer and communication infrastructures. Companies have been facing a dilemma when moving in to new systems in most of the cases to support more holistic approaches where it is required a higher level of integration among company information systems. There is a clear trend to move from an "island" based enterprise ICT systems' organization or technology systems' integration to a more process oriented integration (Depke, 2002). This is a result from a paramount effort led by companies like Sun, IBM, Microsoft, Oracle and SAP to contribute for an integration of different but similar methodologies, technologies and tools. These efforts are contributing to unify approaches promoting enhanced level of reutilization and a clearer and competitive industry of software components assembled to generate enterprise applications/systems.

The ITS-IBus is aligned to this strategy considering that it aims to promote a clear set of interfaces to be adopted in an OEM model (original equipment manufacturer). The objective is to facilitate the plug (assembling) of systems from different producers into complex integrated systems overall contributing to the enterprise business processes. More than another middleware bus, ITS-IBus is an open initiative to promote a generalized interoperability among ITS ICT subsystems in order to develop holistic intelligent transport systems. One example is the challenge to establish interoperability among electronic fee collection systems (EFC) in motorways from different countries. To achieve this goal there is a number of interoperability challenges that need to be solved. On the one hand, the communication between OBUs from different suppliers must interoperate with RSE installed in the involved country's motorways or in other facilities within a toll. On the other hand, the involved organizations need to establish complex collaboration processes able to offer car drivers a virtual toll service. In order to discuss the ITS-IBus strategy in this extended scenario the involved companies and some relations among them are shown in (Figure 1). This is a simplified scenario from what we might see in a near future when all the motorways around the Europe implement similar solutions. In this scenario the questions are:

- What would be the global architecture of a pan-European EFC framework;
- Market strategy - Tier players from customers (ITS infrastructure users) to car embedded endpoint (OBU in the case of DSRC technology)
- Processes - Operators' and cooperative business processes and information models and management strategies
- Technology - Technological infrastructures, system components, quality of services and life cycle technology management.

Beyond the above dimensions, other not less important are not focused on this discussion like the social and organizational aspects.

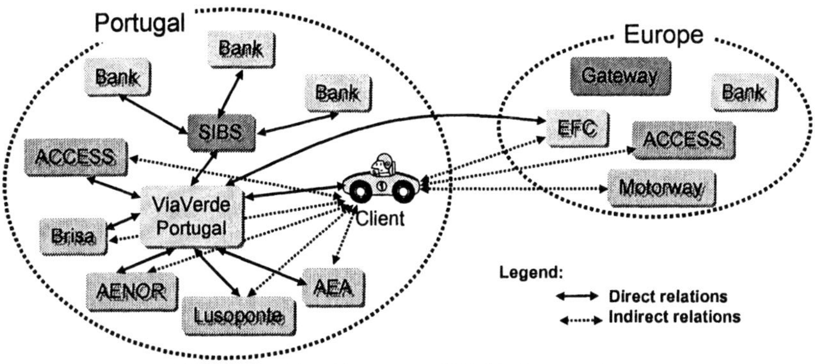

Figure 1 – Pan-European EFC infrastructure

The global architecture of a pan-European EFC framework is considered as a top-level goal to motivate the need for a holistic approach pointing to a tight collaboration among European ITS players. The second vector aims to discuss market strategy considering company arrangements according required services to commit global EFC framework. This involves players from ITS operators (Brisa, Aenor, AEA, Lusoponte), car parking infrastructures (Access), payment gateways (SIBS) and bank service providers, important to consolidate payment operation with ITS customers. In Portugal the car driver (Client) has a business relation with Via-Verde Company. The Via-Verde offers access services to different motorways, car parking and to other facilities (services) on the behalf of the various infrastructure operators (indirect relations). In the future, a European car driver might drive through European countries crossing motorways from different operators and parking in different areas without care about local payment. It will be invoiced in a month basis with a detailed report of all transactions. To achieve this there is a need to understand the collaboration model among the involved companies in order that each one can comply with a predefined and agreed quality of collaborative services. This involves different type of services from money transference involving payment gateways and banks, motorway toll transactions, car parking transactions from others. To achieve this purpose, the normalization forums are of paramount importance because they are contributing to standardize information models and more recently service interfaces based on the web services definition language (WSDL) and XML schema for the underlying data types involved. The interactive financial exchange (IFX) forum is an example of such an initiative to promote interoperability among payment service providers and users. Another forum, the open financial exchange (OFX) is also promoting interoperability through a specification for the electronic exchange of financial data using the Internet as communication platform. Nevertheless these and other initiatives are somehow divorced from others with similar objectives but in other application domains like the information exchange between business applications hosted by open application group (OAG), the object management group (OMG) for a more ICT oriented standards, from many others.

There is a lack of interoperability between technology and business areas partially explained by the fast evolution of ICT during the last decades. Even for ICT experts it is not easy to make decisions about the operating system to adopt (Linux or Windows), the runtime framework for distributed applications (Java/JVM or C#/CLR) from other more specialized frameworks, most of them presenting complex challenges when necessary to integrate them towards more holistic solutions. Even if the W3C XML framework that has emerged from the largest interoperable platform the World Wide Web (WWW or Web) has contributed to facilitate interoperability at process and technological levels, many obstacles exist.

Beyond this more technology oriented discussions, there is also a paramount effort to unify organizational (business) processes as a key measure to make possible enterprise collaboration. An important contribution is the process handbook under development at MIT as a comprehensive framework for organizing large amounts of useful knowledge about business (Malone, 2003).

Under these challenges, the ITS-IBus open initiative aims to contribute to promote interoperability at both technological and process level through a flexible and advanced interoperability framework based on existing open standards. The strategy is not to force a fast change from existing solution and well established system providers but rather involves them in the definition of common interfaces and promoting the reutilization of open frameworks able to assemble a diversity of systems to develop solutions easier to project, develop, maintain and evolve. These more technical requirements follows other more business ones those that must guide the design of a successful holistic technological infrastructure. Actually, the business models are the prime discussion in order to establish the underlying trusted business relations from which the collaborative relations are derived. The collaborative relations can be formalized through specific contracts that regulate all the collaborative processes specifying also the information exchange and auditing procedures.

As an example, we can consider a motorway operator responsible for an infrastructure with a number of electronic fee collection tolls. When a car crosses, a toll (entering or leaving it if in a closed infrastructure) the operator registers the transaction and, as soon as possible, delivers it, possibly in a lot, to the client's service provider. For the client's service provider, there is a need to access all the transactions (client's identification, a location and a time stamp). This information will support client's invoicing with a detailed report explaining each transaction. For conflict resolution, some evidences are necessary to consolidate the stored transactions even if in some circumstances it might be necessary to make a deeper information analysis considering an extended time window to detect exceptional client's patters. This is what happens with other services offered in a diversity of domains. There is a need for entities, the client and the service provider to establish a clear and trusted framework. To guarantee this it is also necessary to extend trustiness to all the intervening partners. If something goes wrong, the client claims to the service provider and he must receive from it all the answers to the open questions. Even if some information is missing it is important to consider that all the responsibility is of the service provider.

3. INTEROPERABILITY STRATEGY

There is a complex challenge which companies are facing when their business processes depend on multiple partners with competing business interests. On the one hand, the underlying business processes need to be unified or at least, some mapping needs to be established. On the other hand the technological framework considering all the ICT infrastructures from operating systems, middleware and networking systems, even if interoperability is being facilitated by some convergence achieved from a set of open initiatives, it continues to be a real problem when ICT life cycle management costs are considered. From different authors, the right strategy to cope with interoperability is not to promote radical substitutions but rather promote smooth transitions where legated systems are considered as valuable assets; at least until an accepted migration road map is established.

Therefore, the ITS-IBus open initiative is grounded on the following premises:

- Systems exist in different technological stages and forms, with focused objectives and other overlapping different application domains;
- Systems based on standard software or based on software and dedicated hardware establishes groups of heterogeneous systems overlapping or not relevant functionalities and characteristics.

We define a system as an ICT unit based on software or software and dedicated hardware able to work standalone or as an assembled component of a more complex system. This definition of system aims to unify in a broad sense to other concepts that fall in our system's definition. A system can be either an EIS (Enterprise Information System), an application, a software component, any piece of software or software and hardware able to be considered as a closed box with a clear behavior (outputs) under different environment conditions (inputs), (Figure 2).

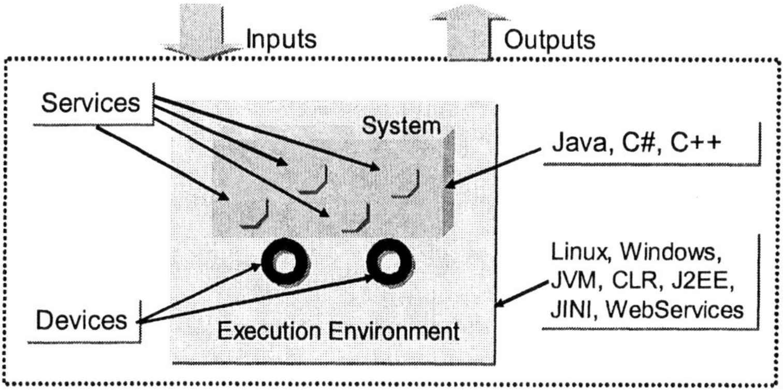

Figure 2 – General model of a company ICT system

The inputs can be configuration variable, information from sensors, user inputs, any information that might contribute to condition system's behaviour. The output can be any information that is needed by other system or user to realize some activity or task. In some extent, the proposal follows the service-oriented architecture (SOA) where all the execution units take the aspect of services with predefined contracts. In addition, in ITS-IBus the toll technological system at different levels are invited to adhere to such a service based framework. From our experience, there exist measurable advantages for both toll technology providers and motorway manager companies as end-users to adhere to a common interface. Furthermore, the strategy might be extended to a set of secondary business processes involving the collaboration with other companies. As an example, a motorway company might need to collaborate with a bank to make electronic payment cards debits to common customer accounts. Nowadays this collaboration is being done with a technology not adapted to the new collaboration needs mainly those related with time. It is not acceptable that some change in customer payment card like revocation or information update, take more time than a few minutes or at maximum a few hours to get information updated in all the tolls requiring such information.

In order to cope with distributed business requirements, ITS-IBus framework proposes a service-based environment where each functionality or group of functionalities are available through services. In the general architecture of a typical toll infrastructure (Figure 3) a lane management system (LMS) coordinates a group of systems through their services. A toll is managed by another system, the toll plaza management system (TPMS) responsible for all the operations management at toll level. The overall tolls are coordinated by a toll management system responsible for the supervision of the entire motorway technological infrastructure. All the systems implement specialized services plugged through an open interoperability bus made of a set of open frameworks offering reusable services like directory, publish and discovery, messaging, authentication and authoring, information security, persistency, user presentation facilities, from others.

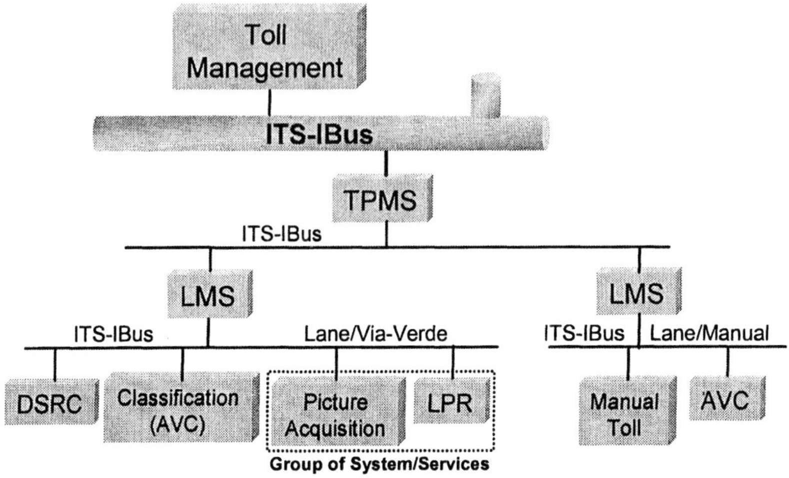

Figure 3 – General architecture of a toll technological infra-structure

There is an ongoing effort from Brisa and involving potential technology suppliers to promote an open framework and service interfaces as an open initiative. This might contribute to enforce a convergence of the presentations or interfaces for systems belonging to a same class. It does not make any more sense to admit that competitive factors are on systems' interfaces or in specific functionalities not available in any other competing system. The trend for companies, willing to adopt well established and proved technologies (some open technologies) under the crescent pressure to get more integrated (holistic) infrastructures, is contributing to support such necessary consensus.

As an example for a Via-Verde lane a DSRC collects car identifiers and according classification and a validation of contractual conditions the LMS decides for the collection of a picture of the car and the plate character string if available the LPR service. On a toll, the DSRC (Dedicated Short Range Communication) system is the component that is responsible for controlling the communications with the vehicles where an on board unit is installed to enable motorway client identification (Osorio, 2003). Both DSRC and LPR system services follow an open interface developed under ITS-IBus initiative. For the license plate recognition where there is a lack of standards applied to this problem, a research activity is being conducted to promote an open standard interface for an automatic vision system (Broggi, 2000). The main objective is to develop a special class of systems offering different automatic vision related services namely long-term traffic statistics (Abrantes, 2002), the detection of dangerous car maneuvers, license plate recognition (Chang, 2004), obstacle detection on high speed railway lines or in risky motorway or rout sectors.

To integrate all these dedicated systems the ITS-IBus adopts a peer-to-peer strategy considering that (business) processes are executed into systems that present services to other processes or service implementations (Figure 2). Everything requiring computational actions is represented by services presenting an open interface and following a set of rules that make them ITS-IBus enabled. The services are plugged into systems that behave like an execution containers presenting a minimal set of service to the ITS-IBus (Osorio, 2003). A system might implement at least the "plug and play" service in order to be identified by other ITS-IBus enabled peers.

A first prototype of a lane management system (LMS) and a toll plaza management system (TPMS) was developed. This first approach uses the JINI framework to create the service community. Following a previous work, the JXTA is also being evaluated to extend the service community to Internet domain getting advantage from the mechanisms offered by this framework to cross firewall and routers/NAT company barriers.

4. CONCLUSIONS

The ITS-IBus initiative involves different projects aiming to create the necessary consensus in different but complementary application domains. The initial focus was on toll technologies namely at those supporting the Via-Verde service model. In this area an effort was done to develop a peer to peer framework based on services implementing specialized functionalities of a toll infrastructure involving systems like a DSRC for car automatic identification and a LPR to automatic plate recognition as an enforcement system. Nevertheless, the extension of the Via-Verde model and services to other facilities like car parking and gas stations added new challenges namely those related with the involved companies that indirectly contribute to the Via-Verde pervasive services. These companies use different technological systems and processes organization what is contributing to a number of difficulties to establish an advanced collaborative model based on distributed business processes. In some areas like car parking infrastructures, there is a need to further normalization efforts in order to promote the adoption of open solutions and a shift to ITS-IBus peer service strategy. There are a number of monolithic and proprietary solutions to manage car-parking infrastructures that need to move to open solutions in order to get pluggable to the ITS-Ibus and to cope with Via-Verde business model requirements.

This requires the development of specialized systems as open components pluggable to an open infrastructure and able to support the execution of the (business) processes at different levels of the ICT framework. At company interoperability level, more efforts need to be done to establish a common understanding for related processes and an execution model able to cope with collaborative business model requirements.

5. ACKNOWLEDGMENTS

This work was partially supported by BRISA group, through the research and development BRISA / ITS-IBus and Brisa / Automatic Vision projects. The work is being developed by the research groups GIATSI and the Signals Processing groups from ISEL-DEETC in collaboration with DID/NID, the Innovation and Development Department of BRISA. We also acknowledge the valuable contributions from Miguel Machado, Bruno Basílio, Rui Lopes, Joaquim Pereira, Rui Gonçalves, Francisco Monte and Ricardo Cruz.

6. REFERENCES

1. Abrantes A., Marques J., Lemos J., Long Term Tracking Using Bayesian Networks, IEEE International Conference on Image Processing (ICIP), Rochester, 609-612, vol. III, S, 2002.
2. Beymer D., McLauchlan P., Coifman B., Malik J., A Real-time Computer Vision System for Measuring Traffic Parameters, In Proc. Computer Vision and Pattern Recognition (CVPR), 1997.
3. Broggi, A., Ikeuchi, K., Thorpe, C., Special issue on Vision Applications and Technology for (Broggi, 2000) - Intelligent Vehicles, IEEE Transactions on Intelligent Transportation Systems, June, 2000.
4. Brownsword Lisa L. Carney D. J., Fishe D., et al - Current Perspectives on Interoperability, Technical Report, CMU/SEI-2004-TR-009, 2004.
5. Chang, S., Chen, L., Chung, Y., Chen, S., Automatic License Plate Recognition, IEEE Transactions on Intelligent Transportation Systems, March 2004.
6. Depke R., Engels G., Langham M., Lütkemeier B., Thöne S. - Process-Oriented, Consistent Integration of Software Components, Proc. of the 26th Int. Computer Software and Applications Conference (COMPSAC) 2002, Oxford, UK, IEEE, Aug. 2002.
7. Gomes J. Sales, Jacquet G., Machado M, Osório A. Luís, Gonçalves C., Barata M. - An Open Integration Bus for EFC: The ITS IBus, in ASECAP2003, 18 - 21 May 2003 in Portoroz, Slovenia
8. Malone, T. W., Crowston, K. G., Herman, G. (Eds.) - What Is In the Process Handbook? Chapter in: Organizing Business Knowledge: The MIT Process Handbook. Cambridge, MA: MIT Press, September 2003.
9. Muehlen M., Nickerson, J.V.; Swenson, K.D.: Developing Web Services Choreography Standards – The Case of REST vs. SOAP. Decision Support Systems 37, Elsevier, North Holland, 2004.
10. Osório A. L., Abrantes A. J., Gonçalves J. C., Araújo A.; Miguel J. M., Jacquet, G. C.; Gomes, J. S. - Flexible and Plugged Peer Systems Integration to ITS-IBUS: the case of EFC and LPR Systems, PROVE'03 – 4th IFIP Working Conference on Virtual Enterprises, published by Kluwer Academic Publishers, ISBN: 1-4020-7638-X, pages 221-230, 2003-b.
11. Osório, A. Luís; Barata, M. Martins; Abrantes, A. Joaquim; Gomes, J. Sales; Jacquet, G. Costa - Underlying ITS Business Processes with Flexible and Plugged Peer Systems: the Open ITS IBus approach, PROVE'03 – 4th IFIP Working Conference on Virtual Enterprises, published by Kluwer Academic Publishers, ISBN: 1-4020-7638-X, pages 221-230, 2003-a.
12. Pezzini M. - Leveraging Legacy Assets: Web Services Through Integration, in the Gartner European Symposium, Fortezza da Basso Florence Italy, 10–12 March 2003.
13. Roy T. Fielding, Richard N. Taylor. Principle design of the modern Web architecture. ACM Transactions on Internet Technology (TOIT), pp. 115-150, May 2002.
14. Vernadat, F.B. - Enterprise Modelling and Integration: Principles and Applications. Chapman & Hall - London, 1996.2. Smith, Adam. "An Inquiry into the Nature and Causes of the Wealth of Nations". In Classics of Economics, Charles W. Needy, ed. Oak Park, IL: Moore Publishing, 1989.

26

ANALYSIS OF REQUIREMENTS FOR COLLABORATIVE SCIENTIFIC EXPERIMENTATION ENVIRONMENTS

Ersin C. Kaletas, Hamideh Afsarmanesh, L. O. Hertzberger
University of Amsterdam, Informatics Institute, THE NETHERLANDS
{kaletas, hamideh, bob}@science.uva.nl

Scientists face a number of challenges when performing their complex experiments. Collaborative Experimentation Environment (CEE) addressed in this paper is a support environment for scientific experimentations, with an emphasis on supporting joint multi-disciplinary projects and collaborations. In order for a support infrastructure to help scientists tackle the challenging characteristics of their experiments, it must properly address their requirements. This paper presents the results of characterization and requirements analysis for CEEs towards supporting the collaborative activities of scientists, and introduces the extensions necessary for the VLAM-G scientific experimentation environment.

1. INTRODUCTION

Among the main challenging characteristics of emerging experiments in e-science domains, one can mention the complexity and diversity of experiments, the size and heterogeneity of data generated by these experiments, and the need for collaboration among heterogeneous and autonomous sites when performing joint experiments.

Several solutions have been proposed to support scientists with their complex experimentations. *Science Portals* (Ashby, 2001), (Pierce, 2002) only provide a single point of access with simplified interfaces to a specific set of resources that are of importance to a certain scientific community. A *Problem Solving Environment* (PSE), on the other hand (Allen, 2001), (Schuchardt, 2002) is a system that provides all the computational facilities needed to solve a specific target class of problems in a certain problem domain (Gallopoulos, 1994). Finally, a *Virtual Laboratory* (VL) (Afsarmanesh, 2002), (Messina, 2002) provides a generic electronic workspace for distributed collaboration and experimentation in research, to generate and deliver results using distributed ICT (Vary, 2000). It supports an aggregation of people who pursue a related set of research activities and share resources, where the resources including the people may be geographically distributed and associated with different institutions (Messina, 2002).

Collaborative Experimentation Environment (CEE) addressed in this paper is a support environment for scientific experimentations, which refers to a virtual laboratory in its broadest sense, with an emphasis on supporting joint multi-disciplinary projects and collaboration, specifically information sharing among organizations and scientists. It is an integrated solution and support environment that addresses different aspects of experimentation and that supports scientists during the entire life cycle of experiments.

In order for a CEE to help scientists tackle the challenging characteristics of their experiments, it must properly address all their requirements. In a CEE, there are different types of users that perform different activities. Consequently, each of these users has different needs and expectations that mainly reflect the major activities they perform within the CEE. A detailed **characterization** of the CEE allows for the identification and characterization of both the different types of CEE users and the activities that they perform. Such a characterization leads to the identification of **user requirements**. Furthermore, a proper fulfillment of user requirements in turn puts a number of **ICT requirements** on the necessary base CEE infrastructure, which also need to be carefully analyzed.

This paper presents the results of characterization and requirements analysis for CEEs towards supporting the collaborative activities of scientists. In the remaining of this paper, first a characterization of CEE is provided, and the performed use case analysis is described. The paper then provides the results of a detailed analysis of requirements, with the focus on collaboration-related requirements. The collaborative extensions planned for the VLAM-G experimentation environment are introduced next. Finally, the paper presents its conclusions.

2. CEE CHARACTERIZATION

The Collaborative Experimentation Environment (CEE) is characterized in this section by distinguishing its major constituents; namely *experiments*, *users*, *data*, *functionality*, and *infrastructure*.

2.1 Experiment Characterization

The focus here is on the characterization of the life-cycle of a typical e-science experiment, which consists of three 'recursive' phases (**Figure 1**). During the **design phase**, the aim of the experiment is usually formulated as a question, and the methodology to answer this question is mapped to an experiment design. In the **execution phase**, the experimental procedure designed in the first phase is executed. It may include laboratory activities, using an instrument, or data gathering. During the last phase of **result analysis**, the data generated by the experiment is analyzed and interpreted by scientists. Several analysis tools can be used during this phase.

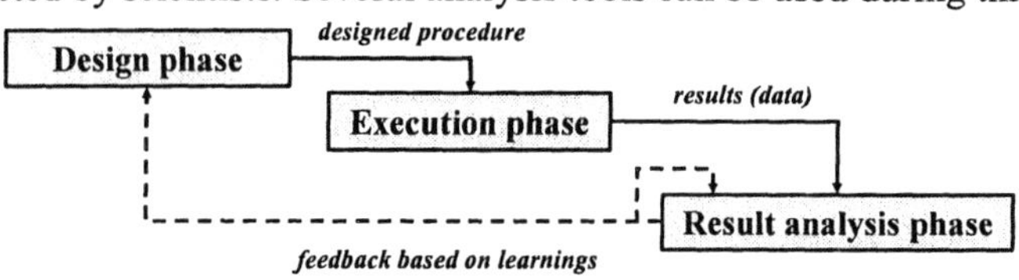

Figure 1. Life-cycle of a typical e-science experiment

The analysis phase is of ad-hoc nature and intuitive. Learning from the analysis results, the scientist may decide to use different analysis methods/algorithms on the same experiment results, or may decide to make a new experiment. The recursive nature of the experiment life-cycle comes from this last point.

2.2 User Characterization

Four target user groups are distinguished for the CEE (**Figure 2**). **Scientists** are the actual users of the CEE. A scientist is typically associated with an e-science domain (e.g. molecular biology). Inexperienced scientists usually follow a pre-defined procedure when making their experiments, while experienced scientists can also define new, customized procedures. **Domain experts** are scientists who have extensive knowledge and experience on a given e-science domain and on the experiments being performed in that domain. They are responsible, for instance, for modeling experimental information, designing experiment procedures, defining protocols to be used for certain activities in the laboratory, and defining parameters to be used for certain hardware and software. Another type of user for the CEE constitutes the **developers of support tools**. **Administrators** are responsible for the tasks related to the proper management and operation of the CEE, such as resource management, infrastructure maintenance, and user management.

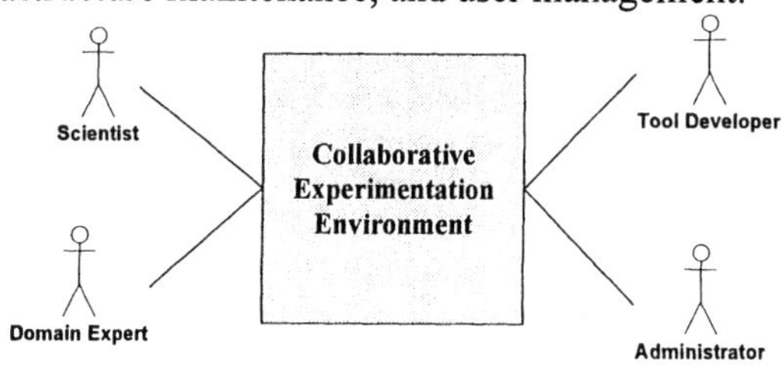

Figure 2. Users of the CEE

2.3 Data/Information Characterization

This section outlines the important aspects of data/information handled in a CEE (**Figure 3**). Large **size** of generated data is one of the most common characteristics of e-science experiments. However, the composition of generated results differs for each experiment. For instance, microarray experiments generate smaller size of results in comparison to material analysis experiments, but they are performed more frequently. **Storage** of data depends mainly on its size, structure, and usage. Large and/or unstructured data sets are generally stored in files, while structured data and/or data that needs to be queried are stored in databases. **Manipulation** of scientific data also varies from one experiment to another. Information generation can be step-wise over time, or at once. Information access can be on-demand basis for a single element, in the form of aggregate queries, or as data scanning. Information is usually **modeled** differently at each organization, following a quick-and-dirty approach, without considering standards, compatibility or possible future extensions. Scientific data is heterogeneous by nature. Among different types of **heterogeneity**, one can mention model/paradigm heterogeneity, data definition and/or manipulation language heterogeneity, semantic heterogeneity, and system heterogeneity. **Interoperability** is important to support sharing and exchange of data among collaborating centers. In line with the different types of heterogeneity, interoperability must address syntactic interoperability, semantic interoperability, and system interoperability.

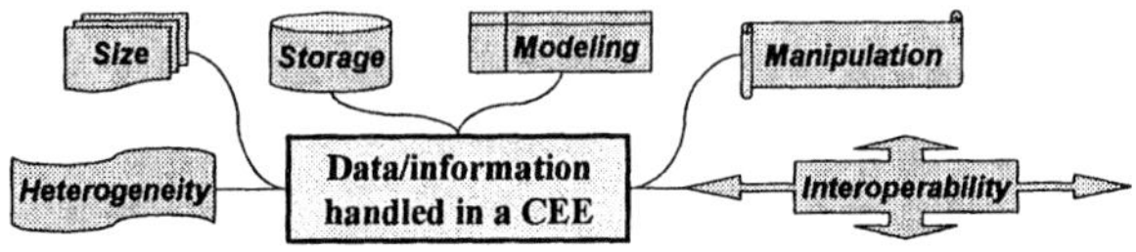

Figure 3. Characteristics of data/information handled in the CEE

2.4 Required Functionality Characterization

The following outlines the main functionalities required from a CEE (**Figure 4**): **Experiment management** (i.e. definitions, models, and mechanisms for managing an experiment during its entire life-cycle); **data/information management** (i.e. mechanisms for storage, querying, retrieval, and modification of wide variety of experiment-related information); **resource management** (i.e. efficient and coordinated management of resources needed and used during experiments); **user management** (i.e. definition and manipulation of users, roles, and their access rights); **security provision** (i.e. mechanisms for authentication of users and authorization of their requests); and **collaboration support** (i.e. support for collaborative activities among scientists, such as resource sharing, knowledge and experience sharing, and cooperative work among remote users).

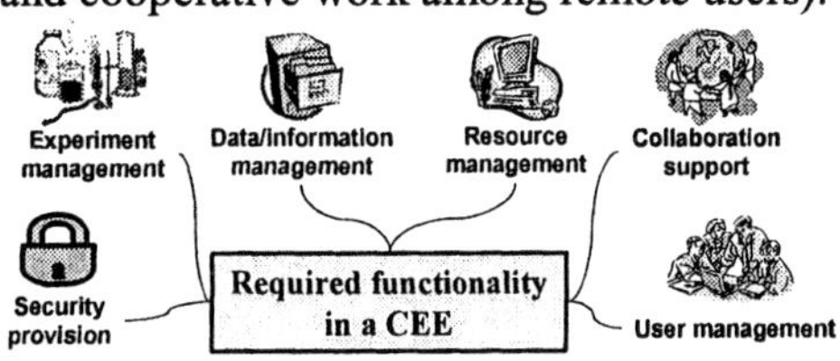

Figure 4. Required functionality from the CEE

2.5 Required Infrastructure Characterization

The infrastructure provided by the CEE must provide the necessary **computing facilities** for the analysis of large data sets (e.g. clusters of personal computers, virtual clusters), high-bandwidth/high-speed **networking facilities** for the transfer of data and/or processes among distributed computing, storage, or visualization facilities, and **software environment** that is open for adding new resources and scalable for coping with increasing number of users and workload.

3. USE CASE ANALYSIS

Use case modeling is a technique used to describe what a system should do. The primary components of a use case model are use cases, actors, and the system modeled (Eriksson, 1998). An *actor* is a person that interacts with the system. *Use cases* correspond to the main activities that an actor performs when interacting with the system. The *system* here corresponds to the CEE.

In addition to the users of the CEE (described in Section 2.2), another 'actor' of the CEE is the **ICT developer**, who develops the base infrastructure. The use cases identified for different CEE actors are provided in **Figure 5**. The use cases presented in this figure are high-level, mainly because e-science experiments are of ad-hoc and intuitive nature, where scientists may follow different routes and perform different activities within a use case.

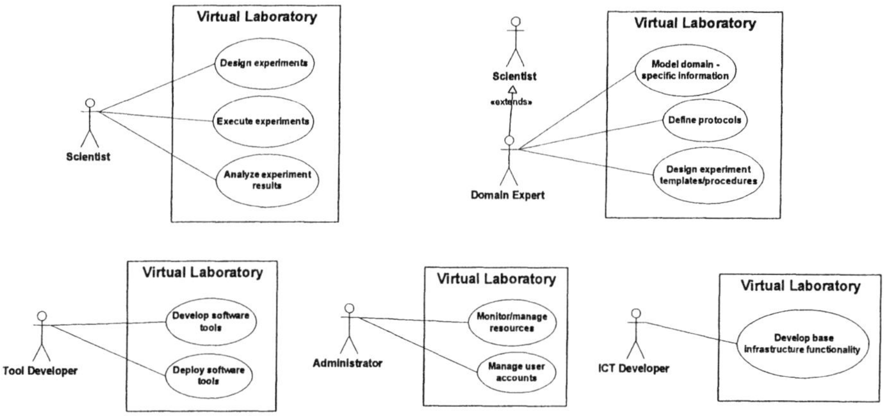

Figure 5. Use cases for CEE actors

4. CATEGORIZATION AND ANALYSIS OF REQUIREMENTS

4.1 Classification of Requirements

Based on the use case analysis, this section classifies all requirements into two categories, namely *user requirements* and *base ICT infrastructure requirements* (**Figure 6**). The former group is further classified into four groups corresponding to the different types of CEE users. The latter group is further classified into two, namely general CEE requirements and information management requirements. Identification and analysis of the base ICT infrastructure requirements constitute a first step towards providing a solution to user requirements. Therefore, in **Figure 6**, this relation is represented with block arrows from user requirements to base ICT requirements. This section presents the results of performed requirements analysis.

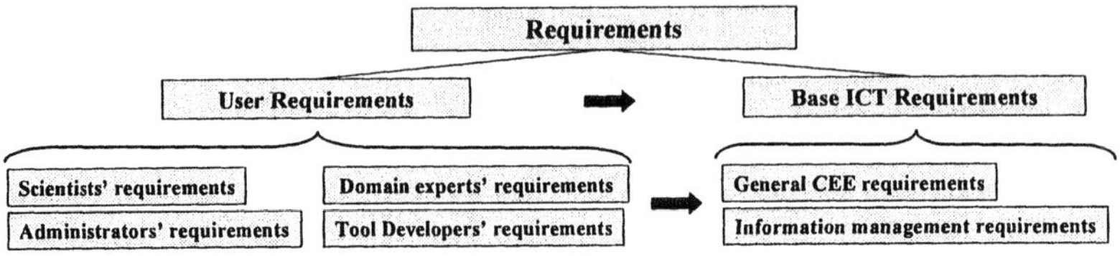

Figure 6. Classification of requirements

4.2 Analysis of User Requirements

The requirements analysis presented in this section aims to answer the following question: *What do the different CEE users require to properly perform their activities within the CEE?*

Scientists' requirements include the following:

- A proper means for clearly and sufficiently formulating the experiment objective/question in an experiment design
- A repository for experiment designs, results, and any other related information
- Necessary infrastructure for the execution of experiments
- Availability of both generic and most commonly used problem solvers
- Flexibility in the analysis phase that can cope with the ad-hoc and intuitive nature of analysis

In addition to scientists' requirements, **domain experts'** requirements include:

- A proper means for modeling standardized activities as protocols, and standardized experiment designs as templates, and a repository to store these
- Well-defined methodologies for modeling highly heterogeneous experiments and related data/information

Below are the **tool developers'** requirements:

- A design philosophy adopted by the CEE for software tools
- Well-documented and easy to understand CEE APIs

Finally, **administrators'** requirements include the following:

- Mechanisms for managing and monitoring the CEE resources
- A proper means for managing user accounts

Moreover, all users require graphical, easy and convenient to use, uniform interfaces for all their activities within the CEE, which are at the same time easy to customize to personal preferences.

Collaboration Related User Requirements

- Mechanisms for sharing and exchanging data and information, instruments, or other resources with collaborating partners in a secure environment
- Various technologies for cooperative work, such as video-conferencing or display sharing, supporting joint experimentation

- When needed, a proper means for easily adding new resources, stopping, re-starting, disconnecting a resource from the pool of resources, and enabling/disabling access to resources for a specific user or for a group of users
- Mechanisms for creating/removing user accounts, defining, updating, removing user roles, and defining and enforcing access rights for various resources

4.2 Analysis of Base ICT Infrastructure Requirements

In the previous section, needs and expectations of different CEE users when performing their activities in the CEE were presented (i.e. user requirements). A proper fulfillment of these user requirements in turn puts a number of requirements on the base ICT infrastructure for the CEE. Requirements analysis for the base ICT infrastructure presented in this section aims to answer the following question: *What functionality and facilities must be provided by the base ICT infrastructure to properly fulfill the requirements of different CEE users?*

General CEE Requirements

General CEE requirements are classified into the following categories: *Infrastructure requirements*, *functionality requirements*, *interface requirements*, and *architectural/technological implementation requirements*. Requirements in the first two categories were already outlined in Section 2 during CEE characterization; therefore, this section focuses on the remaining two categories.

User interfaces in general act as the entry-point of users to the underlying environment, while programming interfaces act as the entry-point of applications. **Interface requirements** include:

- User interfaces must hide the technical details and complexity of the underlying experimentation environment while supporting any possible usage of functionality provided by this environment
- User interfaces must allow an organized working environment, and ease the management and usage of diverse data and available resources and help the scientist to easily find what is where
- The CEE infrastructure must provide platform independent, uniform, well-documented, and easy-to-understand programming interfaces
- The programming interfaces must support the interoperation of domain-specific tools both with the CEE software environment and with other tools

Architectural design plays an important role on the scalability, openness, flexibility and manageability of the overall system. Following are the **architectural/technological requirements** related to the implementation of the base CEE infrastructure:

- The infrastructure must adopt a technical and architectural design philosophy for software development, information management, and resource management. The philosophy must be complemented with well-defined methodologies for each of these activities.
- The architecture must be open, flexible, and scalable to support interfacing with other systems, to improve, extend, or customize the provided functionality when

needed, to sustain a certain level of performance, to support collaboration, and to develop a number of monitors for managing and maintaining the system.
- The system implementation must exploit the existing and emerging standards as much as possible. However, compatibility of a technology with the CEE philosophy and methodologies, and its openness for any future improvements/extensions must be considered beforehand.

Information Management Requirements

In this subsection, the focus will be on the information management requirements for the base ICT infrastructure for CEE. Information management requirements are classified into the following categories: *Modeling requirements*, *storage requirements*, *manipulation requirements*, *security requirements*, *interoperability requirements*, and *implementation requirements*. The last category is already addressed as part of the general requirements; therefore it will not be included here.

Following are the **modeling requirements** for information handled in a CEE:
- Data models must be capable of properly representing the various types of information handled in the CEE.
- Data models must support modeling and representation of various types of experiments with different experiment flows and at any level of detail.
- Data models must be generic to achieve *uniformity* in representing both heterogeneous experiment types and heterogeneous data types.
- Schemas in the developed data models must be evolvable. They must be flexible for future changes, extendible for future extensions, open for customization to specific domains. Furthermore, the schemas must be compatible with the philosophy adopted by the CEE.

Storage requirements focus on the availability of databases for different types of information about experiments, e.g. for templates for the most common types of experiments, descriptions of previously made experiments, and descriptions of the most common techniques, protocols, etc. used in different types of experiments.

Requirements related to the **manipulation** of information include the following:
- Storage, access and manipulation mechanisms for various types of information must be developed, that are uniform within and across disciplines.
- Provided mechanisms must efficiently utilize the generality and expressiveness of the developed data models.
- Mechanisms for arbitrary queries must be provided.
- Mechanisms must be provided for version control.

Requirements related to the **security** of information are enumerated below:
- Mechanisms to define access rights for data security and information visibility must be made available to any user that owns some information in the CEE.
- All provided information management mechanisms must consider and enforce the access rights that are defined for the information that they manipulate.

Applying standards is among the **interoperability** requirements. In case of accessing multiple data sources, mechanisms to help/assist administrators to resolve model/paradigm heterogeneity or semantic heterogeneity must be provided.

Collaboration Related Base ICT Infrastructure Requirements

The base CEE infrastructure must support the collaborative activities among scientists. In specific, the VL infrastructure must address the following collaboration related requirements:

- Necessary infrastructure and mechanisms must be developed to enable sharing of resources, such as availability of a resource management system, information system for up-to-date status information, mechanisms for adding/removing a resource to/from the pool of shared resources, definition and maintenance of user account mappings, and definition and enforcement of usage rules. User and programming interfaces supporting all these mechanisms must be provided.
- Necessary data models, tools, and functionality/mechanisms must be provided to enable and ease the transfer of knowledge and experience sharing; for instance by making experiment templates, designs, protocols defined by expert users available to novice users, or through on-line (virtual) discussion environments among scientists. Such tools may utilize various technologies to support collaboration among partners: synchronous such as video-conferencing, display sharing/simultaneous visualization, joint sessions, etc. or asynchronous such as data exchange, sharing an instrument in another organization, etc.
- A proper infrastructure must be provided for coordination of joint distributed activities and for secure and authorized sharing of resources, which also considers the autonomy of collaborating organizations. Necessary data models and functionality/mechanisms must also be developed for the definition, management, and enforcement of collaboration rules.

8. VLAM-G COLLABORATIVE EXPERIMENTATION ENVIRONMENT

The Dutch VLAM-G (Grid-based Virtual Laboratory Amsterdam) project (Afsarmanesh, 2002), (Afsarmanesh, 2001) provides the main context for the work presented in this paper. VLAM-G is a multi-disciplinary virtual laboratory environment that provides the required generic environment for multi-disciplinary research in experimental science domains. VLAM-G allows its users to perform multi-disciplinary, collaborative experiments in a uniform, integrated environment, complement their in-vitro experiments with in-silico experiments, define customized experimental procedures and analysis flows, reuse generic software components, and share hardware, software, storage, networking resources as well as knowledge and experience.

The architecture of the VLAM-G and interaction among its components are shown in **Figure 7**. **Front-End** is the user environment of the VLAM-G, which presents the VLAM-G functionality to its users in a uniform way. **Session Manager** manages the active user sessions, and is responsible for coordinating the interactions among the VLAM-G components. The distributed computing and networking resources on the Grid are made available to VLAM-G users through the **Run Time System** (RTS), which provides an API to encapsulate the Grid computing code within a simple interface. **Module Repository** is a persistent storage for binaries of software entities to be executed by the RTS. **VIMCO** is the information management platform of VLAM-G, and provides the necessary mechanisms for the manipulation of different types of experiment-related information.

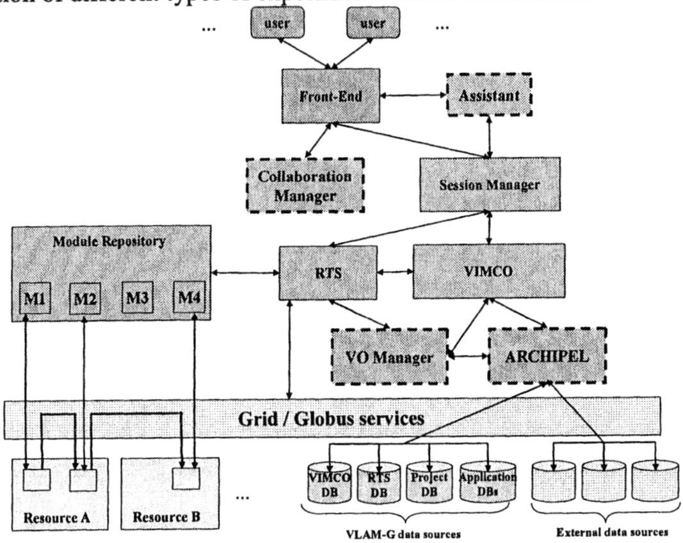

Figure 7. VLAM-G architecture extended with collaboration components

As mentioned earlier, collaboration is one of the main characteristics of e-science experiments. Emerging scientific experiments are evolving towards collaborative efforts involving several partners from different disciplines, different organizations, and different countries. With the increasing complexity and cost of scientific experiments, sharing expertise and sharing resources have become two of the most important motivations for collaboration. VLAM-G already addresses some of the main issues related to collaborative experimentation, such as multi-disciplinary projects, sharing (parts of) experiments, and basic mechanisms for controlling the collaboration (e.g. sharing policies to ensure the semantic consistency of the shared information and basic access rights). In addition, VLAM-G addresses sharing of software and hardware resources.

However, some of the collaboration requirements that were identified in this paper still need to be fulfilled. These requirements are integration of heterogeneous data from autonomous sources, setting and enforcing rules and regulations for a proper collaboration among partners within the context of a virtual organization, and supporting cooperative work among scientists. **Figure 7** shows the initial ideas on extending the VLAM-G architecture with four collaborative components, namely *Archipel*, *VO Manager*, *Collaboration Manager* and *Assistant.*

Archipel is a generic federated information management framework being designed within the context of the VLAM-G project, supporting uniform access to a variety of heterogeneous and distributed information sources. The VO support infrastructure in VLAM-G, called **VO Manager**, is at the design stage (Kaletas, 2004), and it will make use of the other VLAM-G components; for instance, it will use the Archipel for sharing data resources and controlling access to these resources, or Grid for enforcing the sharing and access policies on hardware/software resources, as well as sharing the Grid security credentials needed to use these resources. **Collaboration Manager** will enable simultaneous collaborative design and execution of experiments through cooperative work environments (e.g. chatboxes). Finally, the **Assistant** will assist users during their experiments, for instance, by suggesting the most efficient software to perform a specific task.

10. CONCLUSIONS

In this paper, the CEE solution to support scientific experimentations was characterized and different types of CEE users and the activities that they perform within the CEE were identified and described. Each of these users performs different activities in the CEE, hence they have different needs and expectations from the CEE. In addition to user requirements, requirements for the base ICT infrastructure underlying the CEE were presented, with particular attention on collaboration requirements. Analysis of requirements showed that many requirements are related to each other. User requirements in turn impose a number of requirements on the base ICT infrastructure for CEE. The ICT requirements represent a first step towards providing a solution to user requirements. ICT developers must address these requirements to provide the necessary environment and functionality to CEE users.

As these requirements point out, there are several different aspects that need to be addressed related to collaboration, including VO support infrastructure, federated data access and integration, and cooperative work. With the existence of a collaboration support infrastructure, a number of organizations can join together, sharing their resources and skills towards reaching common goals. As applied to the scientific collaboration domain, VO paradigm can assist organizations in pursuing a common goal, for instance, tight collaboration towards solving scientific problems, where the sub-tasks are distributed among different organizations and the distributed multitasking is coordinated by the VO Manager. As the base necessity in the VO, it will be possible to share (access) privileges on all kinds of resources, from hardware and software to data and information. Furthermore, this sharing and collaboration is regulated by pre-defined sets of rules and policies, which are agreed upon by all collaborating partners. These agreements in form of contracts will further increase the trust among partners, and help them to advance their collaborations.

11. REFERENCES

1. Afsarmanesh H, Kaletas EC, Benabdelkader A, Garita C, Hertzberger LO. A Reference Architecture for Scientific Virtual Laboratories. Future Generation Computer Systems 2001; 17 (8): 999-1008.
2. Afsarmanesh H, Belleman RG, Belloum ASZ, Benabdelkader A, van den Brand JFJ, Eijkel GB, Frenkel A, Garita C, Groep DL, Heeren RMA, Hendrikse ZW, Hertzberger LO, Kaandorp JA, Kaletas EC, Korkhov V, de Laat CTAM, Sloot PMA, Vasunin D, Visser A, Yakali HH. VLAM-G: A Grid-based Virtual Laboratory. Scientific Programming 2002; 10 (2): 173-181.

3. Allen G, Benger W, Dramlitsch T, Goodale T, Hege HC, Lanfermann G, Merzky A, Radke T, Seidel E, Shalf J. Cactus Tools for Grid Applications. Cluster Computing 2001; 4 (3): 179-188
4. Ashby JV, Bicarregui JC, Boyd DRS, Kleese-van Dam K, Lambert SC, Matthews BM, O'Neill KD. A Multidisicplinary Scientific Data Portal. In Proceedings of the 9th International Conference and Exhibition on High-Performance Computing and Networking, 2001, pp. 13-22.
5. Eriksson H-E, Penker M. UML Toolkit. Wiley Computer Publishing, 1998.
6. Gallopoulos E, Houstis E, Rice JR. Computer as Thinker/Doer: Problem-Solving Environments for Computational Science. Computing in Science and Engineering 1994; 1 (2): 11-23.
7. Kaletas EC, Afsarmanesh H, Hertzberger LO. A Collaborative Experimentation Environment for Biosciences. (To appear in) International Journal of Networking and Virtual Organizations 2004.
8. Messina P. The Emergence of Virtual Laboratories for Science and Engineering. iGrid2002 Presentation. http://www.igrid2002.org/ppt/Paul_Messina.ppt
9. Pierce M, Youn C, Fox GC. The Gateway Computational Web Portal. Concurrency and Computation: Practice and Experience 2002; 14 (13-15): 1411-1426.
10. Schuchardt KL, Myers JD, Stephan EG. A Web-Based Data Architecture for Problem Solving Environments: Application of Distributed Authoring and Versioning to the Extensible Computational Chemistry Environment. Cluster Computing 2002; 5 (3): 287-296.
11. Vary JP. Report of the Expert Meeting on Virtual Laboratories. United Nations Educational, Scientific and Cultural Organization 2000; Technical Report CII-00/WS/01.

27 A KNOWLEDGE MANAGEMENT BASED FRAMEWORK AS A WAY FOR SME NETWORKS INTEGRATION

Gerardo Gutiérrez Segura
Laboratoire de Productique et Informatique de Systèmes Manufacturiers
ggutierr@bat710.univ-lyon1.fr
Véronique Deslandres
Laboratoire de Productique et Informatique de Systèmes Manufacturiers,
deslandres@bat710.univ-lyon1.fr
Alain Dussauchoy
Laboratoire de Productique et Informatique de Systèmes Manufacturiers
dussauchoy@bat710.univ-lyon1.fr
FRANCE

This paper initially introduces specificities of small and medium enterprises (SME) networks, then deals with the recent evolution of the Knowledge Management (KM). In coherence with this evolution, we describe a knowledge management process based on a community of practices which can be applied to these types of groups. The main conclusion is that knowledge management projects provide a good manner for SME networks to facilitate and increase the collaboration rate as well as to share knowledge, allowing them to make collaborative work more efficient.

1. INTRODUCTION

A considerable number of computing and social sciences research testify the application of knowledge management (KM) on a large group of enterprises. However in small and medium enterprises (SME) it is also necessary to capitalize the knowledge that could disappear. This need is even more important for strategic SME networks.

The case of SME networks is particular in the sense that generally little time is planned for coordination and collaborative work. By way of consequence, decisions and schedules are made without sufficient consultation and sometimes in a hurry. In this case, some projects of knowledge management are well-adapted. We have been working for six years on the contribution of information technologies (IT) for the co-operation and knowledge division within SME strategic alliances. This research fits into the two GRECOPME projects (GRECOPME, 2000), partially financed by the Rhone-Alpes France's region in which SME's are numerous. These multi-field projects are controlled by l'Ecole Nationale des Mines de Saint-Etienne, and gather teams from Lyon1 and Lyon2, from l'INSA of Lyon, l'IUT of Roanne and the Saint-Etienne University.

This paper initially introduces specificities of SME networks, then deals with the recent evolution of the Knowledge Management (KM). In coherence with this evolution, we describe a KM process which can be applied to this type of networks in order to enhance partner integration and ensure therefore a better collaboration.

2. THE UNIQUENESS OF SME NETWORK

2.1 Presentation of small and medium enterprises networks

For more than a decade, many companies have been aiming alliances in order to develop their activity within a network. Various levels of co-operation and integration can be observed, creating a virtual company. The motivations which lead to firms co-operation are various: offer expansion, introduction on new markets, strategic alliance against competition, etc. For small and medium enterprises, the motivation is still different and can be regarded as defensive or pro-active. Among defensive networks, the objective can be size effect compensation, gap filling, leader's retirement anticipation, etc. The pro-active networks, rarely seen, are created with the strategic aim of expand the offer, to compete, to co-operate on innovation possibilities within a given sector (GRECOPME, 2000).

In SME networks, co-operation intensity strongly depends on the degree of confidence acquired between the companies. We established that these networks had their own life cycle, with three different phases:

- Confidence Construction
- Co-operation Test(s)
- Alliance stabilization (fusion, fragmentation or new alliance)

The coordination and running modes of SME networks are made complex due to the fact that they are not controlled by a single manager but by a group of leaders, even if each enterprises preserving activities with its own customers. Before the final phase of the structure stabilization, the network integration is always considered as potentially reversible by the leaders. But in fact, the more thorough is the co-operation, the more it takes an irreversible character (e.g. company specialization) and the more the company depends on the network. We note that the latter point effectively contradicts leaders' autonomy research. In fact, the firm that comes in a network is rarely a neutral process, and in many cases the integration process itself has to be managed carefully.

As an example, an IT project development aiming at support some of the network activities can be an interesting integration vector, especially if the computer culture of the companies is sufficient (GRECOPME, 2000). Because we have encountered many different cases of IT project, various architectures for the co-operative have been proposed in (Bienner, 1999) and (Gutierrez, 2001). However, the installation of an information system is based mainly on a strong confidence, and it is necessary to wait, in particular in non well-stabilized alliances, until the network succeeded and structured many co-operative activities (objective which is not always reached).

SME networks are organizations where the KM can constitute an undeniable integration vector towards co-operating work. Therefore, the setting up of several

SME network will produce a vast stock of knowledge that will have to be managed in a suitable way. However, at the beginning of such an organization, SME don't especially agree to share information, documents or knowledge with the new partners who may become competitors. We consider that for some SME networks, it can be convenient to set up a co-operation system at the very beginning of the network. Of course such a system will be better accepted in firms which already have privileged technological culture (e.g. software and computing service companies SCSC). Our approach is based on the aim to provide this confidence by using a structure based on the KM adapted rather to the managerial problems (little time for coordination and co-operation works, a little confidence etc.) of this kind of networks, i.e., we plan to set up a knowledge management framework in the two first phases of the above-mentioned life cycle.

3. THE KNOWLEDGE MANAGEMENT

3.1 The weak points of the knowledge management

The main critic that one can make about these tools are that: (1) the majority of them are limited to a particular KM function, (2) the tools do not take into account the company field, and also the standards used, (3) some of these tools were developed specifically for a given company – in general, for a large group and his partners –, this is not very appropriate for other companies, and especially (4) the success of the KM project associated with the use of a tool is never guaranteed. The large groups even doubt about the effectiveness of this type of project.

The methods of acquisition and capitalization were very criticized when they consider knowledge as an isolated object, not located, described apart from any context and giving their interpretation by a random user (Quéré, 1999). Other methods were developed for the knowledge located, but the complexity of the systems of representation used made the update and the evolution adaptation a very complex and difficult activity (Lucier, 1998).

Overall with regard to the evaluation of the KM projects, certain weaknesses are now well-known, in particular the low visibility of the sources of profitability. Sometimes it's necessary to wait two years to get the benefit from such a project, when it is not purely and simply given up because of the lack of effort of maintenance or maladjustment to the evolution of the company. On the actors level, one notes an increased mobility of the personnel and thus a loss of generalized sincerity, the two phenomena being induced by the market and harming particularly the implication of the actors needed for the development of projects related to the human capital (Meissonier, 1999).

Thus, although our perception of the KM models had evolved since its beginnings, no model economically valid shows how knowledge is connected to the tasks and the performance (Malhotra, 2002). For this reason the role of the leadership is perceived like a weak point in the economy of the knowledge, as well as the role carried to the attention (Von Krogh, 2000): "The managers [...] must learn cognition. On one hand, there is the problem of confidence (based on the exchanges) but also the problem of the attention paid to the individuals". The concept of attention (regard) incorporates and extends the confidence. The exchange of tacit knowledge is thus almost impossible without attention. Researchers study for example how to conceive training schemes which would support a behavior of attention. According to our experience, the problem relating to the appropriation by the users of the KM system set up exists indeed, and can be more or less important. Even if the experts having taken part in the project were qualified, the users do not grant confidence in the results obtained only if they estimate that the system does not block their autonomy of decision (Exworth, 2003).

4. A KM approach for SME Networks integration

4.1 Application of KM in SME network

When we talking about of Knowledge Management for SME's, one can notice that it is not widely used in this kind of companies, mostly because small structures can't follow up with big organizations on the technological level. Nevertheless, a small percentage of SME have some KM activities, and a traditionally accepted limit is that KM concerns companies of more than 40 employees [Lim, 2000]. Indeed, current literature states that KM is more needed by big firms rather than by the small ones, due to the fact that in SME, the problem of knowledge exchange and sharing is lessened by the size of the companies (you know immediately who to ask to), the versatility of employees (making them be concerned by more various subjects) and physical proximity (easier meeting possibilities).This is true for an small enterprise, but the SME networks are out of this context : in an organization, the partners almost don't know each other, are not close to each other, and they are not always disposed to individually expose their knowledge.

Because of this domain's evolution and of our experience with small businesses networks, the benefits of KM is essentially located in the implementation of procedures – along with technological networks according to the information technologies (IT) culture of the network - , making the knowledge sharing and the collaborative work easier: Interactions and representations exchange, turning hidden knowledge into explicit one. There is also a need for supports for explicit knowledge management (diffusion and explanation of a given representation) as well as for tools designed for creating new knowledge.

The debate about the added value of IT to the organization and particularly to the management of knowledge is regularly brought back on the front scene (Beckman, 1999). Even if the fact that information systems (IS) have a weak impact on the effectiveness of decision making, is widely accepted as soon as only the technical dimension is concerned (data processing) (Simon, 1980), it is still required to evaluate its contribution once all the needs have been defined and that the attention capacity limits of the users have been identified. This attention capacity seems strongly to be related to the IT culture of the company to us, and the IS will play a crucial role for small software and computing service companies; this is the reason why we have chosen to work with this kind of companies.

4.2 Software and Computing Service Companies (SCSC)

In our opinion, SCSC are a fertile ground for Knowledge Management. We have many arguments to defense this affirmation. Firstly, technology is in the heart of these companies. Information technologies (IT) characteristics like evolution capacities and obsolescence imply a special and primordial role to knowledge. Indeed, one should not let himself to be outdated by some always evolving technology. One must then continually keep his knowledge and skills up-to-date. The counterpart of evolution capacity is definitely obsolescence. A technology can be quickly seen as out-of-date and be replaced with another one. In order not to be put on the technological fringe, it is required for consultants to keep on learning new technologies, and to acquire new knowledge and know-how. Moreover, we consider it is easier for an IT-oriented company to develop KM as it cannot be conceived without an "IT Tools" dimension. We understand easily that for the consultants who are working daily with the new computing technologies it is not hard to convince to KM tools.

4.3 Communities of practice (CoPs)

In the literature, many authors recognize that many enterprises are using community of practices without being aware of it. The community of practice is a group of individuals who have specified subjects in common interest, which need to interact around problems, which develops an expertise on a field, and which is implied in the objective of collective training (Wenger, 2001). In the beginning, this concept was proposed in response to the technological domination of the efforts in KM works.

It was to make knowledge something alive rather then the reduction of a stored and solidified structure, something that pertains to a community able to maintain, develop, and to share its knowledge. It is the understanding and the management of the tacit knowledge who brought certain authors like Wenger (Wenger, 2002) to base CoPs on the relational and social nature of knowledge. This approach has been successfully used in the private sector over the past decade and now being applied in the public sector (Snyder, 2003). By the other hand Community Portals also exist including a set of functionalities (Schneider, 2002), such as the management of contents, interactions, of the community as well as different mechanisms of piloting, including for example the reputation system.

We have chosen communities of practice, which according to our opinion seems provide the appropriate environment for the generation and the transfer of knowledge of work between network partners.

4.4 A community practices instead of Knowledge Management system.

In the current marketing strategy, the software editors do not hesitate to qualify any new functionality which allows the management of documents in KM, the same for KM for the SME. Although the information management and the knowledge management share the same supports (documents, plans, diagrams), their evolution, their objectives and their operating ways are completely different (Malhotra, 1998).

Our method consists in the development of a community of practice where the actors will share information, ideas, documents, etc. with the aim of increasing their own knowledge level, and ensure the correct broadcast of technical information to the persons which need it. In other words, rather than study the wealth of information in the channel of communication, it is interesting to make an analysis of the dynamics of the organization. This is particularly true within SME networks which support the co-construction of strategic directions especially based on exchanged information. The work on knowledge management resulting from human resources recommend that everybody reached the knowledge level at the time that everybody connect actors who communicate. More precisely, we consider that knowledge is the result of the interaction between information (a procedure, information of a customer, an opinion of a colleague) and a person. Indeed by the interpretation process according to the actors that the information is transformed in knowledge.

For the creation of our CoPs, we based ourselves on the four stages of development of the communities of practice proposed by Wegner (Wenger, 2001):

The field: Naturally, it is essential to know on which field the sharing of knowledge of the community is centered. For this point we chose the computing field, in particular on the development of web sites and office applications on which all partners of the group are concerned.

The operation: Rather than an organization of operation of the community we must cultivate it (Prax, 2003). Indeed the evolution of the community practices is organic. However a minimum of organization seems to need. It is necessary that people should be engaged to the community to give him a minimum of stability, points of reference and that will be the reference persons. Being that we work with a SME network geographically away.

The actions: A community usually shares common activities and also some own projects. These types of activities, technological tools, external sources to exploit, etc., should be defined. In the SCSC network that we are working with, the information objectives and knowledge exchanges has been defined with the partners. We identified two main objectives for the starting the community. These objectives were related to new information technologies as well as some projects actually being pursued by the network. Indeed, our purpose is to share a special kind of knowledge, in particular relationship knowledge which concern the actors of the fieldwork. Example the different kinds of knowledge that will be shared where several partners of the group take part: information about the new software for the Internet site development, the way of how protecting a computer against a virus propagated by the Net, but also about information about how to manage the project of setting-up a ERP in a SME society etc.

The tools: They correspond to the whole physical and technological device that facilitates the running of the community. The objective of our approach is to make the most of the powerful technologies to be easy to use for the participants of the community, it permitted us the specification of the computing tools, support tools (a Wiki and a forum) as well as the establishment of use recommendations. Indeed, this community is based on a Wiki and a forum. The Wiki is a piece of a server of applications which permits to create as well as to publish Web contents using as a tool a navigator Web. For instance, Wiki permits the collective creation of documents hypertext : it is about " Open Editing " in reference to Open Source (of the opened editorial content). Wikis are often created in co-operation with other internauts. Anybody can modify a page (a minimum of computer knowledge is required). Every participant has liberty to edit wiki. In our method, we implemented the Wiki and then we only allowed access to members of the community.

The main use of the Wiki is the publishing in order to share knowledge in relation to the new domains among members of the group (and therefore, the knowledge of the actors increases very quickly) so that each one of the partners knows domains of expertise of the others. Activities of administration made in the beginning of the project aimed mainly to launch demands regarding to some domains so that members publish in the Wiki these knowledge, information, experiences, links towards the interesting sites, etc. on the required theme. Thus, as administrators we first created the basis of a page on the theme to develop on the Wiki, and then we sent an e-mail to the group to invite them to participate.

It took for the forum a couple of months to start. Actually, the main function of the forum in the beginning of the activities was to be the support of communication of the group (an e-mail of the group). However, it became afterwards a real forum of sharing on different domains. Indeed, some questions answered to questions but launched other questions in order to complete, so participants answered to these new topics. These first activities allowed the group to introduce themselves (especially about the other participant expertise), but to our opinion, the most important thing was the creation of a confidence climate that grew up dramatically in a short time thanks to the constant communication and sharing relations. For example: activities as asking information directly to other partner on the forum as well as to express points of view in relation to answers or themes treated on the Wiki became very current for all participants.

5. CONCLUSIONS

We have introduced a methodology for a KM system within SME networks which can run at the beginning of co-operating activities. The objective of this research is to propose a solution closely suitable for the problematic of SME networks towards knowledge sharing. We estimate that the KM field for the SME networks is not explored enough by the scientific communities. The objective of this research is to stimulate the integration and co-operation to increase the confidence (which has an very low level at beginning of activities) within such organization, especially using the existing and scattered knowledge capital of the network.

This research was at first based on the analysis of the problematic of integration and sharing existing in these organizations. Then we applied the feed-backs of related works from the knowledge management community. We are aware that the mere installation of certain technologies will not guarantee the success of projects of knowledge management in SME networks, because the management and strategic approach remain fundamental. Nevertheless a global co-operative technological culture is appearing due to everyday internet use by professionals. Our purpose is that the KM within SME network can be applied gradually, aiming to develop a learning organization (Jacob, 2000). Due to their structure -a group of small size organizations -, to co-operation imperatives not always precisely formalized, and to the necessary up-date regarding computer technologies, SME networks provide an auspicious natural setting for the use of the new knowledge economics.

6. REFERENCES

1. Beckman, T. (1999) "The Current State of Knowledge Management", in Knowledge Management Handbook, edited by Jay Liebowitz, CRC Press, Boca Raton, Floride, 1.1-1.21
2. Bienner, F. and FAVREL J. (1999) ''Organization and management of a distributed information system shared by a pool of enterprises''. Acts of the conference IEPM'99. Glasgow, Juillet 1999.
3. Exworth, M., Wilkinson, E. K. McColl, A., Moore, M., Roderick, P., Smith, H. and Gabbay, H. (2003) "The role of performance indicators in changing the autonomy of the general practice profession in the UK", Social Science & Medicine, Vol. 56, Issue 7, 1493-1504.
4. GRECOPME (2000) Vincent L. et al, ''Groupement d'Entreprises Coopérantes : Potentialités, Moyens, Evolutions'', Rapport du projet de la Région Rhône-Alpes 1997-2000
5. Guffond, J.L. ; Leconte, G. (1998) ''Logistique de chantier, modes d'organisations et outils de pilotage - le cas de l'activité de construction'', Acts of the second international meetings of research en logistique, "Logistique et interfaces organisationnelles", édité par N. FABBE-COSTES et C. ROUSSAT, Marseille, January 27-28.
6. Gutierrez-Segura, G. (2001) ''ERP pour les groupements de PME/PMI'', Rapport du DEA ISCE (Informatique et Systèmes Coopératifs pour les Entreprises) Univ. Claude Bernard LYONI.
7. Jacob R. et S. Turcot (2000) ''La PME «apprenante » : Information, connaissance, interaction, intelligence'', Rapport de veille, projet Globalisation et PME innovante, Université du Québec à Trois-Rivières, Juillet, 113p..
8. Lim, D. and Klobas J. ''Knowledge management in small enterprises'' The Electronic Library, volume 18, nombre 6, 2000. http://www.emerald-library.com
9. Lucier, C.E. and Torsilieri, J.D. (1998) "Why Knowledge Programs Fail: A C.E.O.'s Guide to Managing Learning", article web visible sur http://www.it-consultancy.com /extern/extern.html (visible le 26 Juin 2003).
10. Malhorta, Y. (1998) "Deciphering the Knowledge Management Hype", Journal for Quality & Participation, Special issue on Learning and Information Management, 21, 4, 58-60.
11. Malhorta, Y., (2002) "Why Knowledge Management Systems Fail? Enablers and Constraints of Knowledge Management in Human Enterprises". In Holsapple, C.W. (Ed.) Handbook on Knowledge Management 1: Knowledge Matters, Springer-Verlag, Heidelberg, Germany, 577-599.
12. Meissonier, R., (1999) ''NTIC et processus de décision dans les PME-PMI'', rapport de recherche WP n°561, IAE Aix-Marseille
13. Prax, J.Y. ''Le manuel du knowledge management: une approche de 2e génération''. Paris: Dunod, 2003.
14. Nonaka, I. and Takeuchi, H. (1995) ''The Knowledge-Creating Company'', Oxford, Oxford University Press
15. Querel, L. (1999) ''Action et cognition situées'', Conférence Publique 17 juin 1999, Montpellier
16. Scheinder, D. (2002) ''Portail pour les communautés de pratiques'', TECFA (Technologies de Formation et Apprentissage) FPSE (Faculté de Psychologie et des Sciences de l'Education) Université de Genève.
17. Simon H. A. (1980) ''Le nouveau management, la décision par les ordinateurs'', Economica, Paris. Lakatta EG, Cohen JD, Fleg JL, Frohlich ED, Gradman AH. Hypertension in the elderly: age- and disease-related complications and therapeutic implications. Card Drugs Ther 1993; 7: 643-54.
18. Snyder W. M. et al (2003) ''Communities of practice: A new tool for Government Managers'', http://www.businessofgovernment.org/pdfs/Snyder_report.pdf
19. Von Krogh, G., Ichijo, K. and Nonaka, I., (2002) ''Enabling Knowledge Creation: How to Unlock the mystery of tacit Knowledge and Release the Power of Innovation'' Oxford university press, N.Y.
20. Wenger, Etienne. (2001). ''Supporting Communities of Practice: A Survey of Community-oriented Technologies''. Published as "shareware" and available at www.km.gov under "Group Documents", then "Documents and Resources."
21. Wenger, E. McDermott, . R. et W. Snyder (2002). ''A guide to managing knogledge : Cultivating Comunities of Practice'', Harvard Business School Press.

28 COLLABORATIVE E-ENGINEERING ENVIRONMENTS TO SUPPORT INTEGRATED PRODUCT DEVELOPMENT

Ricardo Mejía[1], Joaquín Aca[1], Horacio Ahuett[2], Arturo Molina[1],

[1]Centro de Sistemas integrados de Manufactura - ITESM
[2]Centro de Diseño e innovación de productos- ITESM
Ave. Eugenio Garza Sada 2501 Sur
Monterrey, N.L. 64849 MEXICO
(52–81) 8158-2032/86, fax: (52–81) 8328-4123
rimejia@itesm.mx, aca@itesm.mx,
horacio.ahuett@itesm.mx, armolina@itesm.mx

Nowadays global product development tasks are executed by different facilities usually at different geographically location, where design and manufacturing teams must work remotely. This situation requires three major issues to be tackled: (1) implementation of a collaborative Integrated Product Development process among the different companies participating in the Product Life Cycle activities; (2) establishment of environments that foster the coordination and cooperation among engineering groups; (3) integration of software tools that allows the exchange of information and knowledge among engineers in an effective and efficient manner. A reference model for integrated product, process and manufacturing systems development is described and a methodology to implement Collaborative e-Engineering Environments is proposed to provide a model to transfer the e-engineering concepts to the industry. A case study is described that applied the proposed methodologies to set-up a collaborative environment for high tech product development using low-cost technologies.

1. INTRODUCTION

Global companies have been forced to define and standardize their product development and manufacturing processes, in order to coordinate at a global level all their activities related to an "Integrated Global Product Development". This concept means the integration of all the activities, methods, information and technologies to conceive the complete Product Life Cycle [Tipnis 1999]. At the same time the globalization of the industrial activities and decentralization of many manufacturing processes leads companies to work in relation to very distant collaborators. Due to these trends, organizations are constantly seeking better methods for improving productivity and effectiveness in the accomplishment of engineering tasks, primarily through the use of information technology. Examples include the need to reduce the cost of designing new products and to significantly shrink overall development life cycles [Bochenek and Ragusa 2001]. For these reasons the creation of collaborative e-Engineering environments has been a key challenge in information technology.

Information technology must be designed, implemented and integrated to enable people to collaborate and coordinate their design and manufacturing activities. But technology is not enough; a well defined development process where activities, information/knowledge, techniques and partners involved has to be modeled and visualized in order to have a reference for the global development process. Integrated Product and Process Development (IPPD) is a management technique that simultaneously integrates all essential acquisition activities through the use of multidisciplinary teams to optimize the design, manufacturing, and supportability processes [OUSD 1998]. There are methodologies to support the implementation of IPPD concepts, among them, [Lee et. al. 2003], [Mervyn et. al. 2003], [Song et. al. 2001], [Swink et. al. 1996] and [Yan and Zhou 2003]. However important considerations are: (a) integration level of methods and tools proposed is restricted to specific development activities and (b) the absence of one methodology able to integrate complete product life cycle.

This paper describes a methodology that has been used to design, integrate and execute collaborative e-Engineering environments based on the concept of IPPD. Two case of study related to the application of the reference model is presented to demonstrate its applicability in real situations.

2. DESIGNING AND CONFIGURING COLLABORATIVE e-ENGINEERING ENVIRONMENTS

A process development model is the basis to design, configure and implement the collaborative e-Engineering Environment. Figure 1 depicts the methodology used:

I) Determine the Company process Requirements

Identify the company requirements for product, process and/or manufacturing system development. Once the process has been defined, an AS-IS process model is built. This model should include information regarding the activities, information/knowledge, human and technological resources and organizational issues (practices, procedures, responsibilities). Sometimes for new processes implementation, an AS-IS model might not exists, therefore the methodology begin with the creation of a TO-BE model.

II) Assessment and model TO-BE of the development process

The AS-IS model represents how a process is currently executed. An assessment is carried out based on four views: process, information/knowledge, organization and resources. Afterwards, the TO-BE process is modeled, including all the proposed modification to have a more efficient/effective model. This TO-BE process model will be used to define a Workflow in further stages.

III) Design and integration of environment and applications

Four main steps must be undertaken in order to integrate collaborative engineering environments:

- Modeling the workflow (MODEL): Workflow modeling allows the analysis and visualization of the whole development process. The workflow logic is defined using a Workflow Management System (WFMS) to describe all the elements of the process. It is important to mention that the key elements to describe in a workflow are: activities, flow of information, people, and decisions.

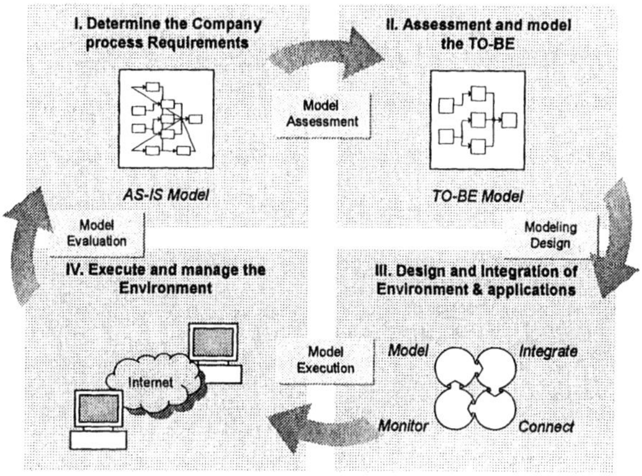

Figure 1 – Methodology for designing and configuring Collaborative e-Engineering Environments

- Selection and Integration of e-Engineering applications (INTEGRATE): The different applications required in a collaborative environment includes: 1) Functional: function oriented systems that support engineers in specific tasks, for example CAD, CAM, CAE, KBES, and Rapid Prototyping tools; 2) Coordination: coordination systems to support sequencing of activities and flow of information., for example workflow and project management; 3) Collaboration: collaboration systems to foster cooperation among engineer, i.e. CSCW - Computer Supported Cooperative Working; and 4) Information Management: product and manufacturing information management systems, knowledge based management systems. The combination of application will be aligned to the company specific needs.
- Connect external applications using standards and web protocols (CONNECT): in a collaborative engineering environment there is a need to connect external applications that customers or suppliers are using. Therefore the required communication standards and web protocols must be defined. Some of the applications includes: marketing information exchange (e.g. Web pages, e-catalogues), Manufacturing / Production systems (e.g. e-RFQ, ERP, MES – Manufacturing Execution Systems), and other CAD/CAM/CAE systems.
- Definition of performance measures and monitoring techniques (MONITOR): the decision of "what" to monitor in the process is established. Performance measure could include: time, costs and resources usage. This is important to allow managers to coordinate, track and control the process at the execution

phase. If the WFMS allows the application of business process intelligent tools depth analysis of process behavior and performance can be carried out.

IV) Execute and manage the Workflow

The workflow is executed once all the stages from the collaborative engineering configuration methodology are covered. The execution of a Collaborative Environment based on a WFMS allows process analysis that can be carried out in order to improve the process. The process management tracks events and data from the Workflow environment execution. It provides both real-time and historical tracking of what is occurring in the workflow engine. All these information can help to improve the process. Based on the analysis and suggested modification a new potential TO-BE model (the currently process in execution is the AS-IS process) and maybe new design improvements can be proposed to improve the business process.

3. A PILOT PROJECT FOR COLLABORATIVE HIGH TECH PRODUCT DEVELOPMENT

The Northern Mexican Region has been seen the establishment of a large number of US manufacturing facilities looking for advantages in Mexican manufacturing expertise and close to border localization. The tasks of design and manufacturing of products are being executed by different facilities of large corporation usually at different geographically locations. Problems faced between US and Mexican companies are (a) lack of collaboration during early stages of life cycle product development; and (b) deficient knowledge of manufacturing process capabilities. To demonstrate how the collaborative product development can be attained an Engineering Collaboration Environment for High Tech Products pilot project was developed.

A methodology was configured to develop High-Tech Products (Mechatronic products) between US and Mexican companies. The product to develop was a communication device for children with special needs for communication. The target was to complete a functional prototype in six months. The responsibilities and resources to ensure the project execution were delegate to students at ITESM. Students from Electronic Department of ITESM were responsible to design the Printed Circuit Board (PCB) of the device and students from Mechanical Engineering Department were responsible to design and fabricate the prototypes of the Housing and fabricate the prototype of the PCB in the Manufacturing Research Center facilities at ITESM.

Step I and ***Step II*** were undertaken during the configuration of a new process for High-Tech products development. For this reason the proposed development process is directly a TO-BE model. A detail description of the process model was created specifying specific activities, methods, tools and people.

For ***Step III)*** a selection of freeware tools was a priority. The purpose of this exercise was to explore a broader range of collaboration tools, because past experiences in designing and implementing Collaboration Environments (Web portals and Groupware) had shown that collaboration tools already integrated in those systems were limited [Aca et al. 2003]. As a consequence, several tools were tested, and it was concluded that commercial public domain tools will be used. The following applications were evaluated: MSN™, Yahoo™, and NetMeeting™. All of these applications provide a high variety of collaboration tools. For this specific pilot project, the MSN applications were selected to demonstrate the concepts. Then, the following aspects were considered to design an environment with free-ware tools:

MODEL: In this pilot project, it was necessary to create a shared space, using MSN groups, to be used as an integrated platform (environment). However in this kind of environments, there is a lack of workflow tools. Therefore the flow of activities and the methods and tools required for each activity, were implemented using a set of working spaces were each phase of the development process included all the workflow components.

INTEGRATE: **Functional:** the engineering stand-alone applications used were: Mechanical Desktop, Design explorer and Electronic Workbench. Several Web based tools were developed: MAS[1], SMT-Advisor[2], and C) Ducade[3]. **Collaboration**: chat, calendar and forums were used to foster collaboration; and the MSN Messenger ™ provided instant messaging, applications sharing, audio / video and whiteboard. **Coordination**: the stand alone application of MS Project was used. **Information Management**: based on MSN group's technology (files uploads, pictures publishing and customized html frames) was used as a repository of data/information.

CONNECT: No connections to ERP or SMEs systems were required in this prototype. Exchange information was achieved through file sharing and uploading in the environment.

MONITOR: A simple set of performance indicators were defined because the embedded coordination system was not able to track and control the development process. Indicators as milestones and dates were managed, but on-line tracking was not possible.

Step IV) The execution of the activities using the e-Engineering collaboration environment was carried out to demonstrate the development of a mechatronic product (Figure 2).

METHODOLOGY

REF	Activities	Methods	Tool
1	Conceptualization	Competitive Intelligence	Internet
2	Project Planning	Gant Diagram	MS Project
3	Benchmarking	Parametric Analysis	Internet
4	Patent Analysis	-	Internet
5	Market Requirements	QFD	QUALISOFT
6	Concept Generation	Morphological Matrix	CREAX
7	Design Variables	-	Pro-E, SPEED DE, SMT
8	Performance Analysis	FMEA Design	Pro-E CADMOULD
9	BOM & Drawings	Product Decomposition	Pro-E Design Explorer
10	Preparation & Purchasing	-	-
11	Construction & Assembly	FMEA Process	-

COLLABORATIVE ENVIRONMENT

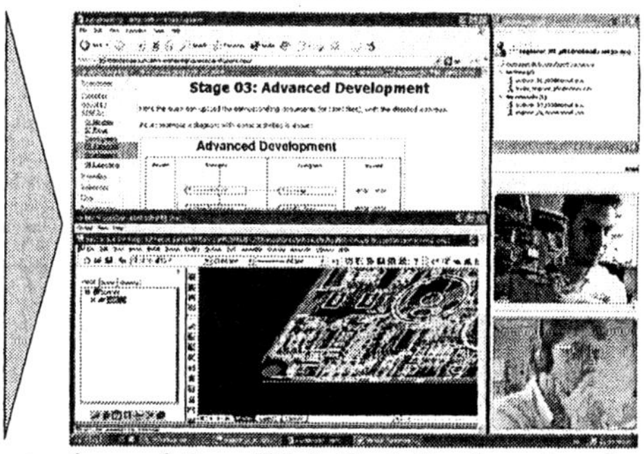

Figure 2 – Freeware collaborative tools used for a High Tech Product development pilot project

[1] http://cybercut.berkeley.edu/mas2/
[2] http://csim.mty.itesm.mx/grupos/smt
[3] http://spiderman.me.berkeley.edu/ducade/

4. CASE STUDY: COLLABORATIVE ENVIRONMENT FOR DRY-FREIGHT VAN DEVELOPMENT

A collaborative e-Engineering environment was design and implemented to support a Mexican company in developing new product. The collaboration was carried out between engineers in two different cities in Mexico (Monterrey and Cordova)

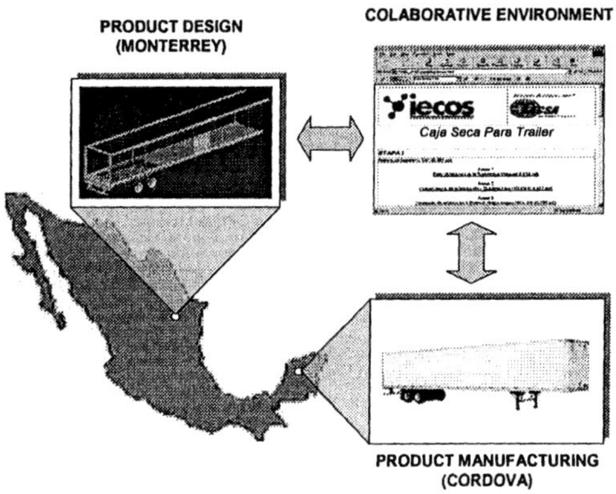

Figure 3 – e-Engineering collaboration for dry-freight van development

The objective of the project was to design and fabricate a Dry-Freight Van in a period of five months. Therefore, two processes were defined: Product Development (Dry-Freight Van) and Manufacturing Process Development. The concept of the collaboration was that the company located in Monterrey was in charge of the Dry-Freight design and the company in Cordova was responsible for the manufacturing process and fabrication (Figure 3).

Step I) Two processes were defined to set up the e-Engineering Collaborative Environment: Product Development and Manufacturing Process Development.

Step II) To design and fabricate a Dry-Freight Van it is necessary to develop a new process that allowed engineers from Monterrey to carry out concurrent activities with the engineers in Cordova, while sharing information and knowledge regarding the manufacturing capabilities and capacities of the company in Cordova. A project coordinator was defined, and leaders were assigned for both teams. The set of standard formats and reports were elaborated to allow an efficient exchange of information among teams; this was very important because even when both companies used the same CAD tool, but versions were different.

Based on the experience with the pilot project, the same freeware technologies were considered in the integration of applications to configure the collaborative e-Engineering environment ***(Step III).*** However some special considerations were essential because it was an industrial development. First of all, due to information confidentiality, a dedicated server was assigned to manage the collaborative environment, therefore any information exchange would be sent through a secure

server. Even so, the communication between companies was performed using NetMeeting™. The environment was designed with following considerations:
MODEL: In this project, it was necessary to create shared spaces. A simple *Password Accessing Webpage* was created, with a structure that emulated a workflow management system, according to the development process defined for the Dry-Freight Van design.

Figure 4 Interactions using the e-Engineering Collaborative Environment

INTEGRATE: **Functional:** stand-alone applications were used such as Mechanical Desktop, AutoCAD and Spreadsheets. The information was exchanged in AutoCAD format. **Collaboration**: NetMeeting™ was used because its communication capabilities were more than enough (instant messaging, applications sharing, audio / video and whiteboard). **Coordination**: MS Project was employed to manage the project. **Information Management**: webpage using XTML were created mainly to store and exchange information between the engineers (with restricted access to partners involved see figure 4).
CONNECT: No external applications were required.
MONITOR: timeframes and specific product deliverables were key performance indicators used to monitor the process. The control of the process was managed using MS Project software by updating project information in the shared webpage.
The execution of the Dry-Freight Van design (***Step IV***) was carried out according to the schedule planned in the project. Nevertheless the fabrication of the Dry-Freight Van was not possible due to financial problems with the company in Cordova.

5. CONCLUSION

Typical problems in product development scenarios at different geographical locations are: the absence of a structured product development process, lack of collaboration during early stages of life cycle product realization and miscommunication between design and manufacturing engineers. This research explored how a structured methodology can be used to design and configure collaborative engineering environments to suits the requirements of company focusing on specific issues such as: actual development process, available knowledge; human resources capabilities, and technological constraints. A pilot project and an industrial case study were presented to demonstrate how the methodology has been used to create such collaborative environments. The execution of the engineering activities using the e-Engineering collaboration environment enable the following: (1) improve the collaboration activities among engineers based on a structured development process (2) facilitate the coordination and exchange of information using the shared spaces (3) make it possible the interactions of teams located at different geographic locations and (4) improve the engineering tasks by using a set of functional tools. However further research in relation of how the collaborative activities are performed are needed because a collaborative engineering environment supported only by information technology is not sufficient to improve global product development. There is a need to understand better of how coordination, collaboration and cultural issues restraint engineers from a company to achieve a successful e-Engineering practice.

6. ACKNOWLEDGMENTS

The authors acknowledges the Chair in Mechatronics from the Instituto Tecnológico y de Estudios Superiores de Monterrey (ITESM - Campus Monterrey) for the support in the development of this research.

7. REFERENCES

1. Aca J., Mejía R.,Velandia M., García E., Galeano N., Ahuett H., Molina A., Wright P., "Integrated Product Development in Virtual Enterprises Supported by Web-based Applications", in Process and Foundations for Virtual Organizations, L.M. Camarinha-Matos, H. Afsarmanesh (Eds.), Kluwer Academic Publishers, 2003, pp. 361-368.
2. Bochenek, G.M.; Ragusa, J.M.; "Virtual collaborative design environments: a review, issues, some research, and the future". International Conference on Management of Engineering and Technology. PICMET '01. Portland, 2001
3. Lee, R. S., Tsaib, J. P., Kaoc, Lind, C. I. and Fane, K. C., 2003, "STEP-based product modeling system for remote collaborative reverse engineering", Robotics and Computer-Integrated Manufacturing, Volume 19, Issue 6, Pages 543-553.
4. Mervyn, F., Senthil Kumar, A., Bok, S. and Nee, A. Y. C, 2003, "Developing distributed applications for integrated product and process design", Computer Aided Design, Article in Press.
5. OUSD, 1998, "DoD Integrated Product and Process Development Handbook", Office of the Under Secretary of Defense (OUSD), Washington, DC - 20301-3000, August 1998.
6. Song, P., Tang, M. and Dong J. 2001, "Collaborative model for concurrent product design", IEEE Sixth International Conference on Computer Supported Cooperative Work in Design, 12-14 July 2001, Pages 212-217.
7. Swink, M., Sandving, J.C. and Mabert, V., 1996, "Customizing Concurrent Engineering Processes: Five Case Studies", Journal of Production and Innovation Management, Volume 13, Pages 229–244
8. Tipnis V.A., "Evolving Issues in Product Life Cycle Design: Design for Sustainability", Chapter 13, in Handbook of Life Cycle Engineering: Concepts, models and technologies, Edited by A. Molina, A. Kusiak and J. Sanchez, London, Kluwer Academic Publishers, pp. 399-412. 1999.
9. Yan, P. and Zhou, M, 2003, "A life cycle engineering approach to development of flexible manufacturing systems", IEEE Transactions on Robotics and Automation, Volume 19, Issue 3, June 2003, Pages 465–473.

29

APPLYING A BENCHMARKING METHODOLOGY TO EMPOWER A VIRTUAL ORGANISATION

Rolando Vargas Vallejos[1]; Jefferson de Oliveira Gomes[2]
[1] *Universidade de Caxias do Sul, BRAZIL (rvvallej@ucs.br)*
[2] *Instituto Tecnológico de Aeronáutica, BRAZIL (gomes@ita.br)*

This paper describes an experience applying a benchmarking methodology in a Virtual Organisation (VO) called Virfebras. A peculiar characteristic of Virfebras is that enterprises' members are competitors, but entrepreneurs are convinced that they have to work co-operatively in order to achieve common business-related goals.

The creation of Virfebras went through several phases. One of these phases, benchmarking, is considered important to supply strategic technological information and to empower entrepreneurs' behaviour, increasing trust between them. The present work describes the benchmarking methodology developed for seven mould and die enterprises, which are part of a Virtual Organisation called Virfebras.

1. INTRODUCTION

Mould and die manufacturing occupies a key position in the industrial value-added chain. The effectiveness of this sector exerts considerable influence on the competitiveness of production companies (Eversheim & Klocke, 1998).

According to Eversheim and Weber (2000), mould and die external boundary conditions are high pressure of time and cost as well as high quality standards caused by fierce competition, new technological developments and lack of qualified personnel. Internal boundary conditions are a complex production system of "one-of-a-kind" tools for a high product spectrum that are disturbed by a high percentage of alteration orders, repair orders and rush orders. Because of these conditions work environment of mould and die industries is always in turbulence.

To minimise this work environment turbulence, some concepts, philosophies, techniques, methods and tools are being used. One of these methods is *benchmarking*, which is used for measuring cost structures, processes and technological performance of enterprises, and to provide them with strategic information, which will lead to highest competitiveness (Kiesel, 2001).

In the present work a benchmarking methodology for the Virfebras Virtual Organization (VO) was developed and applied. The methodology was adapted from the successful German experience of the "Aachener Werkzeug- und Formenbau" in the mould and die sector. This experience is a mutual business effort of the Fraunhofer Institute for Production Technology (IPT) and the Laboratory for

Machine Tools and Production Engineering (WZL) at RWTH Aachen. Virfebras is composed of nine mould and die industries, and its characteristics are described below.

2. THE VIRFEBRAS VO

Virfebras is a VO that resulted from the partnership between the University of Caxias do Sul (UCS), nine mould and die industries, a Brazilian agency for supporting small and medium size companies (SEBRAE-RS), and the State of Rio Grande do Sul government. These companies, which have common market interests, decided to take part in a research project co-ordinated by UCS with the purpose of learning how to build a cooperative environment using ICT (Galelli et al., 2001).

Initially, Virfebras adopted the concepts of Virtual Enterprise (VE) and VO proposed by Camarinha-Matos and Afsarmanesh: *"A VE is a temporary alliance of enterprises that come together to share skills or core competencies and resources in order to better respond to business opportunities, and whose co-operation is supported by computer networks. A VO is a concept similar to a virtual enterprise, comprising a network of organisations that share resources and skills to achieve its mission/goal, but not limited to an alliance of enterprises"* (Camarinha-Matos & Afsarmanesh, 1999). As the concepts of VO and VE were new, one of the major challenges was to set up a VO without previous knowledge of this area. This way, entrepreneurs and professors, until now, have been discussing concepts and operational issues of VO and VE and applying them to their own companies.

Virfebras was created in 1999 and is located in the city of Caxias do Sul, south of Brazil. Whenever an order is submitted to the group, a VE is created, with one of the companies being the co-ordinator (VE-C), and other companies being the members (VE-M). The VE-C takes responsibility on the technical and legal aspects of the order. When the mould(s) and/or die(s) are delivered to the customer, and there are no more issues to deal with that order, the VE is dissolved. This way, within the VO, several VEs may exist at the same time, with one specific company being co-ordinator of one or more VEs, and member of others. It is worth mentioning that every company keeps its identity, and is also allowed to do business alone.

The creation of Virfebras went through several phases, namely *training and education*, *technology set up*, *market strategy*, *benchmarking*, *identification of shareable resources*, *organisational structure*, and *operational issues* (Galelli et al., 2001). In the present work, one of these phases, benchmarking, is considered important to provide strategic technological information, and to empower the entrepreneurs' behaviour increasing trust between them. The authors state that benchmarking should be considered as a strategic activity to increase the integration and performance of enterprises that are involved in a VO.

3. BENCHMARKING

Learning from the practices of others is part of human nature. We apply this principle intuitively at home, at work, in the society, wherever we are.

Benchmarking follows this same basic principle, trying to systematise and apply it in organisations with the purpose to supply them with strategic information.

There are several types and models of benchmarking. In the present work we will explore the competitive benchmarking, characterised by its application between competitors. The idea is that competitors have, if not the same practices, very similar ones; they have the same problems and common solutions too.

Benchmarking is a continuous and systematic process to evaluate products, services and processes against competitors, or renowned organisations considered world leaders in their field (Spendolini, 1993; Zairi & Leonard, 1996). The working definition is the search for industry best practices that lead to superior performance. According to Zairi and Leonard (1996), benchmarking is used at the strategic level to determine performance standards considering four corporate priorities: customer satisfaction, employee motivation and satisfaction, market share and return on assets, and at the operational level to understand the best practices or processes that help others achieve world-class performance.

Benchmarking is an opportunity for an organisation to learn from the experience of others. Even considering benchmarking an experience that stimulates self-questioning organisation processes will bring benefits, because a constructive crisis and challenging ideas and practices is beneficial per se (Boxwell, 1996; Zairi & Leonard, 1996).

4. BENCHMARKING FOR MOULD AND DIE INDUSTRIES

Mould and die manufacturing occupies a key position in the industrial value-added chain. A recent benchmarking work in Europe involving approximately 50 mould and die makers in Europe and South America (Eversheim et al., 2001) identifies some common characteristics of high performing enterprises. In addition, five factors were identified that have a high degree of correlation with those high performing enterprises. These factors are listed below:

a. Focus on the processes. Enterprises had clearly defined core competencies on processes and had developed specifics market niches.
b. A higher effort was extended to project and design in the production life cycle. Project planning and engineering attention was intensified before shop floor activities.
c. Continuous investments and improvement on their chosen core competencies.
d. Increased machine tool utilization reducing set up times. CNC programming was rigorously developed using integrated resources and methods.
e. Highly motivated workforce. Employees enjoy their work, and care about company performance.

The study found that companies that excelled at these practices had experienced superior performance and efficiency, more than 25% lower lead-time to produce moulds and dies.

Based on that experience and adapting it to some Brazilian market realities, a benchmarking methodology was developed for the enterprises that are part of the Virfebras VO. The question was: if benchmarking helps mould and die enterprises to be more competitive, how will affect the entrepreneur's relationship in a VO?

5. THE VIRFEBRAS BENCHMARKING PROJECT

The mould and die enterprises that form the Virfebras VO, in almost four years, have learned how to work in a co-operative environment. Currently, with this strategy, these enterprises offer a broader range of quality services to their customers, including lower costs and lower time-to-market.

The benchmarking project was an initiative of seven enterprises of Virfebras, which wanted to compare and evaluate their technological resources and processes, identifying their practices between the "best and worst practices" of the group through specific technological performance parameters.

The benchmarking methodology is divided in five phases called: *planning* (identification of technological performance parameters, elaboration of the benchmarking questionnaire and its application), *creation* and *analysis* of Virfebras database, *integration* (interpretation and discussion of the technological performance parameters with the group), *action* (planning actions to increase lower performance parameters of the group and of each enterprise, and apply them), and *checking* (the processes performances).

To analyse the technological performance parameters based on the "Aachener Werkzeug– und Formenbau experience" (Klocke & Bilsing, 2002), the "analysis of pairs" methodology was applied. This method is used to verify the number of prevalent qualitative parameters, in a context where it is not possible to establish a comparison numerically. In that way it is possible to establish a ranking of weights for certain technological characteristics.

To exemplify this methodology for the analysis of the technological characteristics of CNC milling machines for the mould and die sector, it is possible to link some parameters, as listed below:

1. Spindle power and speed; 2. Machining area; 3. CNC; 4. Work piece pallet; 5. Tool changer; 6. CAM interface; 7. Number of machines per operator; 8. Integrated measuring system; 9. etc..

Table 1 – Matrix for the "analysis of pairs" methodology considering a specific context (mould and die sector, roughness process, etc.) for milling machines

	1	2	3	4	5	6	7
1. Spindle power and speed							
2. Machining area	1						
3. CNC	1	2					
4. Work piece pallet	1	2	3				
5. Tool changer	1	2	3	5			
6. CAM interface	1	2	3	6	6		
7. Number of machines per operator	1	2	3	7	7	6	
8. Integrated measuring system	1	2	3	8	5	6	7
9. etc.							

With the information of Table 1 it is possible to establish a ranking of "weights" for certain technological characteristics (Table 2). To calculate the qualitative parameters' "weight", can be used the following equation:

$$\text{Weight} = [4(N_i - N_{min})/(N_{max} - N_{min}) + 1] \qquad (1)$$

Table 2 – Ranking of "weights" for certain technological characteristics

Characteristics	Number of citations in the "analysis of pairs"	Characteristic's weights	Characteristic's usefulness	Weight x Usefulness
1	7	5	2	10
2	6	4	2	8
3	5	4	2	8
4	0	1	0	0
5	2	2	1	2
6	4	3	2	6
7	3	3	2	6
8	1	2	1	2
Total				42

For the usefulness the criteria are: 0 – they don't use the characteristic, 1 – they use it sometimes, and 2 – they use it frequently.

Considering the characteristic's weight and usefulness we obtain a total value that is divided by the maximum factor acquired for that analysis. This result is multiplied by 5 to obtain the technological factor:

$$K_{tech} = 5 \times [Total_{used}/Total_{disp}] \quad (2)$$

6. RESULTS

The first result of this benchmarking methodology is summarised in more than 100 benchmarking figures describing several technological performance parameters that measure the organisational and technological performance of the mould and die enterprises that compose Virfebras.

Based on the technological factors of the European benchmarking project, the following associations were established with the Virfebras VO:

a. **Focus on the processes.** The Virfebras enterprises have not clearly identified their core competencies, and developed their market niches yet. Some companies had defined their competencies to manufacture injection moulds, others stamping tools, but without a specific process specialization. The moulds and dies are mostly of medium size and weight. The request for surface quality and dimensional tolerances are uncritical.

b. **Moulds design / manufacturing lead-time.** The companies apply lower time project and process planning comparing to the shop floor activities, which leads to a high effort to finishing and machining process stages (Figure 1).

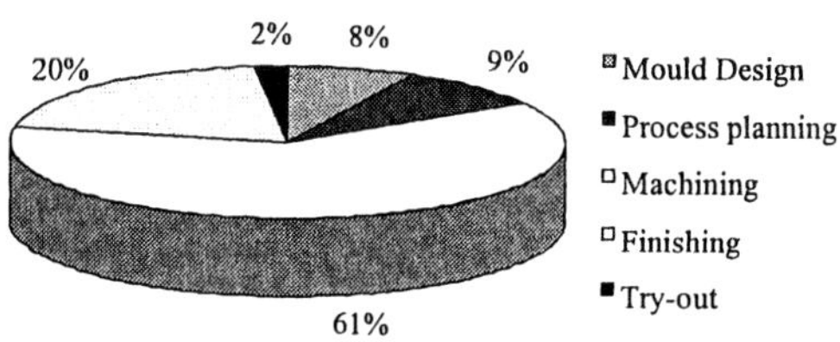

Figure 1 – Percentage of Virfebras moulds design / manufacturing lead-time.

In addition, the analysis shows that about 60% of the total costs are dedicated to tool production, which means high percentage personnel cost and low automation grade. Despite high personal cost, the number of specialists is low and, therefore, the projects do not usually have innovative solutions.

c. **Machine equipment.** During the efficiency evaluation of the machines the five main manufacturing processes in mould and die industry, respectively milling, sink and wire *Electrical Discharge Machining* (EDM), grinding, turning, were examined. For the evaluation some features were determined, related to a technological factor (K_{tech}), thus enabling an evaluation independent of the type of operation. To exemplify this analysis some figures describing the milling and the EDM processes are showed.

d. **Milling machines.** The enterprises within the Virfebras VO have a generic spectrum of milling machines, which covers all areas of different milling operations, roughing processes up to the *High Speed Cutting* (HSC) milling, except HSC for non ferrous (Figure 2).

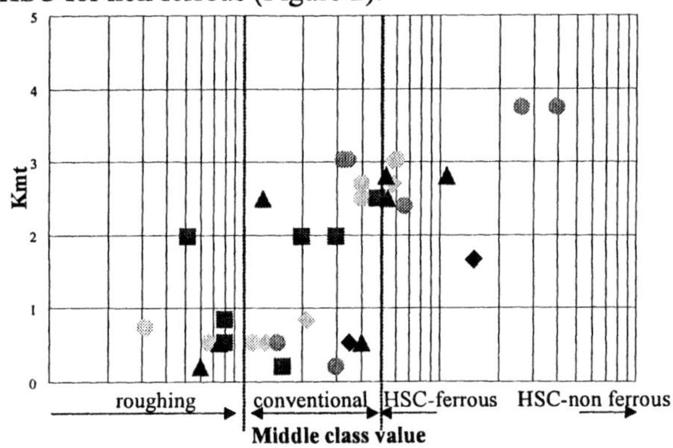

Figure 2 – Technological characteristics of Virfebras milling machines.

e. **EDM machines and processes.** The sink-EDM evaluation presented a low efficiency. The main reason for this is the low machine running time and low automation process. However, the wire-EDM evaluation presented a better efficiency, due to strong investment in new machines and better machine running time (Figure 3).

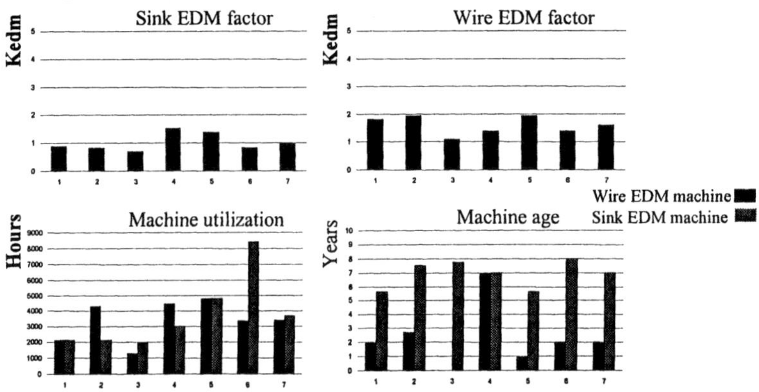

Figure 3 – Analysis of Virfebras EDM machines and EDM processes.

6.1 Comments

After the creation and analysis of the Virfebras database an integration workshop was organised with the aim of understanding and discussing the technological performance parameters with the entire group. In that workshop some common problems were identified, common for the entire group and for some enterprises that were grouped in sub-groups. By the end of this activity an action plan was established to increase lower performance parameters of the group. The action plan was extended to each enterprise identifying its own weaknesses.

The above-mentioned planned actions are being currently implemented. To improve the efficiency of the milling and CNC programming technology, specific work is being implemented, directly on the shop floor. Routines for tool life monitoring procedures to calculate the best cost / benefit relation of different tools were developed. Milling strategies for different surface features (pockets, ramps, cavities, etc.) were also created, and a study of suitable machining parameters, considering three different work-piece materials is being carried out.

6.2 Contribution for the VO context

One important contribution for the VO context is related to the entrepreneurs' behavioural change. Since the application of the questionnaire, analysis and activities after the initial benchmarking results, the entrepreneurs and employees change their behaviour from an initial posture of hiding information, to a frank exposure of confidential technological issues.

Some common technological problems that could be solved sharing experiences and best practices of the benchmarks were identified. The benchmarks accept to receive the visit of the other employees in order to show their best practices. With this experience they have the intention to share their knowledge using ICT through the VO participants. Additionally, were identified other kind of problems that could be solved by joining competencies and efforts in a co-operative way.

With this benchmarking experience the entrepreneurs are now convinced that working as a VO, not only aiming at reaching business opportunities, but also increasing their performance sharing their best practices, mainly in technological aspects, is a good alternative to become more competitive in the global market.

7. CONCLUSIONS

With the benchmarking project more than 100 benchmarking analyses were obtained, describing technological performances that could influence the competitiveness of the mould and die enterprises composing Virfebras.

The authors of this work believe that one result of this benchmarking project reflects in terms of organisational and technological strategies. Some of them are already being adopted by the enterprises within Virfebras.

Another important result of this work is the behavioural change that entrepreneurs went thorough. In the beginning of the project it was common to observe that entrepreneurs and employees used to hide information from the competitors, as time passed, they realised that this behaviour should be replaced by a new one, more suitable for a co-operative environment. They started sharing information and learned that they usually have, if not the same problems, at least similar ones. If they started sharing solutions, every company could benefit from this exchange process. Nowadays entrepreneurs usually share information, once in a while one of them invites the other partners to visit its industry to show its resources and internal processes. The intention is to use the ICT in order to speed up the knowledge sharing. Actually, with this initiative Virfebras VO is studying the possibility to implement a Knowledge Management program.

Finally, it has been possible to observe a significant behavioural change in all Virfebras participants; one of the major changes is related to trust. The evolution of the entrepreneurs' behaviour when dealing with internal or particular issues is remarkable. That's why, the authors of the present work, state that benchmarking should be consider as a strategic activity to increase the integration and performance of enterprises that are involved in a VO.

Acknowledgements

The authors would like to thank SEBRAE-RS for its support to the benchmarking project. Also, we are very grateful to the Virfebras enterprises whose participation has been fundamental for the success of this project.

8. REFERENCES

1. Boxwell, R. Jr., Vantagem competitiva através do benchmarking. São Paulo: Makron Books, 1996.
2. Camarinha-Matos L. M., Afsarmanesh H. The virtual enterprise concept. In L. M. Camarinha-Matos, H. Afsarmanesh (Eds.), Infrastructures for Virtual Enterprises – Networking Industrial Enterprises, Kluwer Academic Publishers, 1999, p. 3-14.
3. Eversheim, W., Geschke, H.-J., Bilsing, A., Deckert, C., Westekemper, M. Vom Komponentenfertiger zum Systemlieferanten - Strategische Ausrichtung des Werkzeugbaus der Hettich Umformtechnik. In: Form + Werkzeug, 2001.
4. Eversheim, W., Klocke, F. Werkzeugbau mit Zukunft – Strategie und Technologie, Springer Verlag, Berlin, 1998.
5. Eversheim, W., Weber, P. The right strategy to success – Strategic orientation in die and mould manufacturing, presentation of the colloquium "Werkzeugbau mit Zukunft", Aachen, 22.-28. September 2000
6. Galelli, A., Costa, C. A., Vallejos, R. V., Graciolli, O. D., Luciano, M. A. "A Virtual Organisation for the Mold and Die Industry in Brazil". In: 5TH World Multiconference on Systemics, Cybernetics and Informatics/ISAS - SCI 2001. Orlando, FL: International Institute of Informatics and Systemics, 2001. v. III, p. 303-308.
7. Kiesel, H. The Development Tendencies at the Tooling Shop of Daimler-Crysler do Brasil Ltda. Presentation at the 6th International Seminar "High Technology", Piracicaba, Brazil, 2001.
8. Klocke, F.; Bilsing, A. Technologisches Benchmarking im Werkzeug- und Formenbau, WT. Werkstattstechnik 92, Nr.11/12, p.595-599.
9. Spendolini, Michael Jr. Benchmarking. Tradução de Kátia Aparecida Roque. São Paulo: Makron Books, 1993.
10. Zairi, M., Leonard, P. Practical Benchmarking: The Complete Guide. London: Chapman & Hall, 1996.

30 A CONTRIBUTION TO UNDERSTAND COLLABORATION BENEFITS

Luis M. Camarinha-Matos, António Abreu
New University of Lisbon
Quinta da Torre – 2829 Monte Caparica, PORTUGAL
cam@ uninova.pt, ajfa@ fct.unl.pt

The identification and characterization of collaboration benefits is an important element for the wide adoption of the collaborative networks paradigm. Departing from some categorization of intuitive advantages of collaboration, this paper introduces an approach for the analysis of benefits in collaborative processes for enterprises networks. The potential application of some indicators derived from this analysis is also discussed in VO breeding environment (VBE) context.

1. INTRODUCTION

In most literature on Virtual Enterprises / Virtual Organizations there is an intuitive assumption that these forms of collaborative networks bring clear advantages to its members and represent even a survival factor in turbulent socio-economic scenarios. However, in spite of this assumption, it is also frequently mentioned that the lack of objective measurements, clearly showing the benefits of such organizational forms, is an obstacle for a wider acceptance of this paradigm.

What will my organization benefit from embarking in a collaborative network? Will the benefits compensate for the extra overhead and even the risks that collaboration implies? These are questions that many SME managers ask when the issue of collaboration is brought in.

It is, in fact, difficult to prove the advantages of (dynamic) collaborative networks in contrast to more traditional organizational forms in terms of improved performance. Being able to measure the performance of a collaborative network as a whole and the performance of each of its members could represent an important boosting element for the wide acceptance of the paradigm. However performance indicators tailored to collaborative networks are not available yet [6].

Performance measurement depends on the premises of the measurement system used. Collaborative networks challenge the premises of the methods developed in the past, therefore the applicability of existing measurement systems in this area is questionable.

Before establishing a new set of indicators it is necessary to analyze in more detail the *basis* of collaboration and its benefits. Understanding the nature of collaboration benefits is also important as a way to ensure that every member of the network understands the measurements in the same way. This is also a requirement for goals alignment in order to facilitate the coherence of members' goals with the measurements.

This paper introduces some discussion of the nature of collaboration benefits as a contribution to a future identification of a set of performance indicators that are suitable for collaborative networks.

2. SOME BACKGROUND

A number of theories focused on different perspectives of cooperation have been proposed in various disciplines. Some relevant examples include:

- **Resource Dependence theory** – which is concerned with the arrangements between enterprises to reduce uncertainty and dependency from products, services, tangible and intangible resources and competencies, to contribute to the creation of their offerings to customers. From this theory point of view, cooperation is explained as an attempt of the enterprises to adapt to their environments to enable the procurement of necessary resources while at the same time maintaining acceptable power-dependency relationship [5,11].
- **Transactions cost theory** - Transactions costs are generally defined as being the cost for gathering information, negotiation and contracting, and physical transaction of *objects* through a defined interface. According to this theory, enterprises and markets are alternative governance structures that differ in their *transactions costs*. From this point of view cooperation is explained as an organizational "hybrid" form between the market and the enterprise [14].
- **Game theory** – A mathematical framework designed for analyzing the interaction between several actors whose decisions affect each other. An interactive situation is described as a *game* including an abstract description of the players (actors), the courses of actions available to them, and their preferences over the possible outcomes. From this perspective, cooperation processes take place when the total utility of acting in conjunction is greater than the sum of utilities for each participant considered individually [1,8].
- **Complexity theory** – Complexity theory deals with systems that show complex structures in time or space, often hiding simple deterministic rules. A complex system can be understood as any network of interacting agents (processes or elements) that exhibits a dynamic aggregate behavior as a result of the individual activities of its agents. Some important characteristics of complex systems include: non-determinism, limited functional decomposability, distributed nature of information, and emergence and self-organization. Emergence is in fact one of the most important properties of complex systems, what makes this paradigm an appealing approach for the analysis of advanced collaborative networks [2,7,9,12].
- **Contingency theory** – Contingency theory is concerned with the identification and understanding of the enterprise structure in different conditions (or contingencies). Various forms of organization can coexist depending on different conditions. These conditions depend on internal factors that are specific to each enterprise but also external factors like: the environment uncertainty and the

distribution of resources. This theory considers a cooperation process as a fast way for an enterprise to quickly adjust its structure to an environment with high uncertainty [10].

Although offering some structuring elements, these theories are mostly "enterprise-centric" (except the theory of complexity and game theory) and lack an inter-organizational focus.

Some other more "network-centric" contributions can be found in various works from the sociology area dealing with "social actors networks". In this area concepts such as *prominence* of actors in a network, *centrality*, *prestige*, etc. and approaches to compute them have been suggested [13]. These approaches are perhaps more abstract, lacking some economic and practical focus, but can be used as a source of inspiration to analyze collaborative networks of enterprises.

From the traditional literature on virtual enterprises / virtual organizations, a number of variables related to the identification of collaboration benefits have been suggested (Table 1).

Table 1 – Cooperation variables and associated target goals

Cooperation variable	Target goal
Costs	Share costs
Risks	Share risks
Dependence	Decrease the dependence level in relation to third party
Innovation	Increase innovation capacity
Market position	Defend a position in the market
Flexibility	Increase flexibility
Agility	Increase agility
Specialization	Increase specialization
Regulation	Establish proper regulations
Social causes	Share social responsibility

Table 2 shows, for each target goal, some examples of associated (intuitive) advantages of collaboration.

Table 2 – Example of some associated advantages

Target goal	Example of some advantages associated to collaboration
Share costs	• Have access to new markets and/or businesses without the need to make high investments. • Share R&D costs. • Ability for SMEs to compete with large competitors.
Share risks	• Companies operate in changing environments and with limited, therefore imperfect, knowledge. Consequently in some cases the level of uncertainty may have a negative impact on the decision-making processes. Sharing knowledge among several partners allows a reduction of this uncertainty level. • When several partners are involved in a collaborative project there is a partition of the responsibilities among them (co-responsibility). • In some cases solidarity mechanisms can be established among partners. • Also enabling the competition of SMEs with large companies.

Decrease the dependence level in relation to third party	• All companies depend on others to some extent for products, services, raw materials, tangible and intangible resources and competencies. Through cooperation companies can reduce this dependence by creating privileged links to other firms in an attempt to reduce transaction costs that arise when uncertainty increases. • Also enabling the competition of SMEs with large companies.
Increase the innovation capacity	• Increase the capacity of generating new ideas through the combination of the existent resources and diversity of cultures and experiences (critical mass). • Emergence of new sources of value. • Reduction of the life cycle of the products and technologies. • Possibility of developing more robust products fitting the customers' expectations and therefore contributing to an increase of the quality.
Defend a position in the market	• Achievement of economies of scale by sharing resources. • Establishment of defensive coalitions with the purpose of building entry barriers in order to defend themselves against a dominant firm or a new player. • Establishment of offensive coalitions with the purpose of developing competitive advantages and strengthening their position by diminishing the other competitors' competitiveness. • Increase the negotiation power in relation to suppliers and/or customers that are outside of the collaborative network. • Also enabling the competition of SMEs with large companies.
Increase flexibility	• Share of resources and combination of skills among partners. • Use the core competences from other partners. • Increase the adaptation capacity towards several business environments simultaneously. • Offer a broader range of products / services. • Grow for new segments in a stable way reaching a larger stability.
Increase agility	• React in a short period of time to a business opportunity through the establishment of more agile procedures. • Increase the interoperability between several processes and products (establishment of norms)
Increase specialization	• Let companies concentrate their resources on the critical activities.
Establish proper regulations	• Definition of rules to avoid opportunistic behaviors and to avoid conflicts. • Increase common culture of trust.
Share social responsibilities	• Obtain recognition from others (intangible value). • Develop social responsibility. • Altruism. • Reinforce values that are common.

From a macro-level, these potential benefits can be regarded from two perspectives:

- *Survival capacity* – Reflecting the capacity of an actor (e.g. company) or a group the actors to stay in operation "alive" when confronted by forces, which tend to destroy them.
- *Performance capacity* – Reflected in the capability of an actor or groups the actors to better accomplish their tasks.

One question is then whether each of the above potential benefits of collaboration is more relevant to a situation of survival or performance improvement. In order to identify possible answers, a small survey (45 respondents) was conducted by email, involving industry and academia experts from Portugal, Italy, Spain, Germany, UK, Denmark, Turkey, Austria, USA, Canada, and Japan. Fig. 1 shows an excerpt of the used questionnaire.

Fig. 2 summarizes the collected answers. The adopted scale considers the following:

- *Strong relationship* - When the distribution of most answers in relation to the variable is in the interval of 75% to 100% of relevance.
- *Moderate relationship* - When the distribution of most answers in relation to the variable is in the interval of 25% to 50% of relevance, or in the interval of 50% to 75% of relevance.
- *Weak relationship* - When the distribution of most answers in relation to the variable is in the interval of 0% to 25% of relevance.

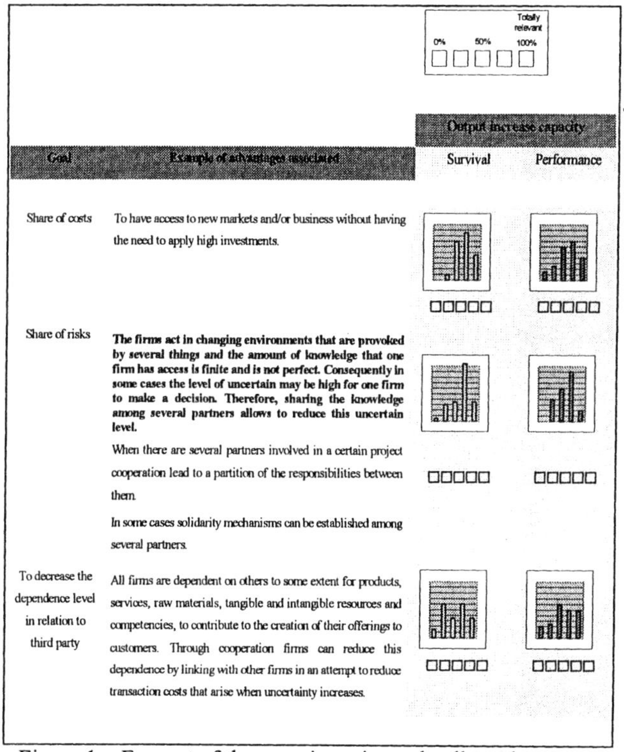

Figure 1 – Excerpt of the questionnaire and collected answers

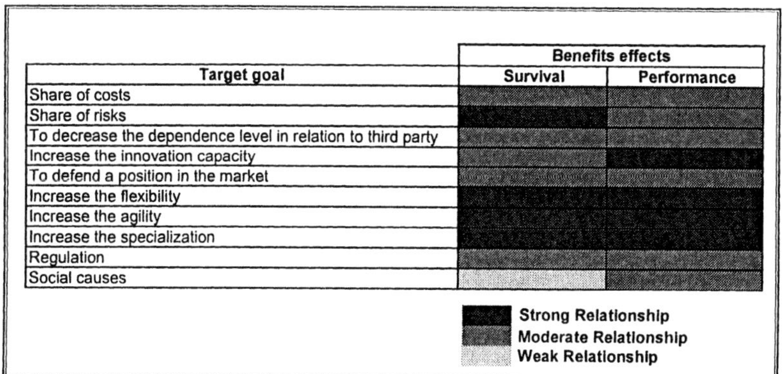

Figure 2 – Contribution of benefits to survival and performance increase

From these results one can conclude there is a clear (intuitive) perception that cooperation benefits are related to the two strategic goals – survival or performance increase.

It is also visible that if the primary goal of a company is to stay "alive" it would likely be motivated to find cooperating partners with the purpose of sharing risks. On the other hand, if the strategic goal is to improve performance, the motivation for partnership will be more related to increasing innovation capacity. Increasing flexibility, agility, and specialization are equality important.in both cases.

3. BENEFITS ANALYSIS

3.1 Task performance benefits

For the purpose of the following discussion, let's consider Task Performance Benefits (TB) as the benefits that result from the performance of a task in the context of a collaborative process. A collaborative process is understood as a set of tasks performed by the collaborative network members towards the achievement of a common goal (e.g. the business goal that motivates the creation of a Virtual Enterprise). For reasons of simplicity we consider a level of granularity of tasks such that each task is performed by a single member of the network (single actor).

The term benefit is used with the same meaning as net profit. In the following analysis benefits are assumed as abstract quantifiable measurements. The actual meaning of a benefit depends on the underlying value system. For instance, in the context of networks of enterprises it most likely represents a measure of economic benefits while in the context of a NGO it could represent a more abstract indication of social prestige or peer recognition. In general this concept represents a combination of multiple variables (as discussed in previous section). How to combine those variables into a single value is not addressed in this paper.

Let $TB_{ji}(t_{jl})$ – benefits for actor a_i as a result of the performance of task t_l by an actor a_j.

When $i = j$ this represents a *self-benefit* (Fig. 3.a); otherwise it is a *received benefit* (perspective of a_i) or *contributed benefit* (perspective of a_j) (Fig. 3.b).

In the context of a collaborative network the total *self-benefits* for a given actor a_i is given by the sum of the self-benefits obtained from all tasks performed by this actor:

$$Self\ Benefits\ (SB_i) = \sum_{l=1}^{L} TB_{ii}(t_{il})$$

where: t_{il} – description of a task t_l performed by actor a_i

L – total of task performed by actor a_i

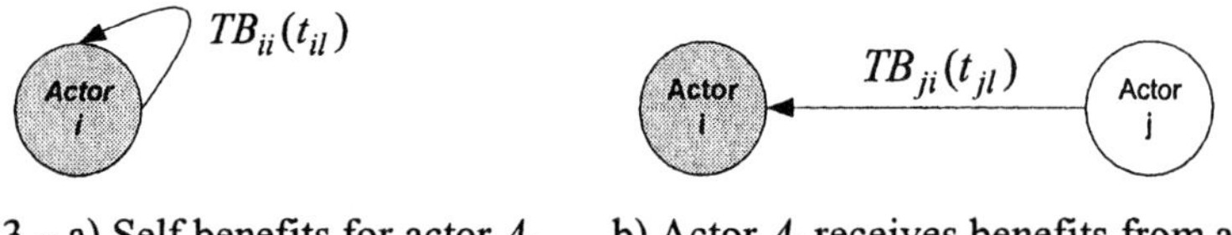

Figure 3 – a) Self benefits for actor A_i b) Actor A_i receives benefits from actor A_j

The total of the benefits an actor a_i receives as a result of the performance of another actor a_j is given by:

$$Received\ Benefits\ (RB_{ij}) = \sum_{l=1}^{L} TB_{ji}(t_{jl}) \quad i \neq j$$

where: t_{jl} – description of a task t_l performed by actor a_j

L – number of tasks performed by actor a_j

And the benefits received (external benefits) by an actor a_i as a result of the performance of all actors involved in the cooperation process is given by:

$$External\ Benefits\ (EB_i) = \sum_{j=1}^{N} RB_{ij} \quad i \neq j$$

where: N – Number of actors involved in the collaborative network.

The external benefits, i.e. what an actor perceives as direct benefit of collaboration, shall be > 0. One actor might accept a non-positive value for some collaboration processes, but in the long run the result needs to be positive in order to keep it interested in collaboration.

The total benefits for a_i are: *Total individual benefits (TIB_i)* = $SB_i + EB_i$.

From the network point of view, the total received benefits are:

$$Total\ Received\ Benefits\ (TRB) = \sum_{j=1}^{N} EB_i$$

Similarly, from the contributor point of view we can define: Benefits contributed by an actor a_i to its partner a_j as a result of all tasks performed by a_i:

$$Contributed\ Benefits\ (CB_{ij}) = \sum_{l=1}^{L} TB_{ij}(t_{il}) \quad i \neq j$$

where: t_{il} – description of a task t_l performed by actor a_i

L – total tasks performed by actor a_i

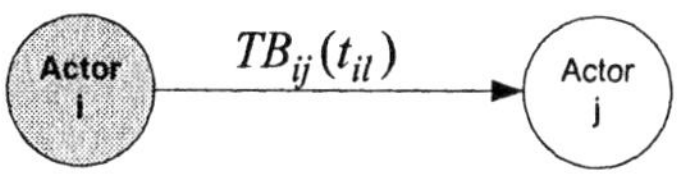

Figure 4 – Actor a_i *contribute* benefits to actor a_j.

And the sum of benefits contributed (social contributed benefits) by an actor a_i to all its partners as a result of its performance in the cooperation process is given by:

$$Social\ Contributed\ Benefits\ (SCB_i) = \sum_{j=1}^{N} CB_{ij} \quad i \neq j$$

where: N – Number of actors involved in the collaborative network.

In a sustainable collaboration network, at least in the long term, SCB_i shall be >0, otherwise the actor would be considered selfish.

Total benefits resulted from an actor a_i:

Individual Generated Benefits (IGB_i) = $SB_i + SCB_i$

From the network point of view, the total contributed benefits are:

$$Total\ \ Contribute\ d\ \ Benefits\ \ (TCB) = \sum_{j=1}^{N} SCB_i$$

Obviously, the total received benefits = total contributed benefits, i.e. *TRB* = *TCB*. In a sustainable collaborative network these benefits shall be greater than 0.

3.2 Task dependence

There is a task dependence when the realization of a task by one actor, and therefore the respective benefits, depends on other agents that are not involved in the execution but have an influence on that execution. An example of task dependence occurs when an actor with a good reputation in the market is present as member of a collaborative network and this fact helps others to acquire a contract (task) that otherwise would be lost.

This task dependence (or influence from some actors) can be modeled as an enabling factor with a value between 0 (inhibitor) and 1 (enabler). The benefits resulting from a dependent task are therefore conditioned by this enabling factor:

$$Dependable\ Task\ Benefits\ DTB_{ij}(t_{im}) = TB_{ij}(t_{im}).\times \prod_{d=1}^{K} D_{di}$$

where: K – actors that influence task t_{im}

It shall be noted however that this expression does not properly model all dependency situations. For instance, it does not capture the cases in which the influences of two or more actors are addictive. What if two actors with positive influence "compensate" for one with negative influence? This formula gives predominance to the negative influence (any value of D_{di} less than 1 represents some form of negative influence).

One possibility is to consider different types of dependencies (Fig. 5):

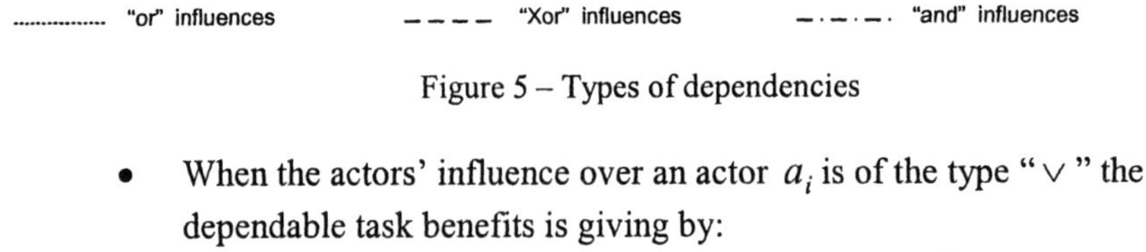

Figure 5 – Types of dependencies

- When the actors' influence over an actor a_i is of the type " $\vee$ " the dependable task benefits is giving by:

  $$Dependable\ \ Task\ \ Benefits\ \ DTB_{ij}(t_{im}) = TB_{ij}(t_{im}) \times \sum_{d=1}^{K} D_{di}$$

  where: K – Number of actors that influence task t_{im}
- When the actors' influence over an actor a_i is of the type " $\wedge$ " the dependable task benefits is giving by:

$$Dependable\ Task\ Benefits\ DTB_{ij}(t_{im}) = TB_{ij}(t_{im}).\times \prod_{d=1}^{K} D_{di}$$

- When the actors' influence over an actor a_i is of the type "xor" the dependable task benefits is giving by:

$$Dependable\ Task\ Benefits\ DTB_{ij}(t_{im}) = TB_{ij}(t_{im}) \times \max\{D_{di}\}$$

It is also important to distinguish between *influences* during the execution of a task and influences during the acquisition of a business opportunity (that is acquired will imply the execution of several tasks). In this discussion we are considering the first case of influences.

If we consider that tasks performed by an actor can be divided in two groups – independent and dependent - the self-benefits for a_i can then be represented by:

$$Self\ Benefits\ (SB_i) = \sum_{l=1}^{L} TB_{ii}(t_{il}) + \sum_{m=1}^{M} DTB_{ii}(t_{im})$$

where: L – independent tasks performed by a_i
M – dependent tasks performed by a_i

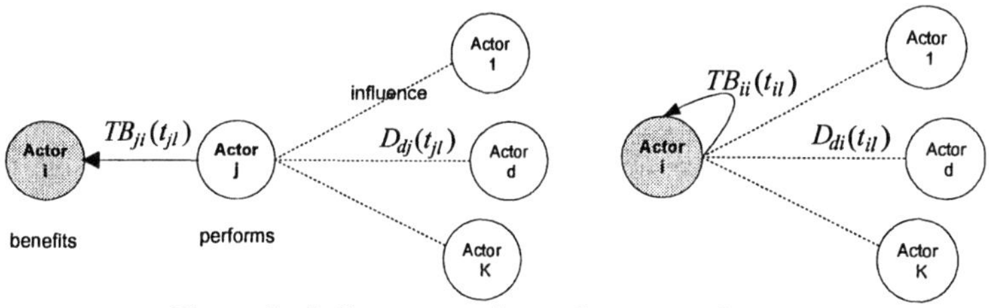

Figure 6 – Influences on the performance of a task

Similarly for received benefits:

$$Received\ Benefits\ (RB_{ij}) = \sum_{l=1}^{L} TB_{ji}(t_{jl}) + \sum_{m=1}^{M} DTB_{ji}(t_{jm})$$

where: L – independent tasks performed by a_j
M – dependent tasks performed by a_j.

3.3 Some cooperation indicators

In this section, some basic indicators of the cooperation process are introduced.

- ***Individual contribution index*** – normalized contribution of an actor to the collaborative network:

$$Individual\ Contribution\ Index(ICI_i) = \frac{Social\ Contributed\ Benefits\ (SCB)_i}{Total\ Contributed\ Benefits(TCB)}$$

- ***Apparent individual contribution index*** – an indicator based on the number of contribution links (i.e. the *out degree* of the actor in the graph representing the cooperation benefits):

$$Apparent\ contribution\ index\ (ACI_i) = \frac{N^{o}\ out\ links\ leaving\ a_i}{N-1}$$

where: N is the number of members of the collaborative network

This index gives an apparent and simple do compute measure of the involvement of an actor as a contributor to the collaboration process. An actor with an ACI close to zero is not perceived as a good contributor to the network (although the real value of its contribution is better expressed by ICI).

- ***Individual external benefits index*** – normalized external benefits received by an actor:

$$Individual\ \ external\ \ benefits\,index\,(IBI_i) = \frac{External\ \ Benefits\ \ (EB_i)}{Total\ \ Received\ Benefits\ (TRB)}$$

- ***Apparent individual benefits index*** – an indicator based on the number of received contribution links (i.e. the *in degree* of the actor in the graph representing the cooperation benefits):

$$Apparent\ \ benefits\ \ index\ \ (ABI_i) = \frac{N^{\circ}\ links\ \ arriving\ \ at\ a_i}{N-1}$$

This index also expresses the ***popularity*** or ***prestige*** of the actor [13] in the sense that actors that are prestigious tend to receive many external benefits links.

4. APPLICATION POTENTIAL

The existence of a *VO breeding environment* (VBE) is considered by many authors as a pre-condition for the effective establishment of dynamic virtual organizations [3], [4]. A VBE represents an association or pool of organizations and their related supporting institutions that have both the potential and the will to cooperate with each other through the establishment of a "base" long-term cooperation agreement. When a business opportunity is identified by one member (acting as a broker), a subset of these organizations can be selected and thus forming a VE/VO. Various VE/VOs can coexist at the same time in the context of a VBE. A breeding environment, being a long-term networked structure, presents the adequate base environment for the establishment of cooperation agreements, common infrastructures, common ontologies, and mutual trust, which are the necessary facilitating elements when building a new VE/VO. In other words, VBE represents a group of organizational entities that have developed a *preparedness* for cooperation, in case a specific opportunity arises. Industry *clusters* or industry districts are examples of such breeding environments.

In this context, the definition of a cooperation benefits model and a set of indicators can be a useful instrument to the VBE manager, to a VE/VO broker, and to a VBE member. Let's suppose a record of the past cooperation processes, represented as collaboration benefits graphs (performance catalogue), is kept at the VBE management level. Using simple calculations as illustrated in previous sections, and some simple statistics / data mining (performance and link analysis), it is possible to extract several macro and micro indicators regarding the performance of the VBE and its members as a collaborative structure. These indicators can be determined for a particular collaboration process (a particular VE/VO occurrence) or over a period of time (average values) and can be used in decision-making processes, such as planning a new VE/VO.

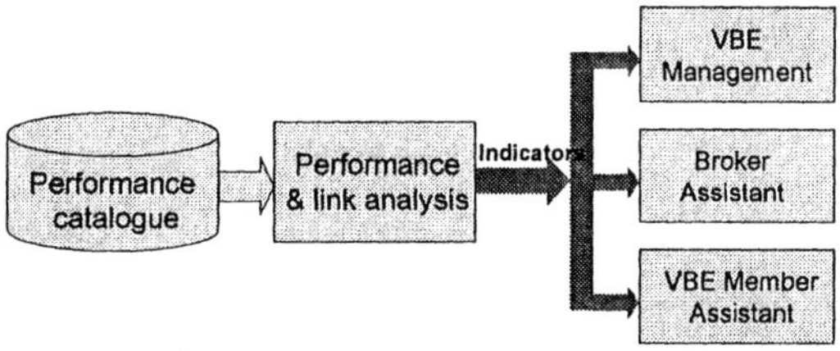

Figure 7 – Framework to support VBE to extract support indicators

For instance:

At the VBE management level:

Global indicators (e.g. cohesion level, identification of closely-related sub-groups / cliques that work well together, average total benefits for all past collaborative networks /VOs, detection of "parasites"/ego-centric) or member specific indicators (e.g. average benefits generated by each member and their variance).

At the broker's level:

Indicators that may help in: partner selection for a specific VO being planned (e.g. average individual contribution index), in the analysis of the planned VO (cohesion, level of uniformity of the external benefits index, global benefits of the VO), etc. For instance, if the benefits in a particular VO are mainly self-benefits it means the level of (explicit) collaboration is low (the work could be done in isolation). For partners selection it is also important to analyze the history of dyads (an actor a_i might be more effective when collaborating with a specific actor a_j than with any other in the VBE). This analysis can be extended to groups larger than 2 elements (cliques).

At the member's level:

A member may ask questions such as: Shall I get involved in this consortium? Was my participation in this collaborative process beneficial to me? What is my level of "popularity" or "prestige"? What is the balance of my interactions with a specific member (dyad relationship)? Have I got reciprocity, in the past, from the potential members to be involved in the same VO?

It shall be noted that other attributes besides the benefits can be recorded associated to the benefits collaboration graph (e.g. failures, delays in delivering results). The above discussion is only an illustration of the approach being followed in a research initiative trying to contribute to the creation of a framework for VBE management. Methods developed in the Social Networks area, combined with a system to monitor and keep track of performance history, are particularly useful here. Further developments [4] and validation of the approach are nevertheless necessary.

For instance, the concepts of centrality and prestige defined in the Social Network Analysis area and typically measured with basis on the outdegrees and indegrees, respectively, need to be discussed in the VBE context. Being "central" or "prestigious" in the (limited) universe of a VBE does not necessarily mean any extra "visibility" to the outsiders (potential customers or originators of the business opportunity), but it certainly has something to do with the internal power relationships.

The assignment of values to each arc of the benefits graph might not be an easy task (when we want to record the history of performances). On the other hand, if the purpose is the elaboration of a simulation model to study emerging behaviors then the approach is easier to adopt as the actual values of such arcs will be parameters of the simulation process.

5. CONCLUSIONS

Reaching a better characterization and understanding of collaboration benefits is a key pre-condition for a wide adoption of the collaborative networks paradigm in its various manifestation forms. This understanding is also a base for the establishment of proper performance indicators to be used in decision making processes at various levels: VO breeding environment management, VO brokering, and VO breeding environment membership.

Some preliminary steps in this direction, inspired in the Social Networks analysis, were presented. Initial results illustrate the applicability of the suggested approach. Further steps are necessary towards the elaboration of the drafted analysis framework as well as its validation.

The ECOLEAD integrated project recently started in the context of the 6th framework program of the European Commission provides the context for the continuation of this work.

6. REFERENCES

1. Axelroad, R. (1984). The Evolution of Cooperation, Basic Books.
2. Bar-Yam, Y. (1997). Dynamics of Complex Systems, Addison-Wesley.
3. Camarinha-Matos, L. M.; Afsarmanesh, H. (2003). Elements of a VE base infrastructure, J. Computers in Industry, Vol. 51, Issue 2, Jun 2003, pp. 139-163.
4. Camarinha-Matos, L. M.; Afsarmanesh, H. (Ed.s) (2004). Collaborative Networked Organizations – A research agenda for emerging business models, Kluwer Academic Publishers, ISBN 1-4020-7823-4.
5. Dussauge, P. and B. Garrette (1999). Cooperative Strategy - competing Successfully through Strategic Alliances, John Wiley& Sons LDT.
6. Evans, S.; Roth, N.; Sturm, F. (2004). Performance measurement and added value of networks, in Collaborative Networked Organizations – A research agenda for emerging business models, Kluwer Academic Publishers, ISBN 1-4020-7823-4.
7. Eve, R. A., S. Horsfall, et al. (1997). Chaos, Complexity and Sociology - Myths, Models and Theories, SAGE Publications.
8. Myerson, R. B. (1997). Game Theory Analysis of Conflict, Harvard University Press.
9. Pavard, B. Complexity Paradigm as a framework for the study of Cooperative Systems http://www.irit.fr/COSI/summerschool/bpstudy.pdf - visited on (22-04-2004)
10. Penã, N. A. and J. C. F. Arroyabe (2002). Business Cooperation, Palgrave macmillan.
11. Pfeffer, J. and G. R. Salancik (1978). The External Control of Organisations: a Resource Dependence Perspective, Harper Row - New York.
12. Sterman, j. D. (2000). Business Dynamics - Systems Thinking and Modeling for Complex World, McGraw-Hill.
13. Wasserman, S. and K. Faust. (1994). *Social Network Analysis - Methods and Applications*. Cambridge University press.
14. Williamson, O. E. (1985). The Economic Institutions of Capitalism: Firms, Markets, Relational Contracting, New York: Free Press.

31

PREDICTIVE PERFORMANCE MEASUREMENT IN VIRTUAL ORGANISATIONS

Marcus Seifert, Jens Eschenbaecher
BIBA Bremen, sf@biba.uni-bremen.de
GERMANY

Predictive Performance measaurement is a decisive task for the further evolution of virtual organisations. The methodologies developed so far have a strong focus on either supply chains or extended enterprise orientied structures. Current trends in manufacturing enterprises change from long-term supply chains to dynamic network co-operation where both the structure and the entities of the network are dynamic and created with respect to the actual customers' order. An important management task within this kind of co-operation is to identify the most suitable partners and to build up the best performing virtual team within a short time frame to fulfil the customers' wishes. This paper discusses a predictive performance measurement approach as planning tool for virtual organisations to anticipate the performance of a planned virtual team.

1. OPTIMISATION POTENTIALS WITHIN VIRTUAL TEAMS COMPARED TO TRADITIONAL SUPPLY CHAINS

Long term co-operations between companies and their suppliers as well as the continuous improvement of more or less stable processes have been the main characteristics of the industrial production for a long time. Related concepts such as supply chain management were applicable in those cases where market needs and products are relatively stable and where the competitiveness is mainly based on the continuous optimization of the established process chain. Not the single company competes on the market but the whole supply chain as the provider of a wide variety of processes to offer a complex product faces the competition (Boutellier 1999, S. 66).

Current trends in manufacturing enterprises are changing from long-term co-operations between suppliers to ad-hoc co-operations related to the specific needs of dynamic customers' order (Hieber 2001, p. 2). The structure of these kinds of temporary networks is well known as the concept of virtual organisations (Camarinha-Matos 2004). The duration of collaboration in a virtual organisation consisting of a variety of independent companies is often limited to one certain project and the network has to be re-built for the next project. Virtual teams consisting out of distributed members collaborating in project teams will become the inevitable path of future (Gassmann and Zedtwitz 2003).

The main characteristic of a network collaboration in virtual teams compared to traditional supply chains is the short operational phase. While supply chains are established to operate over a long time, virtual teams are configured to realize at least one customers' order. Figure 1 shows the life-cycle of a network in virtual organisations (bottom) compared to long-term co-operations in supply chains.

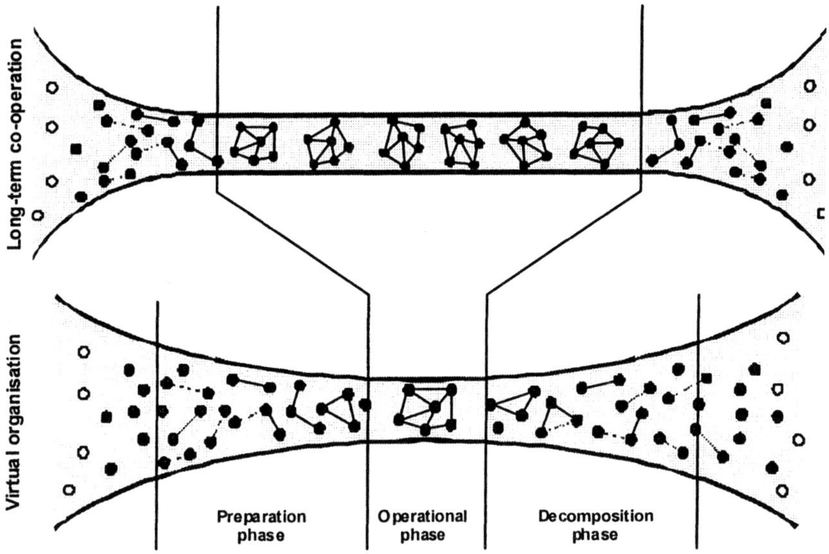

Figure 1 Life cycle of a enterprise network within the virtual organisation compared to long-term co-operations (Thoben/Jagdev 2001)

The identification and configuration of the best performing collaborative network to fulfil a specific customers' order during the preparation phase of a virtual organisation is an important management issue. Apart from the intra-company processes, especially the inter-company processes are the main success factors of a high performing co-operation due to competitiveness requires the integrated optimisation of the whole process chain including all resources. In "stable" supply chains with longer-term oriented co-operations, this optimisation can be performed continuously. An appropriate method to support this kind of process improvement is the performance measurement (PM) (Mertins 1995, p. 10). The continuous monitoring and evaluation of the current process performance enables the identification of weaknesses and is the base for the optimisation of the process chain. In this context, the performance measurement can be seen as retrospective method to monitor existing processes which enables the ex-post reaction on measured syptoms.

In virtual organisations, this continuous, long-term process optimisation by applying a "rear-view" evaluation is impossible. The reason is the dynamic character of virtual organisations where the process chain and the collaborating companies are configured specific to the current customers' order. Considering the extreme case, the best possible performance of a virtual organisation has to be ensured from the first order due to the existence of this specific collaboration is related to only this certain order. In consequence, already the partner selection during the preparation phase of the life-cycle determinates the future performance of the virtual organisation during the operational phase – athough partners can of course be replace if not performing. This means that the prospective performance evaluation of a planned configuration of a virtual organisation during the preparation phase is an important planning task with a high impact on the future performance potential of future inter-organisational team. One criterion during this preparation phase is the qualifying examination of potential partners within the virtual organisation to build a order-specific team to evaluate their capability to contribute to the order specific tasks.

2. PERFORMANCE MEASUREMENT AS METHOD TO QUANTIFY A PROCESS PERFORMANCE

The concept of PM bases on the definition of key performance indicators (KPI's) which represent the performance of the business processes to be measured in a quantitative or qualitative way (Böhnert 1999, p. 92ff.). Using KPIs to support manufacturing operations has its roots back to the late 19th century where Frederick W. Taylor introduced time and motion studies to manage production lines and warehouse operations (Lapide 2001, p. 287). Since value chains are today distributed, companies have spent their efforts to re-engineer and to improve their supply chains. During the last years, a couple of concepts have been developed to measure the performance of stable supply chains (eg. Balanced Scorcard, Logistics Scoreboard, Economic Value-Added EVA). Within this chapter, the SCOR standard to model and to evaluate stable supply chains is introduced which represents currently the most extended approach which is related to Supply Chain Management. It is discussed, which aspects of this approach are also applicable for virtual organisations and which requirements an appropriate performance measurement system for virtual organisations can be derived.

2.1 The SCOR Model as method to evaluate Supply Chains

The SCOR (Supply Chain Opertations Reference) approach developed in the late ninetees by the Supply Chain Council is a methodology to model and to evaluate Supply Chains. The value chain is described as a sequence of standard processes which are namely "make", "source", "deliver", "plan" and "return processes. The contribution of each participant in the value chain can be described as at least one of these processes, which leads on the top level to the general architecture of the supply chain. This top level can be specified on a level of process categories (level 2) and details process elements (level 3). It is obvious that SCOR also focuses on supply chains to describe and to monitor existing value chains with the objective to reach an optimisation.

The benefit of the SCOR model is that it provides standardised processes, which allow to model the whole inter-company value chain with one single method – of course if all network partners agree to this standard. Due to the SCOR approach can be seen as quasi-standard, some of the often-implemented software tools for performance measurement base their models on the SCOR approach. Examples for these tools are SAP APO, Cognos or SCORwizard. SCOR intends to model the supply chain from the suppliers' supplier to the customers' customer. The performance is described by KPIs belonging to five attributes, which are quality, assets, flexibility, responsiveness and costs.

2.2 Requirements for a performance measurement within virtual organisations

All existing approaches base their method on the belief that processes are relatively stable and most of the approaches focus on the evaluation of intra-company processes. They still rely on the idea to learn from the past to improve the future by measuring the actual performance. But only the continuity of the involved partners on the one hand and the high stability of the installed processes in an ordinary supply chain on the other hand allow an ongoing process improvement on the base of a retrospective performance evaluation.

Regarding the introduction of a performance measurement system in the virtual organisation, there is the lack that there are no approaches available which are able to measure the inter-company processes and which calculate the networks performance (Hieber 2001, p.2). Furthermore, the dynamic character of virtual organisations denies the measurement of past processes to initiate a future improvement due to the missing stability of the network. In a time-restricted co-operation with the objective to fulfil one single order, it is necessary to perform from the very beginning. This means that a performance measurement approach has to be initiated already in the initiation phase of the network to ensure that the operation phase will perform in the best way. This environment where possible partners of the planned network are identified and selected from a pool of potential companies, the so called breeding environment (Camarinha-Matos 2004), has still not been recovered potential and should already been supported by a performance measurement approach. Figure 3 shows the principle, how a virtual organisation can be generated dependent on the available potential partners and the required competencies.

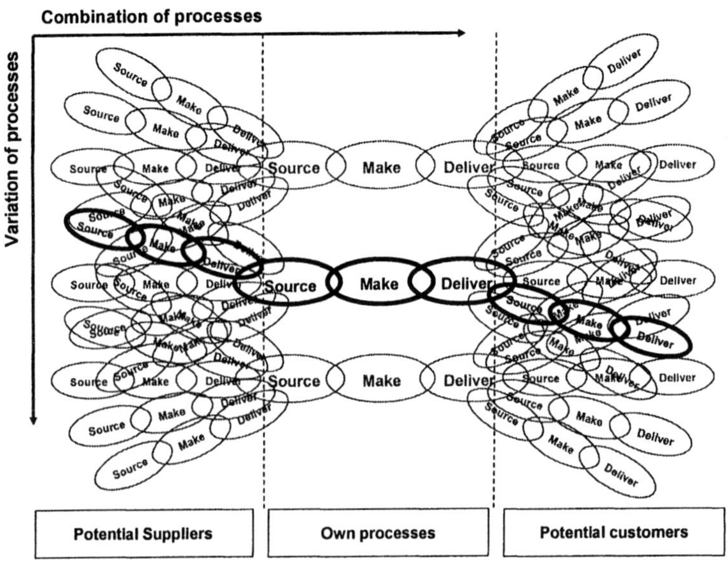

Figure 2: Variation and combination of processes (SCC, 2002)

Each product requires a specific process chain which means a specific combination of processes. Within the virtual organisation, there are multiple providers for each process available which allows to generate different variants of the process chain in terms of the involved partners. The main task within the preparation phase of a collaborative network is to identify that combination and variation of the value process chain which will deliver the best possible performance.

Combination of processes
The breeding environment provides the sum of all available processes from all potential partners for a virtual organisation. According to a specific order, the appropriate selection of processes has to be taken. This selection is in a first step independent from a company and describes the general architecture of the process chain.

Variation of processes
The breeding enivronment usually provides the possibility to select between different process owners which are the potential partners to build up the required process chain. To identify the best performing virtual organisation for a specific order means to be able to anticipate the performance of the temporary process chain.

It is obvious that there is isolated experience regarding the single performance of a potential partner available but not for each possible variation of the process chain with different partners. To ensure the best performing network, it is necessary to anticipate the performance of a planned network. Due to the temporary existence of this network, the traditional approach of continuous process improvement is not applicable. It has to be ensured that the generated virtual organisation represents the best possible selection from the breeding environment. To react flexible and with high quality on customer requests, a decision support within the initiation phase of the virtual team is very important.

Currently, the actual performance measurement approaches do not support this predictive performance measurement. When defining a predictive performance measurement for virtual organisations, a common methodology to measure the performance of all companies within the virtual organisation is required. This covers the following aspects:

Usage of the same process model for all partners within the virtual organisation to enable the modeling of distributed process chains
Up to now, the most completed and detailed model to describe industrial processes is the SCOR model of the Supply Chain Council. Due to its world wide availability and the large community supporting this initiative, it is appropriate to model inter-company process chains. The model itself can also be used to model the processes within in the virtual organisation although some adaptation is necessary. The difference is that the generated "supply chain" within the virtual organisation is dynamic and only temporary existent.

Definiton of a common set of KPIs to ensure the comparability between the partners
The SCOR methodology provides a wide range of KPIs to quantify the processes defined within the process model. The KPIs have been developed to evaluate the whole supply chain. Most of the indicators are company specific which enables the evaluation of each participating company. But the application of the model to the

whole process chain provides an overview on the performance of the whole value chain. Exact definitions in terms of formulas are not provided by the SCOR model and most of the indicators (especially units) have to be defined by the user.

Access to performance data of all potential partners for the main contractor who composes the virtual team
The isolated evaluation of a companies' performance does not support the evaluation of the whole process chain. To be able to quantify the performance of a virtual organisation, the performance data (KPIs) for all participants in the virtual team has to be accessible for the performance manager. An appropriate way to collect and to share information is a we-based system on the Internet. Due to the confidentiality of performance data, the privacy and security of the data has to be ensured for the industrial application.

Methodology to support the search for the optimal virtual team
Basing on the performance data of each partner in the virtual organisation, a method has to be developed to predict the performance of a virtual team and to compare different variants of a process chain.

3. SCIENTIFIC APPROACH FOR A PREDICTIVE PERFORMANCE MEASUREMENT CONCEPT

The concept bases on the assumption that the isolated processes provided by each single company within the virtual organisation are relatively stable. The process chain within a virtual team is composed by combining and integrating these isolated processes. The stability of the single processes which are the entities of the whole process chain allow to develop a performance measurement approach basing on two main tools and a central database. **A web-based database** collects the performance data calculated by the monitoring tool and provides it to the planning tool for a process chain simulation. The **monitoring tool** supports the measurement of the processes of each company involved in the virtual organisation and delivers the necessary data for the planning tool. The **planning tool** contains the process modeler to model a process chain on the base of the SCOR methodology. Developing and comparing the performance of the possible variants using the database, the tool proposes a virtual team with the probable best performance. Both tools as well as the overall approach are described in detail in the following.

The monitoring tool
The monitoring tool provides a real-time performance measurement for a companies' processes. Each company which is part of the breeding environment gets access to the tool with an own profile where it configures its processes and selects indicators. On the base of the actual processes, each company is able to monitor its own processes while all partners use the same methodology, which is the SCOR model. Basing the whole performance management within the VO on one model ensures the comparability between the partners. The data collection can be done in two different ways:

- The manual collection of each KPI according to the selected acquisition period: The tool identifies continuously the maturity of each indicator and asks for the actual values
- The automated data integration: An XML interface enables the user to link the tool with an ERP database to extract the necessary values automatically according to their maturity. The interface has to be specific to the available ERP source. Prototypes of these interfaces have been developed during the European project APM (Automated performance measurement) IST-1999-10279.

The results of the KPIs are stored in a central web-database which is normally located at the main contractor. This part of the tool covers the traditional process monitoring and enables an ongoing improvement of the own core processes. All KPI's are private and they are only visible for the owner of the data. The main contractor can access the data to calculate and to compare the to-be scenarios. This rule ensures the privacy of the performance data for each participant within the VO. Figure 4 shows the application of the monitoring tool described by the steps a) and b).

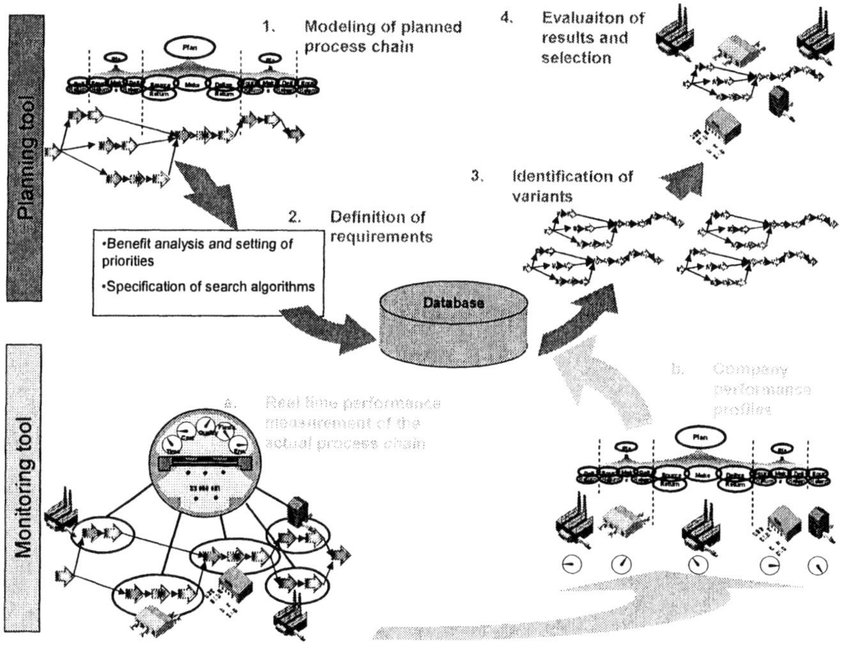

Figure 3: Concept for a predictice performance measurement system

The planning tool

The aim of the planning tool is to support the main contractor to generate different scenarios of a process chain by combining possible partners in the process chain. The comparison is executed by accessing the performance data from the different partners, which allows to estimate the prospective performance of the whole process chain. In a first step, a model of the whole process chain has to be developed by the main contractor. This company-independent model is the starting point for the simulation part where the database is used to identify possible companies for each process. All possible variants of the process chain gained from the database are compared on the base of the stored actual KPI values. This approach allows to simulate the performance of a virtual organisation which sometimes never co-operated in this certain configuration before by using the actual performance data of each partner. Figure 4 shows the necessary steps for this simulation where the first step is the modelling of the planned, company independent process chain. The second step is to define the priorities in terms of costs, quality, responsiveness, assets and flexibility which determinates the selection of the possible partners. The third step identifies possible variants for a virtual team to run the required processes which are compared and evaluated on the base of their KPIs in the forth step. The tool itself is browser-based and uses the PHP script.

4. CONCLUSIONS

Existing performance measurement approaches still base on traditional supply chains and there is a lack of considering the growing influence of virtual organisations which require a interorganisational focus of the performance measurement. The main constraint for the establishment of an interorganisational performance measurement approach is of course the trust between the partners within the breeding environment. The agreement to share performance related information within the network and the understanding that the common output of the network determinates the customers' satisfaction are the main drivers for the success of an interorganisational performance measurement approach.

6. REFERENCES

1. Boutellier, R.: Konkurrenz der Logistikketten, in Logistik Heute, May 1999
2. Shields, M.-G. (2002) ERP-Systeme und e-business schnell und erfolgreich einführen. Wiley-VCH Verlag GmbH, Weinheim 2002.
3. Hieber, R. (2001) Supply Chain Mangement, A collaborative Performance Measurement Approach, VDF Verlag, Zürich 2002
4. Thoben, K.-D, Jagdey, H., (2001) Typological Issues in Enterprise Networks
5. Böhnert, A. (1999): Benchmarking: Charakteristik eines aktuellen Managementinstruments, Hamburg 1999
6. Lapide, L.: What about Measuring Supply Chain Performance?, AMR Research, http://lapide.ASCET.com, 15.11.2001
7. Camarinha-Matos, Luis M. (2004) New collaborative organiszations and their Research needs, in Processes and Foundations for virtual organizations. Kluwer Academic Publishers, Bosten / Dordrecht/London, S. 3ff.
8. Gassmann, O., von Zedtwitz, M (2003). Organising Virtual R&D teams. In: R&D management, 33, pp. 243-262.
9. Mertins, K: Benchmarking: Praxis in deutschen Unternehmen, Springer , Berlin 1995

32

MULTI LAYERS SUPPLY CHAIN MODELING BASED ON MULTIAGENTS APPROACH

Samia Chehbi, Yacine Ouzrout, Aziz Bouras
{samia.chehbi, youzrout, abouras}@univ-lyon2.fr
FRANCE

This paper proposes a strategic multi layers model based on multi agents approach for supply chain system. It introduces a formulation and a solution methodology for the problem of supply chain design and modeling. In this paper we describe and analyze the relationships among main entities of a supply chain, such as suppliers, producers, and distribution centers, in the aim to design the agents and define their behavior. We also study, how these relationships can be formulated in a multi layer model. Finally, a generic multi agent model is illustrated.

1. INTRODUCTION

The most popular research topic in the field of supply chain (SC) management is the formulation of strategic and efficient model. This can be opted by different manners, by using artificial intelligence tools or integer programming methods (Wu, 2001). The problem is commonly arises in the evaluation of some parameters characterizing SC state. A number of production producers supply a collection of distribution centers with multiple products, which, in turn, supply customers with specified demand quantities of different products. The challenge is to determine the number, location, capacity, and type of convenient actors to minimize the total cost of the SC. The mathematical problem of formulation in production context exists since a long time where some works, like the one of Goeffrion and Graves (Goeffron, 1972) described a multi-commodity single-period production-distribution problem and solved it by *Benders Decomposition*. Recently, Hong Y. et al. (Hong, 2003) has developed a proved method based on constraints to design a strategic production-distribution model. Other works have been published recently under this theme, like (Dong, 2003), where Dong J. et al. analyze the formulation and design a demonstrated mathematical model based on lemmas and theorems. Most efforts in these works consider SC activities separately and proceed by studying SC as a linear model and try to represent it globally. We provide another view to model SC, considering it as a non linear system with a high level of complexity, and we try to apply technical tools, classically used to resolve complex systems design and modeling. In the next sections, we describe briefly some SC features, then, we detail our formulation steps and describe the parameters, variables and constraints used to design the multi layers system. After, we proceed by giving some key issues in SC design and modeling using multi agent systems and finally, we give an example of dimensioning supply chain problem.

2. FORMULATION OF SUPPLY CHAIN ORGANISATION

Today, competition in global markets with heightened expectations of customers has pushed the enterprises to invest and make more importance of their SCs. Consequently, to reduce cost and improve service levels, effective SC strategies must take into account interactions at the various levels in the SC. To improve their SC performance, firms must focus on understanding most information and relationships nature of all the partners and SC actors. Having an idea of a model projecting their SC, these firms can improve many strategic decisions like forecasting operations, predicting customers' demands and decreasing warehouse stocks costs.

This paper shows how theoretical programming formulations can be applied to SC design problem, and focuses on the fact of dividing SC into upstream and downstream parts by considering the manufacturer as the reference mark. The main objective in the proposed approach is to evaluate an objective function (to maximize) and a cost function (to minimize). We assume that the SC's flow concerns a single family of product.

Actors in classical SC models (Simchi-Levi, 2000) are not organized, and interactions are ignored. This leaded us to propose in a precedent work (Chehbi, 2003) a multi-layers reorganization in order to facilitate the SC evaluation's phase. Hence, any SC can be transformed into a multi-layers architecture (figure 1).

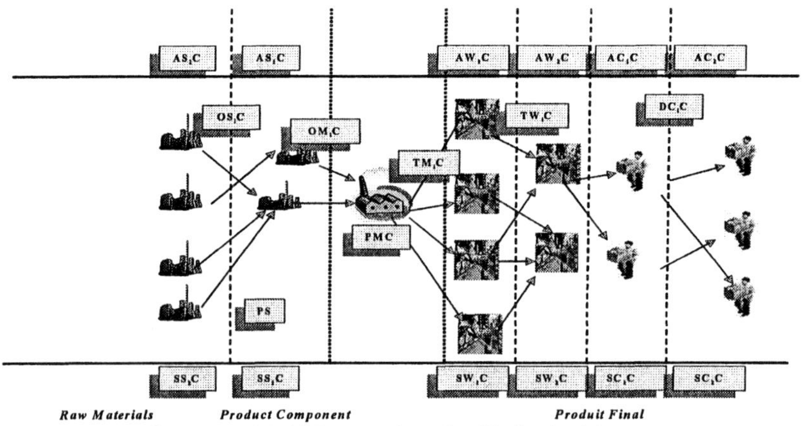

Figure 1 – Multi Layers Supply Chain Architecture

2.1. Supply Chain Parameters

Before formulating the model, we introduce the basic parameters notations and definitions. In this study, we use the following indices: $s \in S$, a set of candidate suppliers; M the single manufacturer; $w \in W$, a set of warehouses and distribution centers; $c \in C$, a set of customers; $pc \in PC$, a set of product components needed for production; $p \in P$, the single type of product characterizing the SC. Using these notations, we describe some considered costs as follows:

Added Cost: It is the cost obtained when introducing an actor in the chain. Hong et al. (Hong, 2003) used the same principle in their formulation of a logical constraints model for SC. They call it the fixed cost to open and operate an actor. Each actor has its proper *added cost* in the SC. Relating to each actor type, we distinguish: *ASiC* (*Added Supplier 'i' Cost*), *AWiC* (Added Warehouse *'i'* Cost) and *ACiC* (Added Customer *'i'* Cost).

Action Costs: Signifies the internal cost evaluated for each actor. We distinguish two types of costs; the first is the *Production Cost* of one unit of the final product in the case of the main manufacturer (*PMC*), or a unit of a product component in the case of a supplier PS_i. The second type concerns the *Storage Cost* of a unit of the final product or its components (SS_iS for suppliers, SW_iC for warehouses and *SS_iC for customers*).

Interaction Costs: Interactions between supply chain actors play an important role in the total cost. Along the upstream supply chain, we define *TMiC* (Transportation Cost Between the manufacturer and its *'i'*th customer), *TWiC* (Transportation Cost between warehouse*s*) and *DCiC* (*Distribution Cost* between customers). In the downstream chain, we define *OMiC* as the cost of materials ordered by the main manufacturer to its ith supplier and *OSiC* between manufacturers to deliver the product.

2.2. Evaluation of objective function

Before providing the total cost function, we define other notations related to each actor location in the multi layer architecture. A supplier i located in a level n is indicated by s^n_{Ln} with L_n the number of suppliers in the layer n. The same rule is applied to the other actors, so we have as a result the indices matrix (*Supply Chain Matrix*) described below.

$$\begin{bmatrix} s_1^n & \cdots & s_1^1 & w_1^1 & \cdots & w_1^p & c_1^1 & \cdots & c_1^m \\ s_2^n & \cdots & s_2^1 & w_2^1 & \cdots & w_2^p & c_2^1 & \cdots & c_2^m \\ s_3^n & \cdots & s_3^1 & w_3^1 & \cdots & w_3^p & c_3^1 & \cdots & c_3^m \\ \cdots & \cdots & \cdots & \cdots & \cdots & \cdots & \cdots & \cdots & \cdots \\ s^n_{L_n} & \cdots & s^1_{L_1} & w^1_{L_1} & \cdots & w^p_{L_p} & c^1_{L_1} & \cdots & c^m_{L_m} \end{bmatrix}$$

Using these notations and all parameters evolved before, we give the total cost of supply chain as the sum of *added, action* and *interaction costs*. By evidence, the *total added cost* is the sum of *added supplier costs, added warehouses costs* and *added supplier costs.*

Total Added Supplier Cost $= \sum_{i=1}^{L1} ASiC + \sum_{i=1}^{L2} ASiC + \ldots + \sum_{i=1}^{Ln} ASiC$, *Total Added Warehouse Cost* $= \sum_{i=1}^{L1} AWiC + \sum_{i=1}^{L2} AWiC + \ldots + \sum_{i=1}^{Lp} AWiC$, and *Total Added Customer Cost* $= \sum_{i=1}^{L1} ACiC + \sum_{i=1}^{L2} ACiC + \ldots + \sum_{i=1}^{Lm} ACiC$

After evaluating *added costs* of each type of supply chain actors, we sum them to have the total added cost in the chain. To evaluate *total action cost,* it is important to notice that there are probably actors which are not producers or storage centers; we interpret this by introducing a binary coefficient q^i_{Lj} to the supplier i belonging to the jth layer, whether it is a producer $\{q^i_{Lj} = 1\}$ or not $\{q^i_{Lj} = 0\}$.

Total Production Cost = Production Costs of suppliers + Production Cost in the main manufacturer and Total Storage Cost = Storage Cost of suppliers + Storage Cost of warehouses + Storage Cost of customers.

$$Total\ \ Action\ \ Cost = \sum_{i=1}^{L1} PS_i . q^i_{L1} + \sum_{i=1}^{L2} PS_i q^i_{L2} + ... + \sum_{i=1}^{Ln} PS_i q^i_{Ln} + +PMC + \sum_{i=1}^{L1} SS_i C.q^i_{L1} + \sum_{i=1}^{L2} SS_i Cq^i_{L2} + ... + \sum_{i=1}^{Ln} SS_i Cq^i_{Ln} + \sum_{i=1}^{L1} SW_i C.q^i_{L1} + \sum_{i=1}^{L2} SW_i Cq^i_{L2} + ... + \sum_{i=1}^{Lp} SW_i Cq^i_{Lp} + \sum_{i=1}^{L1} SC_i C.q^i_{L1} + \sum_{i=1}^{L2} SC_i Cq^i_{L2} + ... + \sum_{i=1}^{Lm} SC_i Cq^i_{Lm}$$

Total Action Cost = Sum of ordered costs between suppliers + sum of ordered costs between the main manufacturer and its suppliers + sum of transportation costs between the main manufacturer and its customers + sum of transportation costs between warehouses + sum of distribution costs between customers.

$$Total\ \ Interactin\ \ Cost = \sum_{i=1}^{L2} OS_i C + ... + \sum_{i=1}^{Ln} OS_i C + \sum_{i=1}^{L1} OM_i C + \sum_{i=1}^{L1} TM_i C + \sum_{i=1}^{L2} TW_i C + ... + \sum_{i=1}^{Lp} TW_i C + \sum_{i=1}^{L1} DC_i C + \sum_{i=1}^{L2} DC_i C + ... + \sum_{i=1}^{L(m-1)} DC_i C$$

By consequence, the objective function is given by: *Min F = Total Added Supplier Cost + Total Added Warehouse Cost + Total Added Customer Cost + Total Action Cost + Total Interaction Cost.*

3. MULTI AGENT SYSTEMS MODELING

This section presents an issue for modeling the dynamic behavior of the proposed SC multi layers model. Our aim is that to obtain an efficient tool of simulation which can be applied to quantify the flow of SC information. With this described model, we think be capable to determine strategic policies are effective in smoothing and reducing variations in the SC. In most recent works in this topic, SC and enterprises networks have been a fertile area of multi agent simulations. That's because there is a growing need to developing decentralized efficient tools aiding to more performed management tools.

Referring to (Ferber, 1995), a multi agent system is a collection of, possibly heterogeneous, computational entities, having their own goals and problem-solving capabilities. Won et al. (Dong, 2002) suggest a set of interactive agents for Harbor SC network. Lin et al. (Lin, 1998) present multi agents architecture to model and simulate SC information system, they propose a shared environment based on agents simulating orders processes. Researches on agents-based SC management can be divided into three types: (1) Agent-based architecture for coordination, (2) agent-based simulation of SCs and (3) dynamic formation of SCs by agents. Our current work is a combination between the two first types of researches. It proposes an agent-based architecture doted of decision making agents to insure collaboration between SC parts. Based on various designs for multi agent systems in the literature (Dong, 2002) and many previous researches, we try to design an agent-based SC model, described statically and dynamically via three types of agents (figure 2). An agent type called '*controller*' to model the SC dimensioning, a set of agents to model SC dynamic, divided into physical agents representing tangible existing objects (such as wholesalers, customers, etc) and logical agents defining a virtual agent for each layer. They are doted of information functions used to control the information flows and manage the interactions as described bellow:

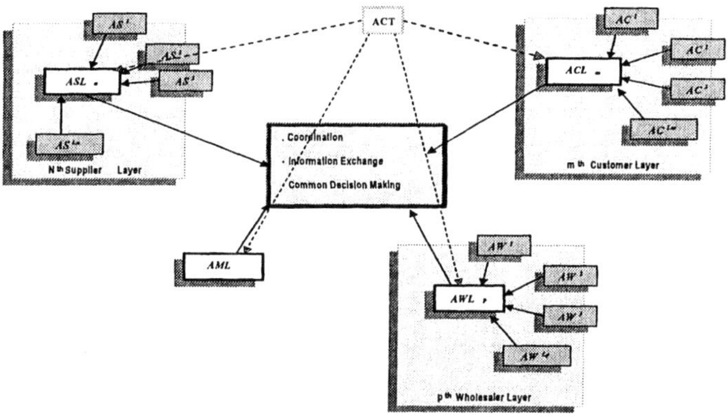

Figure 2 – Multi-Layers Agent SC Model

3.1 Agent-Actors

For each SC entity, we define a specified agent. So, AC_j^i means the ith customer agent in the jth customer layer. Applying the same signification, we have AS_j^i for suppliers and AW_j^i for wholesalers. They are designed with classical standardized functionalities to communicate, negotiate and send or receive requests from other agent-actors of the same layer and the agent-layer (see figure 3, (a)). Three principle modules are implemented in the agent-actor; each one insures a specific task: The communication module serves as a reception-sending filter with the other agents of the same layer. There may be several types of information exchanged, such as products demand, negotiation messages, asking for a shared information, etc. The knowledge management module contains all parameters perceived by the agent and a part of the database. For example, action costs are the most important variables existing in the internal database. Moreover, this agent has the possibility of asking for external data from other agents. To coordinate between previous modules, a coordination module is added.

3.2 Agent-Layer

Each layer is managed by a logical cognitive agent responsible for reactive agents' management in its same layers. We define one agent manager for the manufacturer called *AML* (Agent Manufacturer), one another for each customer layer (ACL_i), one agent for each layer (ASL_i) and also for each wholesaler *layer* (AWL_i). It has to insure and maintain the minimization cost of its layer when it receives the order from the controller agent (Figure 3, (b)). Interactions between these agents enable the flows of products and information within a layer and to other layers that are immediately adjacent to it in the SC. In addition to contain the same modules implemented in the Agent-Actor, we define the decision making module designed to propose negotiated decisions to Agents-Actors. Its internal data base comprises interaction and added costs related to its layer.

3.3 Agent-Controller

We define one controller-agent (*ACT*) in the system, designed to evaluate strategic decisions for SC dimensioning. All formulation part is implemented in the decision making module of this agent. It must communicate continuously with all agents-layers of all the chain in order to update its information (figure 3, (c)).

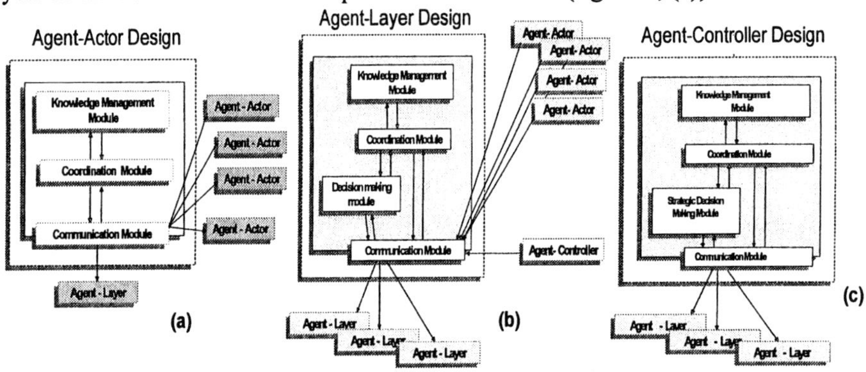

Figure 3 – Agents design

4. ILLUSTRATIVE EXAMPLE

In this section, we present an example of a simplified supply chain illustrated in (figure 4). The chain is related to the product *P,* assembled and delivered by the main manufacturer *Man.* In the manufacturing process, we distinguish an intermediate product component *IP* assembled and delivered by a secondary center of assembly *D*. There are three first suppliers *{A, B, C}* for three raw materials *{M1, M2, M3},* one storage center *E* for the product component and the raw material *M3* in addition to three storage centers *{G, L, M}* for the final product *P*. We also have six final customers *{N, O, P, Q, R, S}*.

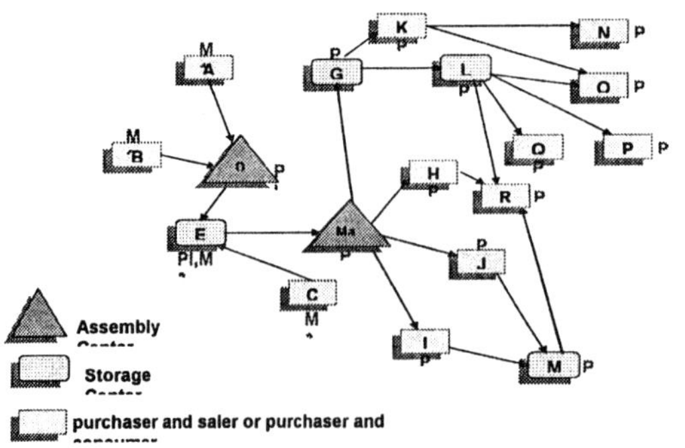

Figure 4 – SC description – Illustrative Example

4.1. Multi Agent Model

In order to decompose the chain into layers, it is important to extract a hierarchic tree to distinguish and define the various levels. We take as reference mark the manufacturer of the final product and then we advance in the hierarchy in the two directions (customers and suppliers) (figure 5). In this example we can divide the chain into six layers *{S1, S2, S3, C1, C2, C3}*.

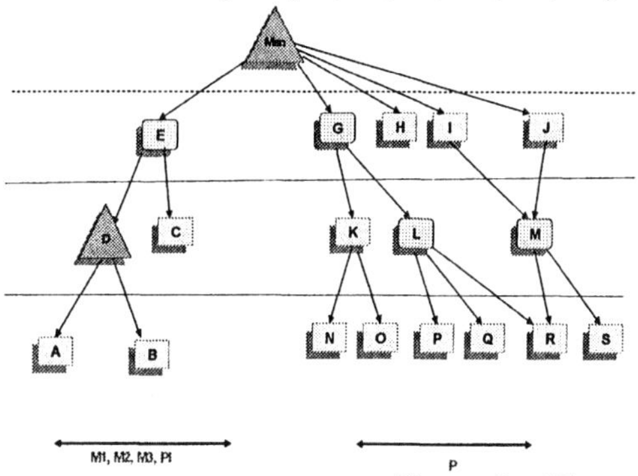

$$\begin{bmatrix} S_3^1=A & S_2^1=D & S_1^1=E & C_1^1=G & C_2^1=K & C_3^1=N \\ S_3^2=B & S_2^2=C & \phi & C_1^2=H & C_2^2=L & C_3^2=O \\ \phi & \phi & \phi & C_1^3=I & C_2^3=M & C_3^3=P \\ \phi & \phi & \phi & C_1^4=J & \phi & C_3^4=Q \\ \phi & \phi & \phi & \phi & \phi & C_3^5=R \\ \phi & \phi & \phi & \phi & \phi & C_3^6=S \end{bmatrix}$$

Figure 5 – Hierarchical tree

Each element in the matrix is represented by one reactive agent containing its actor's information, and each column in the matrix is represented by one cognitive agent manager to coordinate between layers and make partial common decisions, in addition to another cognitive agent defined to take strategic decisions for supply chain dimensioning. By consequence, our multi agent organization contains seven reactive agents related to six cognitive agent-managers in each layer and one cognitive agent-controller. In order to clarify the use of each agent, we describe a simple example of a scenario showing the problem of supply chain dimensioning.

4.2. Dimensioning problem description

We assume that the main manufacturer *Man* is located in France; its storage center *G* is located in USA and it has to know if it is profitable to open a secondary center of assembly in USA instead of delivering products to USA with a high cost of transportation. We suppose that in the state St_1, the agent controller *AC* has estimated the total cost of the chain at a value CSt_1. We suppose also that the state of the chain in the case of adding the secondary center of assembly in USA is called St_2, were the manufacturer buys product components *{PI, M3}* in USA and assembles them in the added center (figure 6).

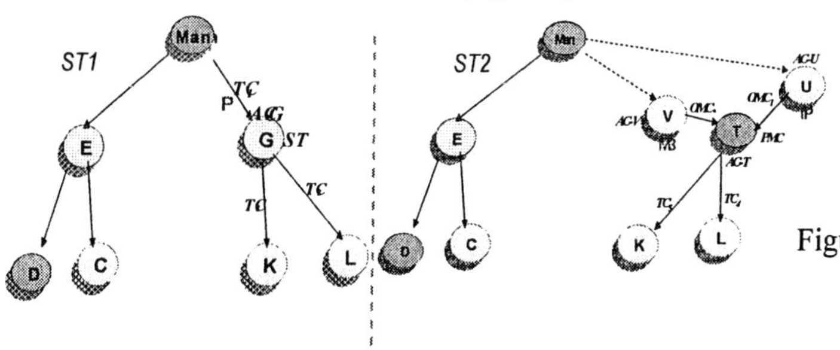

Figure 6 – Illustration of dimensioned supply chain problem

Each agent-actor sends to the agent-manager of the specific layer the values of estimated added costs *AC* of the actors *{U, V, T}* respectively, in addition to the production cost of the secondary manufacturer *T {PMC}*. The agents-managers, in turn, send these information in addition to the interaction costs $\{OMC_1, OMC_2, TC_4, TC_5\}$ to the agent controller. This one collects all cost values, calculates the total cost CSt_2, compares with the previous cost CSt_1 and finally decides which state is profitable for the manufacturer; the state St_1 $(CSt1<CSt2)$ or St_2 $(CSt_1>CSt_2)$.

5. CONCLUSION

Although there is a wealth of literature and research on modeling of strategic SC design, there is an apparent lack of theoretical consideration of SC constraints. In this present paper, we formulate a strategic SC model based on ordered layers and including pertinent constraints. We have considered the SC with a high complexity, in order to justify our choice of strategy in designing it as complex system. We have expressed relationships among actors via inter-relations between layers. We have extensively discussed our parameters and variables representation. We presented our multi layers model as a parallel organization of non linear sub systems of layers where interactions. In addition, we propose a multi agent issue for SC modeling. This paper aims at introducing a representation for building a mathematical model based on the constructive idea in constructing a multi layers model. The principle in designing this model is that to minimize global system cost while satisfying all customers' demands and to design learned agents to simulate SC environment and their actors' behavior. We think our work can open a novel way in proceeding of SC problem resolve by using decentralized tools.

6. REFERENCES

1. Chehbi S, Derrouiche R, Ouzrout Y, Bouras A. Multi Agent Supply Chain Architecture to Optimize Distributed Decision Making. The 7th World Multiconference on Systemic, Cybernetics and Informatics (SCI), Floride-USA. 2003.
2. Dong J, Zhang D. Multitiered Supply Chain Networks: MultiCriteria Decision-Making under Uncertainty. New-York-*USA*. 2003.
3. Dong W, Hie S.K, Nak H. K. Combined Modeling with Multi Agent System and Simulation: It's Application to Harbor Supply chain Management.The 35th Annual Hawaii International Conference on System Sciences (IEEE). USA, 2002.
4. Ferber J. '*Les Systèmes Multi Agent: Vers une Intelligence Collective*'. Inter Edition, Paris, 1995.
5. Geoffrion A.M, Graves G.W. An Interactive Approach for Multi-Criterion Optimization, with an Application to the Operation of an Academic Department. Management Science, 19(4) :357-368, 1972.
6. Hong Y, Zhenxin Y, T.C. Edwin C. A strategic Model for Supply Chain Design with Logical Constraints : Formulation and Solution'. Computer & Operations Research, 30(14) :2135 – 2155, 2003.
7. Lin F, Shaw M. J. Reengineering the Order Fulfilment Process in Supply Chain Networks. International Journal of Flexible Manufacturing Systems, (10:3), pp.197-229, 1998.
8. Simchi-Levi D, Kaminiski P, Simchi-Levi E. Designing and Managing the Supply Chain: Concepts, Strategies, and Case Studies. International Editions, 2000.
9. Wu T.T, O' Grady P. A Network Based Approach to the Design of Supply Chains. Arizona-USA, 2001.

33 A FORMAL THEORY OF BM VIRTUAL ENTERPRISES STRUCTURES

Rui Sousa; Goran Putnik
Production and Systems Engineering Department
University of Minho, School of Engineering
Guimarães, PORTUGAL
Email: rms@dps.uminho.pt
Email: putnikgd@dps.uminho.pt

Formalisms are a tool commonly used in many engineering areas and, as expected, are also being used on virtual enterprises research. However the use of formalisms is not enough to ensure correctness and ambiguities absence on the developed projects. Only with a background formal theory is possible to achieve that goal. This paper presents a formal theory of the structural aspect of virtual enterprises according to the BM_Virtual Enterprise Architecture Reference Model (BM_VEARM) developed at University of Minho – Portugal. The theory is generated and represented by an attributed context-free formal grammar accepting some pre-requisites as input and producing as output canonical structures of virtual enterprises compliant to those pre-requisites. The formal theory of BM_VEARM virtual enterprises structures is in fact the formal language generated by the defined grammar.

1. INTRODUCTION

Contrarily to some speculations the use of formalisms doesn't mean that a formal theory is behind. For the case of first-order theories this claim is proved in (Sousa, 2003), using mathematical first-order logic concepts, and implies that formalisms by their own are not enough to ensure correctness and to avoid ambiguities.

It is commonly accepted that only with solid theories it is possible to achieve the desired rigour on developing projects. Research on virtual enterprises (VEs) is an area of investigation whose importance is rapidly increasing as VEs are seen, especially by the scientific community, as the new paradigm for the factories/enterprises of the future. It is obvious that a formal theory of VEs would be of extreme importance for the investigation on this area.

This paper introduces a formal theory, generated by an attributed context-free formal grammar, of the structural aspects of VEs according to BM_Virtual Enterprise Architecture Reference Model (BM_VEARM).

The concept of theory is rigorously defined by mathematical logic as a set of some formulas with some special characteristics (Mendelson, 1987; Ebbinghaus *et al.*, 1996; Keisler, 1996). Those formulas are obtained from a given alphabet of symbols, using some derivation rules (calculus of formulas) and they constitute a language. Thus a theory is a language but, obviously, a special language. The formal grammar presented in this paper was specially developed to synthesize strings of symbols which are formulas compliant to the theory definition coming from mathematical logic. Hence, this grammar generates a language which is in fact a theory. The theory concept has as background other formal concepts from mathematical logic involving not only the syntactical viewpoint, but also the semantic perspective (e.g. structure, interpretation and model). With the developed grammar, and given some pre-requisites, it is possible to generate canonical structures of VEs compliant to BM_VEARM reference architecture.

The paper is intended to be introductory and self-contained regarding the grammatical principles involved, and its structure is as follows. Section 2 provides the basics of formal grammars, arising from theory of languages. A generic definition of formal grammar and the Chomsky's classification for formal grammars are presented. Attributed grammars are also referred as they are the truly powerful grammars. A simple example, already interpretable in the manufacturing systems structural aspects area, is provided. The fundamentals of BM_VEARM developed at the Production and System Engineering Department, University of Minho, Portugal (Putnik, 2000), are provided on section 3. Comprehensively more emphasis is dedicated on the structural aspects of VEs. On section 4 it is introduced the attributed context-free formal grammar G_{BM} responsible for the generation of the formal language L_{BM} which is a formal theory of BM_Virtual Enterprises structures. On section 5 some conclusions are outlined along with some perspectives of future work.

2. FORMAL GRAMMARS

A grammar is usually known as a set of rules allowing the creation of words and sentences over a given alphabet. The formal grammar concept goes a bit further by including the alphabet itself on the definition. Many similar definitions can be found in literature (Salomaa, 1973; Denning *et al.*, 1978; Hopcroft and Ullman, 1979; Lewis and Papadimitriou, 1981; Mikolajczak, 1991; Révész, 1991; Pittman and Peters, 1992), all based on Chomsky's definition (Chomsky, 1959). Adapted to the notation used in this paper we have:

Definition 1: A formal grammar G is a four-tuple $G=(V_T, V_N, S, R)$ where V_T is a finite set of terminal symbols, V_N a finite set of non-terminal symbols ($V_T \cap V_N = \varnothing$), S is the initial symbol ($S \subseteq V_N$) and R is a finite set of rewriting rules.

Each rewriting rule, or production, is an ordered pair (α,β) usually denoted as $\alpha \rightarrow \beta$ showing how the word $\alpha \in (V_T \cup V_N)^+$ can be rewrite as $\beta \in (V_T \cup V_N)^*$. The word α must contain at least one non-terminal symbol. Recall that if V is an alphabet then V^* represents the set of all the words, including the empty word λ, that can be constructed with the symbols of V and $V^+ = V^* \backslash \{\lambda\}$.

Example 1: Consider a grammar $G=(V_T, V_N, S, R)$ where $V_T = \{m, \mapsto, //,), (\}$, $V_N = \{S\}$ and $R = \{S \rightarrow m,\ S \rightarrow S \mapsto S,\ S \rightarrow S//S,\ S \rightarrow (S)\}$.

Two possible words of terminal symbols generated by this grammar are:

$$S \Rightarrow S \mapsto S \Rightarrow m \mapsto S \Rightarrow m \mapsto (S) \Rightarrow m \mapsto (S//S) \Rightarrow m \mapsto (m//S) \Rightarrow m \mapsto (m//m) \quad (1)$$

$$S \Rightarrow S \mapsto S \Rightarrow (S) \mapsto S \Rightarrow (S//S) \mapsto S \Rightarrow ((S)//S) \mapsto S \Rightarrow ((S \mapsto S)//S) \mapsto S \Rightarrow$$
$$\Rightarrow ((m \mapsto S)//S) \mapsto S \Rightarrow ((m \mapsto m)//S) \mapsto S \Rightarrow ((m \mapsto m)//m) \mapsto S \Rightarrow$$
$$\Rightarrow ((m \mapsto m)//m) \mapsto m \quad (2)$$

Each symbol $\Rightarrow$ represents a derivation step and corresponds to the application of one of the available productions. A derivation process ends when all the symbols of the word are terminal symbols. From the manufacturing systems structures perspective, words obtained by derivations (1) and (2) can de interpreted as different machine compositions (see Figure 1).

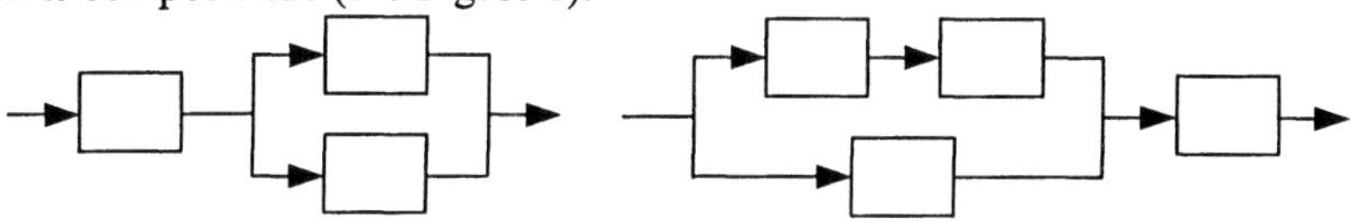

$m \mapsto (m//m)$ $\qquad$ $((m \mapsto m)//m) \mapsto m$

Figure 1 - Machine compositions generated by *G* grammar

Based on their productions type, formal grammars are classified in four classes: unrestricted (type 0), context-sensitive (type 1), context-free (type 2) and regular (type 3). This classification is known as Chomsky's hierarchy. To overcome some limitations of formal grammars the concept of attributed grammar was introduced by (Knuth, 1968). In an attributed grammar each symbol may have none, one or more attributes, addressing thus, besides syntactical aspects, the semantic viewpoint. Consequently the definition of each production must be extended with assertions or predicates about the involved attributes. To illustrate this concept the grammar *G* from example 1 will be transformed into an attributed grammar *G'*. The distinction between different machines can be accomplished using a finite number i of m_i symbols, instead of a single symbol m. Each machine can be characterized by its production rate p_r in parts/h. Thus p_r will be an attribute of each m_i symbol and also of symbol *S* which represents the entire system (see Table 1). This attribute is not applicable to the remaining alphabet symbols.

Table 1 - Symbols with attributes for grammar G'

Symbol	*Description*	*Attribute*	*Description*
m_i	machine i	p_r	machine production rate
S	system	p_r	system production rate

Now the productions of the new grammar *G'* are completed with assertions about the p_r attribute (see Table 2). Note that superscript identifiers are introduced whenever symbol instance distinction is necessary.

Table 2 - Productions and assertions for grammar G'

Production	*Assertion*
$S^{(1)}\rightarrow m_i$	$p_r(S^{(1)})=p_r(m_i)$
$S^{(1)}\rightarrow S^{(2)}\mapsto S^{(3)}$	$p_r(S^{(1)})=\min(p_r(S^{(2)}), p_r(S^{(3)}))$
$S^{(1)}\rightarrow S^{(2)}//\, S^{(3)}$	$p_r(S^{(1)})=p_r(S^{(2)})+ p_r(S^{(3)})$
$S^{(1)}\rightarrow(\, S^{(2)})$	$p_r(S^{(1)})=p_r(S^{(2)})$

Example 2: Consider the attributed grammar $G'=(V_T, V_N, S, R)$ where $V_T=\{m_1,\ldots,m_i,\mapsto,//,),(\}$, $V_N=\{S\}$ and R contains the productions of Table 2.

Recalling the derivation (1) of example 1, but including now instance identifiers and distinct m_i symbols we may have:

$$S^{(1)}\Rightarrow S^{(2)}\mapsto S^{(3)}\Rightarrow m_1\mapsto S^{(3)}\Rightarrow m_1\mapsto(S^{(4)})\Rightarrow m_1\mapsto(S^{(5)}//S^{(6)})\Rightarrow m_1\mapsto(m_2//S^{(6)})$$
$$\Rightarrow m_1\mapsto(m_2//m_3) \qquad (3)$$

Besides the showed generation of machine compositions, G' can also determine the production rate p_r of the generated system, based obviously on the individual machines production rates which in this case are set, for instance, to 20, 18 and 16 parts/h for m_1, m_2 and m_3, respectively. Formally system p_r calculation is done using the assertions associated to the applied productions, starting from the last derivation step because system p_r is an synthesized attribute (Pittman and Peters, 1992; Sousa, 2003). Thus in the last derivation step it is used the production $S^{(6)}\rightarrow m_3$ implying that $p_r(S^{(6)})=p_r(m_3)=16$ parts/h. The previous derivation step applies production $S^{(5)}\rightarrow m_2$ and thus $p_r(S^{(5)})=p_r(m_2)=18$ parts/h. The fourth derivation step uses production $S^{(4)}\rightarrow S^{(5)}//\, S^{(6)}$ leading to $p_r(S^{(4)})=p_r(S^{(5)})+ p_r(S^{(6)})=18+16=34$ parts/h. The third derivation step applies $S^{(3)}\rightarrow(\, S^{(4)})$ and thus $p_r(S^{(3)})=p_r(S^{(4)})=34$ parts/h. The second derivation step uses the production $S^{(2)}\rightarrow m_1$ implying that $p_r(S^{(2)})=p_r(m_1)=20$ parts/h. Finally the first derivation step applies $S^{(1)}\rightarrow S^{(2)}\mapsto S^{(3)}$ and consequently the production rate of the generated system is $p_r(S^{(1)})=\min(p_r(S^{(2)}), p_r(S^{(3)}))=\min(20,34)=20$ parts/h. Although simple this example illustrates the high potential of attributed grammars when compared with traditional grammars.

The language generated by a grammar is the set of all the words of terminal symbols generated by that grammar.

Definition 2: The language generated by a formal grammar $G=(V_T, V_N, S, R)$ is $L(G)=\left\{p\in V_T^* \mid S\overset{*}{\underset{G}{\Rightarrow}} p\right\}$.

Symbol $\overset{*}{\underset{G}{\Rightarrow}}$ denotes derivation in many steps according to the productions of G.

Mathematical logic defines language as the set of all the formulas obtained from a given alphabet according to a set of rules (calculus of formulas). From all those formulas some, under certain circumstances, may constitute a theory (Mendelson,

1987; Ebbinghaus *et al.*, 1996; Keisler, 1996). Thus, and without further justification, we can say that a mathematical logic language may potentially include one or more theories. Hence if a formal grammar generates words that can be considered as formulas, then that grammar is a potential theory generator. This subject is deeply investigated in (Sousa, 2003).

3. FUNDAMENTALS OF BM_VEARM ARCHITECTURE

The BM_Virtual Enterprise Architecture Reference Model (Putnik, 2000) is based on a multilevel hierarchical model (Mesarovic *et al.*, 1970) and supports four crucial characteristics for VEs: integrability, distributivity, agility and virtuality. To achieve the first characteristic BM_VEARM includes an integration mechanism concept. The use of wide area networks supports the distribution of the VE resources. Agility and virtuality are provided in BM_VEARM through the broker concept. Figure 2(a) represents the elementary hierarchical BM_VEARM structure which works as a building unit in the synthesis process of VEs. Figure 2(b) shows an example of a VE structure synthesized according to BM_VEARM. Both diagrams on Figure 2 are logical representations with a high abstraction level. From the implementation viewpoint, integration mechanisms are usually embedded in the adjacent blocks (i.e. control level and resources management).

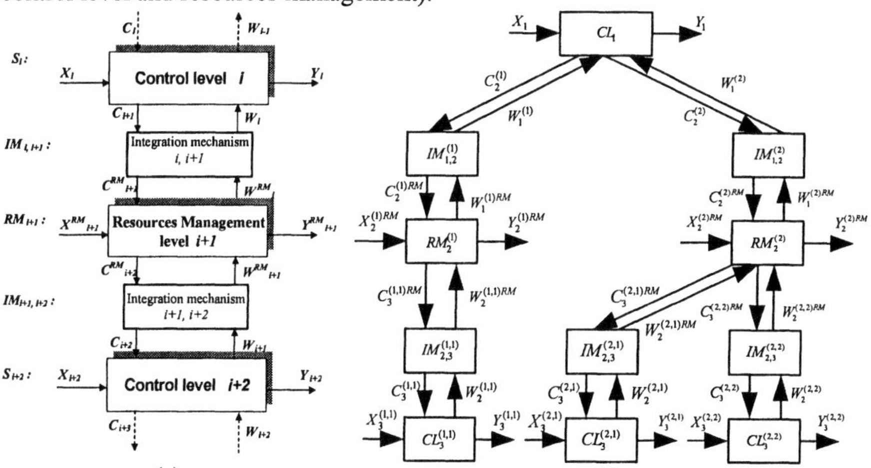

(a) (b)

Figure 2 - BM_VEARM (a) elementary structure (Putnik, 2000) (b) VE instance

Based on this perception the incoming grammar for VE synthesis may include only two types of basic blocks: c_i - control level and r_j resources management. The complete description of BM_VEARM can be found in (Putnik, 2000).

4. A FORMAL THEORY OF BM_VEARM VIRTUAL ENTERPRISES

This section presents a context-free attributed grammar, denoted as G_{BM}, able to generate VEs structures according to BM_VEARM. As seen before two fundamental terminal symbols are necessary: c_i – to represent control level blocks and r_j – for resources management blocks (see Figure 3). Due to space limitations is not possible to include here all the symbols, attributes and assertions of G_{BM}. This is the reason why definition 3 and derivation 4 only refer to the syntactical aspects of G_{BM}. However the entire development process can be found in (Sousa, 2003).

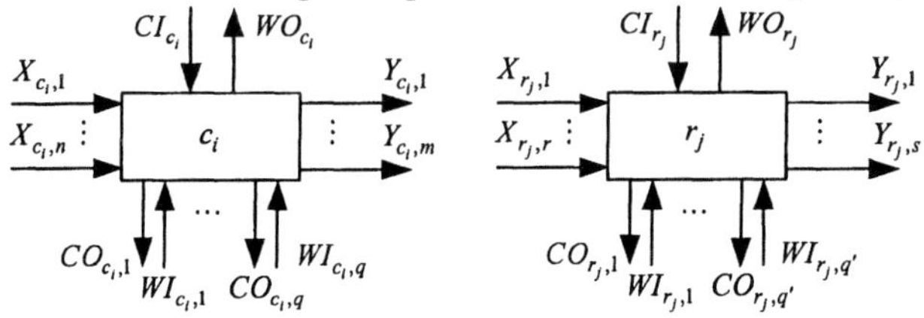

Figure 3 - Basic blocks for G_{BM} (Sousa, 2003)

Definition 3: $G_{BM}=(V_T,V_N,S,R)$ is an attributed context-free grammar where $V_T=\{c_1,\ldots,c_{n_c},r_1,\ldots,r_{n_r},S_{eq},\equiv,\downarrow\uparrow,),(\}$, $V_N=\{S,A,B\}$ and $R=\{S\to c_i(\downarrow\uparrow A)\equiv S_{eq}$, $A\to r_i(\downarrow\uparrow B)$, $A\to AA$, $B\to c_i(\downarrow\uparrow A)$, $B\to BB$, $B\to c_i\}$ dedicated to the synthesis of virtual enterprises according to BM_VEARM.

Figure 4(a) represents the so-called "BM_VEARM minimal system". The VE instance of Figure 2 (b), now with embedded integration mechanisms, is shown in Figure 4(b) and can be synthesized from the following G_{BM} derivation:

$$\begin{aligned} S &\Rightarrow c_1(\downarrow\uparrow A)\equiv s_{eq} \Rightarrow c_1(\downarrow\uparrow AA)\equiv s_{eq} \Rightarrow c_1\left(\downarrow\uparrow r_1(\downarrow\uparrow B)A\right)\equiv s_{eq} \Rightarrow c_1\left(\downarrow\uparrow r_1(\downarrow\uparrow c_2)A\right)\equiv s_{eq} \\ &\Rightarrow c_1\left(\downarrow\uparrow r_1(\downarrow\uparrow c_2)r_2(\downarrow\uparrow B)\right)\equiv s_{eq} \Rightarrow c_1\left(\downarrow\uparrow r_1(\downarrow\uparrow c_2)r_2(\downarrow\uparrow BB)\right)\equiv s_{eq} \Rightarrow \\ &\Rightarrow c_1\left(\downarrow\uparrow r_1(\downarrow\uparrow c_2)r_2(\downarrow\uparrow c_3c_4)\right)\equiv s_{eq} \end{aligned} \quad (4)$$

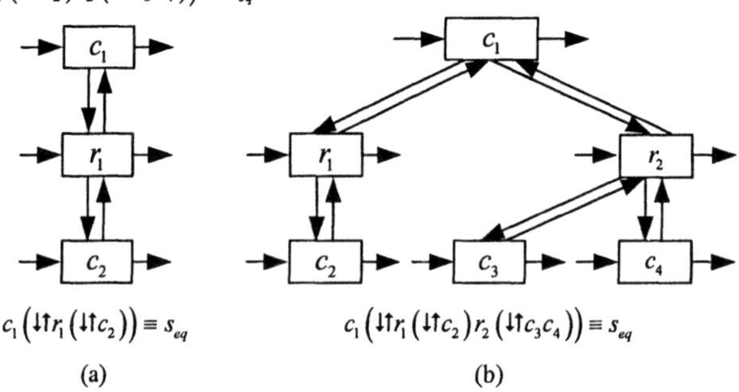

Figure 4 - BM_VEARM (a) minimal system (b) VE instance

As seen before the set of all the words generated by a grammar is a language.

Definition 4: $L_{BM}=L(G_{BM})$ is a formal language for VEs structures compliant to BM_VEARM.

Every word generated by G_{BM} ends with '$\equiv S_{eq}$' being thus a formula. Therefore the language L_{BM} is a set of formulas. According to mathematical logic if those formulas are satisfiable by a given interpretation (and closed under consequence) then they will constitute a theory (Ebbinghaus *et al.*, 1996). The structural interpretation of G_{BM} terminal symbols as control level blocks, resources management blocks, hierarchical connection, etc., satisfies all the formulas of L_{BM}. Thus we can claim that L_{BM} is a formal theory of VEs structures compliant to BM_VEARM.

5. CONCLUSIONS

The importance of the virtual enterprise (VE) paradigm at present and near future seems to be obvious. It seems also consensual that investigation on this area must have a solid theoretical background otherwise sustainable research won't be possible. Following this line of thought this paper presents an important contribution to the establishment of the referred theoretical base.

It is shown how formal grammars, and specially attributed grammars, can be used to deal with some aspects of VEs – structural aspects in this case - in a completely rigorous manner.

It is presented the attributed context-free formal grammar G_{BM} responsible for the generation of the formal language L_{BM}. L_{BM} is not just another formal representation language used, in this case, in the VEs area. Due to the development process of G_{BM}, the language L_{BM} can be used to represent VEs structures but it is also a theory of VEs structures compliant to the BM_Virtual Enterprise Architecture Reference Model (BM_VEARM), providing other potentialities. Although not detailed here, due to space limitations, the inclusion of attributes associated to the symbols of G_{BM} grammar constitutes the true power of this approach. For example we can define how many blocks (control level and resources management) are available and how many inputs and outputs each one of them has, and let G_{BM} synthesize VEs instances compliant to those predefined requisites. Furthermore with simple modifications G_{BM} can be used not only to synthesize VEs structures but also to recognize that kind of structures.

The exploitation of the equivalence grammars-automata will lead to the specification of a pushdown automaton equivalent to the context-free attributed grammar G_{BM}, allowing thus the development of application tools. This work is already running and a very simple prototype tool (not yet based on G_{BM}) was already developed by two computer science students.

The Formal Theory (FT) presented in this paper is not a general FT of VEs, but only the FT of a specific aspect of VEs - the structural aspect - compliant to BM_VEARM. BM_VEARM is a reference model and others may exist. The grammatical approach proposed could be applied to other reference models, implying that a specific grammar should be constructed for each model. What to do with these FT of particular VE models and aspects? Unify them in a more general FT or leave them as they are resolving only specific problems? These, and other related issues, are open questions. This paper is also a contribution to these questions.

6. REFERENCES

1. Chomsky, N. (1959). "On Certain Properties of Grammars." *Information and control* **2**: 137-167.
2. Denning, P. J., Dennis, J. B. and Qualitz, J. E. (1978). *Machines, Languages and Computation*, Prentice-Hall, Inc.
3. Ebbinghaus, H. D., Flum, J. and Thomas, W. (1996). *Mathematical Logic*, Springer.
4. Hopcroft, J. E. and Ullman, J. D. (1979). *Introduction to Automata Theory, Languages and Computation*, Addison-Wesley Publishing Company.
5. Keisler, H. J. (1996). *Mathematical Logic and Computability*, McGraw-Hill International Editions.
6. Knuth, D. E. (1968). "Semantics of Context-free Languages." *Mathematical Systems Theory* **2**: 127-145.
7. Lewis, H. R. and Papadimitriou, C. H. (1981). *Elements of the Theory of Computation*, Prentice-Hall International Editions.
8. Mendelson, E. (1987). *Introduction to Mathematical Logic*, Chapman & Hall.
9. Mesarovic, M. D., Macko, D. and Takahara, Y. (1970). *Theory of Hierarchical, Multilevel, Systems*, Academic.
10. Mikolajczak, B., Ed. (1991). *Algebraic and Structural Automata Theory*, North-Holland.
11. Pittman, T. and Peters, J. (1992). *The Art of Compiler Design - Theory and Practice*, Prentice-Hall International, Inc.
12. Putnik, G. (2000). BM_Virtual Enterprise Architecture Reference Model, *in Agile Manufacturing: 21st Century Manufacturing Strategy (A. Gunasekaran)*, Elsevier science Publ: 73-93.
13. Révész, G. E. (1991). *Introduction to Formal Languages*, Dover Publications, Inc.
14. Salomaa, A. (1973). *Formal Languages*, Academic Press, Inc.
15. Sousa, R. (2003). Contribuição para uma Teoria Formal de Sistemas de Produção. Tese PhD. Departamento de Produção e Sistemas, Universidade do Minho.

34 A DISTRIBUTED KNOWLEDGE BASE FOR MANUFACTURING SCHEDULING

Maria Leonilde R. Varela[1], Joaquim N. Aparício[2], Sílvio do Carmo Silva[3]
[1]*University of Minho, School of Engineering, Dept. of Production and Systems*
Email: leonilde@dps.uminho.pt
[2]*New University of Lisbon, Faculty of Science and Technology, Dept. of Computer Science*
Email: jna@di.fct.unl.pt
[3]*University of Minho, School of Engineering, Dept. of Production and Systems*
Email: scarmo@dps.uminho.pt
PORTUGAL

This paper describes a distributed knowledge base for manufacturing scheduling. A great variety of scheduling problems can occur in manufacturing. For solving different problems, usually different methods are required. The distributed knowledge base enables sharing information about scheduling problems and corresponding solving methods in a widened search space. These methods can be remotely available and accessible through the Internet. Running several methods enables obtaining alternative results for a given problem, consequently, contributing for a better scheduling decision-making. An important aspect is that end users and scheduling methods' providers alike can easily and continuously update this knowledge base.

1. INTRODUCTION

Competitive organizations are operating today in global and worldwide markets. Therefore, the competitiveness of enterprises and the quality of working life, in today's knowledge-based economy, are directly tied to the ability of effective creation and share of knowledge both, within and across organizations.

Manufacturing scheduling is a complex task that involves a wide range of knowledge. Scheduling problems are often complicated by large numbers of constraints, interrelating tasks, resources and events external to a manufacturing system. Moreover, slight differences in the manufacturing environment originate distinct problems, which even though being closely related, require different solving methods to be applied.

The effective and efficient resolution of those problems begins with the identification of suitable scheduling methods for solving them. When there are alternative methods to solve a problem alternative solutions can be obtained, which should be evaluated against specified criteria or objectives to be reached. Thus, users are able to properly solve a problem, through the execution of one or more scheduling methods, local or remotely available and accessible through the Internet, and, subsequently, select de most suited solution obtained.

This work attempts to offer new possibilities for carrying out manufacturing scheduling, following the approach of solving problems through a web-based decision support system. The system follows a peer-to-peer computing model, which permits sharing scheduling knowledge by means of a distributed knowledge base (DKB). This distributed scheduling repository enables accessing knowledge arising from an extended range of contributors and, therefore, providing a widened search space. This infrastructure is based on the principles of virtual organizations (VO) [1, 2].

The system permits the characterization of each problem to be solved and, then, the access to corresponding solving methods. For problem identification, a problem classification model that includes a set of parameters is used. This model enables specifying problem classes to which real problem instances belong [6, 7] and for which, hopefully, suitable solving methods can be found in the DKB. The data representation model for scheduling problems and related concepts is based on XML (extensible markup language), which is used as a specification language for scheduling data representation and processing on the Internet [6, 7].

This paper is organized as follows. The next section briefly describes the nature of scheduling problems and the underlying classification model. Section 3 presents the web system's distributed knowledge base (DKB) for supporting the scheduling decision making process, by any end-user who whishes to solve a problem, and describes the underlying peer-to-peer framework. Moreover, a document type definition (DTD) and the corresponding XML document about scheduling methods specification are shown in order to better explain the DKB updating process. Finally, in section 4, some conclusions are reached.

2. MANUFACTURING SCHEDULING PROBLEMS

Manufacturing scheduling focuses on the efficient allocation of one or more resources to tasks over time. It is an important activity to be performed for a company to achieve competitive production, which usually means to deliver products on time and to use resources efficiently. "Good" orderings to perform a series of given tasks have to be found, whereby specific objectives shall be optimized.

In order to perform the scheduling process it becomes necessary to clearly specify the problem to be solved. Manufacturing scheduling problems have a set of characteristics that must be clearly and unequivocally defined.

Due to the existence of a great variety of scheduling problems, there is a need for a formal and systematic manner of problem representation that can serve as a basis for their classification. A framework for achieving this was developed by Varela et al. [6, 7], based on existing notations available in the literature. This framework allows identifying the characteristics of each problem to be solved and it is used as a basis for an XML-based problem specification model developed [6, 7].

The referred framework for problem representation includes three classes of notation parameters for each corresponding class of problem characteristics, in the form of $\alpha|\beta|\gamma$. The first class of characteristics, the α class, is related to the environment where the production is carried out. It specifies the production system type (α_1) and, eventually, the number of machines that exist in the system (α_2). The second class allows specifying the interrelated characteristics and constraints of jobs and production resources, which are expressed by the β (β_1 ... β_{14}) parameters, and also the performance criterion, which is the third class (γ). Some important processing constraints are imposed by the need for auxiliary resources, like robots and transportation devices and/or the existence of buffers, among others factors. The evaluation criterion, the third class of parameters, may include any kind of performance measure, namely multi-criteria measures [6, 7].

An example of use of this notation is "F2|n|Cmax" which reads as: "Scheduling a set of n independent jobs, on a pure flow shop (F), with 2 machines, in order to minimize the maximum completion time or makespan (Cmax). Due to the absence of some characteristics in this problem characterization, it is assumed that they are defined by default. Thus, for example about the possibility of job preemption (pmtn), the jobs are non-preemptable, similarly no job arrival times are specified (r_j), which means that we are in presence of a static scheduling problem, with all jobs being ready to be processed at the same time, let us say at time zero. Moreover, the jobs are independent, as no precedence relations where defined (simple precedences, prec, or other type of precedences) and they have arbitrary processing time lengths, because no processing time restrictions are specified.

Good schedules strongly contribute to the company's success. This may mean meeting deadlines for the accepted orders, low flow times, few ongoing jobs in the system, low inventory levels, high resource utilization and, certainly, low production costs. All these objectives can be better satisfied through the execution of the most suitable scheduling methods made available through a distributed knowledge base, which enables searching for appropriate methods to solve each particular problem.

3. DISTRIBUTED KNOWLEDGE BASE

The last tendencies show that computing environments are characterized by increasing heterogeneity, distribution and cooperation, where distributed knowledge bases play an important role [9].

Knowledge usage in computer systems directly depends on knowledge representative schemes. The standardization of extensible markup language (http://www.w3.org/XML/) on the Internet gives new opportunities in such direction. XML provides general markup facilities that are useful for data interchange. The web system described in this paper is based on XML modeling and related technologies.

The system is able to quickly assign methods to problems that occur in real world manufacturing environments and solve them through the execution of one or more appropriate implemented methods that are local or remotely available and accessible through the Internet.

The selection of one or more specific scheduling methods for solving a given problem is made through a searching process on the distributed knowledge base. The matching process, between problems and methods is performed by a built-in prolog search engine, which was developed using the SWI-Prolog V.5.2.1. free software tool available at http://www.swi-prolog.org/.

Figure 1 shows a general outline of the system's architecture.

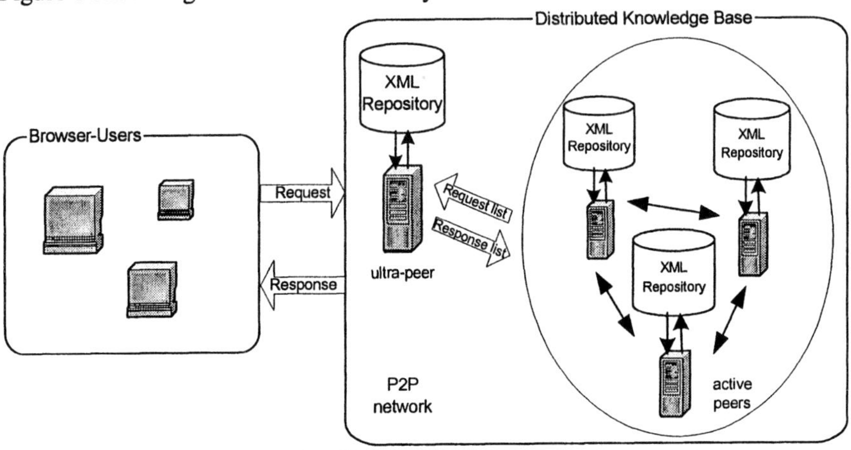

Figure 1 – Web system architecture

Figure 2 illustrates the main system processes, which also include knowledge insertion, about scheduling problems and solving methods, and correspondent information searching. Users can make requests for visualizing scheduling problem classes and methods' information or even browse information about other concepts presented by the system. The data can be shown in different views, using existing XSL (extensible stylesheet language) documents, adequate for each specific visualization request.

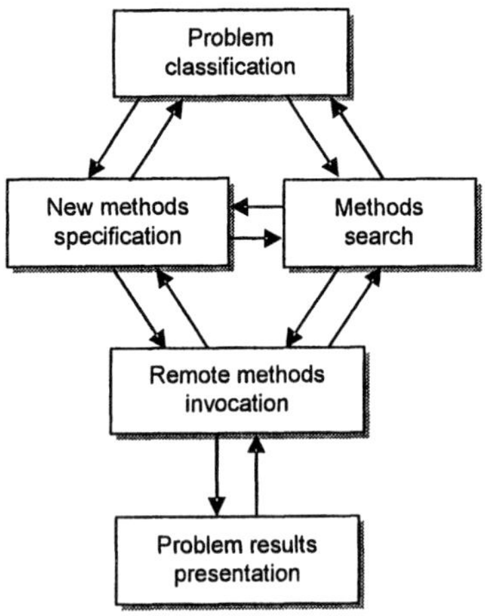

Figure 2 – Main system processes

The system has been designed and implemented as a web service (http://www.w3.org) using the XML-RPC (extensible markup language – remote procedure call) protocol [8] and will be available soon through the http://www.dps.uminho.pt/ web site. In a web service a certain method accepts as input a problem definition and returns a result in some particular form. Different methods' implementations may provide results in different forms, and the system must have a description of them in order to format them according to the problem output to be returned to the client as the very last step of the service.

The system enables different ways of problem results presentation and storage. The result from running a method implementation on the given problem instance can be delivered to the client as an XML file and/ or can be transformed into some more expressive output, like a Gantt chart or other data representation formats, like tables.

3.1 Peer-to-peer network

In networked environments, distributed knowledge bases and intelligent brokers, for information retrieval from specialized servers and knowledge repositories distributed across the Internet, enable to establish high-quality problem solving, through knowledge and resource sharing. In this context, emerging peer-to-peer technology and appropriate networks, suite well to the increasingly decentralized nature of modern companies and their industrial and business processes, whether it is a single enterprise or a group of companies [4]. The P2P framework provides the capabilities that allow users, or peers, to directly interact with each other [4, 5].

The web application described in this paper follows a P2P computing model. A set of peers, contributing with a local knowledge base component, composes a DKB as a P2P network. The P2P network has the capability of allowing a direct-interaction between the peers, which turns the computing environment decentralized, namely in terms of storage, computations, messaging, security and distribution. One of the greatest benefits of this P2P network, in the context of this work, is to easily support the concept of community. Consequently, it is possible for peers to organize themselves into groups that can collaborate with each other in order to achieve certain goals. One of the main goals aimed at in this work is the collaborative improvement of the resolution of manufacturing scheduling problems. This is possible through the DKB for manufacturing scheduling by accessing several different scheduling approaches. This is achieved by providing a mechanism that allows the members of the P2P network to share their scheduling knowledge and scheduling methods.

As previously referred the DKB scheduling system is based on the principles of VO. In this VO, each peer contributes with a local knowledge base (KB) in the P2P network. Each one can then be seen as a VO member or partner interested on solving scheduling problems. Whenever a member stores knowledge in its KB component, he/she is automatically contributing to the enrichment of the whole distributed knowledge repository, which is available to all the members of the VO.

Some peers of the organization can also act as ultra-peers. These special peers have the additional functionality of owning the list of the peers that belong to the VO. Such list contains information about the VO members and a flag that indicates their current state, which can be active or non-active. An ultra-peer also serves the purpose of configuring the P2P network as an open system, allowing any external user to join the organization, or as a closed system in the sense that only the nodes belonging to a certain company or domain can join the organization.

Each active peer is continuously listening for requests from other peers or from browser-like users. When a request reaches a peer, it firstly asks to one of the ultra-peers for the list of other active peers. Next, it propagates the request to all the peers of that list. Once the replies have been returned from the contributing active peers, the compiled results are presented to the user in order to fulfill the request, as previously illustrated in Figure 1.

At any time, external users can join the VO and configure themselves as active peers. This can be easily done, by just installing a set of common components that compose the interface for accessing the network and it's DKB. When a new (ultra)peer joins the VO it sends a request with its address to the ultra-peer(s) that represent the root domain of the P2P network, which is guaranteed to be always available. The ultra-peer(s) register the new (ultra)peer address, which is dynamically broadcasted to the remaining ultra-peer(s) of the current list. The VO members can join and stay connected or disconnect and leave the P2P network whenever they want, which configures very dynamic features to this VO model.

3.2 Updating the knowledge base

The web system enables introduction, validation, and transformation of manufacturing scheduling data. These processes are mainly controlled by DTD and XSL documents stored in the distributed knowledge base and all the scheduling information is stored in XML documents, which are validated according to associated DTDs, before being put in the corresponding knowledge base [6, 7].

At each peer the knowledge base can be continuously improved with new problem descriptions and available solving methods.

In the Internet many implementations may exist for a given method. From the point of view of the web system two implementations of the same method may differ if, for example, they differ on its outputs. Unfortunately, not all implementations work in the same way. Therefore, for the system to be able to match problem instances to resolution methods and to retrieve and use implemented methods available, in a programmatic way, they must also be described within the system. This description must include, among other things, the uniform resource locator to the running method and its signature, which, in turn, includes the definition of the parameters that are necessary for its invocation (inputs) and its output format.

The scheduling methods and their implementation details are described by a given DTD. Listing 1 illustrates this DTD for specifying the methods' information, such as their signatures, which are subsequently used to invoke the methods as web services and for other relevant information retrieval.

```
<!ELEMENT methods (method)*>
<!ELEMENT method
(id,name,url?,problem_class,method_class?,reference,complexity?,protocol?,signature?,gantt?)>
<!ELEMENT id (#PCDATA)><!ELEMENT name (#PCDATA)>
<!ELEMENT url (#PCDATA)><!ELEMENT problem_class (#PCDATA)>
<!ELEMENT method_class (#PCDATA)><!ELEMENT reference (#PCDATA)>
<!ELEMENT complexity (#PCDATA)><!ELEMENT protocol (#PCDATA)>
<!ELEMENT signature (input,output)>
<!ELEMENT input (param | array | matrix)+><!ELEMENT param (#PCDATA)>
<!ATTLIST param name CDATA #REQUIRED type CDATA #REQUIRED control (submit)
#IMPLIED>
<!ELEMENT array (item+)>
<!ATTLIST array from CDATA #FIXED "1" to CDATA #REQUIRED control (submit) #IMPLIED>
<!ELEMENT item (#PCDATA)>
<!ATTLIST item name CDATA #REQUIRED type CDATA #REQUIRED>
<!ELEMENT matrix (item+)>
```

```
<!ATTLIST matrix lines CDATA #REQUIRED columns CDATA #REQUIRED control (submit)
#IMPLIED>
<!ELEMENT output (param | array | matrix)+><!ELEMENT gantt (#PCDATA)>
```

Listing 1 – DTD document sample about methods specification

Many scheduling methods may be more or less adequate to solve a given class of problems. In the methods distributed knowledge base the system records the scheduling method(s) that can be used for solving a certain problem class. Searching for the adequate methods, for a given problem, is performed by matching the problem details with the methods' characteristics, a process performed by the built-in prolog engine of each peer. The methods are usually available in the knowledge base of the peers belonging to the VO but they can also be found in other sites not belonging to the community.

Listing 2 shows a sample of the XML document about scheduling methods. It illustrates the information related to the implementation of the Johnson's Rule [3] for solving problem instances belonging to the F2|n|Cmax class described in section 2. This document is validated against the corresponding DTD, previously shown in Listing 1, before being put in the corresponding knowledge base component.

```
<?xml version="1.0" encoding="UTF-8"?>
<!DOCTYPE methods SYSTEM "methods.dtd">
<methods>
    <method>
        <id>32</id><name>Johnson</name><url>http://localhost:6002/RPC2</url>
        <prob_class>F2|n|Cmax</prob_class><method_class>Sequencing Rule</method_class>
        <reference>Johnson,1954</reference>
        <complexity>Maximal Polynomially Solvable</complexity><protocol>XML-RPC</protocol>
        <signature>
            <input>
                <param name="n" type="integer"/>
                <param name="m" type="integer" control="submit"/>
                <matrix lines="m" columns="n">
                    <item name="job" type="string"/><item name="machine" type="string"/>
                    <item name="p" type="double"/>
                </matrix>
            </input>
            <output>
                <param name="Cmax" type="double"/><param name="sequence" type="string"/>...
                <matrix lines="m" columns="n">
                    <item name="start" type="double"/><item name="finish" type="double"/>...
                </matrix>
            </output>
        </signature>
        <gantt>yes</gantt>
    </method>...
</methods>
```

Listing 2 – XML document sample about methods specification

The inputs include the definition of a parameter n, for the number of jobs to be processed, a parameter m, for the number of machines, and a set of three items organized as a matrix structure, which represent the job name, the machine name and the processing time p of each job on each machine. There is also the definition for the method's output following the same lines. After a method definition has been inserted in a local KB it becomes immediately accessible to any further information retrieval. For the example given, after the insertion of the Johnson's method definition any methods search that is a match for the F2|n|Cmax problem class will include this method in the search results. The methods' definitions are also used in the automatic generation of interfaces for methods invocation and corresponding inputs introduction and subsequently outputs presentation.

4. CONCLUSIONS

In this paper a web system based on a peer-to-peer (P2P) infrastructure and a distributed knowledge base (DKB) is presented. The DKB is spread through a set of members (peers) forming a virtual organization. These members can store information about methods for solving different kinds of manufacturing scheduling problems, as well as their implementations. Each peer, in a decentralized computing environment, is able to directly interact with each other, as well as with any other external user, in order to enable executing diverse scheduling functions, including the ability to represent different scheduling problems, searching for appropriate solving methods and running methods that are made available and accessible by the DKB, through this P2P network. As running different methods enables obtaining different solutions for a problem, the system contributes for a better decision-making process, enhanced by means of the collaboration among the peers forming the virtual organization.

5. REFERENCES

1. Camarinha-Matos, L. M.; Afsarmanesh, H., The Virtual Enterprise Concept, In: *Infrastructures for Virtual Enterprises, Networking Industrial Enterprises.* Kluwer Academic Publishers, 1999, pp. 3-14.
2. Camarinha-Matos, L. M.; Afsarmanesh, H., Design of a Virtual Community Infrastructure for Elderly Care, In: *Collaborative Business Ecosystems and Virtual Enterprises, PRO-VE'02*, Sesimbra, Portugal. Kluwer Academic Publishers, 2002, pp. 439-450.
3. Conway, R. W.; Maxwell, W. L.; Miller, L. W., Theory of Scheduling. England: Addison-Wesley Publishing Company, Inc., 1967.
4. Papazoglou, M.P.; Krämer, B.J.; Yang, J.; Leveraging Web-Services and Peer-to-Peer Networks, In: *Proceedings of Advanced Information Systems Engineering, 15th International Conference, CaiSE*, Klagenfurt, Austria. June 16-18, 2003, pp. 485-501.
5. Terziyan, V.; Zharko, A.; Semantic Web and Peer-to-Peer: Integration and Interoperability in Industry, Industrial Ontologies Group, MIT Department, University of Jyvaskyla, Finland (http://www.cs.jyu.fi/ai/vagan/papers.html).
6. Varela, L. R.; Aparício, J. N.; Silva, C. S., An XML knowledge base system for scheduling problems, In: *Proceedings of the Innovative Internet Computing System Conference, I2CS'02*, Kuhlungsborn, Germany. Springer-Verlag in the Lecture Notes in Computer Science series, 2002; 61-70.
7. Varela, M. L. R.; Aparício, J. N.; Silva, S. C., Developing a Web Scheduling System Based on XML Modeling, In: *Knowledge and Technology Integration in Product and Services – Balancing Knowledge and Technology in Product and Service Life Cycle, BASYS'02*, Cancun, Mexico. Kluwer Academic Publishers, 2002, pp. 61-70.
8. Varela, M. L. R.; Aparício, J. N.; Silva, S. C., A Scheduling Web Service based on XML-RPC, In: *Proceedings of the 1st Multidisciplinary International Conference on Scheduling: Theory and Applications, MISTA'03*, Nottingham, UK. ASAP, The University of Nottingham. 2003, pg. 540-551.
9. Wu, J.; Distributed System Design. New York: CRC Press, 1999.

35

EFFICIENTLY MANAGING VIRTUAL ORGANIZATIONS THROUGH DISTRIBUTED INNOVATION MANAGEMENT PROCESSES

Jens Eschenbaecher
University of Bremen and Bremen Institute of Industrial Technology and Applied Work Science (BIBA) at the University of Bremen, Hochschulring 20, D-28359 Bremen, GERMANY, esc@biba.uni-bremen.de

Falk Graser
University of Bremen and Bremen Institute of Industrial Technology and Applied Work Science (BIBA) at the University of Bremen, Hochschulring 20, D-28359 Bremen, GERMANY, grs@biba.uni-bremen.de

Competition in the future will be characterized by an increasing meaning of customer-driven mass customization. This challenges enterprise networks to develop innovative and more flexible structures than before. The kind of organization that is considered to meet those challenges in the best possible way is the Virtual Organization. To successfully stand the competition, Virtual Organization need clear governmental structures guiding them though their collaborative processes. This paper discusses the Distributed Innovation Management (DIM) concept as an instrument to successfully govern innovation processes in Virtual Organization.

1. INTRODUCTION

Virtual Organizations (VO) offer a dynamic organisational form to meet the challenges of future competition and better distributed innovation management performance. Distributed Innovation Management is defined as the process of managing innovation within and across groups of organizations joining to co-design and co-produce products and co-service the customer's needs (Duschek 2002). Their temporary, flexible and dynamic nature supports the necessity to integrate different enterprises quickly for realizing common business objectives [Sydow 2001]. Whatever these business objectives may be, their common denominator is their innovative character. Creating new ideas, transforming them into a product or service, and bringing them successfully to the market is a challenge that is difficult to manage already within single enterprises where several players within one singular organization need to be streamlined to a common objective. Within a collaborative network, the success of innovation processes depends highly on an efficient network government: Several players within several different organizations, sharing different processes, company cultures, and information systems need to be harmonized to successfully realize an innovative idea (Gassmann and Zedtwitz 2002). Developing and implementing an innovation management system in a VO is a crucial process, not only for the reasons just mentioned, but also for time-to-market reasons. Already today, competition leaves insufficient time for organizations to iteratively optimise new processes within a non-competitive environment; vice versa processes must work reliably at once to ensure achievement of the companies' common objectives.

This paper will take a three step approach to discuss how the concept of Distributed Innovation Management can be applied for governing a Virtual Organization. First, it will expose the basic concepts of the Virtual Breeding Environment as an incubator for VO and Distributed Innovation Management, and will eventually give an integrated life-cycle schema especially regarding these two concepts. Second, it will derive recommendations for successfully implementing Innovation Management Processes, and third, it will state a couple of findings summarizing and prospectively reviewing the paper's contents.

2. STREAMLINING ENTERPRISES FOR SUCCESSFUL INNOVATION

The Virtual Breeding Environment

A Virtual Breeding Environment creates a community of occasionally collaborating companies. The community ensures that the partners apply methods and procedures to ensure a certain quality standard. The term virtual breeding environment was recently developed by Camarinha-Matos and Afsarmanesh (2003). Basically, a VBE supports the exploitation of local competencies and resources by an agile and fast selection of the most adequate set of partners for each innovation project. Consequently if a business opportunity has been identified by one VBE member a virtual organisation can be created rather quickly. More information about the breeding environment can be found in (Camarinha-Matos 2004).

State of the art

The concept of Innovation Management is crucial for companies and collaborative networks. Companies are also required to collaborate with other organisations, because many of them do not possess all the required skills or necessary resources to innovate (DiMAN, 2002). However, innovation activities often cannot reap the desired fruits their implementation promises. Most researchers agree that between 50-80% of innovation fails to have any impact on organisational goals. Some surveys' results are showed in table 2-3.

Source	Strebel, P.	Rothwell, R.	Hammer, M.	Jaikumar	Field, T.	Burnes, B.
Percentage	60 %	80 %	50-70 %	50-75 %	73 %	70-80 %

Table 1: Percentages of failure for innovation projects in organisations [Eschenbaecher 2004]

An organisation typically can invest between 0.01% and 20% of its annual turnover in innovation. The rate of investment can depend on whether the organisation is a corporate "shooting star" where investment can be as high as 20% or "cash cow" where it can be as low as 0.5%. A recent survey of European companies stated that their expenses on innovation have an average of 4%. This fact implies, according to the percentages of failure for innovation projects, that there is a big amount of wasted investment that will not lead to any growth or increase in efficiency. Besides the economic loss, there are also significant consequences within organisations, regarding their culture, for example, increased scepticism among employees and greater resistance to change in the future innovation projects.

Evolution towards Distributed Innovation Management in networks

The state of the art described before clearly indicates the need for good methodological support for Innovation Management (IM). IM has been widely discussed in the past. Throughout the last five decades, the environment in which Innovation Management is embedded changed significantly several times (Möhrle, 2003). This, of course influenced Innovation Management itself, and led to adoptions of the respective methodologies. Pavitt, Rothwell, Dogson and others have put forward more than a decade ago non-linear innovation models, such as the systems integration and networking model, or 5th generation model, that highlight implementation as a non linear process of both explicit and tacit knowledge flows among a network of firms and their suppliers and customers. (Pavitt 2003, Rothwell, 1993, Dodgson 2000); they are depicted in Figure 1.

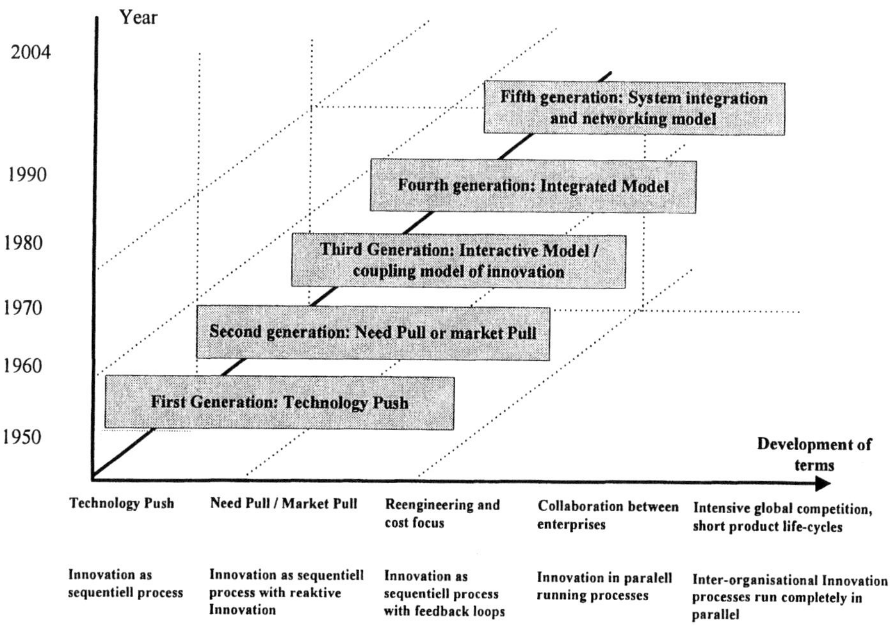

Figure: 1: Evolution of IM

The picture includes a brief description of the dominant trend in each generation with regard to the innovation processes. DIM can be subordinated to the fifth generation "Systems integration and networking model" (Möhrle 2003). Thus, enterprise networks are conceived to undertake innovations in a collaborative and global environment, so achieving outcomes with more efficiency and effectiveness. The final aim is to fulfill customers' needs, adapting to the changing situations within the markets and improving the quality offered by the products and services. For doing this the companies must align their strategies for exchange of knowledge, ICT, processes and people. Additionally they need to have intra-organizational innovation management systems in place which need to be aligned.

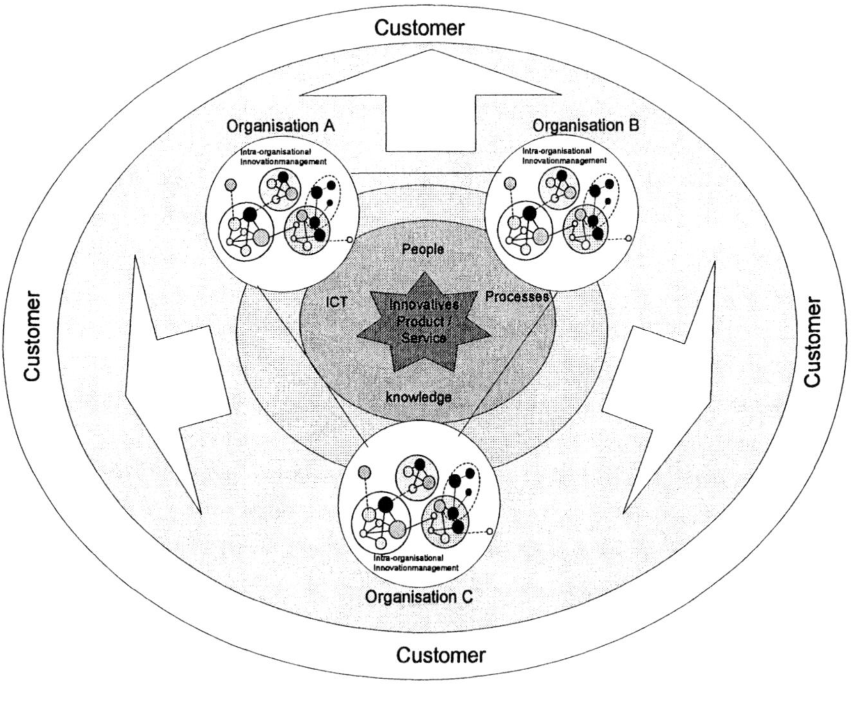

Figure 2: Distributed innovation management in collaborative networks

The network's companies must have intra-organisational innovation management systems which can be adjusted to each other. The authors structure innovation management in four levels. These are individual innovation, project innovation, collaborative innovation, and distributed innovation. The lower levels in the hierarchy are embraced by the distributed concept. In DIM, collaboration is extended beyond the limits of a single organization, cutting across the enterprise network.

An integrated Life-Cycle schema for the Virtual Organization

Figure 3 shows how the concept of the Virtual Breeding Environment is embedded into the life-cycle of a VO. When the VBE created an operable VO, the partners are ready to realize the innovative product and service that is objective of the partners involved. That objective represents the "Virtual Centre" of the life-cycle, to that all actions need to be adjusted.

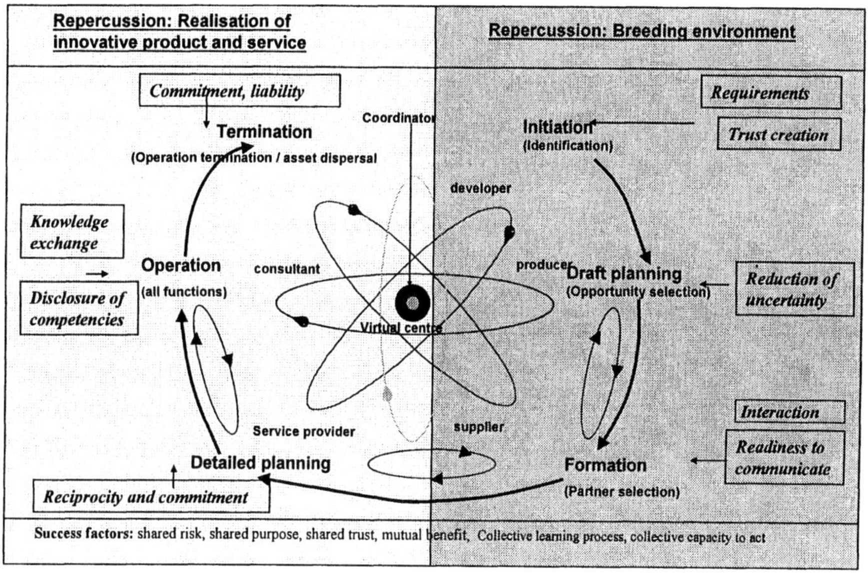

Figure 3: Requirements and success factors in the VO life-cycle

3. RECOMMENDATIONS FOR DIM IMPLEMENTATION

Industrial case studies

The results presented in this section are based on the AIT Implant project and some follow-up empirical studies. Additionally information can be found in Segarra (1999), Eschenbaecher und Cocquebert (1999) as well as recent case studies (Behnken 2004, Gerken 2004). Behnken (2004) illustrates different cases from the aeronautical and furniture industry whereas Gerken shows a distributed innovation management case from Porsche. The cases have clearly shown the need for a methodology which supports the management of distributed innovation processes within collaborative networks.
The case studies show that there is a high interest in tools and mechanisms to better govern innovation projects (compare Eschenbaecher und Hahn 2004).

Collaboration concept

The concept of collaboration has a major impact on distributed innovation management. Collaboration is characterised by three main aspects.

- Co-ordination by using
    - Transparent structure of responsibilities, defined control mechanisms, power structures, structuring and governing phase model, definition of organisational forms of virtual teams, steering

committee as project controlling and external support by consultants or non-team members.

- Communication by using
  - Portal structure to save, open and revise documents, e-mail, calendar, news editor, quick links, status window, category definition
- Cooperation with
  - Subscription opportunity,
  - Web-browser user interface (24 hours availability)

The user process and the users, which collaboratively conduct the innovation process, are in the center of the framework. The model splits the innovation process into specific, discrete phases. Each phase is concluded by a review that delivers one of the two following results:

- Entering the next phase is cleared by accepting the previous' phases results, or
- The previous phase's results are rejected forcing the process back into the previous phase for improvement.

These iterative recursions separate the model from traditional, linear innovation models. Hence, this model is a non-linear one. Within the phase's context, assessment, preparation, achieving and auditing many parallel activities take place. Every phase is finished by a review organised by an independent group of experts deciding whether the activities should be continued or not. This can be an internal management board, reviewers of a funding body or a steering committee of a distributed team. Furthermore the innovation process is separated in the two main stages innovation planning and innovation introduction. These areas are governed by project teams which co-ordinate the project. This methodology shows that conducting a distributed innovation management activity within a network makes a substantial effort in co-ordination, co-operation and communication necessary. The authors propose that the optimal selection and strategic implementation of innovation methodologies describing necessary efforts and suggest an approach for application of innovation methodology in organizations.

Web-based implementation

The extension of the basic conceptual ideas can be conceived as a result of the new opportunities provided by internet-based web-portals. The innovation management methodology has been tested within a large user case (see Eschenbaecher 2004). Altogether 55 organisations participated in a study about their judgement about a web-based distributed innovation methodology (www.expide.org/ecolead). The following figure shows the portal. The main result was that all the testing organisations agreed that a web-based DIM approach is the appropriate way to support innovation management in virtual organisations.

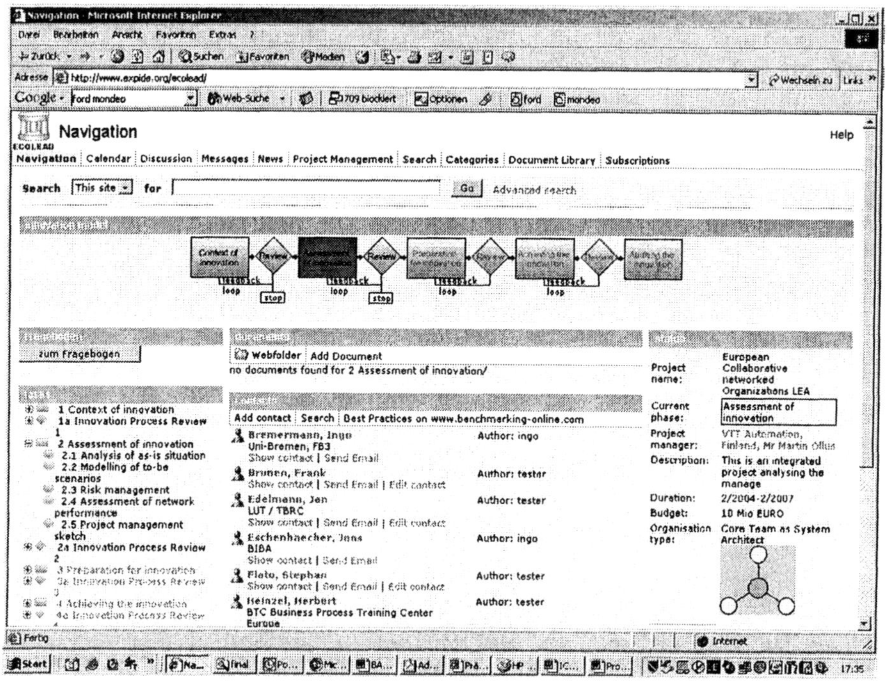

Figure 4: Web based distributed innovation management model

4. SUMMARY AND FINDINGS

Distributed innovation management will become a decisive task for Virtual Organizations and it is not sufficiently implemented. This is based on the large set of requirements shown and the missing common strategy development for the overall network. Furthermore the paper has presented a DIM approach considering the need to create a distributed innovation management system. The DIM portal has been validated by a sufficient user group and has been tested in various case studies.
It is expected that innovation management in virtual organisations will become a major issue in future collaborate networks

5. REFERENCES

1. Camarinha-Matos, Luis, Afsarmanesh, Hamideh: Elements of a base VE infrastructure; to appear in International Journal of Computer Integrated Manufacturing, 2003.
2. Camarinha-Matos, Luis, Afsarmanesh, Hamideh: New collaborative organiszations and their Research needs, In: Processes and Foundations for virtual organizations, Kluwer Academic Publishers, Boston/Dordrecht/London, 2004, p. 3-12.
3. Cascio, Walter: Virtual workplaces: implications for organizational behaviour. In C.L. Cooper & D.M. Rousseau (Eds.), The virtual organization. Trends in Organizational Behaviour, Vol 6, pp. 1-14. Chichester: John Wiley & Sons, 1999.
4. Cormican, Kathryn: Product Innovation Management for Networked Organisations, Galway 2001, PHD-Thesis.
5. Dodgson M: The Management of Technological Innovation, Oxford: Oxford Univ Press, 2000.

Drucker, P. F.: "Innovation and Entrepreneurship: Practices and principles", Heinemann, London, 1985

Drucker, P. F.: „Innovations-Management für Wirtschaft und Politik", Econ Verlag, Düsseldorf und Wien, 3. Auflage, 1986

Duschek, S.: Innovation in Netzwerken – Renten – Relationen – Regeln, Wiesbaden 2002, PHD Thesis.

Eschenbaecher, Jens; Hahn, Axel: Approach for implementing distributed innovation management in collaborative industrial networks, Forthcoming Proceedings of the ICE 2004 conference in Sevilla, 2004

Gassmann, Oliver, von Zedtwitz, Maximilian: Organising Virtual R&D teams. In: R&D management, Vol. 33, No. 3, pp. 243-262.

Hauschildt, Jörg.: Innovationsmanagement, Verlag Franz Vahlen, München, 1993

Hess, Thomas: Netzwerkcontrolling – Instrumente und ihre Werkzeugunterstützung. Deutscher Universitätsverlag GmbH, Wiesbaden, 2002.

Jarvenpaa, Stephan, Leidner, Dieter: Communication and trust in global virtual teams. In: Journal of Computer-Mediated Communication, Vol. 3, No. 4, 1998, p.1-38.

Lorenz, G./Veit, E.: „Die treibende Kraft: neue Technologien", in Staudt, E. (Hrsg.): „Das Management von Innovationen" FAZ, 1986, p.295

Marshall, Paul, McKay, James, & Burn, Janice: The Three S's of Virtual Organisations: Structure, Strategy and Success Factors. In: Hunt & Davnes (Eds.), E-Commerce and V-Business (pp. 171-192): Butterworth Heinemann, 2001.

McDonough III, Eve, Kahn, Kabal & Barczak, Gerd: An investigation of the use of global, virtual, and collocated new product development teams. The Journal of Product Innovation Management Vol 18, Heft 2, 2001 110-120.

Mertens, Peter; Faisst, Wolfgang: Virtuelle Unternehmen, In: Wirtschaftswissenschaftliches Studium, Heft 6, 1996, S. 280-285.

Möhrle, M. G.: "Der richtige Projekt-Mix", Springer-Verlag, Berlin, 1999

O'Sullivan, David.; Cormican, Kathryn.: A Collaborative Knowledge Management Tool for Product Innovation Management, in: Int. Journal of Technology Management, Vol. 26, No. 1, 2003, S. 53-67.

Pavitt K (2003) The process of innovation, SPRU working paper no.89, Brighton: Univ of Sussex

Schuh, Günter, Katzy, Bernhard and Milarg, Klaus: Wie virtuelle Unternehmen funktionieren: Der Praxistest ist bestanden, In: Gablers magazin, No. 3, 1997, p. 8-11

Schumpeter, J.: „Theorie der wirtschaftlichen Entwicklung", 3. Aufl., Leipzig, 1931

Segarra, G. (1999), The advanced information technology innovation roadmap. Computers in Industry, Volume 40, Issues 2-3, November 1999, Pages 185-195

Rothwell R (1994) Towards the fifth generation innovation process, International Marketing Review, pp7-31.

Sydow, Jörg: Management von Netzwerkorganisationen – Zum Stand der Forschung. In: Jörg Sydow (Hrsg.), Management von Netzwerkorganisationen (pp 293-329): Verlag Dr. Th. Gabler GmbH: Wiesbaden 2001.

36 SME-SERVICE NETWORKS FOR COOPERATIVE OPERATION OF ROBOT INSTALLATIONS

Peter ter Horst
Demar Laser B.V., peter@demarlaser.nl
Gerhard Schreck
Fraunhofer IPK, gerhard.schreck@ipk.fraunhofer.de
Cornelius Willnow
Fraunhofer IPK, cornelius.willnow@ipk.fraunhofer.de
GERMANY

A major obstacle for the introduction of industrial robots in small and medium enterprises (SMEs) is formed by the complexity of the robot systems, and the required expertise and qualified personnel. These high requirements for the companies and their personnel could be reduced, if SMEs that dispose over the required skill could provide them to other SMEs as Internet services. This paper describes working procedures, methods and tools for creating such Internet-based SME-service networks.

1. INTRODUCTION

When introducing industrial robots, small and medium enterprises (SMEs) are today faced with the full complexity of robot systems. I.e. for small enterprises the introduction of robot installations already forms a high investment. Additionally, this is accompanied by the need for qualified personnel that is able to perform the planning, operation and maintenance tasks for the robot system.

While larger companies can afford specialist departments for planning and performing robot application, smaller companies dispose only over a small number of persons, and often only over a single person, for performing all robot related tasks (IFR 2001). This consequently overtaxes the companies and their personnel. For these reasons, SMEs often hesitate to introduce robots, even if this is urgently required for productivity and quality reasons.

Different SMEs together, however, dispose over the required types of expertise for efficiently operating robot installations. They could mutually support each other by providing the required skills as services. The spatial distance between the companies could be bridged by Internet.

In pursuit of this approach, the EU CRAFT-Project 'Small and Medium Enterprises - Robotics Service Inter-Network' (SME-Rosin) was started in November 2002, with a planned duration of two years (SME-Rosin Consortium,

2003a). In the project, services and tools for robotics support networks for SMEs are developed.

This paper presents developed company network structures, interaction schemes and support tools for cooperative, balanced integration of automated systems and human involvement.

2. THE EXAMPLE NETWORK

The SME-Rosin consortium consists of European SMEs that form the SME-network shown in figure 1. It is used as an example for developing and testing the required services and tools.

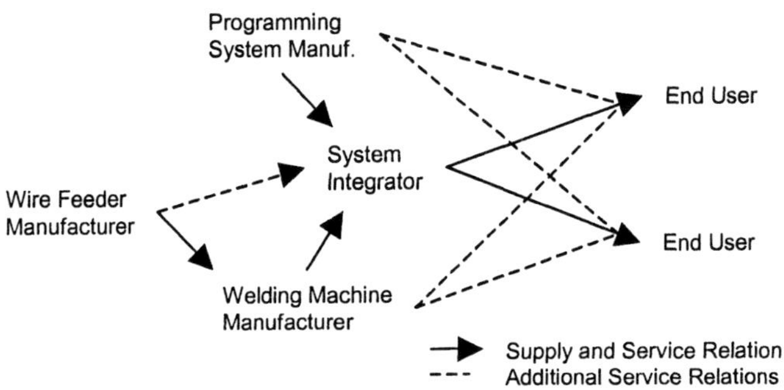

Figure 1 – The example SME-network

The example SME-network includes two typical end user companies. One already disposes over a robot installation and contributes the experience from introducing it and taking it into operation. The other one does not yet dispose over a robot and represents the challenge of introducing a robot.

A supply chain is represented by a system integrator, a manufacturer of welding machines and a manufacturer of wire feeder units. A further supply chain for software tools starts at a programming system manufacturer.

The network shows existing customer/supplier relations. Further services, that are developed in SME-Rosin and that are supported by Internet, enable a closer and more efficient cooperation of the companies.

3. DEVELOPMENT AND VERIFICATION PROCESS

The development of the service processes takes place in development-verification-cycles as illustrated in figure 2. After an initial analysis phase, service processes are defined and required tools are realized. The resulting developments are then verified in test scenarios and close-to-reality pilot installations. Based on the obtained feed-back, the developed processes and tools are improved. Over-all three cycles are performed.

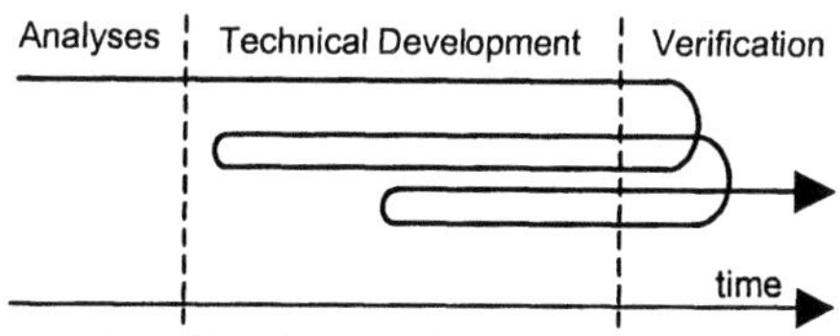

Figure 2 – Development-Verification-Cycles for the service and tools development

4. SERVICE AREAS

Based on the analysis performed during the SME-Rosin project, business processes for the service areas illustrated in figure 3 were developed (SME-Rosin Consortium, 2003b).

The area Production Engineering includes services for the development of manufacturing processes for robot cells. This concerns Product Re-design for automation in robot cells, the development of suitable fixtures, and the development of the processing tool and method, e.g. for arc welding.

The service area Programming includes services for Manual and Automatic Programming of robots. The term 'off-line programming' (OLP) denotes the creation of robot programs with simulation models of robot, work cell and work piece. This stands in contrast to 'on-line programming' what denotes programming directly at the physical robot. Programming as a remote service requires off-line programming. The activities for creating and maintaining the required consistency between the real cell and the simulated cell are summarized as Model Consistency.

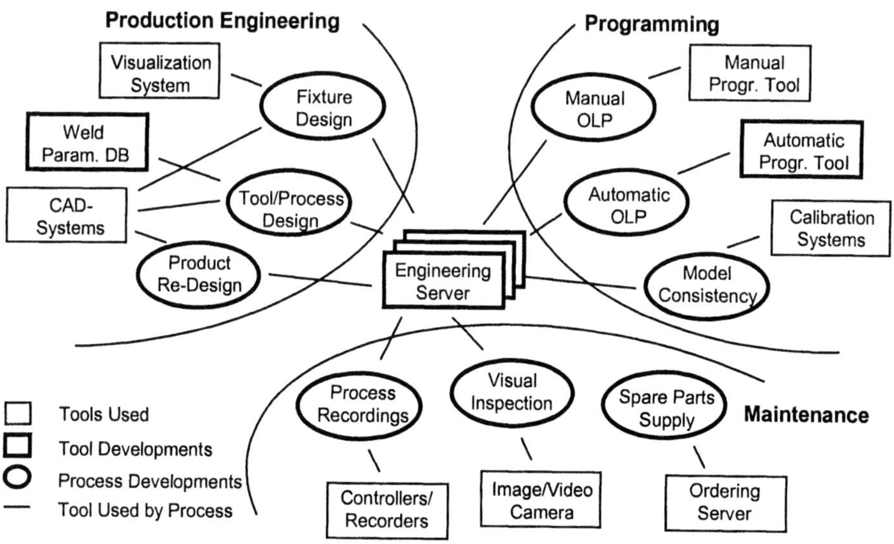

Figure 3 – Service areas, their services and tools involved

The service area Maintenance consists in obtaining Process Recordings for analyzing a robot cell, in Visual Inspection for obtaining pictures and films from a cell and in Spare Parts Supply services.

In practice, the different services and their business processes are often strongly related to each other. Process Recordings may be required for analyzing process performance during Production Engineering, Production Engineering may require Programming for testing designs, Programming services may be required after spare parts replacement, e.g. for re-establishing Model Consistency.

5. THE ENGINEERING SERVER

The central tool for organizing Internet-based services is an Engineering Server, as shown in the centre of figure 3. It organizes the exchange of documents like CAD-drawings, measuring data and robot programs. It provides mechanisms for notification, archiving and project progress control (Berger, Hohwieler 2003).

The SME-Rosin Engineering Server is especially designed for SME-networks. The SMEs are already burdened with the introduction and operation of the robot system. Therefore, the Engineering Server has to be very simple to use and to integrate into existing infrastructures (Kärkkäinen, Ala-Risku, 2003). It has to noticeably facilitate work and, for this reason, its usage has to flexibly fit into actual work procedures and current user needs.

Furthermore, it has to be possible to arbitrarily combine Internet-based activities with conventional means like telephone, fax, e-mail and travelling. Since almost all activities of robot planning and operation are strongly related to physical processes, the service provider has to be well familiar with the cell and has to have a good understanding of the manufacturing process. For a number of activities, direct at-site presence of the service provider is indispensable.

Consequently, activities via Internet have to be executable in combination with conventional means. For this reason, service processes have to be designed in a way that allows to combine conventional means with Internet activities, as required by the actual situation and practical needs. This reflects the flexibility that is a major strength of SMEs.

6. COOPERATIVE INTERACTION SCHEMES

In the course of the project, interaction schemes for the service processes described above are developed. As a typical, but concise example, the interaction scheme for Product Re-Design is discussed in more detail.

The Engineering Server is the central means for project organization between the different involved actors, as illustrated in figure 4. It is run by the Service Provider, which is the organization that provides the services.

For providing a Re-Design service, the Engineering Server is operated by a Designer employed at the Service Provider. The Service Provider, has a Company Interface to the Service User which is also an organization.

An employee of the Service User is in charge for cooperation with the Service Provider in a given project. For the case of a Product Re-Design project, this may be the Sales Engineer in charge of the project. The Sales Engineer at the Service User cooperates with the Designer at the Service Provider via the Engineering Server. Of course, as practical needs demand, also other communication means are involved.

The cooperation between the Designer at the Service Provider, and the Sales

Engineer at the Service User follows the diagram of principle activities in figure 5. It starts with the initiation of the service, followed by a number of service cycles.

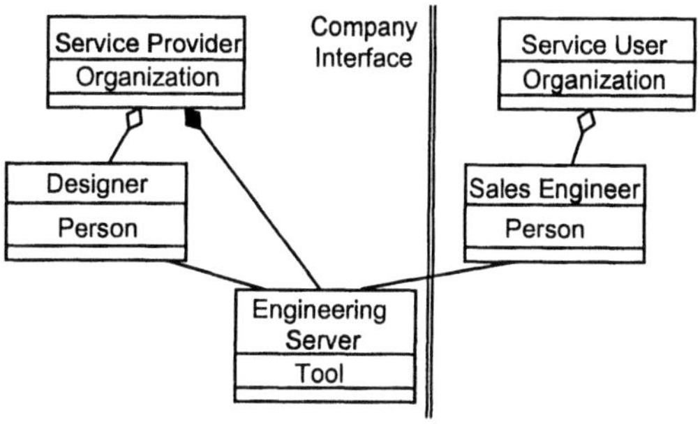

Figure 4 – Engineering Server as central means for project coordination

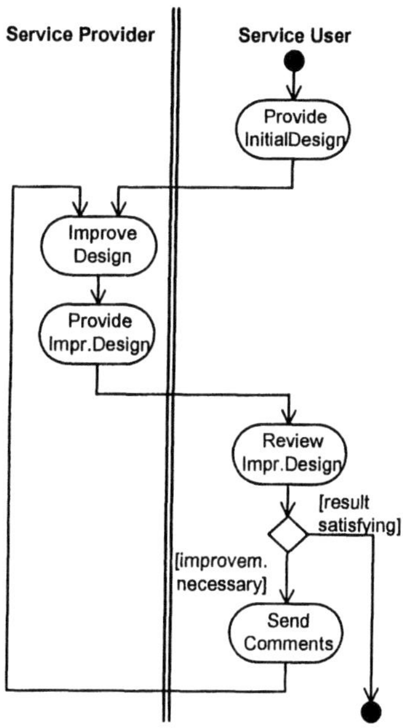

Figure 5 – Basic activity scheme for Re-Design services

After project initiation, the initial design is provided by the Service User and sent to the Service Provider via the Engineering Server. The initial design may consist of CAD-drawings, part lists, material specifications, surface treatments, quality requirements, etc.

For CAD-drawings, standards like Step or Iges could be used. This however may lead to data losses during conversion. Proprietary data formats, e.g. CATIA or AutoCAD files, avoid the losses but need agreements for using the same tool at Service User and Service Provider.

A further relevant technology are graphics viewers. They do not allow to modify data, but allow to inspect designs and they are provided for free. Then Service Providers can use the tool of their choice, while Service Users need no licenses.

Viewers are also available for simulation and animation tools (Visual 2003).

After providing the initial design, the service cycle starts: The Service Provider reviews and improves the design. During this, further interaction may take place. Additional specifications and requirements can be sent, and details may be clarified by e-mail or telephone. Then the improved design is provided to the Service User. This may include design variants for evaluation and selection by the Service User.

The Service User reviews the improved design. If the result is satisfying, then the Product Re-Design cycle is completed. Otherwise, the Service User comments on the actual version and sends the comments to the Service Provider. The comments may consist of text, modified or commented CAD-drawings, a design review by telephone, etc. as it suits best. For exchange of electronic data, the Engineering Server is used. Then, a new development cycle starts.

Please note again, that this development cycle is nominal and may be altered pragmatically as needed. It may be interrupted by conventional means like telephoning, faxing, travelling, etc. where this is more efficient. The improvement of the design may be interrupted by new data from the Service User. The review of the improved design may be interrupted by new ideas from the Service Provider, etc.

Furthermore this nominal cycle can be combined with the Engineering Processes of other service areas. For an implementation of the cycle in industrial praxis, any need and any degree of flexibility has to be possible.

7. USE CASE FOR COOPERATIVE PRODUCT RE-DESIGN

The following use case illustrates the interaction scheme by an example. The example and the names of the involved companies are fictive and used for illustrative purposes only. The example is, however, typical and realistic.

The use case is built on the following scenario: CoMa Ltd. manufactures containers with high volume. The containers have a high number of variants, reaching from differing dimensions to a variety of special equipment like rolls and suspensions. In addition to the number of standard container types, CoMa produces containers also for customer designs. Most of them are derived from standard designs. Sub-assemblies like side parts, bottom parts, back planes and doors are robot welded.

Since the number of customer designs does not justify to employ a skilled designer for automation, CoMa uses services provided by the Re-Design Company ReDeCo GmbH. ReDeCo disposes over models of the robot installation at CoMa. The cooperation in a project may proceed as followings:

Project Initiation

1. A customer of CoMa needs containers that are adapted to the customer's existing internal transportation system. For this, containers have to be adjusted in size and the suspension mechanism is to be modified.
2. During a meeting at the customer, hand sketches for the new design are made, and CoMa is supplied with printouts of technical drawings for the suspension and dimensions.
3. The sales engineer at CoMa creates a new AutoCAD design of the customized container variant, by modifying an existing design.
4. ReDeCo GmbH is contacted by telephone for the new task.
5. CoMa's sales engineer places the new AutoCAD design in a new folder of

ReDeCo's Engineering Server.

6. The designer in charge at ReDeCo is notified by the Engineering Server that the new design is available.

First Re-Design Cycle

7. The designer at ReDeCo fetches the design from the Engineering Server and reviews it. It turns out that the suspensions are placed in a way that the tool collides with them when welding the vertical stiffeners of side parts.
8. With a telephone call, ReDeCo checks with CoMa if it is better to shift the suspensions or to shift the stiffeners. Since the suspensions have to fit to the customers's transportation system, they may not be shifted.
9. Since shifting the stiffeners may reduce the admissible load of the containers, CoMa checks with the customer for the maximal load. The result it sent to ReDeCo via the Engineering Server.
10. The designer at ReDeCo is notified that the feed back arrived and fetches it from the Engineering Server.
11. The designer at ReDeCo shifts the stiffeners in the AutoCAD design. The admissible payload is verified by a tool for stress analysis. The resulting design is placed in the Engineering Server.
12. The sales engineer at CoMa is notified and fetches the new design from the Engineering Server. It is checked for compliance with the customer's specification.

Second Re-Design Cycle

13. The new distribution of stiffeners in combination with the reduced payload suggests to omit stiffeners. This would reduce production costs. CoMa creates a re-design and passes it via the Engineering Server to ReDeCo.
14. The designer at ReDeCo is notified about the feed back and clarifies the reasons for the modifications during a telephone conversation.
15. ReDeCo reviews the design, tests it with the tool for stress analysis and returns it with minor modifications.
16. CoMa reviews the redesign with the customer, who accepts it.
17. The acceptance is sent via the Engineering Server to ReDeCo who uses the data in the Engineering Server for accounting and places the invoice.
18. CoMa passes the design to production, where the robot programs are created (this could be done by a Programming Service Provider).
19. The containers are produced.

8. BUSINESS ADVANTAGES

The central benefit from the developed interaction schemes for Internet services is that robot end user companies are relieved from various tasks in robot application and from the need for highly skilled and specialized personnel. This holds for all service processes of the described service areas.

The given example use case for Product Re-Design services shows that CoMa does not need to employ an experienced product re-designer, CoMa can profit from the experience of ReDeCo in Product Re-Design, and CoMa can benefit from the specialised tools that ReDeCo disposes over, e.g. for computing stress models.

The advantage of ReDeCo is an extended market. Re-Design services can be offered also to companies that had no need for them. The efficient cooperation of both companies is enabled by defined interaction schemes and data exchange mechanism that are well known to the partners. This is supported by the Engineering Server that organises data exchange and project progress control.

9. SUMMARY

The introduction of robot installations at small and medium enterprises (SME) end user companies is faced with the challenge of the complexity of handling robot systems. This requires expertise and a high educational level of the personnel. For facilitating the introduction of robots in SMEs, the EU project 'Small and Medium Enterprises - Robotics Service Inter-Network' (SME-Rosin) was started. It aims at developing Internet-based services that support the planning, operation and maintenance of robots at SMEs by services of SMEs that dispose over the required special expertise.

In the project, services for the areas Production Engineering, Programming and Maintenance are developed. Each service area defines a number of interaction schemes for performing the required distributed cooperation of SMEs. The cooperation schemes flexibly integrate company interaction via Internet with conventional means like telephoning and traveling.

As the central means for project coordination and data exchange, an Engineering Server is developed. It is especially tailored for the needs of SME-networks. It is easy to use and its operation flexibly adapts to current project needs. This reflects the flexibility that forms a typical strength of SMEs.

The cooperation schemes for the services and the required tools are developed and verified repeatedly in a number of development-verification-cycles.

The resulting service concepts were illustrated in this article by interaction schemes for Product Re-design, followed by a close-to-reality use case.

Acknowledgements. The project SME-Rosin (Small and Medium Enterprises - Robotics Service Inter-Network) is funded by the European Commission as FP-5 CRAFT Project IST 2001-55039. The consortium is formed by Industrieausstatter fuer Schweissen und Umwelttechnik GmbH (ISU), Neubrandenburg Germany, Visual Components Oy (Helsinki Finnland), Cooptim Ltd., (Erd (Budapest), Hungary), rs-technologies (Leipzig, Germany), Demar Laser B.V. (Hengelo, The Netherlands), MEBIA GmbH (Ossendorf, Germany) Fraunhofer IPK (Berlin, Germany), ZDIS (Gliwice, Poland).

11. REFERENCES

1. IFR International Federation of Robotics, "European Robotics, A white paper on the status and opportunities of the European Robotics Industry", Prepared by European Robotics Forum (IFR ERF) & European Robotics Research Network (EURON), September 2001
2. SME-Rosin Consortium (a), SME-Rosin Homepage, http://sme-rosin.ipk.fhg.de/, updated continuously
3. SME-Rosin Consortium (b), "Infra Structure: Distributed Work Procedures", project documentation, 4 December 2003, publication planned
4. Berger R., Hohwieler E., "Service Platform for Web-based Services for Production Systems", Proceeding of the 36th CIRP International Seminar on Manufacturing Systems, June 03-05, 2003, Saarland University Saarbrücken, Germany
5 Mikko Kärkkäinen, Timo Ala-Risku, "Facilitating the Integration of SMEs to Supply Networks with Lean IT Solutions", Building the Knowledge Economy: Issues, Applications, Case Studies, IOS Press, 2003
6. Visual Components Oy, "3DCreate (White Paper)", http://download.visualcomponents.net/docs/VC_whitepaper.pdf

37

INFORMATION INFRASTRUCTURES AND SUSTAINABILITY

Rinaldo C. Michelini*, George L. Kovacs°
*PMARLab-DIMEC, University of Genova, ITALY
(michelini@dimec.unige.it)
°Computer and Automation Research Institute and
°Technical University of Budapest, HUNGARY
(gkovacs@sztaki.hu)

Today sustainability is a basic demand. It often contradicts with the needs of an affluent society. The life quality of industrial countries could never lower, if trading of extended *artefacts (*products-services*) will be based on exchanging information-intensive deliveries. The wealth build-up will follow in the knowledge society, fostering eco-consistent behaviours by balancing tangibles decay by intangibles increase. The idea is described by the KILT model, which characterises by the TYPUS metrics. The paper discusses some topics of the prospected scenario, underlining supply chain issues, showing that the ICT options are critical aids to create the required information environment. Some basic trends are sketched, focusing on products-services trading, supplied by extended enterprises, under supervision of independent certifying bodies.*

1. INTRODUCTION

Sustainability (James, 1997) is a new challenge. Until now the industrialized countries, profited by the haphazard consumption of tangibles to widen the *manufacture* market. The *affluent* society tries new offers to out-date the previous ones; wealth generation stresses on quantities, as factories return increases by selling greater amounts of wares, to supply items at prices that greater number of buyers can afford. Market saturation and technology options, recently, turned rivalry to *scope* economy, to supply items at client's satisfaction, exploiting plants' flexibility to manufacture market-driven mixes of items. The change is consistent with simultaneous engineering, which leads to the merging of design and fabrication into *intelligent* organizations (Ettlie, 1994; Michelini, 2001 and 2001a). The welfare growth of industrialized countries, thereafter, aims at the *service* market (Michelini, 2002), by supply chains jointly embedding *commodity* and *utility* provision.

The *extended* artefact (*product-service*) blends *commodity* and *utility*, by information intensive delivery to grant specified *functions*. *Consumables* (raw materials and grown or manufactured commodities), have prevailing birth from non-renewable sources and, as the earth is a closed system, development sustainability shall asymptotically cause wealth downgrading, unless the staple turns to yield value chains into *intangibles* (knowledge, technology, etc. with services, functions, etc. delivery). The result depends on the *information technologies*, believed to be able to strengthen and unfold the paradigm shift to *scope* economy, with changes in habits: *knowledge-* vs. *tangibles*-marketing, leading to eco-consistent progress, without lowering welfare by proper balance. This scenario is worth to be investigated in the following way:

- we shall explore the ***KILT*** model (see later on) to work out sustainable quality features (***TYPUS*** metrics) and coherent support (*net-infrastructures* or *collaborative* networks) needed to fulfil the paradigm shift to eco-consistency (*method* innovation);
- then the analysis considers the technicalities for thrifty achievements through *extended* artefacts and extended enterprises, once the ***TYPUS*** metrics charges consumers for resources decay: backward cycle factories, third-party certifying bodies and *product-service* business are main falls-off.

The *information technologies* are instrumental aids, directly and indirectly affecting sustainability by pervasive provisions. In the switch from *affluent* society (*consumables*) to *thrifty* society (*intangibles*), the technical-scientific patterns only represent necessary prerequisites. Legal-political and socio-economical patterns need to be established, too.

2. FROM AFFLUENT SOCIETY TO THRIFTY SOCIETY

Earlier we used to deal with invested capital (I) and/or involved labour productivity (L); these remained for a long while the only chief factors affecting manufacture delivery (Q). Recent assessments show that the relevant role of a third quantity, knowledge (K: know-how, technology, expertise, etc.), leading to value chains increases up to 40-50 % or more. At the millennium turn, ecology concerned people require to stop free access to non-renewable resources, with profit for manufacturers and purchasers and damage for present and future population. Basic claim is to refund the leftover humans for the tangibles decay (T) diverted along the supply chain. Thereafter, the manufacture delivery (Q) will depend on four independent factors (K, I, L and T) (Michelini, 2002 and 2001a):

$$\boldsymbol{Q} = a_0\, \boldsymbol{K\, I\, L\, T} - a_1\, \boldsymbol{K} - a_2\, \boldsymbol{I} - a_3\, \boldsymbol{L} - a_4\, \boldsymbol{T}$$

where: ***K***, knowledge and technology; ***I***, invested capital; ***L***, labour entry; ***T***, consumed tangibles, as mentioned above.

The ***KILT*** model assumes that the four (scaled) factors have similar effects and that the lack of any of them brings to negative built-in delivery. The direct dependence on an individual factor characterizes clustered companies, which resort to non-proprietary technologies, venture capitals, outsourcing or leased provisions to keep the business work. Four productivity figures, accordingly, appear, and *fair* competition needs *equal opportunity* players.

The explicit dependence on ***T*** and the request to repay for tangibles depletion are coherent with sustainability goals. The challenge aims at drastically lowering downgrading, still preserving welfare: the scenario looks after *knowledge* society surroundings, where *extended* artefacts (*products-services*) are traded by *extended* enterprises (*nested infrastructures*), so that the supply-chain grants a ***K*** factor value, properly balancing the extra costs paid for the *T* factor.

The innovation brought forth with explicit account of the ***K*** factor leads to the *knowledge* society; it does not mean overcoming the *affluent* society. The uneven distribution of wealth among the world countries shall simply modify with higher profits localized within the knowledge-intensive ones. The economic bias will not, repeat the trends arose in the past, by ***I*** or ***L*** factor built-ups, due to ***K***-trade peculiarities, shortly mentioned in section 3. The *thrifty* society foundation establishes on the further explicit account of the ***T*** factor, again, with peculiarities not repeating known patterns.

2.1 The TYPUS Metrics

The explicit inclusion of natural resources spoilage in pricing life-cycle artifacts should be based on worldwide accepted standards. The exact amounts represent a public income, with twofold goal: to remunerate the people not involved by the specific transaction, and to spur thrifty choices either to hinder squandering. The other side, no intention aims at hampering or stopping the progress, rather at modifying the staple in consumables by enhanced focus on renewable (natural) stuffs and re-cycled commodities. The approach requires worldwide withdrawals for tangibles decay, with equivalent tax burden, objectively linked to the life cycle of every traded artefact, including overall provisions and dismissal opportunities. On these ideas, several metrics can be proposed, with figures stated within the acknowledged legal metrology precepts, on condition to have full *visibility* on the artefact life-cycle data and proper *control* on the actual operation falls-off.

In any case, the definition of measurement standards is a preliminary fulfilment. A coherent answer looks after defining a framework, which gives account for the all material-and-energy flows activated along the considered artefact life-cycle and assumes that the net depletion is assessed at the life-cycle end, including side-effects to remove negative impacts and positive contributions due to recycling and recovery. The idea leads, typically, to the ***TYPUS*** (*tangibles yield per unit of service*) metrics. The framework is built on the assumption that most buyers are primarily interested in the *functions* delivered by the instrumental artefacts they purchase, thus a scale based on the *unit of service* is specially relevant to turn users to conservative behaviour, as pricing the *tangibles yield*, more than abstract quality figures, shows that actual needs are favoured.

The *collaborative* network presumed by the ***TYPUS*** metrics is a challenging development where highly structured ICT tools are available. We might look at prospected standards from two viewpoints:

- the short terms preparatory practice, to help introducing *costs* for the actual decay of tangibles;
- the longer terms habits, to foster the agreement on the *scope* of maximizing ***T*** productivity.

2.2 The Collaborative Networked Support

The ***TYPUS*** metrics needs suitable *collaborative* networks to manage the life-cycle data, within transparent and scrupulous facilities. The arrangement basically requires three facts: the marketing of *extended* artefacts, the involvement of *extended*

enterprises and the overseeing of third party certifying bodies. On these conditions, information nested infrastructures are basic aids:
- to provide collaborative forms and behaviours for product life-cycle management;
- to rule conformance assessment and restoration within networked responsible bodies.

Thus, the network has direct links with "conventional" *extended/virtual* enterprise implementations. The *collaborative* network complexity appears highly tangled, as several firms are involved, through competing offers to manage equivalent *product-service* settings. Thus, interlaced *net-infrastructures* shall exist within almost worldwide contexts, and need grant the protected access to the *extended* artefacts' life-cycle databases, from the overseeing certification bodies. This, within the many achievements of the *knowledge* society the ***TYPUS*** metrics (or an equivalent reference) cannot operate without the full visibility on the artefact life-cycle, with all related beforehand, side and afterwards effects.

2.3 The Role of Method Innovation

The *knowledge* society is viewed to the winning path to the *thrifty* society, providing technical aids, directly by the mixed utility-commodity ICT provisions, indirectly by supporting the *extended* artefacts market. The technology-driven issues are not *sufficient* to generate sustainability. Today, the purchasing decisions that favour a *product-service* with lower impacts in resource provision, in life-cycle use, etc., with properties that facilitate reuse or recycling, etc. are qualitative spurs; resource productivity (***TYPUS***) is an hypothesis: no established standard is available; no testing and overseeing body exists. The proposition looks after three aids (Binder, 2001; Graedel, 1997; Giarini, 1998):

- technical-scientific support of innovation by targeted R&D projects
- socio-economical promotion of the appropriateness of eco-consistent behaviours
- political-legal actions by means of the governmental regulation of eco-compatibility

The relevance of the legal and social (beside technical) conditions for *method* innovation stems from the current refusal of alternatives: engineers are *manufacture* economy minded; consumers belong to, possibly, even less receptive areas. The thesis that people is more interested in *using* goods and profiting of *functions*, than in possessing items, is dubious; more questionable that ownerless consumption leads to eco-benefits: leased items, e.g., may age faster than owned one, due to the lessee's irresponsible use in wear-out protection and up-keeping carelessness.

3. EXTENDED ARTEFACTS AND ENTERPRISES

Only some technical aspects are investigated, even if we know that socio-economical and political-legal contributions are fundamental, too. Central role is played by the *extended* artefact (*product-service)*, i.e. instrumental (tangible and intangible) delivery to a client, granting the enjoyment of specified *functions*, according to life-cycle indenture. The *extension* obliges the supplier to the user for conformance assessment at the point of service, both being bound by enacted (safety, environment, etc. protection) rules.

The main actor in the *extended* artefact market is the *extended* (recently often called as virtual) enterprise or *net-infrastructure*, i.e.: factual alliance of partners merging skills, know-how and resources and enabling co-design, co-manufacture, co-market, co-maintain, co-servicing, co-recycle, etc. efforts, to offer *extended* artefacts at purchaser's benefit *and* environment safety (Graedel, 1997). The *extension* provides visibility on *products-services* operation life to support:
- resources consumption and surroundings impact recording;
- third-party conformance assessment and eco-figures certification.

The *extended* artefact and enterprise definitions assume that *method* innovation is the main concern and ***TYPUS*** metrics in included within the economy of *scope* patterns, according to so called, longer terms habits.

With focus on *extended* artefacts (Thoben, 2001) the critical opportunity is to enable the practice of technological *sustainability* (James, 1997; Mirchandani, 1996), by moving welfare generation from a typically *manufacture* market, to a mainly *function* market. The *extended* enterprise case is slightly different, as ICT is critical there. Although the expected advantages of interconnected infrastructures are properly recognized, existing tools suffer drawbacks, as lack of common reference models to be shared as *type*-facility; lack of effective interoperability mechanisms and approaches; lack of eligible protocols and frames, free from non owned details; heavy design and engineering efforts to make proprietary technologies co-operate; rapid software and hardware obsolescence, frustrating provisional goals; actual obstacles in the effective transfer of locally tested instruments; and the lack of viable leadership proposing low-cost linking environments.

3.1 Supply Chain Management Issues

Today the world-wide globalisation and the appearance of extended/virtual enterprises require more than only Supply Chain Management (SCM) for some tasks of a given enterprise. Due to the physically and logically distributed character of the co-operating units (workshops, plants, enterprises, etc.), taking advantage of the existence of global networking, web-based solutions are suggested. There were two EU projects (FLUENT, 1998 and WHALES, 1999) that provided such solutions. (FLUENT, 1998) gives "beyond SCM" workflow/supply chain solutions for distributed (mainly SME) organizations dealing with manufacturing, services, maintenance, etc. The main target firms of (WHALES, 1999) are the distributed, multi-site, multi-firm, powerful organizations (and SMEs), and the goal is to manage complex, one-of a kind products and projects, manufacturing and management as well.

The results provide new IT solutions for managing complex logistic flows, occurring in distributed manufacturing networks with multiple plants and co-operating firms. Networks of this kind are gaining relevance and diffusion, under the impulse of the following main factors:
- emerging virtual/extended enterprise paradigms
- pull-oriented production models, like just-in-time, requiring synchronisation of internal and external flows
- lean/agile manufacturing models, based on horizontal, goal-oriented process chains
- evolving market conditions, calling for business globalization and decentralization of manufacturing facilities.

Traditional SCM implementations refer to a linear, standardized and relatively stable view of the supply chain: "supply chain management is about managing the flow of products and services and the associated information, across the whole business system to maximise value to the end consumer." (Price Waterhouse, 1997).

Recent analyses have pointed out the potential failure behind the traditional, linear logic, especially where revenue increase is pursued instead of cost reduction:

- Cost reduction leads to: standardisation and simplification of supply chain and its operation; minimization of integration costs; definition of "functional silos" independent of each other.
- Increasing revenues means to take advantage of diversification and differentiation, exploiting changes in demand and supply. This means making more money thanks to the supply chain ability to reconfigure itself, to harmonize capacities and to respond quickly as a whole.

To look at the supply chain complexity as a competitive advantage, rather than as a source of costs, means a radical change of perspective in the organization models supported by SCM tools: "For a start, the supply 'chain' is really not a chain at all - it is a complicated web of relationships between demand and supply. The concurrent and multidimensional nature of these relationships creates a complex fabric, woven step by step."(Mirchandani, 1996).

The logical architecture of a new network of an extended enterprise means some nodes equipped with the new system, other nodes are acting as customers, suppliers or subcontractors. Nodes of the latter type can only take part as executors in logistic flows controlled by the flow management nodes. The reason is that these nodes lack the network-level vision and decision support tools to actively participate in the planning and co-ordination of supply flows.

Each node is perceived by the other nodes as an autonomous source of: (i) information on the node and the goods it supplies and consumes (*knowledge* level); (ii) demand/availability signals and allocation decisions (*planning* level); (iii) supply control signals and exceptions (*control* level). Independently of ownership and position in holding hierarchies, nodes in the network are modelled as source and destination of logistics flows. To this purpose, each node is attributed a three-tiered structure including: a Flow Collector, that manages incoming logistics flows, a Flow Dispatcher, that manages outgoing flows, and a Flow Processor, responsible for integration with internal production flows.

Co-operation between nodes is realised through links, each representing a stable relationship for the exchange of a given product between a "supplier" node Flow Collector and a "receiver" node Flow Dispatcher. The Flow Processor is not directly involved in the link, since the flow control is based on a clear separation of logistics decision-making domains. Internal logistics are managed by each node on its own, and are perceived at the network level only through requirements, events and constraints on external logistics flows. A link definition fixes the characteristics of supply flows taking place through the link, in terms of:

- data on the supplied product, including shipping, transportation and delivery parameters
- planning policy applied to the link, in terms of planning parameters, planning method, e.g., "push" or "pull", and planning responsibility, e.g., either the supplier or the receiver, or a third node controlling the flow

- workflow model, i.e., the sequence of messages and events characterizing the nodes interaction during planning and control of supplies over the link.

This way, a high degree of generality and flexibility is reached in modelling the variegated network configurations found in the real world. For example, a node can establish "pull" links with a network of suppliers, keeping a centralized control of suppliers selection and orders allocation. The node product can be delivered to a trading partner on the basis of an inventory replenishment agreement, modelled by a "push" flow controlled by the supplier, and to a customer on the basis of a normal "pull" link. Both types of outgoing flows can originate dependent requirements for the above suppliers network.

3.2 The Backward Cycle Business

The backward cycle deals with parts and materials processing after (partial or total) dismissal of the handled commodity. Nevertheless, as we are concerned by extended artefacts, the information contents are not neutral and two restricting patterns establish: • feedback of forward cycle features, to recognise the appropriate design-for-specifications; • forecast of backward cycle features, to include suitable design-for-recycling specifications.

The backward cycle is simply an option, to be weigh against others, when eco-design becomes a main purpose, so that focus scans on: • planning for quality protection, disassembly, material reuse, etc.; • designing for long-life, rare maintenance, low energy consumption, etc.; • preferring self-tuned rigs, re-used packaging, improved logistics, etc.; • setting optimal effectiveness, pro-active up-keeping, etc. artefacts; • choosing high throughput, material saving, energy recovery, etc. cycles; • making use of recycled, less energy-intensive, renewable, etc. materials. After dismissal, re-conditioning or re-manufacturing are relevant options:

- re-conditioning has the goal to back establish overall conformance to specification, by combined industrial processes addressed to artefacts at their life end; re-conditioning is limited, if re-setting is partial;
- re-manufacturing recovers parts and material with properties matching the original ones, by combined industrial processes applied to dismissed artefacts, and candidates them to new duty-cycles; the issue is limited, if the processed parts do not recover the original characteristics.

When new artefacts are conceived, the backward cycle affects original choices to include re-conditioned and re-manufactured items and to forecast careful set-ups for recycling. However, the integrated design steps are not sufficient, by themselves, to grant economical return; the thrifty society surroundings, actually, establish when the artefacts true price includes the overall cost for materials and energy depletion suffered by the eco-system; thereafter, the world-wide use of the **TYPUS** metrics, or equivalent taxing procedure, will be enabling reference for the backward cycle, at the different ranges of the forward one.

3.3 The Conformance Assessment

Taking the life-cycle into account, alternatives are possible, once extended artefacts are supplied by extended enterprises and third party certification bodies oversee the life-cycle incumbents. The three parties ruling seem to be a good compromise to enhance competition and to balance responsibilities, under real fair-trade conditions.

The picture is coherent with a controlled collaborative network, directly linking an extended enterprise to individual clients, so that the supply chain of each delivered extended artefact is transparently available.

The relevance of the conformance assessment service shows that this new business might grow to large percent of the gross national product of each country, becoming a wealth source of the knowledge society. The involvement of third party certifying bodies needs, of course, proper regulations, enacted by the national authorities, but suitably harmonized to assure worldwide equivalence.

Certifying bodies compete in a free market, being replaced possibly any time, exactly as the partners of an extended enterprise, or the agreements about the extended artefact responsibility might be up-dated. The changes do not interrupt supply chain monitoring, simply request that the new entries are accepted by proper data transfer, and the new duties are assigned.

3.4 The Falls-Off on the Manufacture Market

The build-up of backward cycle enterprises and of eco-certifying bodies can progressively establish on existing patterns, drastically expanding the business domains and enhancing the collaborative network aids, to fully achieve the method innovation of the thrifty society. The technical opportunities are mainly provided by ICT instruments, and correspond to focus, for fixed deliveries on specialising the web links of extended enterprises to individual resources utilization requirements (Price Waterhouse, 1997; Mirchandani, 1996) in a way that, even in front of defective cross-link occurrences, decisive helps establish along finalized patterns, to help the surfacing of filtered knowledge (whether the series of consents verify for the selected tracks), leading to a set of tailored provisions, such as: - interoperability by integration and sharing of federated information; - management of distributed activities, based on self-acting clusters; - supply-chain transparency given by eco-consistency assessment records; - goal-oriented co-operative knowledge problem-solving capability; - sectional bounded and case-driven trust building processes. These and similar contrivances contribute to the coherence of the collaborative network.

The evolution brings to supply chains jointly embedding commodity and utility provision, so that the value of intangibles becomes prevalent as compared to the one of consumables (whether non renewable resources are concerned). The scenario is not new, as the boasted merit of industrial society was the delivery of low price artefacts, as compared with people wealth, based on wide resort to natural raw materials. After a while, this becomes a deceitful virtue: alternative provisions need to be explored for the value chain of artefacts to avoid squandering earth treasures. The knowledge society might be a winning answer, grounded on selling information and assuring high revenue based on intangibles. The out-coming market presents some peculiarities: trading information will never dispossess the dealer of his original know-how; sharing information is based on individual commitment and does not automatically follow from paying for it; developing information is typically non-linear process, grounded on synergic accumulation; augmenting information could be costless, whether built on collaborative settings with additive specialisation; and so forth. The extended artefact supply chain represents a challenging bet, and the falls-off on the manufacture market could bring to the thrifty society, with the many facets we have tried to sketch in this study.

4. CONCLUSIONS

The paper moves from the recently acknowledged KILT model (Michelini, 2001 and 2002b), and - among others - arrives to the TYPUS metrics (Michelini, 2002b). It should be said that, by now, eco-consistency already looks after specially-enacted resource-duty collection systems (Kyoto protocol, carbon tax, etc.), thus the approach is coherently generalised by means of the TYPUS metrics, to establish world-wide taxation settings for fair trade preservation. For sustainability, the affluent society, supported by ceaselessly replacing artefacts, needs evolve to the thrifty society, based on carefully sparing natural resources. This would quite obviously leads to notably lower welfare, unless alternative contrivances are sought to build up wealth. Now, ICT aids are paramount opportunities, which add to established instruments that mankind disposes to protract his progress. The ICT tools characterise the knowledge society, stressing on totally new goods either deeply modified artefacts, by up-graded information contents, so that relevant paradigm shifts apply to the common manufacture practice and to the current consumers' habits. These paradigm shifts are addressed in the paper on, mainly, technical viewpoints, even if legal-economic factors critically affect their feasibility and actual falls-off are properly related to complex issues in the extended artefacts market, generally referred to as *method* innovation. The prospected issues, in fact, depend on exploiting knowledge-driven options by networked set-ups, to manage the supply-chain and embedded information flow, through extended enterprises. The study has the goal to turn the European scientists to the emerging fields of eco-consistency, to take a lead in pioneering objective quantitative assessments of the environment suffered impact, while tangibles are traded to satisfy consumers' requests.

Sustainability growth, indeed, shall first address broadband eco-compatibility goals, assuming that transition to the thrifty society requires changes in habits even before than technical innovation. According to the suggested research lines, the TYPUS metrics should be strictly grounded on scientific principles and technical standards; the visibility of tangibles consumption, then, will be assured by jointly enabling extended artefacts and extended enterprises. This will lead to a different concern in front of natural resources spoiling, with drastic changes of the industrial organizations, supporting the new business of the backward cycle (from dismissed scraps, to recovered materials). The existing welfare, on these ideas, rather than decreases, could widen, recovering by K-growth, the taxes paid for T-decay. The scenario, however, requires:

- technical-and-scientific innovations, to grant the overseeing and the control of artefacts up to dismissal, with life-cycle recording;
- socio-economical assessments, to prove the return on investment of eco-conservative behaviours and the benefits of method innovation;
- political-and-legal changes, to extend the providers responsibility to the point-of-service, for community protection and tax collecting.

The build-up of bylaws, rather than neutral, will be a spur toward sustainability, giving transparency of the performance between competing solutions. The eco-qualified factory, with extended-artefacts, increases its market share, based on information extended infrastructures, advertising the eco-consistency by connected frames, binding, at the points-of-service, suppliers and users with accredited certifying bodies, under world-wide regulation acts.

5. REFERENCES

1. Binder, M. Janicke, M. Petschow, U. Eds.: "Green industrial restructuring: international case studies and theoretical interpretations", Springer Verlag, 2001, ISBN 3-540-67467-5.
2. FLUENT, Esprit IiM-1998-29088: "Flow-oriented Logistics Upgrade for Enterprise NeTworks.", EU project documentations
3. Friend, G.: "The end of ownership? Leasing, licensing, and environmental quality", The New Bottom Line. Vol. 3, 1994, No. 11.
4. Frederix, F.: "Enterprise co-operation leads to extended products and more efficiency", Production and Operations Management Conf., Orlando, USA, March 30-April 2, 2001.
5. Graedel, T.E. Allenby, B.R.: "Design for environment", Prentice Hall, 1997.
6. Giarini, O. Liedtke, P.: "Working in the new (service) economy: the employment dilemma and the future of work", The final draft of the report to The Club of Rome, March, 1998, pp. 140.
7. James, P.: "The sustainability cycle: a new tool for product development and design", J. of Sustainable Product Design, July, 1997, pp.52-57.
8. Ettlie, J.E., Dreher, C.L., Kovács, G.L., Trygg, L.: Cross-National Comparisons of Product Development in Manufacturing, in the book: Advances in Global High-Technology Management, Vol.4, 1994, Part B. (Comparative Perspective of Technology Mngment), JAI Press Inc., pp. 91-109.
9. Michelini R.C., F.Pampagnin, R.P.Razzoli, 2001: "Trends in engineering design and eco-compatibility constraints", 12th Intl. ADM Conf. Design Tools and Methods in Industrial Engineering, Rimini, Sept. 5-7, 2001.
10. Michelini R.C., Kovacs G.L., 2002: "Integrated design for sustainability: intelligence for eco-consistent products-and-services", EBS Review: Innovation, Knowledge, Marketing and Ethics, n° 15, 2002-3 Winter issue, ISSN-1406-0264, Tallin, pp. 81-95.
11. Michelini R.C., Kovacs G.L., 2001a: "Integrated design for sustainability: intelligence for eco-consistent extended-artefacts", Invited Lecture, Intl. IFIP Conf. PROLAMAT: Digital Enterprise, New Challenges: Life-Cycle Approach to Management and Production, Budapest, 7-9 Nov. 2001.
12. Michelini R.C., 2002a: "Information infrastructures and eco-labelling", COVE Newsletter 4, IFIPWG 5.5, pp. 2-8, Oct. 2002, ISSN 1645-0582.
13. Michelini R.C., 2002b: "TYPUS: tangibles yield per unit service metrics for eco-consistent design", EU VI FP-EoI, CORDIS, May 2002.
14. Price Waterhouse©, "Supply Chain Management Practice", In: "Supply Chain Planning for Global Supply Chain Management", November 1997.
15. Mirchandani V., Block J., "Supply Chain Management and the Back Office", Gartner Group© Strategic Analysis Report, September 1996.
16. Suhas, H.K.: "From quality to virtual enterprise: an integrated approach", CRC Press, Boca Raton, Aug. 2001, ISBN/ISSN 0849310156.
17. Thoben, K.D. Jagdev H., Eschenbacher, J.: "Extended products: evolving the traditional product concepts", 7th Intl. Conf. Concurrent Enterprising: Engineering the Knowledge Economy through Co-operation, Bremen, 27-27 June, 2001.
18. WHALES, ESPRIT IST-1999-12538: "Web-linking Heterogeneous Applications for Large-scale Engineering and Services", EU project documentations

PART C

INTEGRATED DESIGN AND ASSEMBLY

38 KNOWLEDGE-BASED REQUIREMENTS ENGINEERING FOR RECONFIGURABLE PRECISION ASSEMBLY SYSTEMS

Hitendra Hirani
Precision Manufacture Group
University of Nottingham
epxhjh@nottingham.ac.uk
Svetan Ratchev
Precision Manufacture Group
University of Nottingham
Svetan.ratchev@nottingham.ac.uk
UK

Reconfigurable Precision Assembly Systems are being developed in response to assembly systems becoming obsolete due to condensed product life cycles being so closely linked with assembly system life cycles. However methods and tools to promote the use of the hardware technology, which is centred on the deployment of reconfigurable modules, are non-existent. The paper presents a knowledge-based requirements engineering approach that gathers user requirements and converts them into system requirements using knowledge rules stored within a database structure. This is illustrated through a case study.

1 INTRODUCTION

Reconfigurable Precision Assembly Systems have been highlighted as one of the visionary manufacturing challenges for 2020 (Bollinger, 1998). Although there are initiatives being undertaken that aim to develop hardware for the concept (Mehrabi, 2002; Koren, 1999; Monfared, 1997; Heilala, 2001; Chen, 2001) there are no schemes that aim to develop methods and tools to disseminate the research to industry.

Requirements specification is the key activity in filling this gap as it forms the first stage of the assembly system design process (Bray, 2002). The user defines a set of user requirements and these are converted into system requirements by the system integrator.

The role of requirements engineering is to provide an abstract solution for a design problem. Moreover *"a good set of requirements defines precisely what is wanted, but simultaneously leaves the maximum space for creative design.* (Stevens, 1995) These requirements have to reflect the customer's expectations of the system.

The interaction between the machine and its environment is the key aspect to consider here and all system properties must be defined using these terms. One method of having concrete requirements to work from is to develop formal models, tools and techniques (Jackson, 1995).

Although there are many commercial tools (International Council on Systems Engineering, 2004; easyweb, 2004) that perform requirements engineering functions, these are mainly tailored to suit software engineering aspects. They chart user requirements declared in natural language and system requirements are defined with reference to these. The tools facilitate functions such as traceability analysis, charting history of requirements and consistency and quality checking, but no knowledge intensive activities are performed.

Knowledge engineering is a key aspect for organisations in the 21st century as the increase in movement of employees between jobs means that knowledge carried by those employees also moves (McCampball, 1999; Bender, 1998). It is in the organisations interest to harness this knowledge within formalised structures where it can be applied to perform some tasks carried out by the workers (Robertson, 2000). For requirements specification this means the semi-automated derivation of system requirements from a set of user requirements.

This paper reports on a research initiative that aims to use assembly system design knowledge to gather user requirements, analyse these requirements and then define system requirements based on these user requirements. This has been implemented through a web-based environment. An overview of the research framework is presented with an outline of the requirements engineering process and the knowledge involved. A case study is included to demonstrate the results.

2 KNOWLEDGE ONTOLOGY FOR REQUIREMENTS ENGINEERING OF RECONFIGURABLE PRECISION ASSEMBLY SYSTEMS

The two areas we are concerned with in this research are those of user requirements specification and system requirements specification, where the user requirements specification consists of a set of business requirements, product definition, part definition, part liaison and other constraints. The system requirements specification comprises the description of assembly tasks and requirements for the specification of assembly modules. The two sets of requirements are owned by two different stakeholders where the user requirements specification is owned by the system user and the system requirements specification is owned by the system integrator. Each is underpinned by the respective stakeholder's knowledge. This knowledge is consolidated by a common knowledge model based on assembly system capabilities. The assembly system capability model will not be explained further in this article as it has been explored in depth in Hirani (2002).

An overview of the ontology is illustrated in Figure 1. The requirements engineering process and a representation of the knowledge are presented separately in this paper to maintain clarity of the boundaries between the knowledge base and the class structure.

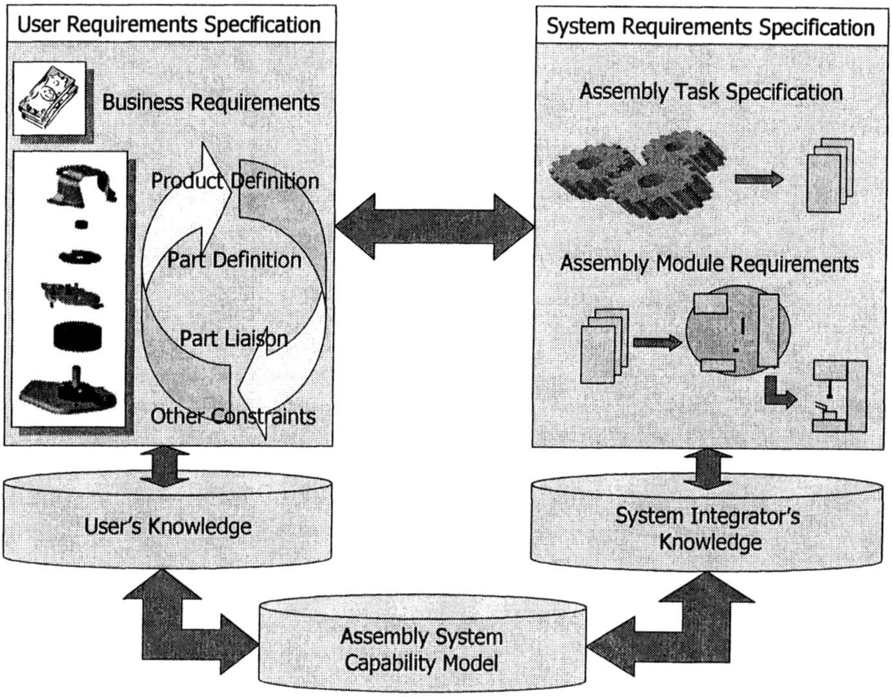

Figure 1: Knowledge Ontology for Requirements Engineering of Reconfigurable Precision Assembly Systems

3 REQUIREMENTS ENGINEERING PROCESS

Requirements engineering for Reconfigurable Precision Assembly Systems begins with the elicitation of user requirements. This entails the specification of business requirements, such as budget, production volumes, delivery timescales, maintenance and training agreements, etc together with a definition of the product(s) to be assembled with their part properties.

A user requirements document is created and sent to the systems integrator, who uses the information to derive a set of task specifications for the finished system to perform. Each task represents the addition of a part to the assembly with handling, feeding and operational properties. These tasks must adhere to the business requirements of the project so they have to be within the universal constraints. All the information derived here is collated to form the system requirements document, which then has to be approved by the system user before any further work is done on the system design. Each item in the system requirements document must be traceable to the user requirement(s) from which it was originally derived. These are properties as illustrated in Figure 2.

For example the Control Architecture is derived from Legacy Systems, Production Volume, Future Modifications, Total Output and System Lifespan user requirements, whereas Packaging is solely dependent on the Product Delivery required. The knowledge model that underpins the decision making is explored next.

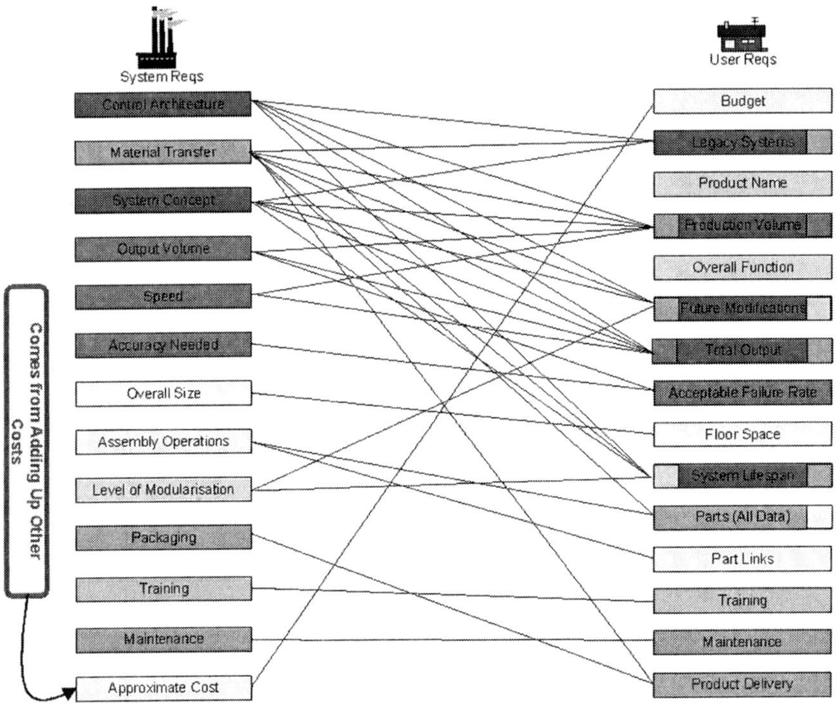

Figure 2: Mapping of System Requirements to User Requirements

4 KNOWLEDGE REPRESENTATION

The knowledge that is needed for requirements engineering for Reconfigurable Precision Assembly System is explored from two viewpoints. The system user's knowledge comprises the knowledge the user has of the business requirements, the products and parts being assembled and the type of liaison that exist between the various parts whilst the system integrator's knowledge includes cost knowledge and technical knowledge on control architectures, material transfer methods and assembly and test operations (see Figure 3).

Knowledge is contained within a database structure at tow levels of abstraction so that it can be easily stored, retrieved and edited as new knowledge is created. Domain knowledge is static knowledge about system properties whereas task knowledge defines the activities that need to be performed by each stakeholder. The link between the domain and task knowledge is facilitated by inference knowledge. This layer describes how domain knowledge should be manipulated for each task and is made up of a series of if then commands for each task. An example of an inference rule is presented in Figure 4.

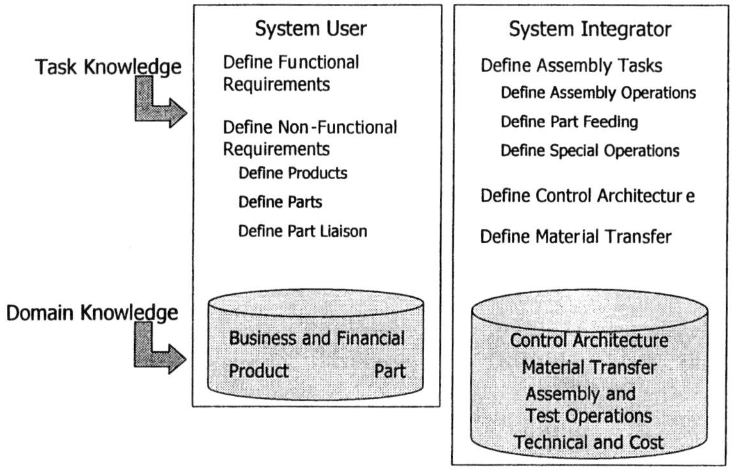

Figure 3: Summary of Requirements Engineering Knowledge

```
;;************************************************
;********** CHOOSE OPERATION *********************
(defrule repeatability
        (cycletime ?c&:(>= ?c 4))
        (partlink ?pl)
        =>
        (if (= ?pl screwed) then (assert (operationtype screwing))
        else
          (if (= ?pl joind) then (assert (operationtype joining)))
        )
)
(defrule move
        (movement ?m)
        =>
        (if (= ?m simple) then
                (assert (machines pn_cylinders))
                (assert (machines linear_drives))
          else
                (assert (machines artic_robot))
        )
)
```

Figure 4: Inference Rule for Choosing Operation

The various forms of knowledge are used to define the type of assembly system required and this information is later used to select physical assembly modules that comply with the requirements for assembly of the product. This has been implemented through a web-based decision making environment (Figure 5).

The environment interfaces with the web through a server, which is behind a firewall for security reasons. The server exchanges messages and code with a JSP Servlet which calls different tasks and inferences represented as Java objects and Java beans. This is backed up by a relational database management system that stores the domain knowledge in the MySQL format. The requirements specification stage of the process has been implemented as a prototype environment. It includes the specification of the business constraints for the project, gathering of product data, data on the parts that make up the product and their connectivity. An example is used to demonstrate the implementation.

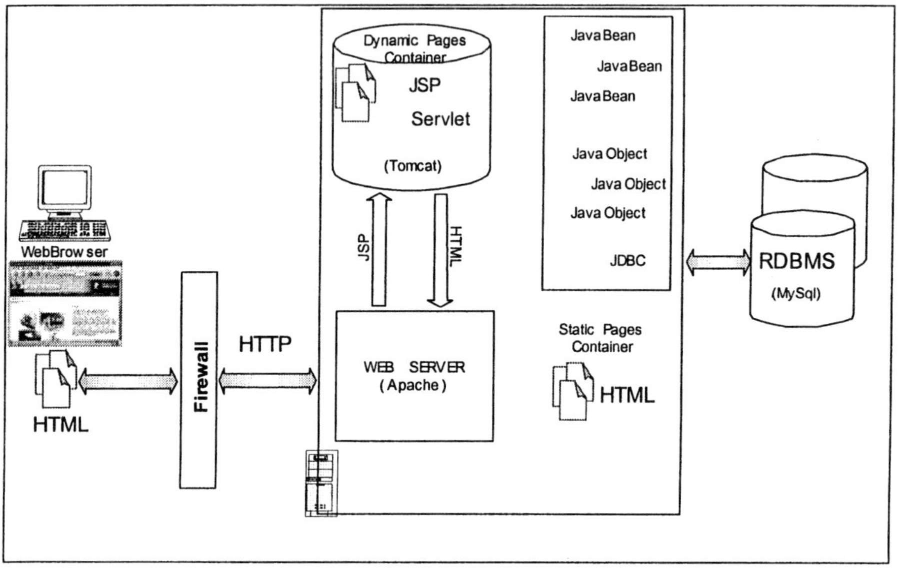

Figure 5: System Architecture for Web-Based Decision-Making Environment

5 CASE STUDY

The requirements specification of a seat recliner mechanism for a car has been performed to demonstrate the use of the system. User requirements for the product assembly have been captured and parsed to form system requirements.

Each part has been described in detail as per the criteria shown in Figure 6 and part liaison characteristics have been declared. These have then been parsed through the system to arrive at a set of task specifications based on inference rules defined within the environment. The result is a set of task descriptions as illustrated in Figure 7.

Each assembly task contains a similar description that can later be used by system integrators to design reconfigurable precision assembly modules that satisfy both the task requirements and the non-functional requirements. These modules would then be integrated to form a Reconfigurable Precision Assembly System.

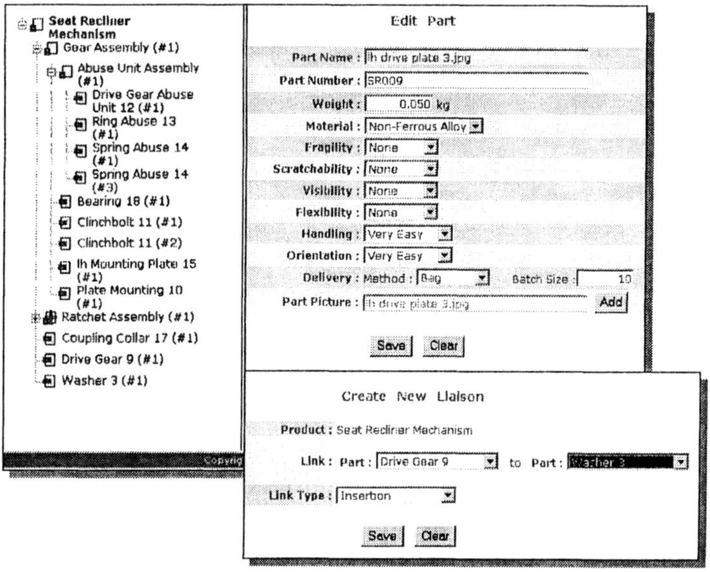

Figure 6: Product and Part Descriptions for Seat Recliner Mechanism

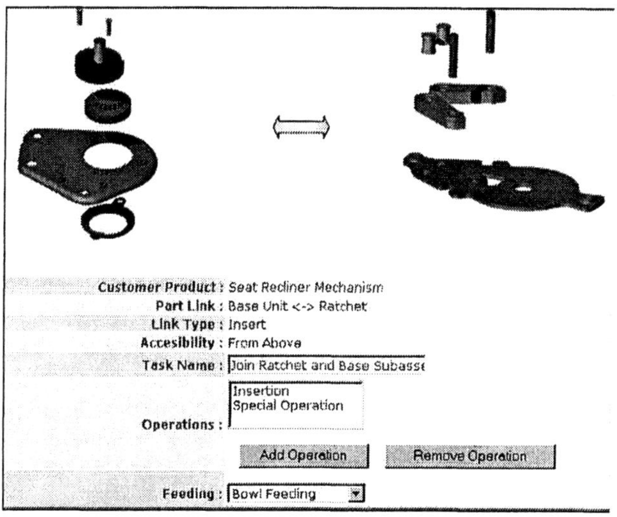

Figure 7: Task Specification for Seat Recliner Mechanism

6 CONCLUSIONS

The paper has presented a framework for the knowledge-based requirements specification of a Reconfigurable Precision Assembly System. This approach centres on users supplying knowledge about the product to be assembled and its constituent parts and how they are related together with some business requirements. These are then parsed through the system using knowledge supplied by system integrators to develop task specifications that satisfy the user requirements.

The method has been implemented through a web-based environment and demonstrated using an example of a seat recliner mechanism in a car. Further work needs to be carried out to expand the knowledge base and to add more functionality to the software and make it holistic.

7 ACKNOWLEDGEMENTS

The reported work is partially funded by the Department of Trade and Industry in the United Kingdom as part of the EUREKA Factory E!2851 E-RACE project the support of which is gratefully acknowledged. The authors would also like to thank TQC Ltd for sharing information on automated assembly and testing equipment and case studies.

8 REFERENCES

1. Bender, S.; Fish, A. (1998): The Transfer of Knowledge and the Retention of Expertise: The Continuing Need for Global Assignments, Journal of Knowledge Management 4:2, pp.125-137
2. Bollinger, J. (1998): Visionary Manufacturing Challenges for 2020, National Research Council Publication, National Academy Press, Washington, D.C. (USA)
3. Bray, I.K. (2002): An Introduction to Requirements Engineering, Pearson Education Ltd, Harlow, Essex (UK) ISBN 0201 767929
4. I-Ming Chen (2001): Rapid Response Manufacturing Through a Rapidly Reconfigurable Robotic Cell, Robotics and Computer Integrated Manufacturing, 17, pp199-213
5. http://easyweb.easynet.co.uk/~iany/other/vendors.htm, 6th April 2004
6. Heilala, J.; Voho, P. (2001): Modular Reconfigurable Flexible Final Assembly Systems, Assembly Automation, 21, 1, pp20-28
7. Hirani, H.; Ratchev, S. (2002): Definitions and Measures for Requirements Engineering of Reconfigurable Precision Assembly Systems, Proceedings of the 31st International Symposium on Robotics, 7th-11th October, Stockholm, Sweden
8. http://www.incose.org/tools/tooltax.html, International Council on Systems Engineering, 6th April 2004
9. Jackson, R.B; Embley, D.W.; Woodfield, S.N. (1995): Developing Formal Object Oriented Requirements Specifications: A Model, Tool and Technique, Information Systems, 20:4, pp273-289
10. Koren, Y., Heisel, U., Jovane, F., Moriwaki, T., Pritschow, G., Ulsoy, G., Van Brussel, H., (1999) Reconfigurable Manufacturing Systems, Annals of the CIRP, 48:2, pp.527-539
11. McCampbell, A.S.; Clare, L.M.; Gitters S.H. (1999): Knowledge Management, The New Challenge for the 21st Century, Journal of Knowledge Management 3:3, pp.172-179
12. Mehrabi M.G; Ulsoy, A.G.; Koren, Y.; Heytler, P. (2002): Trends and Perspectives in Flexible and Reconfigurable Manufacturing Systems, Journal of Intelligent Manufacturing, 13, pp135-146, Kluwer Academic Publishers
13. Monfared, R.P.; Weston, R.H. (1997): The re-engineering and reconfiguration of manufacturing cell control systems and reuse of their components, proceeds of the institution of mechanical engineers conference, 211:B, pp495-508
14. Robertson, M,; Hammersley, G. (2000): Knowledge Management Practices Within a Knowledge Intensive Firm: The Significance of the People Management Dimension, Journal of European Industrial Training, 24/2/3/4/ pp.241-253
15. Stevens, R.; Martin, J. (1995): What is Requirements Management? Proceedings of the 5th Annual International Symposium of the NCOSE, vol 2, pp13-18

39 DEFINITIONS, LIMITATIONS AND APPROACHES OF EVOLVABLE ASSEMBLY SYSTEM PLATFORMS

Henric ALSTERMAN Mauro ONORI
hal@iip.kth.se onori@iip.kth.se
The Royal Institute of Technology,
Dept. of Production Engineering,
Brinellv 68, 100 44
Stockholn, SWEDEN

Europe, as most other OECD areas, is confronted with major potential opportunities in the decades to come. Although often portrayed as threats, the symptoms being denoted in the European economy are, in fact, part of a shift in knowledge and technology infrastructures created by these trends. These current challenges being faced by manufacturing companies nowadays require production systems to become ever more responsive and agile. This is particularly relevant to micro-products, since manual assembly becomes impossible, rendering outsourcing strategies less effective if not deliberately negative. Furthermore, traditional approaches to R&D in this field no longer suffice to cope with the challenges imposed since these imply new business methods, continuous technological evolution,and the increased tendency towards networks of enterprises.

To meet such demands there is a need for new rapidly deployable and affordable (economically sustainable) microassembly systems based on re-configurable, modular concepts that would allow continuous system evolution and seamless reconfiguration. Furthermore, as will be detailed later, one of the required foundations to sustainable assembly system concepts lies within a new way of thinking and working: a methodology that could integrate the various aspects related to the life cycle of the production systems, with particular focus being placed on the re-engineering phase. This article will present some definitions, clarify the basic approach, and outline the serious requirements being posed by such a paradigm: Evolvable Assembly Systems.

1. INTRODUCTION

Modular assembly systems, standardised solutions, and re-configurable approaches have appeared all the more frequently in recent publications. Such terminology indicates that the R&D community has responded to the industrial demands for a more agile *re-engineering phase* (shaded area, fig.1.0). However, since the underlying problems are holistic, and therefore include many different segments of the production equation, the ensuing solutions have only addressed parts of the problem: management, human, design and supply chain issues are not yet well integrated. As given in the preceding figure, the approach given in this article will focus on the re-engineering phase, which is central to the issue of re-configurability or evolvability. Re-engineering is hereby defined as the modification or adaptation of currently available system solutions to fit the new assembly needs. In this respect re-engineering is of capital importance because, in reality, the major part of producing companies have to deal with planned products and existing production facilities. Ideally, they would like to fit any new product, variant, or volume fluctuation into an existing assembly system with as low costs as possible: the re-engineering phase. To date, this has only been a dream. Therefore, if the equipment cannot easily adapt to changing product & market requirements , the overall flexibility is greatly reduced.

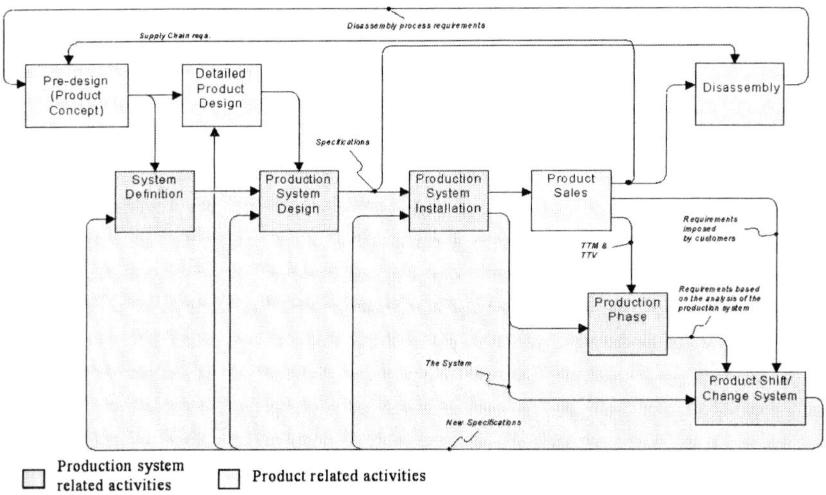

Figure 1 - Simplified View of Product-System Lifecycle

Basically, a radical new way of thinking and working is required: in terms of assembly systems, what is required is not a solution which tries to accomplish all of the envisaged assembly needs within a closed unit (Flexible Assembly systems) but, rather, a solution which, being based on several re-configurable, task-specific elements (system modules), allows for a continuous evolution of the assembly system. In other words, many simple, strictly task-oriented components with standard interfaces are better than few, very flexible but extremely expensive solutions that cannot be integrated within existing systems. The Evolvable Assembly Systems (EAS, [1]) paradigm offers such an approach.

The objective of this article is to clarify the complexity of such an approach, and that it is far more complicated than simply calling a system solution as "modular". Such an endeavour requires a structured approach, and the methods and tools must be collected around a methodology.

2. DEFINITIONS

Modularity has been widely acclaimed in recent publications [2], in fact, it has become a goal in itself, much in the way flexibility was a few years ago. In simple terms, any assembly system which has one module for a given assembly process, cannot be termed as modular: e.g.- a manipulator module, a transport module, and a feeder module, all part of a "modular" system in which no other alternative modules may be interconnected for new or different operations. The misuse of terminology, however, is only counter-productive, since the end-users will only be deceived if the goods are not delivered. In common terms, the modularity denoted to date only refers to a local, mechanical interoperability of a very restricted set of units. The so-called "modules" are often functionally (high-level) specific units with a dedicated, in-house interface. The problems reside in:

- the very weak link between the functionality being enclosed within the "module" and the assembly processes to be accounted for, in detail.
- the non-existent set of classified, formalised assembly processes accounted for by each "module".
- the local, limited, and not well-defined standard being used for the interface.
- the instability of the processes or sub-process being handled by the "module".

In order to counter such misconceptions, a more widely accepted definition of module and modular must be attained. The same may be said for the terms standardised or standardisation, which lie at the core of the misunderstanding of modularity. Even though fairly succesful standards have been derived at in-house level, they may not be regarded as true standards or modular systems since they have not succeeded in becoming as widely used as intended.

These problems with terminology clearly underline the need for a more concerted effort in forming the correct taxonomies and ontologies for this branch of technology. A platform for discussion may be found in the common definitions of these terms[1]:

Module : any in a series of standardized units for use together: an assembly system unit which covers a classified set of formalised assembly operations.

Modular : constructed with standardized units (modules) for flexibility, interconnectability, and variety in use within a specified class of operations.

Standard : An acknowledged degree or level of requirement.

Evolvable : The capability to develop, or arise through, evolutionary processes.

Evolutionary: A gradual process in which something changes into a different and usually more well-adapted form.

[1] Derived from the Merriam-Webster Dictionary; http://www.m-w.com/cgi-bin/dictionary?book=Dictionary

The core issue behind this drive for evolvable or re-configurable assembly systems should be that micro-assembly is *process-driven*; that is to say that the product design may not be miniaturised without serious consideration of the assembly processes that will be required, since a scaling down of existing assembly systems is *not* viable. Therefore, the required micro-assembly processes, which are unstable and practically invisible, dictate a large range of constraints upon the possible product designs (at this stage of events). Hence the need to develop process-oriented concepts, as given by the Evolvable Assembly Systems paradigm [3]. Another important aspect brought forward is that the key issue within any re-configurable system resides in the manner in which the solution caters for the assembly process knowledge. The modularity achieved by such an approach is consequently based on the careful classification, structuring and formalisation of assembly processes and sub-processes.

In order to achieve such solutions, and create a more robust approach, the E-Race[2] project and Assembly Net[3] community have attempted to define the terms and conditions required to attain Evolvable Assembly Systems. Since the endeavour requires the collection of applicable methods, ontologies, and architectures, the formation of a methodology is given the highest priority, which includes control architectures and multi-agent technology [4]. The next sections will now delve into a proposed paradigm, and the article will attempt to clarify the complexity of applying such scientific paradigms into applicable solutions, and the requirements generated.

2.1 The EAS Concept

The proposed EAS vision, first proposed in 2002 [3], aims to provide the business vision, the methodologies and the underlining technologies and educational foundations for developing new rapidly deployable, modular and re-usable, ultra-precision assembly systems that will allow complex, micro-scale products to be successfully assembled in Europe on a competitive and sustainable cost basis. The EAS concept principles have been embraced by the Assembly-Net, E-Race projects, as well as the 6th framework Integrated Project called EUPASS- Evolvable Ultra Precision Assembly Systems. The term evolvable was chosen to pinpoint the creation of a new paradigm and to differentiate this approach, and ensuing methodology, from others: EAS actually embraces two concepts: evolvability and process-oriented systems.

Note that the term *evolvable* is herewith used as an *attribute* of the concept that is to include the modularity aspects within it. Summarising the EAS concept:

- The focus of the EAS approach is on the processes involved within assembly (perspective: entire product lifecycle) rather than on flexibility, technology, or automation issues.

---

[2] E-Race, Eureka Factory (E!-2851-Factory)

[3] *EU Growth Thematic Network on Precision Assembly Technologies for Mini and Micro Products (Assembly-Net, EU GIRT-CT-2001-05039)*

- EAS implies that theoretically very flexible, multi-purpose cells will be replaced by a highly flexible concept consisting of several targeted but not, in themselves, flexible components.
- the focus is not on short to medium-term product changeover scenarios, but on long-term sustainability of the company's capability to maintain in-house assembly.
- EAS focusses on the assembly processes and their classification, stabilisation and formalisation. This is to attain true modularity rather than mechanical interconnectability.
- EAS intends to integrate all activities within a product lifecycle into a single methodology.
- EAS introduces re-engineering as part of the system life-cycle.

Another very important aspect of the EAS is that it introduces the idea of evolution rather than adaptation. Survival-of-the-fittest, legacy systems, and conceptual mutations will have to become part of the EAS scenario, which will discriminate against ineffective solutions to the benefit of innovative ones. This is only possible if the engineering community accepts that the product design and production systems departments can no longer be assumed to be two independent entities or activities. The essence of the EAS concept will include two main components, which will have to be grouped to attain the desired equation:

1. Functional issues, such as process-oriented assembly systems;
2. Quality attributes, such as found within the EAS paradigm.

Work is currently being finalised within dedicated projects, and will be detailed in forthcoming publications. The difficult issue here is to derive the essential variables for correlating the two components given above. The Required Functionality will be a function of the functional components, whilst the Evolvability attribute will be a function of the quality attributes. A non-linear relation will probably ensue. The task is being developed at present, in which the Quality Attributes and Functionalities are being detailed. The point is to try and validate, for example, that the requirements posed by the EAS one is trying to build will be given by a certain level of modularity, which, in turn, will provide a quantifiable level of quality.

2.2 A Potential Application

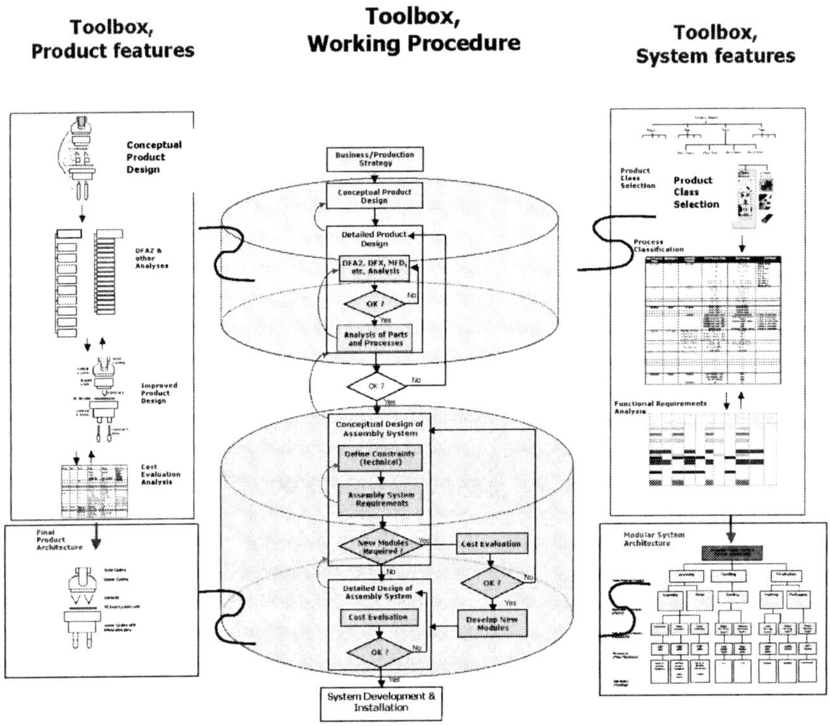

Figure 2 - Working Procedure and Toolbox details for the EAS Approach

The approach could result in a radically new and elaborate working procedure for assembly system developments. The working procedure given as an example reflects the life-cycle given in figure 1.0, and is a step-by-step procedure from fundamental business & production strategies to final product design and ensuing assembly system. As given in this figure, the development toolbox resulting from the application of the EAS methodology may include product-related actions and system-related actions. The shaded areas represent the phases within the working procedure during which the toolbox is most relevant.

It is clear that the EAS toolbox will inevitably require the incorporation of several methods. It will also require the development of structured architectures, knowledge acquisition routines, and data validation schemes. Therefore as it stands, this figure only represents a theoretically possible application. The next section will try to clarify the steps needed to bring such an attempt to application level.

The approach initiates from business and/or production strategies. These are essential, since clear and concise strategic objectives will be interpreted into module drivers: aspects of priority when defining the primary characteristics to be sought after within a module. These may vary, from quality, low-cost, and automated, to high granularity, knowledge transparency, and ease of maintenance. Obviously, a weighting scheme will have to be adopted to classify the priorities/module drivers.

The working procedure given above illustrates the events required when developing the first system (phase 1). Once a system exists, the present architecture is to evolve, hence the working procedure becomes sharper (feedback to product design is based on existing modules) and more focussed on long-term strategies (phase2).

3. APPLICATION ASPECTS

Let us consider the System Features part of the EAS working procedure. The first step is to select a single product type out of a single product class (size, tolerances, no. of parts, complexity). The first stumbling block is that this will require a known taxonomy and ontology. Note that this general classification must be completed and available prior to any analysis. Note also that many products will be of a mixed class nature, such as mini products with micro components: MiniMicro, etc. This precludes that product classes will have to be derived and classified. These classification schemes must then also be applicable to other aspects of the methodology, such as architectures, products, etc.. The main aspect here is to first find a classification scheme that may be applied throughout the product classes and to control, economic and social aspects as well. The zoological classification system[1] has therefore been applied to the proposed manufacturing classification system, which in standard (unexpanded) form may give:

Zoological	Proposed Manufacturing
Kingdom:	Production.
Branch or Subkingdom:	Product Assembly.
Class:	Mega, Macro, Mini, Micro, Nano-Assembly
Order,	Micro-Macro, Micro-Micro, Micro-Nano Assembly
Sub-Order	μClass1, μClass2, etc.
Famil	Assembly, Joining, Handling, Transport, etc.
Genus,	Assembly Processes, Control Processes, etc
Specie	Assembly Sub-Processes, Control Sub-Processes, etc
Individual.	Assembly Modules, Control Modules, etc

The second stage of events in the EAS working procedure requires, for each class, the definition of the processes & sub-processes (from the operations required and other input channels). This may be assumed to be the heart of the System Features part of the approach, since the particular sub-order of product class gives a distinct and unique set of assembly sub-processes.

Main Process Class	Workobject	Assembly Process	Sub-Process Class	Sub-Process	Constraint
Assembly	Part(s)	Assembly	Fit, Type 1	Placement	Always vertical
			Fit, Type 2	Short Insertion	Vertical
					Non-vertical
			Fit, Type 3	Long Insertion	Vertical
					Non-vertical
			Fit, Type 4	Press	Vertical
					Non-vertical
			Connection, Type 1	Side-Fit, stiff component	Non-vertical
			Connection, Type 2	Side-Fit, soft component	Non-vertical
		Drop	Drop	Drop	Always vertical
Joining	Part(s)	Joining	Glue, Type 1	Spot Glueing	
			Glue, Type 2	Seam Glueing	
			Rivet	Riveting	
			Snap-Fit	Snap-Fit	
			Solder, Type 1	Spot Soldering	
			Solder, Type 2	Joint Soldering	
			Weld, Type 1	Friction Welding	
			Weld, Type 2	Laser Welding	
Handling	Part(s)	Grasping	Grasp, Type 1	External Grasp	
			Grasp, Type 2	Internal Grasp	
			Grasp, Type 3	Surface Grasp	
		Feeding	Single Part Feed, Type 1	Mechanical; Vertical	Position & Orientation
			Single Part Feed, Type 2	Mechanical; Non-Vertical	Position & Orientation
			Multiple Part Feed, Type 1	Pattern	Position & Orientation
			Multiple Part Feed, Type 2	Free Placement+Vision	Position (& Orientation)
			Multiple Part Feed, Type 3	Tape	Position & Orientation
			Multiple Part Feed, Type 4	Bulk	Position & Orientation
			Flexible Feed	Mechanical+Vision, vertical	Position & Orientation
Transport	Part/Product	Main Flow System	Product Flow	Conveyor	
		Individual Flow Syst.	Product Transport, Type1	Pallets	
		Independent Flow	Product Transport, Type2	AGVs	
		Internal transport	Product Transport, Type3	Robot	
		Product Fixation	Fixation, Type 1	Fixtures without memory	
			Fixation, Type 2	Fixtures with memory	
		Flow Balancing	Re-Flow & Buffers	Elevators	
			Main buffer	Carrousel	
Quality Control	Part/Product	Function Testing	xxxx	yyyy	
			xxxx	yyyy	
			xx	yyyy	
		Process Control	dddd	iiii	
			eeee	jjjj	
			ffff	kkkk	
			gggg	llll	
		Surface Inspection	hhhh	mmmm	
Finalisation	Product	Marking	Laser Marking, Type 1	nnnn	
			Laser Marking, Type 2	oooo	
			Ink-Jet Marking	pppp	
		Packaging	qq	rrrr	

Figure 3 - Structured Process & Sub-process Classification

These are the sub-processes that, after analysis and classification, lead to the system modules for this sub-order. The problems associated with this step regard the formalisation of the data. Typical issues of importance regard the stability of the sub-processes being classified, the ability to formalise them into mathematical expression for software exploitation, and the validity of the information being supplied. These are considerable issues that need very structured working procedures in themselves, and clarify that system modularisation is far more complex than simply tagging a name onto a system.

Once the sub-processes and associated parameters are formalised and classified, the following step takes the formalisation procedure a step further. This step is particularly tricky since the specific parameter/attributes of each sub-process class are to be compared with one another. The robustness of the procedures exploited in the previous stage will now be put to their test. Furthermore, a method needs to be developed, by which the functionally similar sub-process classes are aggregated into potentially exploitable modules. The resulting sub-division must also be capable to re-iterate which functionalities are grouped into *modules*, which are set as *resources*, and which *attributes* are left to specific tooling (see fig. 4.0).

In other words, many sub-processes will require similar characteristics when viewed from an operational point of view. These may be grouped, such that the main functionality may be given by a module, whilst the detailed aspects resolved by a specific resource (that can be attached/detached from the module; i.e.-gripper, tool, etc.). This will require a specific software-based tool/method. Constraints and related mechanisms obviously need to be developed.

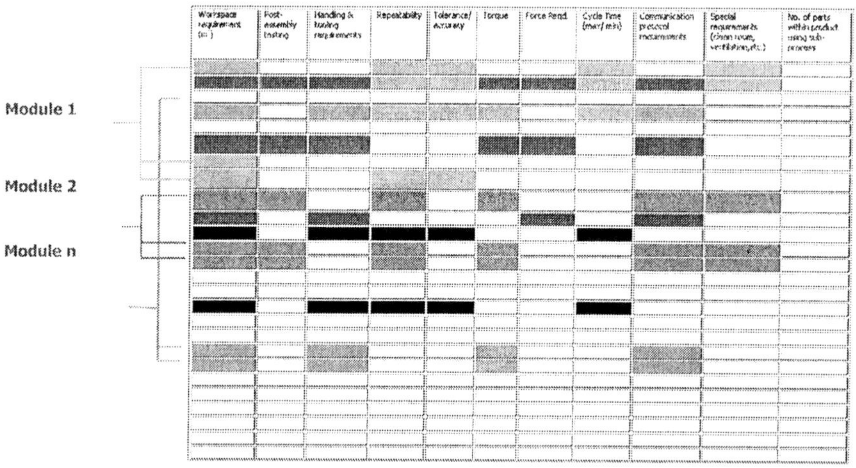

Figure 4 - Sub-division into Potential Modules

The final step in the EAS working procedure, step 5, regards the definition of the Modular Assembly Platform components required: after the first iteration of the EAS procedure, there will be a given system architecture. In this step, the user will pick the modules that correspond to the qualities required by the groupings given in Step 4. However, there may be groups that require entirely new modules. Therefore, there is a need to couple this stage with the final stage in Product Features Analysis: the user/team will have to check whether it is cost-effective to develop a new module, or whether it is better to change the design of the part/product, or even outsource.

Such a Cost Evaluation step may bring new modules to be defined and incorporated into the Module Platform, hence the coupling to a higher-level strategic aspect. Such strategic aspects bring about the issue of having specific "module drivers" set at an early stage in the procedure, an approach already adopted by the Modular Function Deployment product design methodology [5].

4. DISCUSSION & CONCLUSIONS

Figure 2.0 illustrates the potential of fulfilling the requirements given in the EAS working procedure, and also depicts the vital link between processes and modules. However, this is purely theoretical at present.

In order to achieve the solutions mentioned in this article, major efforts are required on several fronts. First of all, as the article points out, there must be some convergence and agreement on the taxonomies, and an assembly ontology should be created. The whole process of developing the assembly system, which will be software-based in a "Development Toolbox", will rely on the correct priority being selected before starting: the module drivers. That is, the user must decide if the tool is to optimise (set priority) for costs , product design, fast ramp-up, or any other "Module Driver". Weighting schemes are not the only requirement to be applied in order to succeed. Other prerequisites may include:

- Definitions, ontologies.
- New cost models for analysis.
- Impact of Social & Management Issues (see fig.5.0).
- Application of solutions in collaborative networks.
- Standardisation/formalisation of Product Classes.

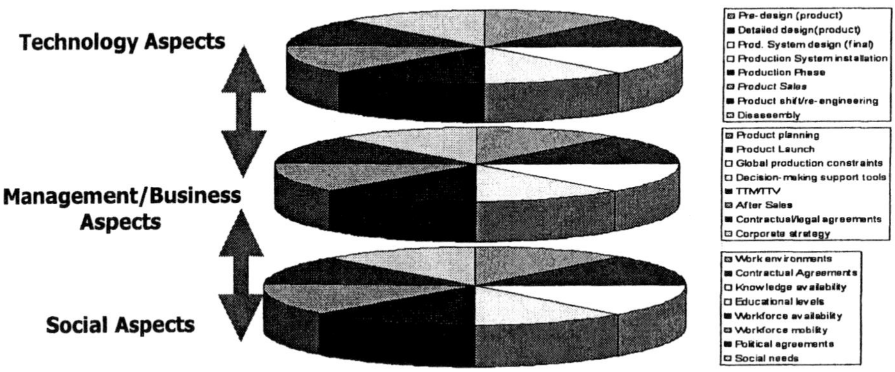

Figure 5 - Concurrent cycles in a product's lifecycle

However, the most important factor remains the collection of the methods required into the so-called working procedure. This includes formalisation procedures, methods, software developments, etc. It is, therefore, important to make absolutely clear that this will not succeed without an appropriate methodology that integrates these issues in an effective manner, a point which was underlined in the Assembly Net Roadmap and other publications ([1],[3]). Note that since the EAS paradigm requires a methodology for successful implementation, it also requires a holistic perspective. This is why social and management issues must also be considered more elaborately. Considering the fact that an eventual implementation may have to recur to multi-agent technology and collaborative networking theories, in which contract management issues arise, these two topics become even more relevant. The article obviously does not present a solution or implementation, but intends to clarify the difficulties behind developing working procedures that result in truly modular system components: the intention is to illustrate the efforts required to achieve evolvable systems, what they represent, and the work that lies ahead. The authors pinpoint that three major initiatives have come together to collaborate around these issues: Assembly Net, E-Race and EUPASS. Assembly Net acts as a dissemination channel and assists the group in finding new collaborating partners.

The E-Race and EUPASS projects are products of such efforts. E-Race is focussed on the software issues related to EAS. EUPASS is an Integrated Project that will attempt to develop EAS systems for two demonstrators at industrial level, and focuses on the architectural, standardisation and hardware issues. The scale of the problem remains significant and the theoretical foundations of many of the issues portrayed in this article will have to be consolidated. Collaboration is also being established with IFIP WG5.5 in order to better incorporate management & social issues. The work described in this article is also detailed in three forthcoming Phd Theses ([6],[7],[8]).

Once again, the point being made is that it is important to clarify that re-configurable, modular, or evolvable systems can only succeed if the processes to be accounted for are classified, stabilised and formalised. Hence, a very focussed approach is strongly recommended: select a well-delimited process for validation/falsification of the ideas. Broad approaches cannot succeed because of the inter-relationships that exist between industrial processes, which complicate the analysis of the core issue, e.g.- re-configurable *production systems* is a doubtful approach since not only manufacturing, assembly, logistic, and sales processes inter-relate on the basis of current solution premises, but sub-contracting, supply-chain, external logistics, marketing and other strategic processes influence the outcome. It is therefore safe to assume that a narrow focus is the only way to scientifically examine, falsify/validate the aspects and potentials behind the EAS paradigm.

4. REFERENCES

1. Onori,M.; Barata, J.; Lastra,J.M.; Tichem,M.(2003); "European Precision Assembly – Roadmap 2010"; Assembly Net, Document DO2c, ISBN 91-7283-637-7, EC Report
2. Gaugel,T.; Bengel,M.;Malthan,D.(2003); "Building a Mini-Assembly System from a Technology Construction Kit"; Proceedings of the International Precision Assembly Symposium (IPAS2003), Bad Hofgastein, Austria.
3. Onori,M.(2002); "Evolvable Assembly Systems – A New Paradigm?"; Proceedings of the 33rd International Symposium on Robotics (ISR); Stockholm, Sweden.
4. Barata, J., & Camarinha-Matos, L. M. (2002); "Shop Floor Re-engineering Using Agents"; Proceedings of the 33rd International Symposium on Robotics - ISR2002, Stockholm.
5. Erixon,G.(1998); "MFD - Modular Function Deployment, A systematic method and procedure for company supportive product modularisation"; PhD Thesis, The Royal Institute of Technology, Stockholm, Sweden; ISRN KTH/MSM/R-98/1-SE
6. Barata de Oliveira, J.A.(2004);" Coalition Based Approach for Shop Floor Agility- A multi-Agent Approach"; PhD thesis, Universidade Nova de Lisboa, January 2004
7. Alsterman, H.(2004);" Strategic Issues for Achieving Sustainable Flexible Automatic Assembly"; PhD thesis, The Royal Institute of Technology, Feb.2004.
8. Lastra, J.M.(2004);"Reference Mechatronic Architecture for Actor Based Assembly Systems"; PhD thesis, Tampere University of Technology, 2004.

40 BENEFITS OF MODULARITY AND MODULE LEVEL TESTS

Patrik Kenger
Dalarna University and the Royal Institute of Technology, SWEDEN
pke@du.se

Many companies implement a modular architecture to support the need to create more variants with less effort. Although the modular architecture has many benefits, the tests to detect any defects become a major challenge. However, a modular architecture with defined functional elements seems beneficial to test at module level, so called MPV (Module Property Verification). This paper presents studies from 29 companies with the purpose of showing trends in the occurrence of defects and how these can support the MPV.

1. INTRODUCTION

Any product or process defect causes losses, and these will be repeated unless a more formalized approach to the problem is taken. Today product tests to detect defects are necessary but they are a difficult phase in industry which results in both extension of lead-times and increased costs, O'Connor (2003). The difficulties are caused by lack of time and knowledge of how to plan and perform the tests in the assembly system; and how to design products which are suitable for, or at least facilitate, tests. In this paper product verification denotes the process of determining whether or not the product at a given phase in the life-cycle fulfils its properties. This definition includes the commonly used word test, but also manual inspection and quality control, and the planning, evaluation and documentation of the verification results. The trend of shorter lead-times and life cycles, Onori (2003), seems to further enhance the difficulties. Since every new variant introduced in the assembly line has its own properties, it also needs its own specific verification. The operators performing the verifications face an impossible task: verifying increasing volumes and variants with the same amount of personnel and equipment. This is the case at one company where the actual verification process has become the bottleneck. One way to handle lead-time and cost of verification is to reduce the verification itself. Although this will cut cost and time, Varma (1995) points out that it is more important to focus on product profitability and verification strategies.

However, modularity has proven to have benefits related to defects and verifications. The goal of this paper is to show the correlation between a modular architecture, module level verification and a potential decrease in design and assembly defects. The discussions in the paper also have a direct bearing upon the correct implementation of re-configurable or evolvable assembly systems, Onori (2003), which strongly rely upon system modularity.

2. MODULAR ARCHITECTURES

To keep up with increased volumes and product variants companies strive to implement a modularized product assortment. Modularization has shown to have numerous benefits, see e.g. Ulrich and Tung (1991), Erixon (1998), Stake (2000), or Baldwin and Clark (2000). In fact many Swedish companies have successfully utilized modularity to stay competitive. Among these are Scania and VBG, see Erixon (1998), ABB, VOLVO, and ITT Flygt, see Stake and Blackenfelt (1998). One studied company shows a potential increase of 6700 variants (theoretically possible variants) and a decrease of 7000 parts after two years of modular implementation, Table 1.

Table 1: Benefits of a modular architecture

Product architecture	Years	Variants	Number of different parts
Integrated	30	300	10000
Modular	2	7000	3000

The product architecture in Table 1 denotes the scheme of the functional elements of the product, Huang (1999), and how these elements are arranged into physical blocks (modules) and the blocks interaction. Huang (1999) describes a modular architecture as the architecture where the functional element is implemented by one block which has few but well defined interactions between other blocks. The integrated architecture is characterized by optimization of a certain performance. The interactions between blocks in an integrated architecture are not as defined as in the modular case, as each block embodies several functions.

However, the company described in Table 1 has a challenge to be faced before the full potential of the modular architecture may be utilized. The company plans to have a minor module storage in which final assembly selects module variants that fit the product the customer asked for. The modules and the module storage will support a potential lead-time decrease of 450%. This decrease though is only possible if defect-free modules are available from storage. The challenge is to verify the increased product variants, made possible by the modular architecture, and to do it on module level. In Figure 1, the challenge is described with 4 modules (1, 2, 3, j) with 5, 3, 7, and 2 variants. If the verification is performed at product level, the 210 possible variants need 210 verifications. This number is reduced to 17 verifications if the product verifications are performed at module level.

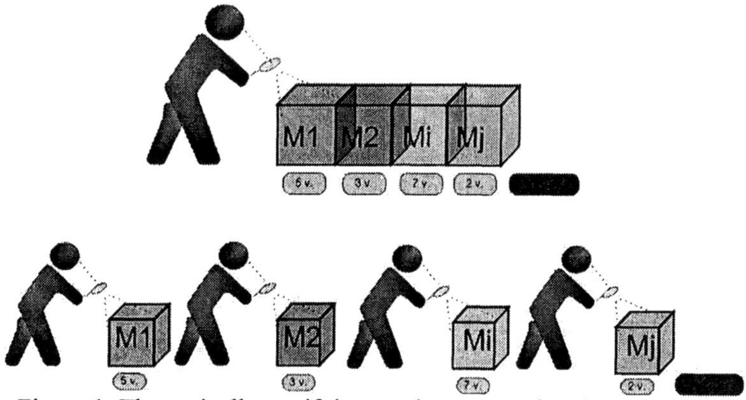

Figure 1: Theoretically, verifying products at product level requires 210 verifications to cover the whole product range, while module level requires 17.

2.1 Modular verification

There are more benefits to be obtained by module level verifications then just a decrease in the number verifications. Researchers and companies studied agree that the product verifications should take place early in the value chain, discussed by Baudin (2002), Robinson et al. (1988), and Nevins and Whitney (1989). The more time spent on embodiment of the product, and the more parts manufactured and added to the product, the greater the value added. At the same time the complexity of the product increases, i.e. more parts are added which give the product more details and functions. The approach to verify the product early in the value chain is specifically beneficial in a modularized product assortment where specified functions and interfaces in each module can be verified already at the module assembly workshop, so called module property verification (MPV), Kenger et al. (2003). As discussed above, a modular architecture has few or one function in each module which in turn simplifies the verification. Benefits of MPV have been discussed by among others Baldwin and Clark (2000), Erixon (1998), Baudin (2002) and Stake (2000). As pointed out in Kenger and Onori (2003), by performing MPVs, detected defects can be repaired at module level where less parts have to be disassembled, spare parts are already available at the module assembly workshop, and no additional assembly or verification tools are necessary since they are also available in the module assembly workshop.

Even though there are several benefits by performing MPVs, there may be reasons for performing the verifications at product level, so called product property verification (PPV). PPV may be more beneficial to perform when the number of defects per product (defect rate) is relative low. Only a final check of the product is performed as a precaution to ensure the compatibility of the parts or modules building up the product. Compared with MPV, there is less number of separate verifications in PPV since one PPV might correspond to several MPV's. That is, at module level each module may need its own verification while it may be enough with a single verification on product level. All in all, it is necessary to measure the benefits of MPV compared to PPV to avoid costly rearrangements at the point of verification which in turn affects both the assembly system and the module design.

3. DESIGN AND ASSEMBLY DEFECTS

A way to measure the benefits of MPV is to analyze occurred design and assembly defects. How often, the causes, and defect trend can show where it would be beneficial with MPV. Here, a defect is a fault that causes the product not to fulfill its properties, i.e. the product does not work or have the intended appearance. Defects themselves are a symptom of poor machines, designs and routines where the defect origin is claimed to always be human. Baudin (2002) and Shingo (1986) point out that verifications to detect defects are a waste of time and resources since it does not add any value to the customers' interpretation of the product. Thereby, the most profitable way to verify is not to verify at all which in turn is related to increase risks of having defect products shipped to customers. Also, the verifications themselves do not contribute to reducing the defects. Case studies, presented below, show that zero defects is an utopia, at the same time demands on verifications increases from customers, standards and governments. Therefore, verifications are necessary but should be performed with a minimum of time and resources. This means that personnel, verification equipment, documentation and preparation have to be optimized and verify the exact demanded properties.

Branan (1991) showed a relationship between defects per million parts and manual assembly efficiency. This relation was further analyzed by Barkan and Hinckley (1994). They show that longer assembly times are related to difficult assembly tasks which increase the probability that a defect may occur. Five assembly factors are also identified related to a qualitative product. (1) Assembly operations, (2) assembly quality control, (3) assembly operation complexity, (4) number of parts, and (5) part defect rate. A relationship between assembly time and the defect rate (defects per product) can also be seen in Table 2.

Table 2: The relation between assembly time and defect rate

Assembly time	Defect rate	Assembly time	Defect rate
< 0.5 hour	< 0.01	0,5 to 1 hour	0.2 to 0.5
<0.5 hour	< 0.01	1 to 2 hours	0.2 to 0.5
< 0.5 hour	< 0.01	2 to 3 days	0.2 to 0.5
0,5 to 1 hour	< 0.01	2 to 3 days	0.2 to 0.5
1 to 2 hours	< 0.01	>2 weeks	0.2 to 0.5
1 to 2 hours	< 0.01	1 to 2 hours	0.6 to 1
5 to 10 hours	< 0.01	1 to 2 hours	0.6 to 1
< 0.5 hour	0.01 to 0.1	> 2 weeks	0.6 to 1
< 0.5 hour	0.01 to 0.1	2 to 3 hours	1.5 to 2
<0.5 hour	0.01 to 0.1	5 to 10 hours	2.1 to 5
0,5 to 1 hour	0.01 to 0.1	2 to 3 days	5.1 to 10
1 to 2 hours	0.01 to 0.1	1 to 2 weeks	5.1 to 10
2 to 3 hours	0.01 to 0.1	> 2 weeks	5.1 to 10
3 to 5 days	0.01 to 0.1	> 2 weeks	5.1 to 10
3 to 5 hours	0.01 to 0.1		

Each assembly time in Table 2 corresponds to a certain defect rate given by a surveyed company. The trend is that longer assemblies results in more defects than shorter ones. The questionnaire was answered by 27 Swedish companies and 2 Norwegian companies. 82% of the responding companies claimed that their products were built up by modules or subassemblies, and 18% that the product has an integrated architecture. Of the same companies, 88,5% answered that their products is mainly assembled manually, and the other 11,5% that they are mainly assembled automatically. At one company a case study was performed and 1600 defects reported over an 8 year period were analyzed. The cost of assembly and design defects in Figure 2, 3, and 4 are designated to specific departments. This means that it is a design defect if the design department is charged for the repair of the subsequent defect. The same goes for assembly defects. Each dot in Figure 1, 2 and 3 represents a customer order of a certain volume. As can be seen in Figure 2 and 3 the trend is that the defect rate decreases as the order volume increases.

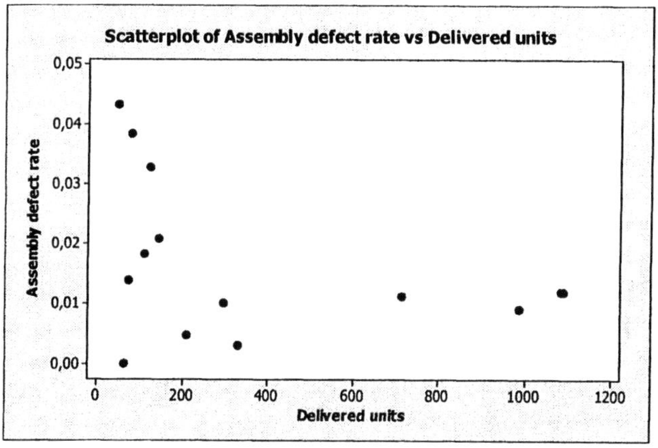

Figure 2: Assembly defect rate and delivered units.

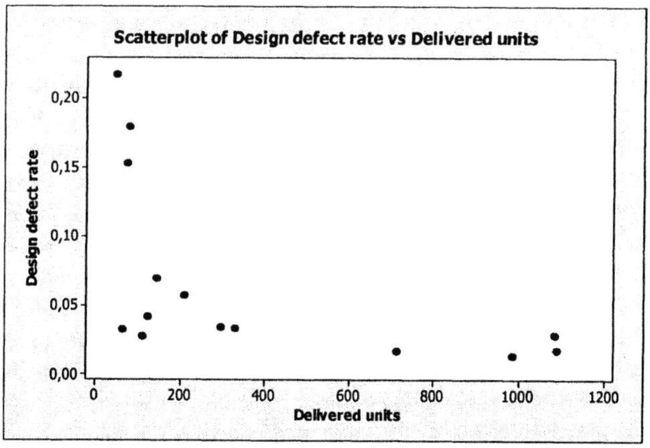

Figure 3: Design defect rate and delivered units.

The design and assembly defects were plotted against each other in order to analyze a possible correlation, Figure 4. The sample correlation coefficient was shown to be

0,83 which is a strong relation, Johnson (2000). This in turn implies that a design which is complex (many parts and many functions) later on also causes the assembly operators to make mistakes.

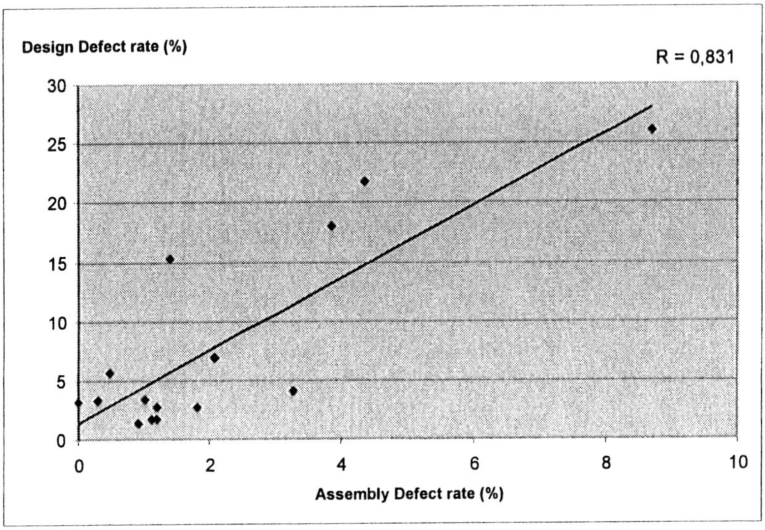

Figure 4: Least square estimate with a sample correlation coefficient R=0,83 which indicates a strong relation between design and assembly defects.

3.1 Difference between design and assembly defects

Even though there is a strong relation between the design and assembly defects, there is a major difference in how they occur. Design defects can be said to occur over a period of time, somewhere during the design process between the gathering of customer needs and the detail design. The design process at the studied company often involves several designers, the design team, who together design the product. Design defects can be detected by the designers themselves, making the design team working as a net which detects defects in time, before the repair becomes much more costly. Design defects, passing the net, can be called consequential design defects, originating from having the "wrong thinking" within the team. Consequential defects denote that the "wrong thinking" which caused the defect, follows the design all the way through the design process. These consequential defects occur even if there is one designer working with the development of the product; each time the designer starts a new working day, or opens up the CAD-software the defect is overlooked and designed "into" the product. Consequential design defects can be difficult to detect since each designed part, sub-assembly, or module may possess a design defect which is not revealed until the product is virtually assembled, simulated or, worse, manufactured and assembled as a physical product. The designer(s) guilty of the "wrong thinking" may therefore not be aware of the design defect until the product actually physically exists. However, frequency of design defects is related to the maturity of the product. Given that the design team is made up of the same members, during the period of design and delivery of the product the design team learns from previous "wrong thinking" and consequential design defects as well as being more and more familiar with the intended function of the product or its parts. This in turn explains the decreased defect rate at higher volumes.

Assembly defects occur more instantly compared to design defects. For example, inserting a part can be done the right or wrong way, where the actual assembly defect (wrong insertion) occurs at the same moment as the part is inserted. The assembly defect is not made by systematically having the "wrong thinking" to the same extent as design defects, but more of the presence and the experience the assembly worker possesses. Although, a systematic pattern is difficult to see in assembly defects, the assembly workers also said that they become more familiar and learn in similar ways to the designers.

After analyzing the defects at the studied company, the result was presented to design and assembly personnel, including the managers. They agreed upon the different ways design and assembly defects can occur.

4. SUMMARY

Verifying products is one of the areas where there is still much to be gained for some companies. It has been shown that defects occur even though preventive measures are taken. However, not only the actual verification is important but also where in the value chain it is performed, how easy it is to detect each defect as well as repairing them.

Modular architectures have numerous benefits, where several of them are related to verifications and defects. Figure 2 and 3 show that as the volume increases the defect rate decreases in design and assembly defects. It was clear from the study that the defect reduction can be explained by learning about the product and being familiar with the work. Modularization gives higher volumes from a decreased number of parts, as well as smaller assemblies, which take less time to assemble. This can have the same effect as having an integrated architecture and high volumes. However, in the modular case the order volume can be one unit and still obtain the benefits of reduced defects.

To decide whether MPV or PPV is the most economical and time efficient way to verify, statistics on defects can serve as a measure. If a relative high defect rate of a certain defect occurs, say assembly defects, then it would most probably be beneficial to perform MPVs. Similarly if a minor storage of modules will be held, to cut the lead-time, MPVs would probably be beneficial. However, if the defect rate is relative low, as well as the cost to repair defects, the PPV approach is suggested. In the MPV case, more verification stations are needed to correspond to one PPV, see Figure 1. Also more test and assembly fixtures (TAFs) are needed in the MPV case, which increases the initial cost when moving from PPV to MPV.

4.1 Further work

The collected data, more than given room for here, will serve as input to develop a prototype of a module assembly line. A TAF is being built to perform MPVs for a proposed module, which shall be part of the assembly line. The prototype line should run long enough to compare numbers from the line today, the defect rate and time to repair, with the prototype line and its defect rates and time to repair. The work is also being proposed to be integrated within an Evolvable Assembly Systems project called A^3-Applied Agile Assembly, to be performed within industry under the leadership of IntRoSys SA, a Portuguese SME.

5. REFERENCES

1. Baldwin, C., Y., Clark, K., B., (2000), Design Rules - The Power of Modularity, Massachusetts: MIT Press, ISBN 0-262-02466-7
2. Barkan, P., Hinckley, M., (1994), Benefits and Limitations of Structured Methodologies in Product Design, Management of Design, Engineering, and Management perspectives, USA: Kluwer Academic Publishers, ISBN 0-7923-9509-3
3. Baudin, M., (2002), Lean Assembly – The Nuts and Bolts of Making Assembly Operations Flow, New York, USA: Productivity Press, ISBN:1-56327- 263-6
4. Branan, B., (1991), Six Sigma Quality and DFA, DFMA Insight, Vol. 2, Winter 1991
5. Erixon, G., (1998), Modular Function Deployment - A Method for Product Modularization, Doctoral Thesis, Stockholm: The Royal Institute of Technology, ISSN 1104-2141
6. Huang, C., C., (1999), Overview of Modular Product Development, Proc. Natl. Sci. Counc., Vol. 24,
7. No. 3
8. Johnson, R., A., (2000), Probability and Statistics for Engineers, 6th Edition: Prentice-Hall, ISBN 0-13- 014158-5
9. Kenger, P., Onori, M., (2003), Module Property Analysis in the Assembly Process, International Precision Assembly Symposium: Bad Hofgastein, Austria
10. Kenger, P., Erixon, G., Lennartsson, S., (2003), Module Property Verification - A Conceptual Framework to Perform Product Verifications at Module Level, 14th International Conference on Engineering Design, 19-21 August, Stockholm, Sweden
11. Nevins, J., L., Whitney, D., E., (1989), Concurrent Design of Products and Processes – A Strategy for the Next Generation in Manufacturing, USA: McGraw-Hill, ISBN 0-07-0463417
12. O'Connor, P., (2003), Testing for Reliability, Quality and Reliability Engineering International
13. Onori, M., Barata, J., Lastra, J., M., Tichem, M., (2002), European Precision Assembly Roadmap 2010, Assembly-Net Report to EC, Assembly-Net (G1RT-CT-2001-05039)
14. Robinson, L., W., McClain, J., O., Thomas, L., J., (1990), the Good, the Bad and the Ugly: Quality on Assembly Line, International Journal of Production Research, Vol. 28 No. 5
15. Shigeo, S., (1986), Zero Quality Control – Source Inspection and the Poka-yoke System, Portland, USA: Productivity Inc., ISBN 0-915299-07-0
16. Stake, R. B., (2000), On Conceptual Development of Modular Products - Development of Supporting Tools for the Modularization Process, Doctoral Thesis, Stockholm: The Royal Institute of Technology, ISSN 1650-1888
17. Stake, R., B., Blackenfelt, M., (1998), Modularity in Use – Experiences from five companies, 4th WDK Workshop on Product Structuring: Delft University of Technology, The Netherlands
18. Ulrich K, and Tung K, (1991) "Fundamentals of Product Modularity", DE-Vol. 39, Issues in Design Manufacture/Integration, ASME
19. Varma, P., (1995), Optimizing Product Profitability – The Test Way, International Test Conference

41 AUTOMATED SYSTEM FOR LEATHER INSPECTION: THE MACHINE VISION

Mario Mollo Neto,
Oduvaldo Vendrametto,
Jóse Paulo Alves Fusco
UNIP mariomollo.cdg@unip.br ,
UNIP vendrameto@unip.br ,
UNIP jpafusco@uol.com.br .
BRAZIL

Nowadays Brazil is the world largest producer of cattle leather and the total exportations in 2002 were around 19 million of pieces or 930 million dollars. In the other hand, the low performance of domestic tanning plants, has compromised the final numbers and as a result the product remains considered as a mere commodity in the international market. The semi-finished products obtained in the very first stages from the process, with low level of value aggregation, represent circa of 60% of exportations.
This paper is a summary of the findings obtained from a research done and aims to present an automated inspection system to classify the wet blue *leather, using image processing and under a quality control system guiding rules..*

1. INTRODUCTION

The globalization effects on developing (or emerging) countries has addressed the label of commodities exporters presented by them. The expectations of sharing the world market, commercializing products in higher levels of aggregated value, motivated by an increasing global competition, has occurred only in the importations side of the so called global products. The gap of knowledge presented by the Brazilian productive system concerning technology, management and quality, jointly with the MNC´s tradition on international trading, weakened the position of national best firms. These ones, without conditions to face properly the fierce competition against their foreign counterparts, were incorporated by rivals or stopped operations or even reshaped their business anyway.
During the 90´s, because the adoption of partnerships, alliances, joint ventures or simply acquisition of Brazilian firms, the participation of foreign firms in the Brazilian GDP increases from 36% (1991) to 53.5% (1999) (Kupfer, 2002).
Changes on processes and operation systems, the use of automation technologies, TI, besides decreasing the participation of Brazilian investmens on the domestic market, co-operate to decrease the employement level and brought other types of

social problems already on the table to be solved. While entire productive networks have been completely absorbed by foreign groups, e.g. autoparts and electrodomestics, conversely there are other remaining sectors virtually untouched as in the case of meat, leather and footwear, wooden-made furniture, gems and other. Probably some of typical characteristics presented by these sectors, e.g. labour intensive processes, lack of organizational control, lack of stable conditions concerning tax legislation and capital-labour relationships, and the general poor quality of buyer-supplier relationships, has pushed the focus of foreign investors away.

Some of the modern (regarded) firms evolved towards higher levels of competitive conditions, working concepts such as world products and making use of very sophisticated technologies and management tools. However, most of the remaining firms stay precariously in the business without investments in technology and, therefore, in a low level of competitive power, still commercializing commodities or semi-finished products. Table 1 shows the potential of sales and development presented by the Brazilian network of meat, leather and shoes business. These numbers could be increased significantly through the adoption of an adequate set of policies by the federal government, aiming to increase the Brazilian share of international market and elevating the employement level as well.

Table 1 - Source: Couromoda Calçados e Artefatos (**) Additioned by the author

Brazilian exportations of meat/leather/footwear - US$				
Product	**2000**	**2001**	**2002**	**Variation % 2000/2002**
Shoes	1,548 bi	1,615 bi	1,449 bi	- 6,3
Eláter	744 mi	850 mi	930 mi	+ 25,0
Components	448 mi	500 mi	519 mi	+ 15,8
Machinery	5 mi	6 mi	ND	ND
Meat (**)	805 mi	807 mi	1,050 bi	+ 30,4
Sum	**3,550 bi**	**4,156 bi**	**~3,948 bi**	**~ + 11,2**

The system proposed is based upon a type of "machine vision", to be developed using new technologies available, to perform activities within the process of classification of *wet blue* leather. Then, its main function would be to classify cattle leather just after the chemical treatment of tanning, looking for increasings on quality standards.

The classification occurs after the very first stage of the production process, on the semi-finished product called *wet blue*. One piece at a time is then extended over one special table and one specialist takes a visual inspection on the leather surface for around 30 and 60 seconds. The specialist then evaluates the quality and assigns the piece a grade between 1 and 7, writing it with a chalk on its surface. According this scale, one piece of leather graded 1 would present the best "value" for quality while the other limit would refer to the worse quality.

The procedure cited above is quite arguable because of its large uncertainty margin and this fact represents the main reason behind the existing commercial problems between tanning plants and their customers, primarily in the case of exportations.

The resulting pieces of leather can be commercialized in a briny or crude state, without being submitted to any inspection of quality up to this phase. Only after the process on tanning plants the piece is converted into leather one and classified, receiving the grade. This late evaluation can bring losses to the tanning plants, because the quality will be known only in the end of the entire production process.
Additionally, the "machine vision" presented and proposed in this paper will help firms to know precisely what they are really doing in terms of quality of good products, creating a value basis to support commercial negotiations in a more realistic way, instead of making use of speculative arguments as up to now.
The new form proposed intends to analyze electronically the leather through the use of one digital optical system, mapping the defects on each piece and obtaining the correspondent classification or grade. Besides, the leather mapping allows the application of more specialized methods to optimize the working area on it, decreasing rejects and the related environmental problems.
One experimental test was done to verify the compatibility of the system proposed and the present visual system in terms of the outputs and the quality of the data to be generated. Despite the fact that the technology proposed still needs a more profound examination before its full adoption by firms, the results suggest that both systems are totally compatible.

2. BACKGROUNDS

Along the last 30 years the research field called "machine vision" has been evolving and developing to play an important role on either side of production processes (using robotics) and quality control of manufacturing operations (AVI - Audio-Video-Interleaved systems).
In the early papers, the main purpose has been to automate the quality control through the use of images (Stapley, 1965; Norton, 1970; Aalderink, 1976; Saridis, 1976; Roland, 1982).
Many researches have been conducted in the last few years to elaborate high speed techniques more efficient to analyze images, trying to emulate the human vision. To do so, besides the images processing system, it is necessary to develop another technique of decision making, very difficult to automate indeed.
In parallel with the academic evolution, the required technology to capture images (e.g. cameras, digitizers) and their subsequent treatment (computers, microprocessors) has been developed in such a way that nowadays it is possible to build a sort of very efficient visual systems to capture images and to take decisions based on information gathered from it.
The development of a system to analyze quality using image processing involves many other important aspects from the choice and implementation of adequate methods up to the elaboration of experiments to verify its suitability.

2.1 Scenario analysis

The research focusing leather issues has been done under a wider range, going beyond the usual activities involving its simple applications as a part of other products or even as a supply in the leather artisan industry.

The innovations already cited in the text revealed new horizons to the leather sector and deserve to receive special attention because the possibility to obtain products with more aggregate value, competitive advantages and increasings on ROI.
Nowadays the leather, as a raw material, is responsible for an important economic sector worldwide, presented a great diversity of products from manufacturing and service firms. Porter (1993, 479-497) presents the competitive advantages that lead Italy into the world leadership of leather industry. Soon after, the author suggested in a lecture the figure 1 above, containing one wide view of the different sectors forming the entire network.
Concerning cattle leather, results obtained from many researches revealed that Brazil has missed an opportunity to gain close to 900 million dollars each year because the poor quality of the leather produced by domestic plants and the unbalance between domestic production and demand (CICB, Brazilian leather plants union, 2000). Usually pieces of leather used as raw materials by manufacturing firms (footwear, purse, belt, Clothes, furniture and car seats) present scratches, scars, punctures and spots over their surfaces.
It is important to point out that 85% of the leather made in Brazil presents some type of defects, and 60% of that occurs within the limits of the tanning plants. The remaining 40% occurs because the bad conditions of logistic factors from the farms to the plants.

a) Figure 2 shows the types of defects presented by the leather made in Brazil. The general lack of secure information about the leather as a product, inside the limits of the farms, represent a huge difficult to be overcome.

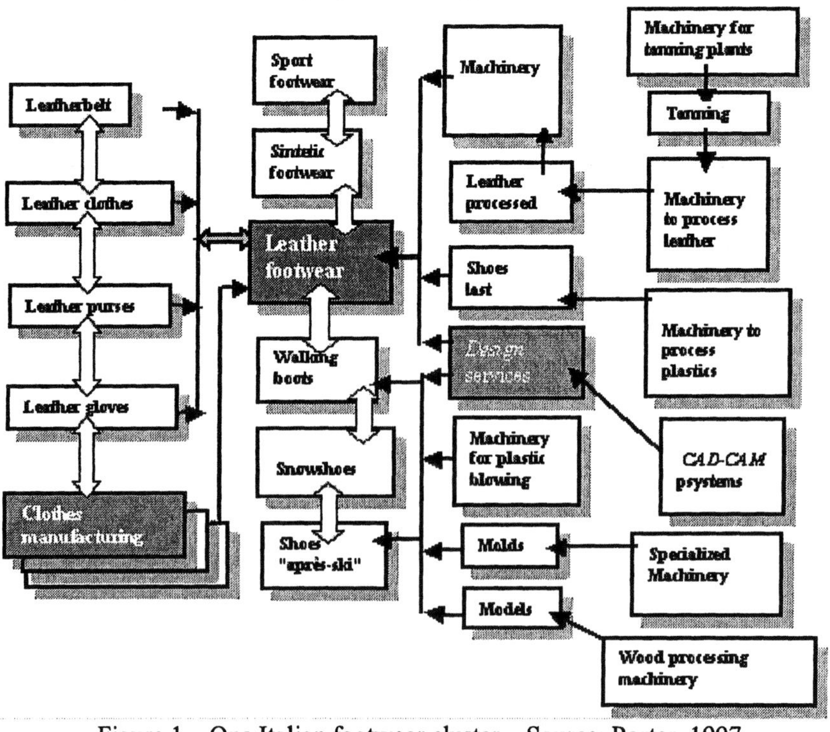

Figure 1 – One Italian footwear cluster – Source: Porter, 1997

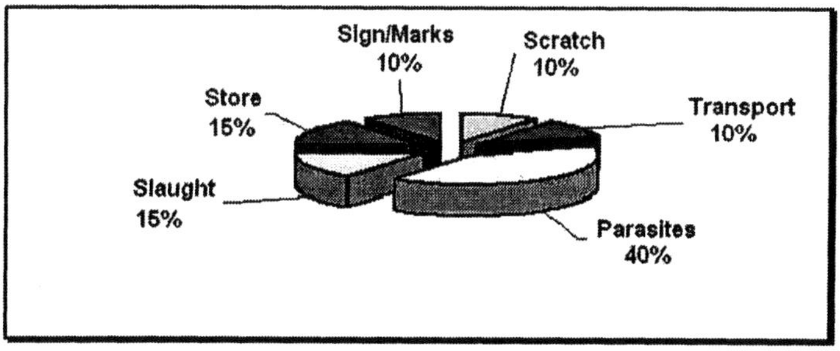

Figure 2 – Defects of Brazilian leather - Source: ABQTIC (1998)

Additionally, three relevant questions need to be considered:

b) The geographic localization and climate conditions of the country, with some regions quite suitable to microbe proliferation and also the emergence of a wide variety of coetaneous diseases. All this stuff requires additional efforts during the entire productive process to obtain the leather.

c) The present practices adopted by the slaughtering plants require corrective actions to improve the product to be sent to the tanning plants, avoiding undesiring rests, usually accepted because of sales criteria (weight).

The treatment process of effluents released by the plants need to be developed more properly to eliminate environmental impacts.

In fact there is a negative cycle to disrupt. The poor profitability presented by the leather sector raises difficulties to answer the question of poor quality. To do so it becomes necessary to provide an adequate flow of investments on new technologies, innovation, management and negotiation processes involving suppliers and customers together. The need of further developments on products and manufacturing plants is quite evident indeed. Efforts in such a way would bring significant results to every firm within the business network, and would increase the incomes from Brazilian exportations.

In so far as the wider and deeper relationships between agriculture sector and other sectors, upright and downright within the productive chain, its connection with other economic segments has been expanded as well. Thus, any macroeconomic or sectorial change affecting one or more structural link of the CAI (Agro industrial Complex) also reverberates on the national economic structure.

Impacts on other sectors would induce side-effects on the Agro industrial Productive Chains (CPA- Agroindustrial Productive Chain). The participation of agribusiness incomes represents something around one third (1/3) of Brazilian GDP (Gross Domestic Product), also a large number of people employed and close to 34,5% of total of domestic sales (Embrapa-gado de corte, 2001).

Although Brazil experiments presently an accelerated process of economic internationalization, the solely economic globalization tends to induce changes on every economic and political aspects of the country. In fact, it is quite difficult to isolate economic events to avoid the natural implications on political and social aspects. To sum up, all of this make necessary to think again about the participation of the federal government in such scenario without one defined array of rules.

It becomes important to point out that, under a type of "free trade" criteria, and in a global sense, one can list three main characteristics to be included as a part of an essential strategic agenda: Organization, representativity and power to negotiate.

2.2 Why to invest on technology?

2.3.1. Economic point of view

The leather and footwear sector is extremely important concerning Brazilian economy, because its volume of exportations and generation of jobs. The country presents the largest commercial herd of the world and also one of the largest groups of cattle breeders, slaughtering and tanning plants. According to information gathered from the Brazilian agriculture office, the industry generates close to 2% of GDP(Gross Domestic Product) and around of 800,000 direct and indirect jobs. (Gostinski, 1997)

Because the high rate of defects and the general lack of uniform batches, the Brazilian tanning plants (per year) embitter a loss of incomes around 320 million dollars. The state of Rio Grande do Sul, responsible for 40% of the total leather manufactured in the country, shows close to 128 million dollars of unrealized sales per year. Regarding the farmers and slaughtering plants, it represents a loss of incomes of 270 million dollars per year. In the same reasoning, the state government suffers a loss of taxes close to 12 million dollars per year.

Thus, it becomes clear that every investment made on technology development would probably bring significant results and increasing on profitability of the firms within the business network.

2.3.2. Manufacturing point of view

The scope of the research presented in this paper encompasses areas of equipments and technologies adopted by the tanning plants (figure 3).

The technology proposed within the "machine vision" needs to be incorporated as another equipment to classify properly (capable of being reproduced) and economically the pieces of leather and aggregating value to the entire process. This type of automation would generate improvements on quality and efficiency, without being dependent on foreign technologies. It could also settle the economic profile of the tanning plants because the optimization of raw material and labour.

3. THE AUTOMATED CLASSIFICATION SYSTEM

3.1. System design

The classification is based upon the determination and mapping of defects presented in the surface of the leather motivated by external agents. Nowadays, the classification is totally performed visually, being highly dependent on the operator's ability, therefore imprecise and subjective (Oscar, 1988). The proposed system divides the batches of leather according 8 standard levels of classification commonly used in the market.

The proposition aims to automate the process through the adoption of a computerized classification, more reliable and capable of being reproduced. The software "embedded" within the machine analyze digitally images obtained from a piece of cattle leather and store it in a data bank with other information about non-conformities or defects such as total area of defects, eccentricity, optical density, and defects per unit (piece of leather). Thus, the information stored would allow firms to compare the results obtained from each inspection against the quality standards accepted in the market.

In general terms, the machine gets the images from a piece of leather using one or more CCD (Charge-Coupled Device) cameras, digitally converted by a microcomputer (PC-like), using a video converter device. The proposed "machine vision" is a composite of one hardware set and another of control software (Roland, 1982) to perform the following functions (see figure 3):

a) Establish the real dimension of the piece to be analyzed
b) Establish the useful area to be used by subsequent processes
c) Track the defects and the influences over the piece under analysis
d) Calculate the total weight of the defects presented by the piece and determine its batch classification
e) Storage of the information gathered from the individual and batches analysis
f) Generate reports containing defects and non-conformities found in each piece or batch of leather

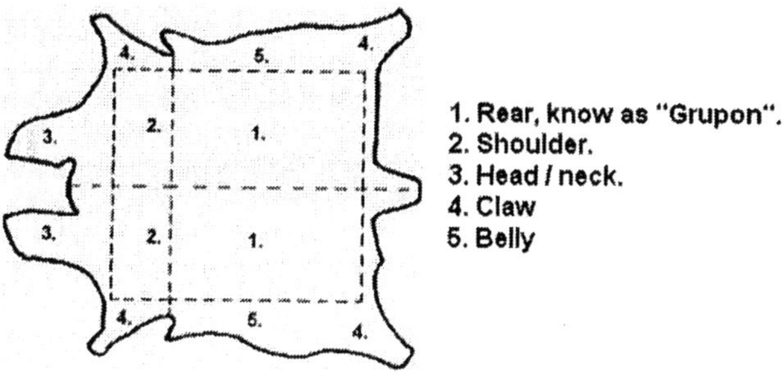

Figure 3 - Source: SENAI tanning school – 2° grau, april, 1988 (Oscar Jacó Schefel and Valmar Silveira dos Santos, quality control of *Wet Blue* and semi-finished leather).

3.2. Physical assembly

The physical application requires the use of a special table to receive the pieces to be surveyed (see figure 4) with a translucent lid to back-of illumination and with an attached metallic structure to install one fixed camera to be used in the analysis to get the quality of the "grupon"(Best part of one piece of leather) and other regions.

The data generated are then processed by the software and compared against a data bank, in such a way to understand the meaning of the data and related evaluation.

3.3. Measurement process

As already cited within the text, the implementation of an accessible and low cost automatic system to control the quality of leather still remains as an important objective to reach.

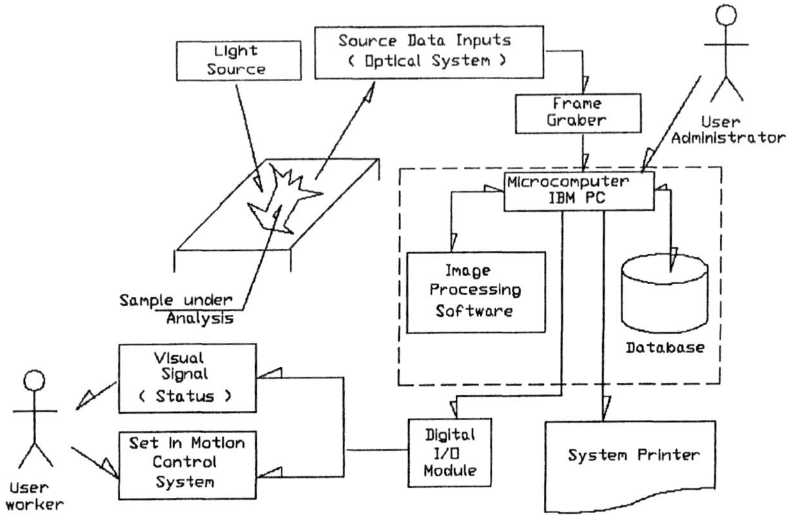

Figure 4 – Automate inspection process of quality leather

To sum up, the process comprises the following activities:

a) Capture images digitally generated by the cameras
b) Put to use one specialist system of visual computation to detect variations on leather characteristics in comparison with standards
c) Link the system above mentioned with the operator through the use of reports, warning or whatever considered more convenient

3.4. Benefits and costs

The proposed system would be able to eliminate or reduce the uncertainty level of the present hand-made classification process, aggregating value to the products being obtained and pushing the pricing negotiations into higher levels.

This system-based process allows more favorable conditions to conduct studies about leather quality, because the availability of data and information about the types and localization of defects found in each batch, species and region. In having plenty of information it becomes possible to implement special policies to decrease the problems of quality through the elimination of the reasons behind them.

4. A PRELIMINARY TEST

The proposed system has been tested in a leather tanning plant located in Franca, one of the Brazilian footwear cluster. The table 2 presented below shows the results obtained from one preliminary test, involving the inspection and classification of a sample with 20 pieces of leather. To do so, one experimental test has been done to classify the sample using two methods, e.g. one using visual classification and the other performed automatically by the system.

After the tanning process, the piece of leather is then divided in two longitudinal halves, from the neck to the tail of the animal. These parts received an identification tag (R-right; L-left side) represented in the third column of the table.

Table 2 – Results obtained for the experiment

Sample	Results	Orient.	Average	Sample	Results	Orient.	Average
1P4	2338,14	L	1902,10	3P6	3099,61	L	
2P4	1922,12	L		4P6	2905,82	L	
3P4	1584,03	R		1P7	4770,02	R	5485,41
4P4	1764,13	R		2P7	5864,31	R	
1P5	2394,97	L	2482,41	3P7	5613,64	L	
2P5	2673,06	L		4P7	5693,69	L	
3P5	2348,67	L		1PR	7901,23	R	7846,46
4P5	2512,97	L		2PR	7865,42	L	
1P6	2976,39	L	3092,76	3PR	7657,94	R	
2P6	3389,25	R		4PR	7961,27	R	

The first column represents the elements of the sample under test, while the second shows the results obtained to each element, through the application of factors pondering the types of defects, their location over the surface and the density of the piece. According the table, the results increase according a logarithmic curve, determining like bands of value to define each level for classification (grade). To check the methods against each other, the sample was divided in five groups according levels of grade well known previously. Thus, the sample presented 4 pieces grade 4, 4 pieces grade 5, 4 pieces grade 6, 4 pieces grade 7 and 4 pieces to be rejected.

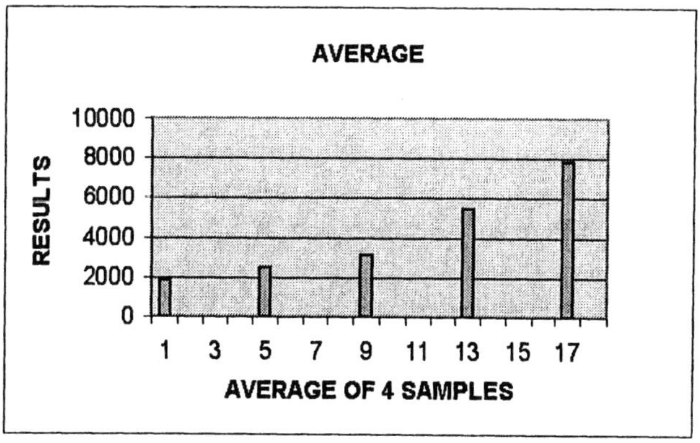

Figure 6 – Compatibility of the methods

After all, the last column represents the average value obtained for the testing results of each group of pieces.

The cycle time to classify and to measure each piece by 1 operator and 4 assistants has been taken close to 43 seconds. To do the same job, in the present stage of development, the automate system lasts around 53 seconds to classify and measure by using of electronic processing devices. However, it is important to take in account that the equipment used to do the experiment was one personal laptop without any special feature to enhance its performance. Thus, it is reasonable to say that using another computer system with high performance should reduce to an estimate value of 30 seconds, suitable to this type of activity.

5. CONCLUSION

As a conclusion from the reasons and the preliminary test presented in this paper, there are many related elements that could justify the adoption of one or more units of the "machine vision" to improve quality of *wet blue* leather produced by the tanning plants, aggregating value to the products been sold in the domestic and foreign markets.

Another important gain refers to the development itself and the knowledge of a new technology like that, adjusted to the economic capacity of Brazilian firms. There are many other possibilities available in the market to perform such a task, however the prices are still too much expensive, besides the fact that usually it generates a type of technologic dependence in terms of adopting one unique machine and its maintenance. Therefore, with the knowledge and range to specify the types of defects presented by the leather, besides one more detailed cost-benefit analysis and study to fulfill the needs to design and produce a "machine vision", it becomes possible to obtain a more uniform classification of leather being produced by tanning plants and increasing the quality level of the material to be used downright by other firms within the productive chains.

6. REFERENCES

1. Aalderink,B.J. and deJonge, M.W.C. (1976) *Automated Surface Inspection of cold-rolled sheet In Proc. International Meeting of Iron and Steel Making,* pp 11-19.
2. AVA (1985) *Machine Vision Glossary*, Automated Vision Association.
3. ABQTIC (1988) *Matéria-prima couro*. Estância Velha.
4. COUROBUSINESS (2000) *Para onde vai o couro brasileiro*. vol. 3, n. 13, p.20-21.
5. Embrapa – Gado de Corte-MS (2001) *Sistemas integrados de produção de peles e couros no Brasil.* Novembro.
6. Gostinski C. (1997) *Brasilian footwear 96/97*. English/Portuguese. Novo Amburgo: Catânia.
7. Norton-Wayne, L. and Hill, W.J. (1970) *The Automated Classification of Defects on Moving Surfaces Second International Joint Conference on Patt.* Recognition, pp 476-478.
8. VENDRAMETTO, O. ; Desenvolvimento e Ruptura: O caso da rede produtiva da carne, couro e calçados. In : FUSCO, J.P.A. (Organizador): Tópicos emergentes em engenharia de produção. Vol.01, São Paulo, Arte e Ciência Editora, 2002, ISBN 85-7473-091-2.
9. Oscar J. S.; dos Santos, V. S. (1988) *Controle de qualidade de couros "Wet Blue" e semi-acabados.* Escola de curtimento SENAI – 2° grau, abril.
10. Porter, M. (1993) *A vantagem competitiva das nações*. RJ: Editora Campus. Pp 479-497.
11. Revista Globo Rural (1987), junho.
12. Roland T. C.; Harlow, C. A. (1982) *Automated Visual Inspection*: A Survey IEEE Transactions on Pattern Analysis and Machine Intelligence, vol. PAMI-4, No 6, pp 557-573, November.
13. Saridis, G.N. and Brandin, D.M. (1979) *An Automatic Surface Inspection System for Flat Rolled Steel.* Automatica, vol. 15, pp 505-520.
14. Stapley, B.E. (1965) *Automatic Inspection of Metal Visible Defects.* Iron and Steel Engineer, november.

42

A SIMULATION BASED RESEARCH OF ALTERNATIVE ORGANIZATIONAL STRUCTURES IN SEWING UNIT OF A TEXTILE FACTORY

Halil Ibrahim Koruca[1]
Ceren Koyuncuoglu[2]
Gultekin Silahsor[3]
Gultekin Ozdemir[4]

[1]*Süleyman Demirel Uni., Dep. of Mech. Eng., 32260 Isparta/Turkey, koruca@mmf.sdu.edu.tr*
[2]*Dokuz Eylul Uni., Vocational School, Dep. of Technical Programs, 35160 Izmir/TURKEY, ceren.koyuncuoglu@deu.edu.tr*
[3]*Funika Textile A.S., Gumusler,20100 Denizli/TURKEY, gsilahsor@funika.com.tr*
[4]*Süleyman Demirel Uni., Dep. of Ind. Eng., 32260 Isparta/TURKEY, gultekin@mmf.sdu.edu.tr*

In textile industry, productivity, flexibility, and quality must be improved and maintained to meet the challenges of an increasingly competitive word market. Simulation has been widely used in the investigation and evaluation of organizational structures and alternatives. In textile industry, meeting the demands of customers or market changes, to make cutting right and standard, to have cutting time short and perfect, the use of new technologies and transformation into progressing organizational structures get importance for productivity. The processing time of right cutting directly affects the cycle time of the end product. In this study, the research of current organizational structure and the use of alternative organizational structures on meeting customer demands on cutting unit of a textile company are made by using ARENA® Simulation Software. Selecting the most productive organizational structure by simulation based approach to evaluate parameters is the purpose of this study and the results are evaluated.

1. INTRODUCTION

In textile manufacturing, Turkey started to seek the ways of Branding for its products to be able to compete with China and Far-East countries having low labor cost. The most important factor that affects the Branding is re-structuring for products of high and standard quality and customer satisfaction. With the increasing quality and productivity understanding in production unit, the importance of right and perfect cutting besides quick operation timing is quite clear.

Simulation modeling and analysis have become a popular technique for analyzing the effects of changes mentioned above without actual implementation of new technologies or assignment of resources. Many manufacturing systems can be easily and adequately analyzed with discrete event simulation models (Banks et. all, 2001). Simulation has long been accepted as an effective approach to design and analyze production structures. In many manufacturing organizations, increasing process flexibility is becoming more important while the reliance on product cost to measure manufacturing performance is being lessened.

Due to restructuring of production, the role of new technologies in a company has increasingly changed over the past several years. Quick change in fashion with the higher and standard quality induced by the use of new technology which is one of the most important factors of international competition in textile and Ready-To Wear (RTW) industry, forces producers to make structural changes in their manufacturing systems. To be able to use the quick change in fashion as a competitive advantage companies must meet the demands in short time by quick production and this obligation brings out the need to employ new technology and organizational change.

This paper considers and explores design of organizational structure and alternatives in a sewing unit. These actions should not simply be related to "new technology and sewing automation", as the achievements which might possibly ensue are mainly obtained by carefully merging the information flow into material flow in such a way that to affect each other when external inputs (commands or disturbances) apply (Acacia et. all, 2003).

A new technology offers many advantages over many advertences over confessional machines, such as higher productivity, more consistence in quality, higher precision, and reduced set-up times

The firms aim at competitiveness, with focus on styling and on critical processing tasks, while the work-intensive phases (e.g. sewing) are eventually decentralized where operators' wage is smaller. Is this an effective set-up? (possibly) yes, on conditions that:

- The market accepts the full amount of ready made suits or dresses, delivered by (large-enough) season's batches (to optimize the productivity on tactical horizons),
- The flexibility is included by 'quick-response' techniques, so that extra items are managed on-process, to personalize size or details (as case arises, on the operation horizons).

These two conditions are consistent with simple rules, such as:

- To aim at work-plans *leanness*, with visibility on cost build-up and quality transfer,
- To focus on the core business and to remove 'intangibles', which make the business with 'little' benefit.

Leanness entails decisions, based on benchmarks, with purport on management tasks (to distinguish administrative or bureaucratic requests) and on technical issues (to plan out product and process innovation). Preliminary step for effectiveness is the setting of performance ranks, at the strategic, tactical and operation levels of the manufacturing engagements, to exploit flexibility through a properly sophisticated govern framework (Acacia et. all, 2003).

2. AIMS OF ORGANIZATIONAL AND TECHNOLOGICAL CHALLENGES

Implementation of high technology production system during the 1980s and 1990s has helped -to same extend- in sharpening the competitive edge of companies by reducing manufacturing lead times and increasing flexibility. However, more potential for improvement lies hidden in the departmental structure of an industrial organization (Zülch et. all, 2001).

The successful implementation of a cutter system is dependent on arriving at a satisfactory solution to interrelated engineering, management, organizational, and human resources issues. Often lower than expected productivity gains were achieved due to a lack of consideration to human aspects in the design, operation, and maintenance of computer-automated technology.

New ways of managing, planning, and implementing cutter systems are needed in sewing unit to deal with rapid development of new technologies such as vision, off-line programming, and system integration, which permit a wide range of applications. For the successful implementation of cutter systems, the connection between product design and manufacturing process design must be understood. Advance planning for the implementation of cutter systems should be made with the objective in mind that it is not merely planning for new technology and equipment, it is planning for human beings.

In the presented analysis herein, particular thought was given to organizational structures because of the large scope for development of the new technologies and workers and the interdependence between disposition strategies and efficiency of the logistic system.

The use of advanced manufacturing technology, new organizational structures and new strategies of human resource management had a significant impact on the demand of labor and the labor market. For those qualified workers, which were underused in the production structures, new opportunities of more demanding work opened. They tried to find jobs which matched with their skills and qualifications. However such changes could not always be managed without any problems; it often took some time, before they could find a better job. So even highly qualified workers became unemployed; this kind of unemployment can be called "search unemployment".

This manufacturing technology will have a profound impact on the ability of the system to react to market needs and the perceptions of customers regarding the processes that should be undertaken, the time scales that the business should operate on and the cost involved. There will also be people issues arising from a technological perspective that sees a requirement to introduce new technology and information management systems. The introduction of new information management technology may also have a profound impact on the organizational structure of the business as communication patterns alter (Bradford and Childe, 2002).

There may be further impacts on the organization structure as decision making moves between people and traditional authority and accountability structures no longer reflect the practice of the business (Bradford and Childe, 2002).

Organizational change is a well established discipline that specializes in the analysis of the business organization, their strengths, their weaknesses and the optimal methods for getting from one state to the other (Bradford and Childe, 2002).

3. RESEARCH METHODOLOGY

Simulation is the main tool used in this study. For the design and control of production systems, simulation has proven to be a powerful tool. However, investigations of the market situation have shown that many companies are not willing to use simulation as a permanent planning tool (Schmittbetz 1998, Zülch et. all, 2002).

Analyzing production systems in a static as well as a dynamic way should support the analysis and design process. This is usually done by applying simulation tolls. For a successful re-engineering process, the production system has to be studied from a different point of the view. For the correct selection of modeling aspects, the global objectives of the company must be studied in detail (Zülch et. all, 2002). During this study it is verified whether the focus should be on information technology or the business processes. Reorganizing the company usually leads first to a business process-oriented approach (Scheer, 1994).

ARENA 7.0 simulation software was used for model constructions and analysis in this study. As the first step, a base model was developed which depicted a system without process variation. Model verification and validation was done by structured walkthroughs of model logic, extensive use of execution traces and by reasonableness of the animation.

4 EXPERIMENTAL ANALYSIS AND RESULTS

4.1 Case-1: Initial Situation of Sewing Unit

There are three things a company must do to compete effectively. They are:

- To provide an efficient well automated manufacturing system which will give the business a distinct advantage over competitors; To focus on the core business and to remove 'intangibles', which make the business with 'little' benefit.
- To determine the order-winning criteria (OWC),
- To control the process in such a way that the product meets the order winning criteria and maximizes profit (Hörte and Ylinenpää, 1997).

By considering these, the company wants to obtain the compatibility goals. Therefore, the company wants to improve the current cutting unit by employing a cutter system or by making some design changes for the minimization of the cycle time of the products, but also the cost is another dominant factor.

The cutting unit can be described as traditionally organized manufacturing system with a function-oriented departmental structure. Figure 1 shows the initial layout of the cutting unit. The company currently employs 22 workers in the cutting room. There are three identical manual spreading machines, which are operated by one worker for each. After the spreading, the fabric is cut according to the product model and numbered for the production control and part pursuit. After the numbering process, related parts are sent to marking table for the marking of the fabric for the designation of the sewing places. After all these operations, control for the cutting errors of the parts is done by two workers and if the parts are satisfied, two workers package them and sent to the sewing unit after the binding operation. Also there are four workers to assist the operations in spreading, cutting, numbering, marking and controlling stations, to prevent the bottleneck occurrence. The company works in one shift for five working days of the week.

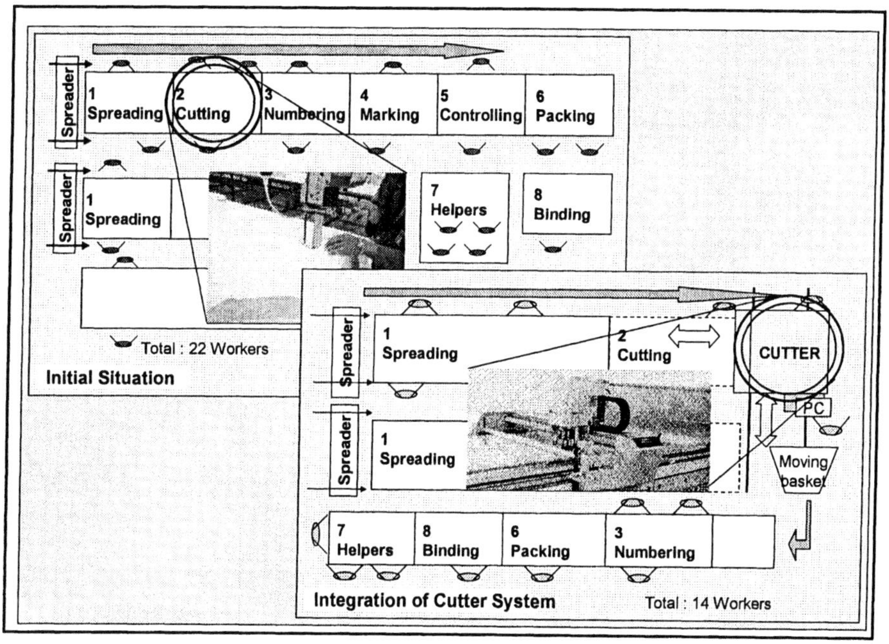

Figure 1 – The current situation of the cutting unit and after the integration of the Cutter System

Since salaries of the employees in textile industry are rather low in Turkey, turnover rate of workers is very high. Because of the learning period for every new worker, new personnel start learning by assisting the existing staff generally. Lying, numbering, marking and packing activities can be fulfilled by new workers, who are capable of doing these activities. In the following alternative models, it is assumed that the qualifications of the workers have been increased as a result of the reduction in the turnover rate of the workers.

Changing the Qualifications for Initial Situation

Variant A: In the first alternative model, it is assumed that the assisting workers are qualified to perform binding activities, so that they can be allocated to those jobs if needed.
Variant B: In this alternative model, it is assumed that the turnover rate of the workers has been reduced; hence workers in the cutting unit can perform all the activities. For instance the workers, who pack the goods, can also control the quality after the cutting activity. But the labor cost per hour will be increased.

4.2. Case-2: Integration of Cutter System

Variant C: It is assumed that the new technology has been integrated to the cutting unit. Based on the assumption that the demand has not been changed, new organizational structures and personnel allocations have been included to the new simulation model. Some activities, such as controlling the quality, are no longer necessary after integrating the new cutter to the unit. Thus, production speed will increase and the number of workers required in the unit will be reduced.

In the new system, 1 qualified worker, who can use the new cutter, 3 semi-qualified workers for lying activities, 2 workers to help the cutting activity, 2 workers for numbering, 2 workers for packing, 3 unqualified helpers and 1 worker for binding activity will be required. Hence, a total of 14 people will be allocated to the new cutting unit.

Variant D: In the last alternative model, the cost affect of working in two shifts in stead of one is analyzed. The labor cost is doubled in the second shift, but doubling the working hours increases the utilization of all the labors and the machine as the resources of the cutting unit.

4.3 Simulation Results

Technological change that renders existing specialized knowledge obsolete unless knowledge of new technologies can be easily transferred to lower-level managers, decisions involving choices of technology will be centralized. The simulation run length is 300 hours and 18 hours period of time is taken as the warm up period for basis, Variants A, B, and C. The simulation run length is 600 hours for Variant D where shift system is simulated.

The simulation results are summarized in Figure 2 for all models comparing to the initial basis. It seems that the results of Variant B where qualification of workers are increased are better than results of other organization models. In Variant B, the production time has been reduced meanwhile production rate and utilization are increased. On the other hand, the unit cost increases because of higher wage of qualified workers.

In Variant C where new technology cutter is used, it is a noticeable result that the unit cost and utilization are in very low levels. In this organization structure, it is possible to have higher utilization and increased production rate by using qualified workers.

In variant D where shift system production has been applied, the unit cost increases as a result of higher wave of workers of night shifts. Therefore, the rate of wages in the unit cost of a product is considered significant. Whereas, the production rate is not as much as high it is expected. The simulation results show that accumulation of products in the production site is a result of the bottlenecks in some work stations. The possible solution for this problem can be using qualified and productive workers. The production times must be balanced in work stations to obtain a continuous flow in the system. Some line balancing studies are suggested on the system for this purpose.

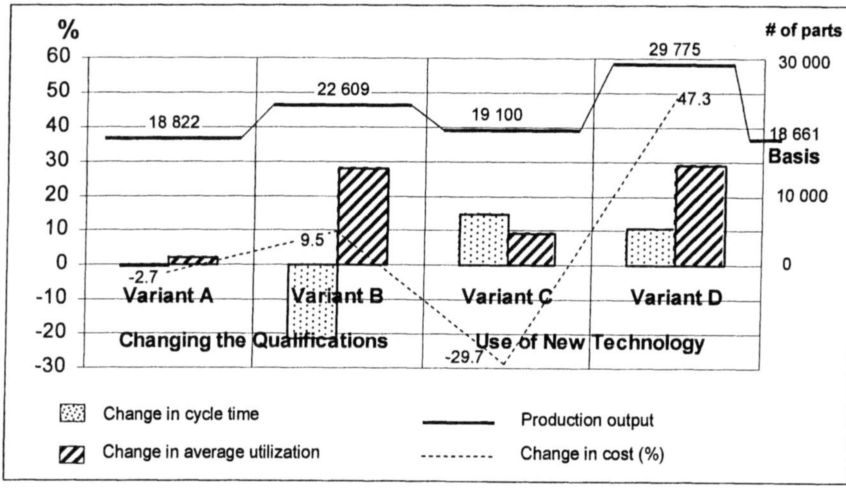

Figure 2 – Simulation results of the studied cases compared to the initial situation

The main issue is to have same qualified worker to work on same work stations for a long time. The location of the company has an important effect on the circulation of workers. The company is located in a city which is one of the popular cities in textile industry in Turkey. Therefore, it is not easy to keep qualified workers in the company for a long time.

5. SUMMARY AND CONCLUSIONS

The cutter system is usually introduced to increase industrial productivity by reducing manufacturing costs, increasing production output or capacity and improving product quality. The introduction of cutter systems can create a reduction in the number of workers, changes in skills and transfers from work locations. The workers who have to move to a less skilled job will fell the effects of cutters systems most directly.

For maximum benefits from the implementation of a new technology such as cutter systems, an understanding of the personnel qualification problems involved is necessary.

For automated cutter systems, there are some differences in the personnel qualifications of prior to cutting preparation flows relative to traditional cutting systems. Therefore, qualified workers are needed for preparation of designed cutting models and to move them to the cutter.

Consequently, the productivity per worker was significantly increased trough the integration of cutting and also less material handling and opportunity for damaging products. There were few options for improving the productivity further through layout changes.

From a people perspective, there was a real scope for improving morale and job satisfaction through cross-training. This could also improve productivity through flexibility.

Acknowledgments

The Authors gratefully acknowledge the financial support of Suleyman Demirel University Research Fund (Project No: 03M-711). Authors would like to thanks Mr. Ali Talebi and management as well as the staff for allowing the authors to access all information used and for contribution to the project. This work was undertaken in Funitex, in Denizli, Turkey.

6. REFERENCES

1. Banks, J., Carson, J. S., Nelson, II B. L., and Nicol, D.M., Discrete-Event System Simulation. New Jersey: Prentice-Hall. Inc., 2001.
2. Accacia, G.M, Conte, M., Maina, D., and Nichelini, R.C., Computer simulation aids for the intelligent manufacture of quality clothing. Computer in Industry 2003; 50: 71-84.
3. Zülch, G. Jonsson, U., and Fischer, J., Hierarchical simulation of complex production systems by coupling of models. International Journal of Production Economics 2002; 77: 39-51.
4. Bradford, J. and Childe, S.J., A non-linear redesign methodology for manufacturing systems in SMEs. Computer in Industry 2002; 49: 9-23.
5. Schmittbetz, M., Simulation wird zum Triebwerk für Innovation. VDI-Nachrichten 1998; 24(52): 18.
6. Scheer, A.-W., Business Process Reengineering, Berlin: Springer, 1994; 5.
7. Hörte, S.-A and Ylinenpää, H., The firm's and its customers' views on order-winning criteria, International Journal of Operations & Production Management 1997; 10: 1006-1019.
8. Porcaro, D. Simulation modeling and design of experiments. Industrial Engineering Solutions 1996; 28(9): 28-30.

43 MODELLING AND SIMULATION OF HUMAN-CENTRED ASSEMBLY SYSTEMS - A REAL CASE STUDY

Anna M. Lassila, Sameh M. Saad* and Terrence Perera
School of Engineering, Sheffield Hallam University, UK
**email: s.saad@shu.ac.uk*

Tomasz Koch and Jaroslaw Chrobot
Institute of Production Engineering and Automation,
Wroclaw University of Technology, POLAND

This paper reports the experiences gained through the use of computer simulation in the modelling of a human-centred assembly line in an automotive manufacturing company. The main themes of the paper include the problem of achieving sufficient accuracy in the description of human operations, the difficulties encountered in data collection, and the modelling of human-centred operations. The importance of an accurate representation of human behaviour for the validity of the developed simulation model in a human-centred system is discussed. Finally, the simulation results and some recommendations for system improvement are presented.

1. INTRODUCTION

In today's highly competitive environment, companies are under constant pressure to improve their production processes. However, correct actions and decisions require accurate information about the system performance. Since the behaviour of many complex systems changes over a period of time, a technique such as computer simulation is necessary for building valid models and generating accurate performance data.

Simulation is a widely used tool for building models of real or proposed systems in order to evaluate their performance in dynamic conditions. The output data of the simulation can be used to identify system bottlenecks and to generate alternative states that may provide the desired performance improvements for the system. The major advantage of simulation is that it enables experiments with the system without disturbing the operations of the real process (Kelton et al., 2004).

Traditionally, system modelling and simulation has concentrated on the technological aspects of the systems (Ehrhardt, 1994) whereas the workers have simply been modelled as resources that perform simple and clearly defined tasks with time dependent availability and varying efficiency (Baines and Kay, 2002). While this approach is sufficient for highly mechanised processes (Ehrhardt, 1994), it can seriously distort the capacity predictions of systems with high human work content (Baines, 2003). Since the competitiveness of many manufacturing companies still depend on the flexibility and responsiveness of humans (Baines and Kay, 2002), an increasing number of researchers have in recent years attempted to develop methods for a more accurate modelling of human behaviour, e.g. Ehrhardt et al. (1994), Cacciabue (1998), Schmidt (2000), Baines and Kay (2002), Baines et al. (2003) and Brailsford and Schmidt (2003). Throught (2004) argued against this trend stating that the non-deterministic behaviour of people is a result of individual characters, opinions, needs and requirements that cannot be modelled probabilistically without a high level of simplification.

In this paper, the problem of accurate modelling of assembly systems with high human work content is highlighted in a case study of a human-centred assembly line in an automotive manufacturing company. The problems encountered during the development of the simulation model using discrete event simulation software Arena are reported. These range from data collection to model validation. According to Jayaraman and Gunal (1997) discrete event simulation is now a standard tool in the automotive industry.

2. LITERATURE REVIEW

Over the last two decades, fully automated factories that would not have to consider the cost or variation problems caused by human involvement have been the ultimate goal of many companies (Braun et al, 1996). However, demand for skilled workers has increased simultaneously with growing automation levels (Arai et al., 1997). This was caused by the increasing complexity and customisation of products, shorter product life cycles and low and unpredictable demand, creating production conditions that can best be dealt with by flexible workers. Even flexible, technologically advanced manufacturing systems relay heavily on the skills of workers (Hitomi, 1996).

Consequently, in recent years academic interest in human factors and their influence on manufacturing system performance has intensified. Sociological insights, for instance on the role of the conditions in which people work (e.g. Bonnes and Secciaroli, 1995) and the characteristics and behaviour of individuals (e.g. Furnham, 1992), are often found to be relevant to manufacturing system development. Hence, current manufacturing research activity is concerned with how to properly include these factors into the system development process e.g. system modelling and simulation.

One of the first detailed methodologies for including human factors in simulation models was proposed by Ehrhardt et al. (1994). Ehrhardt considered strain and stress factors of different manual tasks and their influence on human performance. The significance of human behaviour on manufacturing system performance was emphasised in a study by Fan and Gassmann (1995). They noted that the performance of a manufacturing cell depended greatly on the person working on the cell regardless of the automation level of the cell. Cacciabue (1996) emphasised the importance of a good working environment analysis and of the collection of appropriate information for successful modelling of man-machine interactions.

Later, Schmidt (2000) argued that even though human behaviour is very complex, it can, within limits, be modelled and thus can become deterministic. He proposed a general reference model for modelling human behaviour in any application. His model was applied by Brailsford and Schmidt (2003) to a health care system. Baines and Kay (2002) proposed a methodology for modelling highly simplified relationships between workers, their environment and their subsequent performance. Baines et al. (2003), on the other hand, connected an external human performance model with a discrete event simulation tool in order to study the influence it can have to the performance of a system.

3. CASE STUDY FROM AUTOMOTIVE INDUSTRY

The complexity of trying to model human work with sufficient accuracy was encountered during a simulation and development project of a human-centred assembly line in an automotive manufacturing company. The experiments carried out during the project are shortly described in this section.

3.1 System description

The automotive manufacturing plant, located in Eastern-Europe, has two very similar manual un-phased assembly lines for producing two of their products (A and B), which are produced in more than 300 different versions. The assembly lines, presented in Figure 1, operate independently and differ from each other only by stations 13 and 14, which are present on line 2 only. Consequently, line 1 is unable to process one version of product A. This study mainly examines line 2.

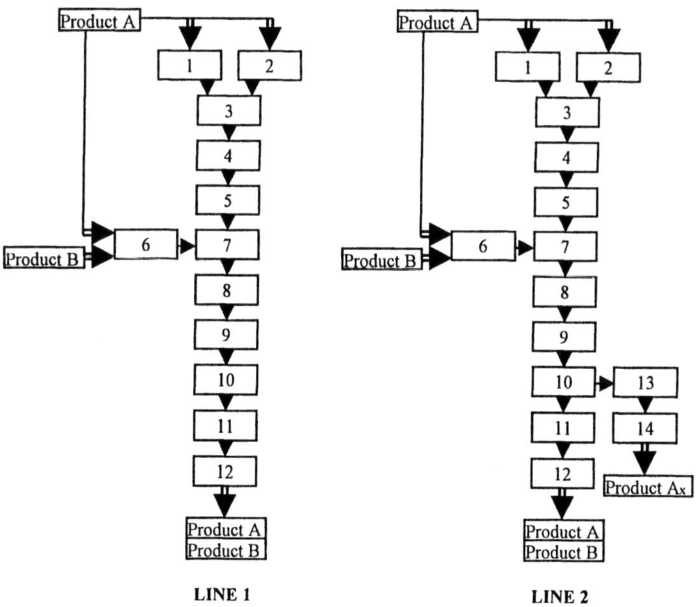

Figure 1 - Assembly lines

Normally, the assembly of products A require work on all stations on the line, except on stations 13 and 14. Those resources are only used by a special version of product A. Products B, on the other hand, use only half of the line starting from station 6 and finishing at station 12. The products are processed in batches according to the orders. Any batch size is possible, starting from only one product up to hundreds. Machines need to be reset every time the product version changes. The length of the changeover depends on the similarity of two consecutive products.

Each station has one operator who performs the assigned tasks either fully manually or with the use of simple machines at some stations. A major part of the work on the line is performed manually as can be seen in Figure 2, which illustrates the estimated manual work content at each station.

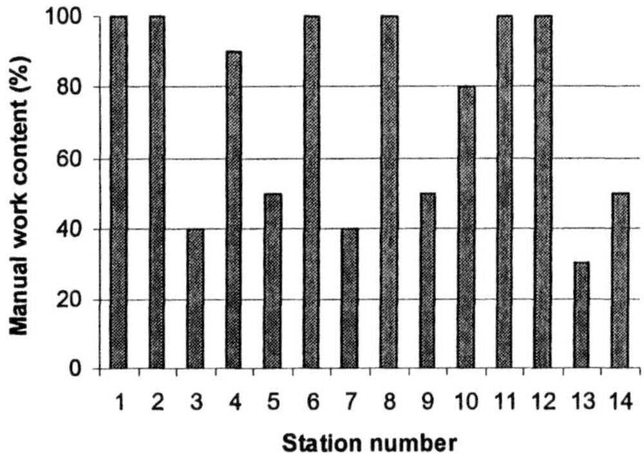

Figure 2 – Estimated manual work content of the stations on assembly line 2

Currently, the workload is not fully balanced between the stations and the processing of a part on consecutive stations may require different times. For this reason, small buffers have been built between stations to hold 5 to 10 products. The flow is improved by encouraging operators to move between the stations to prevent starving or blocking on the line.

3.2 Data collection and process observations

In the past the company had collected very little data about the line performance, and no real measurements of the system had been conducted. However it was estimated that the current line output rate is one product in every 80+ seconds, while the line had been designed for a 44 seconds cycle time. Typically, the system produces around 330 A products or 350 B products in one eight hour shift on one line. The estimated average uptime of a line is 85% and the amount of products requiring rework is around 2%. The setup time between products can reach anything up to 15 minutes.

Since the existing data was not sufficient for the simulation model development, the requirements for additional data were recognised and the measurements were performed on the line. Since the number of product variations assembled on the system exceeded 300 and many of them had very similar processing requirements, ten different products were selected to represent all products processed on the line. The measurements captured the processing, transfer, setup and rework times for each product at each station.

During these measurements the general functioning of the line was observed. In overall, the line appeared very unorganised. Big containers of materials were waiting to be used, most of the buffers were full with unfinished products, people were continuously moving between the stations, regularly a group of people gathered around a machine that was not functioning correctly, and while some operators were working very hard others were idle while waiting for work to be completed somewhere else on the line. In addition, there was an apparent lack of balance in the workload of the stations and standardised work practices. Since most of the work on the line was performed manually by the operators, a large number of people were constantly present and moving around in a relatively small area as encouraged by the management.

The collected data revealed a high variation at least on one station in all time measurements. No particular station was consistently recognised as having a higher variation than others for all products. Figure 3 illustrates the processing time variation for one of the measured products. It indicates that the variation at most stations are just within acceptable limits, however the mode values still vary strongly among the stations. We suspect that this variation was to some degree caused by differently skilled and motivated operators.

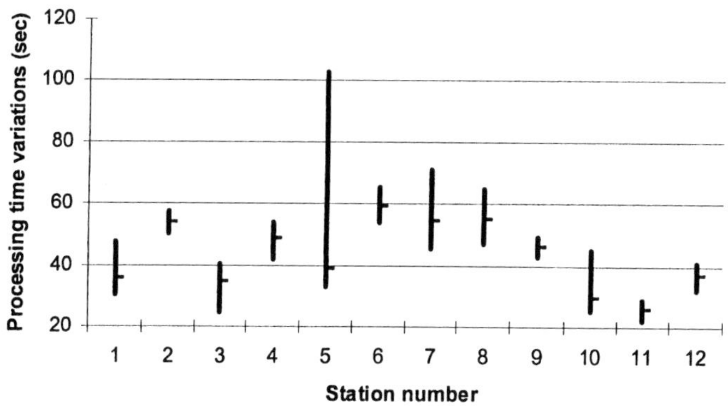

Figure 3 - Processing time variations (min, max, and mode) on assembly line 2 for one product type

3.3 Modelling human-centred processes

The main question encountered during model development was how to model these human-centred processes. The decision was made to represent each operator as a process with limited capacity and scheduled availability. The processing times were included as triangular distributions with the mode calculated over the sample space of the normal values, where normal values were defined as values that regularly occurred during the measurements.

In addition, a separate random breakdown operator was created to model the temporary unavailability of operators caused by off-station work tasks or not work related activities. These activities had a significant influence on the performance of the system as they regularly disturbed the normal functioning of the line by causing starvation and blockage between stations. The effects of running the simulation model with and without the random breakdown operator are illustrated in the next section.

3.4 Model validity and simulation results

The validity of the developed simulation model was evaluated by comparing the performance of the model to the real system. Three separate tests were carried out, the results of which are presented in Table 1. The output and cycle time values obtained from the simulation model were found to be very similar to the estimated values of the real system, differing at most 10.9%. Therefore the tests are suitable for system analysis and experimentation.

Table 1 - Simulation model validation test results

Test no.	Description	No. of setups	Actual system		Simulation model		% difference	
			Output	Cycle time (sec)	Output	Cycle time (sec)	in output	in cycle time
1	Batch of A's	1	~330	~80	345	80	4.5	0
2	Batch of B's	1	~350	~80	372	74	6.3	8.1
3	Production plan (a mix of A's and B's)	3	~310	~80	311	88.7	0.3	10.9

A lead time analysis using the developed simulation model was performed for the same test cases previously used in the validation test. In addition, a fourth test was run using the production plan of test case 3 but removing the random breakdown operators. The results are presented in Table 2.

Table 2 - Lead time analysis using the developed simulation model

Test no.	Average lead time (min)	Average processing time (min)	Maximum transfer time (min)	WAITING (min)
1	39	8.6	1	**29.4**
2	26.7	4.5	1	**21.2**
3	33.4	6.6	1	**25.8**
4	28.5	6.6	1	**20.9**

The results indicate that the temporary unavailability of operators and the machine downtimes together increase the average lead time by around 5 minutes. However, even without the breakdowns waiting in the queues still constitute a major part of the lead-time as can be seen from Table 2. While the average actual processing time for a product during a mixed production is only 6.6 minutes, the average lead-time is over 28 minutes. Therefore, on average, a product spends more than twenty minutes waiting in the buffers between the stations, accounting for more than 70% of the total lead-time. The set-ups that occurred during the last two tests increased the lead-times by only a few seconds.

The long waiting times and persistent high utilisation of the buffers were revealed during the system simulation. Since all buffers had limited capacities, they regularly blocked the upstream processes causing chaos on the line. Figure 4 shows that for products A the buffers at the beginning of the line were full for almost all the time, while the B products filled the output buffers of stations 6 and 7. The average utilisation of resources was around 62%. The utilisation of the resources was clearly higher at the middle of the line. Blocking prevented utilisation to be increased.

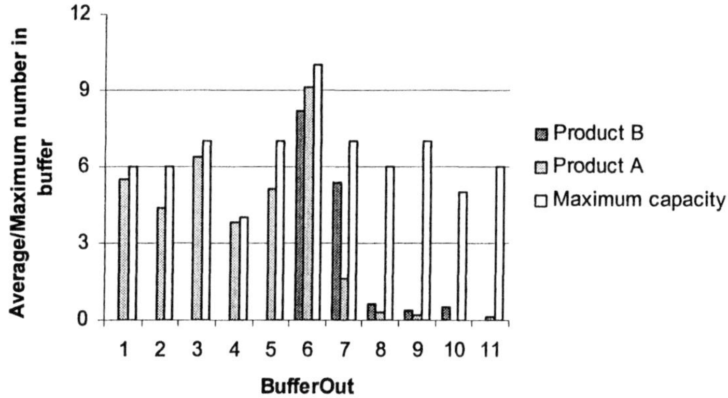

Figure 4 - Average number of products in the buffers compared to the maximum capacities of the buffers

The results suggest that the high variation in the processing times caused by a general lack of organisation and standardisation concerning manual activities resulted in a highly unstable system. The high content of manual work on the assembly line increased the system variation, because the work of the operators was not time-sequenced by the machine times and did not follow defined procedures.

The buffers around these human-centred stations increased the processing time variation even further, since they stopped the continuous flow of the parts and broke the direct connection between consecutive stations. As a consequence, the time-sequencing force of the preceding and following stations was lost. The stations became isolated islands without a common rhythm, and each station was working for its own buffer rather than for the next station. Gradually, the whole line slowed down and the product lead times increased.

To improve the system, first the actual work content of each station needs to be clarified, standardised and measured so that the work load can be rebalanced among the stations in order to minimise variation. Next, the work practices including the operator movements between stations and the system layout should be critically reviewed. In the current chaotic line conditions, the recommendation from system simulation to add a machine to station 7 would not address the actual problem.

4. CONCLUSIONS

This paper has reported on the experiences gained during the simulation model development in an automotive manufacturing company. The paper focused on the human modelling issues from data collection to defining sufficient accuracy to operators. Data collection and simulation revealed a chaotic system with high time variation in every processing stage. The immediate actions for line performance improvement should address organisational practices rather than capacity increases. In order to reduce system idle time and improve utilisation a full line balancing study needs to be carried out. However, at this stage, two types of scheduling scenarios, backward scheduling before the bottleneck station 7 and forward scheduling beyond this station, has been recommended.

The chaotic conditions on the assembly line were caused by the lack of proper planning and control. Appropriate control measures are especially important in systems with high human work content to manage the additional variation human operators create. The rather simplistic modelling approach of humans used in this study provided accurate data as the system itself was chaotic. However, if the product flows were controlled better, the individual characteristics of the humans would become more influential on system performance and a more detailed representation of operator characteristics may be required.

5. REFERENCES

1. Arai E, Shirase K, Wakamatsu H, Murakami Y, Takata M, Uchiyama N. "Role of production simulation for human oriented production systems" In the Proceedings of 14th International conference on Production Research. 1997: 758-761.
2. Baines TS, Kay JM. "Human performance modelling as an aid in the process of manufacturing system design: a pilot study" In the International Journal of Production Research. 2002; 40 (10): 2321-2334.
3. Baines T, Mason S, Siebers PO, Ladbrook J. "Humans: the missing link in manufacturing simulation?" In the Simulation Modelling Practice and Theory, 2003.
4. Bonnes M, Secciaroli G. Environmental psychology: a psycho-social introduction. Sage Publications, 1995.
5. Brailsford S, Schmidt B. "Towards incorporating human behaviour in models of health care systems: An approach using discrete event simulation" In the European Journal of Operational Research. 2003; 150: 19-31.
6. Braun WJ, Rebollar R, Schiller EF. "Computer aided planning and design of manual assembly systems" In the International Journal of Production Research. 1996; 34 (8): 2317-2333.
7. Cacciabue PC. "Understanding and modelling man-machine interactions" In the Nuclear Engineering and Design. 1996; 165: 351-358.
8. Cacciabue PC. Modelling and Simulation of Human Behaviour in System Control. Springer-Verlag Berling and Heidelberg GmbH&Co, 1998.
9. Ehrhardt I, Herper H, Gebhardt H. "Modelling strain of manual work in manufacturing systems" In the Proceedings of the 1994 Winter Simulation Conference. Orlando, Florida. 1994: 1044-1049.
10. Fan IS, Gassmann R. "Study of the practicalities of human centred implementation in a British manufacturing company" In the Computer Integrated Manufacturing Systems. 1995; 8 (2): 151-154.
11. Furnham A. Personality at work: the role of individual differences in the workplace. Routledge, 1992.
12. Hitomi K. "Manufacturing excellence for 21st century production" In the Technovatio. 1996; 16 (1): 33-41.
13. Jayaraman A, Gunal AK. "Application of discrete event simulation in the design of automotive powertrain manufacturing systems" In the Proceedings of the 1997 Winter Simulation Conference. Atlanta, Georgia, USA. 1997: 758-764.
14. Kelton WD, Sadowski RP, Sturrock DT. Simulation with Arena. 3rd Ed. New York: McGraw-Hill Companies, 2004.
15. Schmidt B. "The Modelling of Human Behaviour" Online in http://www.or.uni-passau.de/english/2/Human_Factors.pdf, 2000.
16. Trought B. "Why people are a manufacturing problem" To appear in the Proceedings of the 2nd International conference on Manufacturing Research. Sheffield, UK. September 2004.

44

VERTICAL INTEGRATION ON INDUSTRIAL EXAMPLES

Andreas Dedinak, Christian Wögerer, Helmut Haslinger, Peter Hadinger
ARC Seibersdorf research GmbH
andreas.dedinak@arcs.ac.at, christian.woegerer@arcs.ac.at
helmut.haslinger@arcs.ac.at, peter.hadinger@arcs.ac.at
AUSTRIA

"Vertical integration" is an often used headline in the closed communication from office to machine level The paper will give an overview of state-of-the-art, problems as well as opportunities of concept. Apart from a more general problem description, the paper also reports on results exemplarily in 3 industrial projects achieved by ARC Seibersdorf Research GmbH. The first example covers industrial measurement and testing automation - the most modern test stand for heat meter calibration set up for the PTB laboratories in Berlin. The second one shows the development of a full-automated casting plant for magnesium implementing the safety system. The last one is a high-speed coin sorting system for more than 1000 coin types consisting of control, visualisation, image processing and database.

1. INTRODUCTION

Integration of the main business levels of a production or process orientated enterprise reduces costs, leads to higher flexibility - ending up in prompt time to market and increased quality. As this vertical integration will play a major role in the automation business within the upcoming years, this paper will give an overview over this topic.

First one addresses the issue of the motivation for a structured networking of different business levels. The stimulus is the existing situation which nowadays often leads to problems in not meeting the right delivery date, not meeting the according quality, not being competitive, not being flexible, etc. These problems often arise in historical grown companies, where the individual business levels build their own islands, their own solutions for their own problems with lacking interprocess communication. This situation makes production enlargement difficult, implementation of new technologies expensive, inhibits establishment of standards and makes processes and data difficult to be understood. Moreover, planned and actual status are hard to compare.

1.1 Upcoming Business Trends

The above described lack of in-house business networking is further increased by todays and tomorrows demands on modern business units:

- Product life cycles shortened, even technological cycles shortened
- Orders to stock replaced by short orders to delivery
- Shift form in-house production to integration of sub suppliers
- Decentralised stock and service
- High pricing pressure
- Shift form manual process integration to integrated processes
- Shift from product supplier to system supplier
- Production processes control multiple enterprises
- E-commerce and quality management systems additionally produce enormous quantities of data and require the introduction of a data management throughout the company

These demands can only be met by the horizontal and vertical integration of business units within the company, and exceeding the network to suppliers and even customers.

1.2 The Business Pyramid

When speaking of vertical integration of business levels, the main levels in a production or process-orientated enterprise should be defined:

- Enterprise Planning Level: Uppermost managing level, covers the enterprise management as well finance, logistic and human resources.
- Factory Management Level: covers production planning, production analysis, production optimisation and quality control.
- Production Management Level: covers production control and monitoring, production data concentration, production cell automation, SCADA systems, etc.
- Process Level: covers process control including sensors and actors for signal acquisition and control execution.

In near future the limits between the enterprise levels will become blurred (vertical integration) and all steps within a process chain will be linked (horizontal integration). As a final result, productivity will grow.

2. COMMUNICATION AS THE KEY FACTOR

Analyzing the necessary methods for integrating business islands vertically and horizontally reveals communication as the key factor. The data and communication pyramid has to represent the enterprise pyramid. A sweeping communication network from the sensor level to the managing level is evident, although the demands of such networks differ in individual levels. The requirements on field (production area) related communication networks (busses) are:

- availability very high, as production breakdowns are possible
- changes ins system configuration seldom
- distances possibly high (>100m)
- EMC pollution often high
- mechanical stress possibly high

- temperature stress possibly high
- installation often by unskilled workers
- operation and service by process orientated personnel

In contrast, the request on office related communication networks (busses) are:

- availability medium, as "only" loss of working hours is possible
- changes ins system configuration frequently
- distances small (>100m)
- EMC pollution modest
- mechanical stress modest
- temperature stress modest (often air conditioned)
- installation by skilled personnel
- operation and service by network specialists

We will divide the communication methods in IT related technologies, in Data related technologies and in Field related technologies. Typical representatives of the IT technology are Ethernet, Industrial Ethernet and Wireless LAN (Local Area Network). Typical representatives of the Data related technology are Profibus, Interbus, Foundation Fieldbus, Modbus, Devicenet, etc. Typical representatives of the Field technology are ASI (Actor Sensor Interface bus), CAN (Controller Area Network), EIB (European Installation Bus), LON (Local Operating Network), etc. These networks often serve in their typical application environment and are linked together by converters or bridges connecting the individual levels of communication.

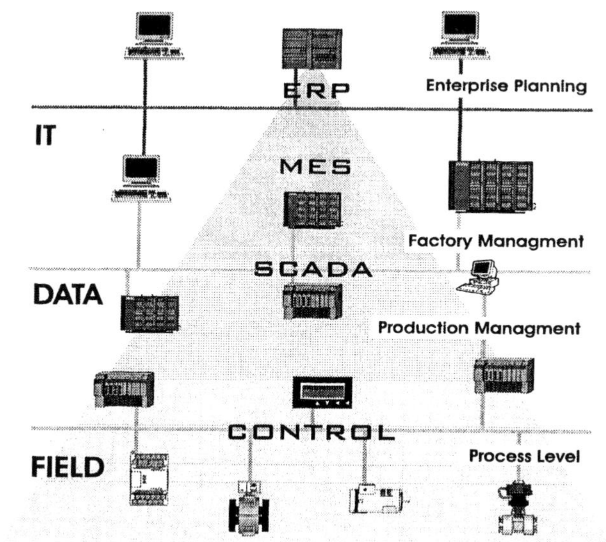

Figure 1 - The communication pyramid corresponds to the enterprise pyramid

This network of networks (busses) is frequently called the communication pyramid (Figure 1), because the members in the lowest communication hierarchy are numerical the largest group (sensors and actors), but have the smallest data rates.

The members in the highest communication hierarchy are numerical the smallest group (servers), but have the highest data rates.

Nevertheless Ethernet and similar technologies penetrate in direction of field level. This results from a rising amount of data in the sensors and actors, as well as from decentralised operation of these devices.

Besides the hardware side of the communication network, the software technology is at least as important. The most important components in non-proprietary communication between software modules of different suppliers in production systems were established under the leadership of Microsoft Corp. and are based on their COM / DCOM ((Distributed) Component Object Model) system. Based on this object model with a clear defined access to the data and property interface of a binary software module, OPC (OLE for Process Control) established a standard in communication of automation systems in an unpredictable speed. All major suppliers of various automation products today support this standard.

Although OPC DA (Data Access) is the world wide standard, OPC further extends this standard to a variety of areas as batch processes (OPC Batch), historical data access (OPC HAD), data exchange between OPC servers (OPC DX) and last, but not least, OPC XML. With XML in general there will be a remedy for the big disadvantages of OPC – the inability to communicate to systems beyond the borders of the Microsoft platform. It remains to be seen whether OPC XML will overcome these ultimate barriers to an unlimited communication between automation products.

3. MANUFACTURING EXECUTION SYSTEMS

As workshop and enterprise level based automation software has been developed since decades without large-scale integration (also developed and operated by people of different educational backgrounds), the modern MES (Manufacturing Execution Systems) software seems to be turning to the missing link capable of bridging the gap of the above mentioned business levels. So far, MES is a loose, but powerful collection of software modules for tasks like production dispatching, resource management, production tracking, maintenance management, production execution, etc. Implementing the modules adapted to the respective business process should result in following benefits:

- Right data at the place of interest
- Real time data at the right moment
- Consistency in production databases
- Higher level of transparency
- Higher level of reliability of decisions
- Observance of schedules
- Observance of quality rules
- Optimised Processes

All these benefits should sum up to:

- INCREASED QUALITY
- SHORTER RESPONSE TIMES
- DECREASED COSTS

Nowadays software producer work on establishing clearly structured frameworks to increase efficiency and usability of these MES modules. Nevertheless today and in the near future these frameworks and their subsystems will work only in a proprietary manner; as yet standards for these MES frameworks are not implemented.

Combining today's opportunities of e – business with the potential of a fully networked production enterprise, the vision of a global market capable of overcoming physical distances and geographic locations by establishing enterprise structures and cooperation's as demanded by products or projects, is no longer a question of technical feasibility.

4. INDUSTRIAL PROJECTS

ARC Seibersdorf research has long time experience in mechanical and electrical automation of production and testing systems. Combining these skills with the knowledge of industrial networking, databases and interfaces for enterprise planning systems the ARC Seibersdorf research GmbH deals with the problems of integrating the workshop level to the management level since several years. Three exemplary projects will be described in short:

4.1 Fully Automated Test Stand for Heat Meters with Integration of the Complete Test Administration

ARC Seibersdorf research was put in charge to build on of the largest and most modern test stand for heat meters in Berlin. Customer was the well-known PTB (Physikalisch Technische Bundesanstalt).

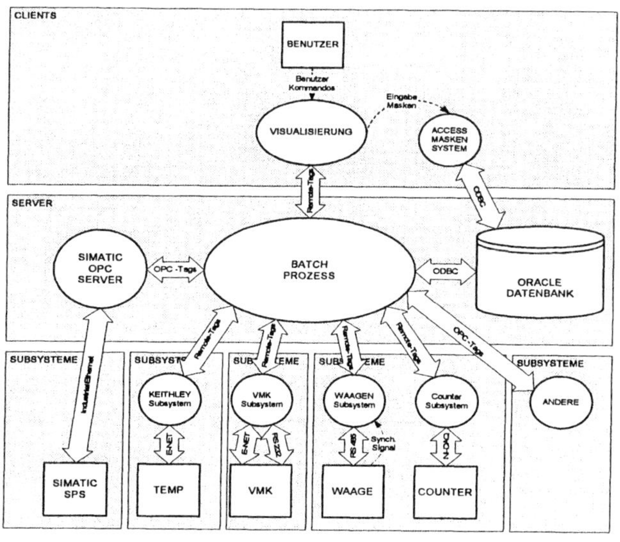

Figure 2 - Systems communication pyramid

In addition to build up the mechanic and measurement system challenging the frontiers of today's realizability, the test stand should be embedded in an overall test administration system. This starts up with the management of customers, test orders, devices under test and ends up in the management for test reports and quality management for the test stand with its subsystems itself (Figure 2). The essential step was the integration of the test stand management database (Oracle or MS SQL server) with its user front-end in the production database, organising the test routines itself and is responsible for the storage of real time trend data from the test process.

In these databases not only the test and test management data is stored, but also the complete setup and control parameters for the control system as well as the calibration data of the test stands gauges.

For exact repeatability of tests, all setup and control parameters of the complete facility can be reloaded to the subsystems automatically by calling a certain date or test order from the database. By additionally generating the necessary documents for quality control automatically, the customer not only owns one of the most modern test stands in measuring technology, but as in terms of data management.

4.2 Control Technology for a continuous casting Plant

Because of the familiar behaviour of magnesium alloys (strong exothermal reaction, explosion hazard) it was necessary to develop the whole processing chain starting from melting, alloying, casting, heat treatment and finally to extrusion of Mg profiles to provide a safe technology for the operators in Mg – direct chill foundries.

One key factor was the automation-, simulation- and control process. For a full automation it was necessary to adapt an existing casting plant, to develop a new launder system from the furnace to the die, new sensors (level, visualization) and to create a control - software witch is also able to control the plant and to log continuously various parameters of the casting process within maintenance via Internet.

4.2.1 Control System

During the casting process of magnesium the prevention of the oxidation reaction is the first goal. According to the potential explosive reaction of magnesium with the cooling water, no staff is allowed near the casting equipment during the process. At the same time it must be warranted that caster and material experts are able to optimize the process online at anytime.

The automation solution (Figure 3) bases on a high performance industrial controller (PLC) and a process visualisation and data management system (HMI - Figure 4). The PLC controls the process parameters, all safety relevant limit values and the communication with the bus linked periphery equipment. The casting plant, the inert gas plant, the melting furnace and the casting plant for massels are individually controlled by an own PLC connected to the central controller. The periphery equipment of the casting plant for massels is connected via profibus with its controller.

4.2.2. Visualisation

Industrial Ethernet networks the HMI system to the control system. In this manner the casting process can be controlled from almost any distance and from any place of the world via Internet. Conditional on the integrated database it is furthermore possible to register process- and automation parameters online to save them for following analysis. This is basis for a further optimisation of the process. Naturally all data can be observed online by customer specified images, evaluated by various statistics and combined to trends or reports. The HMI system is able to allocate parameters by open interfaces like OPC to external evaluation programs.

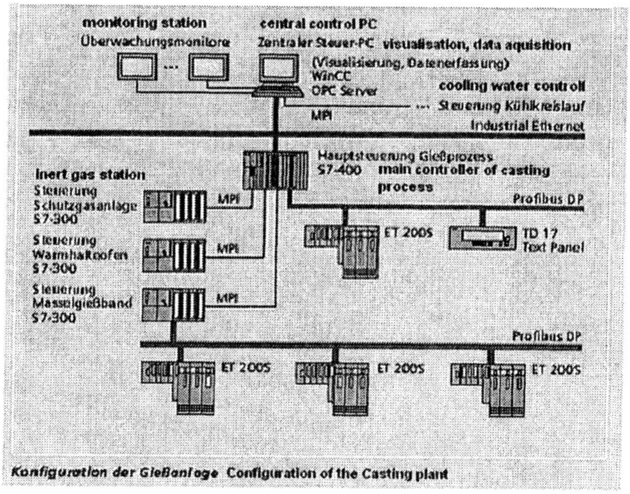

Figure 3 - Control Configuration of the Casting plant.

Besides a recipe administration is integrated. With it is possible to administrate the process parameters of the control of the casting process for various alloys and products easily.

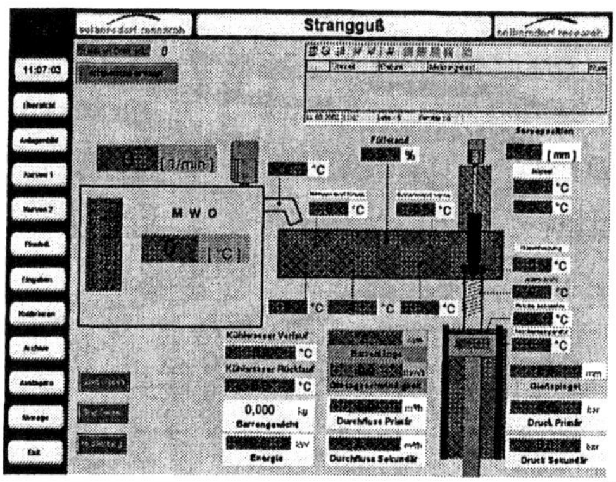

Figure 4 - Visualisation (main panel)

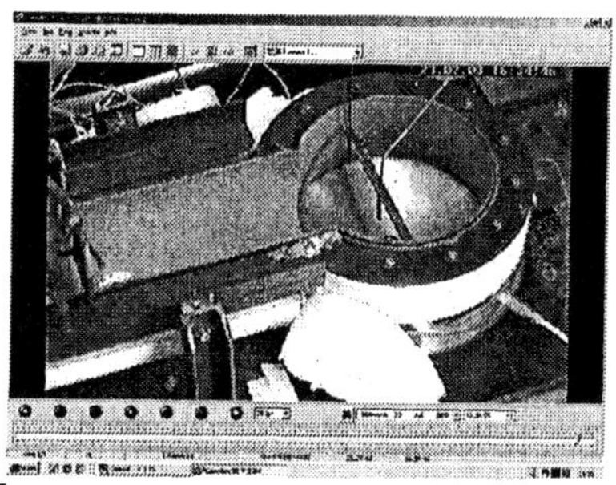

Figure 5 - Live pictures

For a live visualization of the process the control station is equipped with Video monitors and 5 Cameras. They send live pictures directly from the casting pit. These picture are also shown in the HMI system and used for online control and visual analysis (Figure 5).

4.2.3. Telemaintenance via the Internet

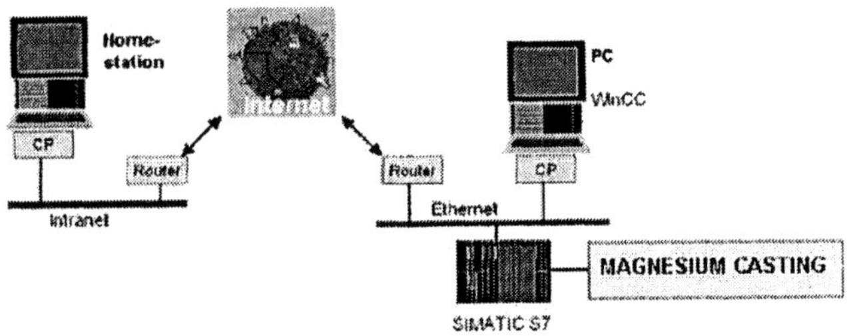

Figure 6 - Telemaintenance via Internet (schematic).

To increase the capacity and improve quality, it's advantageous to connect the magnum-casting machine with the producer. This connection was realisized by Internet (Figure 6). This concept allow to check the status of PLC, report status and alarms, to debug machine diagnostics, to do modifications and enables agreements with customers, to check together material-related machining adjustments and optimisation of production-related parameters from our home station. With this Internet interface it is possible to check these features from any place of the world, providing an Internet connection. Assuming you have the right access privileges to remote control the plant. All drives, sensors, controls and monitoring units are linked to the PC over Ethernet connection and hence these devices can be operated via telematic functional requirements. The communication mechanisms are based on event-driven, high-throughput transfer, resulting in excellent web performance.

4.3 Control station of a high speed coin sorting machine

For sorting more than 1000 different types of coins with nearly 100 percent recognition, a unique high-speed coin sorting machine called "Dagobert" was developed by ARC Seibersdorf research GmbH. Based on image recognition, coins in different mintings from approximately 100 countries had to be classified and sorted with a speed of up to 10 coins per second.

4.3.1 Control System

The high sorting frequency (10 coins per second) and the time consuming calculations that need to be performed for recognizing each coin require a control system with well-defined communication channels. The control station coordinates all sub-systems. Figure 7 shows the control scheme of the system. In order to maximize system security in service, all relevant data are stored redundantly both in the control station and in the database. Thus sorting data are not lost even if the software crashes.

Via a simple and clearly arranged control terminal, the plant can be controlled conveniently. Any error conditions (synchronization and communication errors, fill level overflow, sensor defects, blockages...) are recognized automatically and – if possible – repaired automatically by special error handling routines. If automatic repair is not feasible, the plant automatically changes to fail-safe operation and allows the user to handle the errors without risking loss of any data. Via the graphical user interface, all sorting parameters (box allocation, allocation to the respective benefiting charities, sorting speed, ...) can be adjusted, and all sorting results are monitored.

4.3.2 Visualisation

Via the visualisation at the control station the state of each subsystem can be controlled and monitored. The sorting results are displayed in real-time at the user interface (Figure 8).

The coins delivered to be sorted and counted have been collected and are therefore owned by different charity organizations. For this reason, the sorting results have to be assigned to the respective organization. This assignment is likewise done at the user interface.

4.3.3 Database

In order to provide the highest possible data security and to realize real time behaviour, all sorting data of the current batch are stored in the hardware of the control. An Access database is used as front-end for performing data analysis. This database is updated currently with the real time data of the control. Here, the data of all trained coins and their assignments can be monitored and managed.

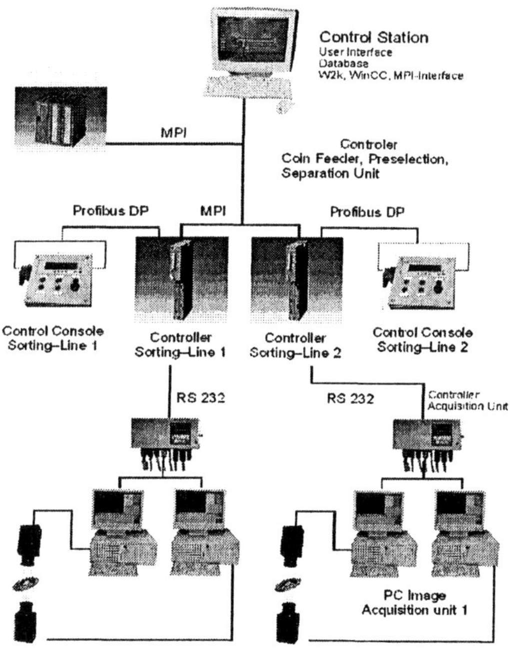

Figure 7 - Control station.

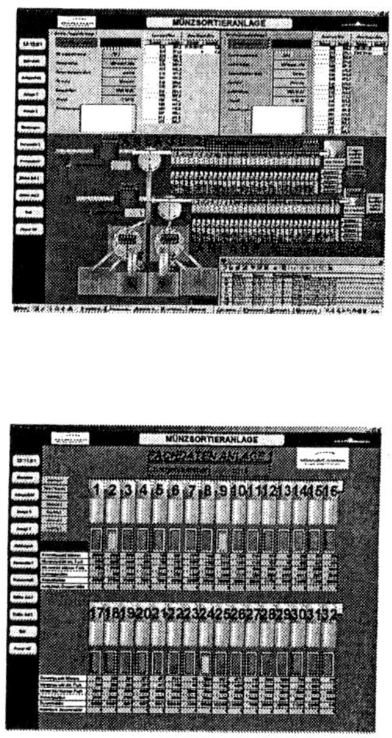

Figure 8 - User Interface

Additionally, a large amount of further information is managed (current exchange rates, processing and delivery state, exchange progress, management of coin containers, ...). A large number of user-definable reports are provided by the database to analyze the entire volume of the collected sorting data (Figure 9).

All data records can be filtered and/or grouped by several parameters (charity organization, processing batch, box number, coin type, currency, ...). At the touch of a button, delivery notes for the respective exchange office can be created.

Likewise, we can calculate estimates for future results can be calculated any time, based on previously obtained results. This statistical report yields information that is important for organizing future coin deliveries.

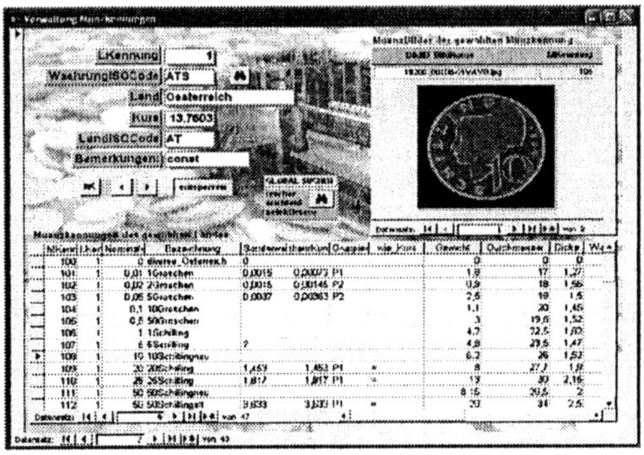

Figure 9 - Coin Administration Tool

5. CONCLUSION

Vertical integration of production systems stands for an automated continuous dataflow throughout all enterprise levels. The enterprise network joins enterprise management and production. It optimises resources and productivity, and assures transparent figures in factory management. It controls logistic and production processes, and assures transparent information in production management. It integrates technologies and data, and generates production data in production process.

On the other hand, increasing pressure to enhance productivity, cost-efficiency, competitiveness, and time to market will further drive the demand for ever more sophisticated integration methods and tools.

6. REFERENCES

1. Arbeitskreis Systemaspekte des ZVEI Fachverbandes AUTOMATION, "Die Prozessleittechnik im Spannungsfeld neuer Standards und Technologien" (German), atp Journal 43, pp. 53-60, 2001.
2. OPC Task Force, "OPC Overview", OPC Foundation 1998.
3. "Inside Look at OPC, XML, NET", Start journal, Volume 5, Number 9, 2001.
4. K.C. Laudon, C.G. Traver, "E-Commerce – Business, Technology, Society", Addison-Wesley, 2001.
5. Siemens AG, "Information Security in Industrial Communication", White Paper, 2003.
6. Siemens AG, "Distributed Automation on the Basis of Industrial Ethernet", White Paper, 2000.
7. Dedinak a., Wögerer Ch., "Automatisierung von Großprüfanlagen am Beispiel eines Wärmezählerprüfstandes für die PTB", (German), White Paper, ARC Seibersdorf research, 2002.
8. Dedinak A., Kronreif G., Wögerer Ch.: "Vertical Integration of Production Systems" IEEE international Conference on "Industrial Technology ICIT'03", Maribor, Dezember 2003.
9. Dedinak A., Koetterl S., Wögerer Ch., Haslinger H.:"Integrated vertical software solutions for industrial used manufacturing and testing systems for research and development", Advanced Manufacturing Technologies – 2004 AMT 2004, London, Ontario, Canada.

45 DECISION SUPPORT WHEN CONFIGURING AUTOMATIC SYSTEMS

Magnus Sjöberg
Ph. D. Student, Dept. Of Production Engineering, Royal Institute of Technology, magnus.sjoberg@iip.kth.se, SWEDEN

Automatic systems are used to varying extent within the manufacturing industry. The challenge is to find the most advantageous applications of automation to the manufacturing system over time. Enquiries concerning automation appear when configuring or re-configuring the manufacturing system. The objective of this paper is to describe and to evaluate existing methods that can be used as decision support when deciding on automation. A participating study was conducted within an automation project. The work procedures within the industry are often based on experience and not on systematic methods. Outgoing from these methods and industrial experience a frame work for a new method is suggested. Issues that are critical to a useful and applicable decision method are pinpointed and discussed.

1. INTRODUCTION

The automation issues are a subset of the configuring of manufacturing/assembly systems, and depending on to what extent, the issues are more or less complex. Configuring or re-configuring a manufacturing system is a many-sided undertaking. The configuring process involves many decisions and engineering tasks to be carried out. Often the tasks are coupled and entail multidisciplinary problems. Depending on the product and the volumes the automation solution might be more or less given. It is said, when investing in technique, that there are only three questions that need to be answered:

- What does the technique do?
- How much does it cost?
- What is the reliability in the answer of the first two questions?

These questions are very unspecific and arbitrary and to be able to answer the questions they must be divided into more precise questions. The first question, *what does the technique do?* can be divided into: What are the abilities and capabilities? What other techniques can be used? Manual or hybrid solutions? How does it affect the system according to system parameters such as cost, quality, delivery and flexibility. The second question, *how much does it cost?* can be divided into: Short- or long term costs/earnings? Intangible/tangible costs/earnings? The third question, *what is the reliability in the answer of the first two questions?* can be divided into: Questions concerning empirical data? Routine? Simulation? Still these questions are not easily answered. The use of methods and decision supports, enable a systematic way to determine the problems. Decision support tools is in this paper a collective term for all concepts with the intention to facilitate the work procedure, when making decision on automation. In this paper the objectives have been to collect and review representative decision supports that are available for engineers concerned with these issues, to suggest a framework for a new method, and to discuss critical issues within this area. No distinction between parts manufacturing and assembly has been made. To a certain level the issues concerned can be seen as common for both assembly and parts manufacturing. This paper is based on a litterateur survey and on a participating study. The ambition has been to cover the different decision support tools, categorise and to analyse them outgoing from user preferences. The literature survey contains sub chapters where different representative decision support tools are described and discussed. The participating study took place at a company, a major producer of robots. The project was conducted within their own manufacturing. The purpose of the project was to enable automatic assembly, for a sub assembly system. Their work procedure was examined. The study serves as start of collecting empirical data of industrial use of decision supports.

2. LITERATURE SURVEY

The survey covers representative samples of different decision support tools. Approximately 100 papers were considered and 17 papers were sorted out and further analysed[i]. The different support tools are divided into the categories: Methodologies, methods, check lists and thumb rules. Further they are described and analysed. Some of the support tools refer to a system solution and others to specific process solutions. Methodologies should be seen as a scientific and systematic work procedure, methods are often included in methodologies. Methods are more specific in their tasks and are often represented by a model, a selection schema or a logically structured diagram. Check lists and thumb rules are, as their name implies, more vague and arbitrary. Automatic systems and automated solution within manufacturing and assembly have huge variations and there is no uniform categorisation and nomenclature. In this paper a division between system applications and single process applications is made. User preferences are in this paper defined as the value of using the decision support. What comes out from using the method; how much effort is needed to get relevant answers, what is the accuracy of the answer?

2.1 Design methodologies

The methodologies analysed in this paper are not of the same character as for example reaserch methodologies, but more of methods with substantial context.

There are some different manufacturing system design methodologies, and parts of them treat the aspects of deciding what process technique to use. The first

---

[i] The papers were collected from the databases: Emerald, Science Direct and ISI.

methodology discussed, Figure 1 (Rao and Gu, 1997) is presented as a manufacturing system design methodology. The methodology is a top down approach where the first step is requirements of the manufacturing system design and the last steps are evaluation and reconfiguration. The steps; selection and design of machines and design of manufacturing system configuration, briefly declare what issues are to be determined, but not how they should be managed and accomplished. Abdel-Malek et al (2000), describes similar system design methodologies, but with different focus, for instance on flexibility or simulation.

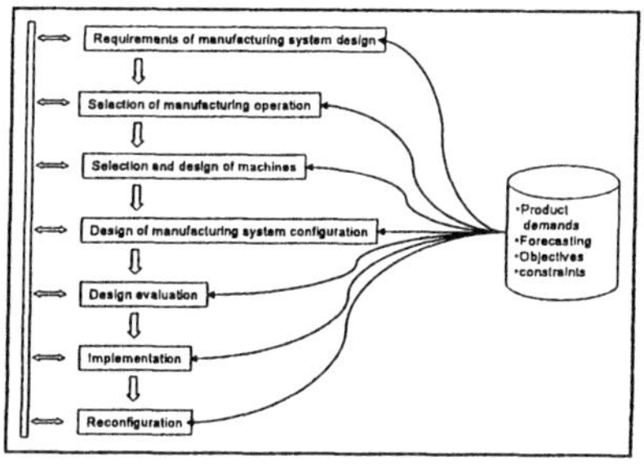

Figure 1. Manufacturing system design methodology (Rao and Gu, 1997)

The methodology described by Borenstein et al (1999) is concerned with selection and comparing between different system configurations and alternatives, Figure 2. The principles which this methodology is based on are:

- Strategy based analysis
- Systemic analysis
- User centred analysis
- Interdisciplinary analysis

As in the previous methodology described, this is a top down approach. The methodology describes all steps in detail and the issues concerned. The above principles from which the methodology is developed, give a strategic perspective instead of a strictly financial which is very common (Burcher, Lee, 2000). Simulation is a requirement to facilitate use of the methodology, and is suggested in other similar methodologies (Pflughoeft et al 1996). There has been considerable research within the area of FMS decisions. Many researchers have considered the issues and decisions about flexibility and the often large investment required to implement these systems.

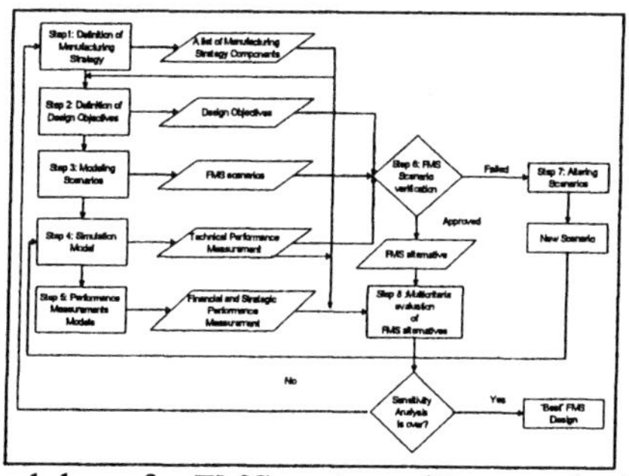

Figure 2. Design methodology for FMS-systems (Borenstein, Becker, Santos, 1999)

Some methodologies do not primarily focus on automation issues, but is more to be seen as the context in which the decision on automation is taking place. The methodologies provide a holistic perspective and a project approach for which they are useful. The focus is mainly on system solutions. However none of the methodologies that have been examined have dealt with the primary issue, whether to automate or not. Often that decision is assumed and approved. It is difficult to evaluate from a user perspective. Methodologies like the above described serve as important project guidance.

2.2 Methods

Methods focus on more specific tasks. The one described below (Boubkri and Nagaraj, 1993) delivers answers to what kind of automation to use, in terms of dedicated or programmable. The factors on which the schema is based are:

1. Annual number of end products
2. Number of variants of the product
3. Life cycle of the product
4. Number of parts in the product

The annual number of products is important when deciding on assembly techniques. Robotic systems play an important roll in some volume ranges. In other there are no economic competitors to manual assembly.

Dedicated automatic assembly emerge where there are large volumes and few variants or single variant production. Thus there is a large span where dedicated systems do not fit the wanted solution. As the number of models increase the demands can not be met by dedicated assembly systems. It is in this range that flexible systems are most feasible. Life cycle of the product also affects the system requirements.

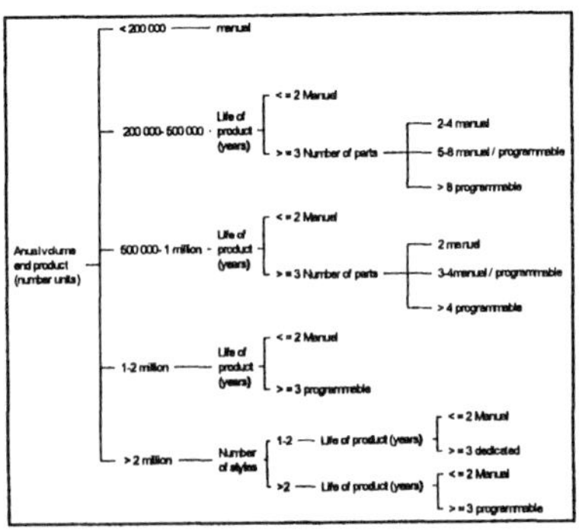

Figure 3. Selection schema (Boubkri and Nagaraj, 1993)

The sum of key indicators; annual number of end products, number of models of the product, life cycle of the product, and number of parts in the product are essential when deciding on what attributes that are important in this selection schema.

This selection method might be useful to give an indication, but it lacks some important aspects. It is not dynamic, and does not consider the changes of technique and cost. Modularisation of the system i.e. system flexibility, is not considered. Due to this the accuracy of methods like this, is changing as function of time. The assembly systems are getting more technically advanced i.e. are capably of managing more tasks, faster and with a higher quality. Further this schema is based on economic assumptions such as short payback times, and lacks the aspects of strategic thinking. The focus is on system solutions.

2.3 Check lists and thumb rules

Checklist and thumb rules are, as their name implies, arbitrary and can be seen as complements to methodologies and methods. What they do is also to pinpoint critical issues but they do not provide any technical solutions. Managers who make automation decisions must consider the following factors (Norman and Frazier,1999):

1. Economic factors
2. Effects on market share
3. Effects on product quality
4. Effects on manufacturing flexibility
5. Effects on labour relations
6. The amount of time required for implementation
7. Effects of automation implementation on ongoing production
8. Amount of capital required

When making decisions concerning in which processes to invest, companies need to satisfy both technical and business perspectives. According to Hill (1995) the following issues have to be considered when configuring the manufacturing system.

1. Decide on how much to buy from outside the company, which in turn, determines the make-in task.
2. Identify the appropriate engineering-technology alternatives to complete the task embodied in each product. This will concern bringing together the make-in components with the bought out items to produce the final product specification at agreed levels of quality.
3. Choose between alternative manufacturing approaches to complete the task embodied in providing the products involved. This will need to reflect the market in which the product competes and the volumes associated with those sales.

Technology alternatives and manufacturing alternatives include the selection of automation technique. The guidance is that it should correlate to the business objectives.

These checklists serve as input to discussions and further analysis. The first list mentioned does not consider the solution as a strategic resource. Without a strategic perspective the issues do often end up in how to evaluate non monetary values. In such cases there can be a risk to overestimate values that easily can be transformed into payoff calculations etc.

Thumb rules that are common when reasoning on automation are:

- If a machine/robot replaces two employees, then it might be economically justifiable
- Does the investment have a payback time shorter than one year, then it's acceptable

This argumentation and thinking are used within industry. The reasoning in terms of replacing employees with machines is essential when deciding on automation, though this thumb rules do exclude many aspects and render a short term economic perspective. Strategic benefits are not considered. Unfortunately these examples are common argumentation within the industry. However they do reflect the environment where the industry exists.

3. PARTICIPATING STUDY

The purpose of the project was to enable automatic assembly, for a sub assembly cell. The work procedure of developing the cell was examined, and serves as start of collecting empirical data, of the industrial use of decision supports.
Conclusions so far is that in this specific case, some decision supports for automation were used and they sort under the category thumb rules. The work procedure was following a project agenda. The project was very well defined, i.e. objective, time and cost were clearly defined. However, much of the work of designing the cell and convert system requirements to cell abilities was based on skilled and experienced personnel. Further the study indicates the need for systematic work procedures such as decision methods. However more studies must be conducted to give relevant input of specific industry requirements on such methods.

4. SUGGESTED FRAMEWORK FOR A NEW DECISION METHOD

A decision method that aims at facilitating decision on automation involves three main areas of different kind:

1. The strategic area where the manufacturing/assembly system is seen as a facilitator of the business strategy.
2. Technical aspects of automation- and manufacturing/assembly system practice.
3. Decision making which involves managerial and communication issues.

The strategic area is of outmost importance since investments in automation technology often end up in justification reasoning. The company must see their production system as an enabler of their business strategy (Skinner, 1969).
The second area: The technology is under constant change and development. Therefore the method must be able to handle these dynamic changes. Strategic and successful production systems depend on many factors. The decision method must

correlate to other production system design issues. The hierarchy, and whether the decision on automation is subordinated to other design issues or not, depends on the situation and context. This must also be considered. Often the third area is neglected, though it is not of less importance. The engineering / production development staff (PDS) must be able to communicate and get acceptance of his / their suggestions. That is both internal engineer-PDS, and external PDS-managerial level. A basic condition for high-quality solutions within the manufacturing/assembly system is knowledge and acceptance at all concerned levels within the company. These are the corner stones on which the method should be based upon. To be able to communicate and generalize the method there must be a structured and well defined taxonomy and categorisation of automation systems. An approach where the strategic requirements and the system capabilities and abilities are mapped is one way to cover that issue.

5. CRITICAL ISSUES

Automation affects many system parameters and the causalities are hard to detect. This tends to a myopic reasoning and exclusion of important aspects. Aspects that affect and are hard to include in common methods are:

1. Life cycle of the product
2. Life cycle of the production system
3. Flexibility (dynamic capability)
4. Non economic and intangible effects
5. Short time economic results dominates the company structure and their way of acting
6. Long time planning is less accurate than short time planning

The concern within industry often is that automation and the implication from the technology involved is no issue, as long as there is a sufficient pay back of the investment. What is considered as sufficient pay back time differs, depending of the situation and company. Thus they might exclude strategic benefits. Strategic benefits are concerned with increased risks. Solutions on automation require a strategic perspective and the industry lacks strategic and systematic work procedures (Axelson et al, 2004). Manufacturing strategy definition, strategy links to competitive priorities (Garvin, 1993) and automation decisions are areas where it is hard to find substantial support. Automation is a wide term. Classification and nomenclature are issues mentioned in this paper. This is an issue concerned with generalisation and appliance of methods. Taxonomy (Bourgeois, et al, 2002), the focus is on assembly and consequently it does not cover other areas.

6. CONCLUSIONS

The methodologies and the category thumb rules and checklists have some common aspects. Neither of them aim towards specific solutions Therefore are more feasible for a wider range of applications. The methodologies examined lacks in the argumentation for the trade off that has to be made and in the support for the decision. A method often supports a certain kind of decision and is therefore delimited in its application area. Generally one can say that methodologies and checklists sustain more accurate over time, and that methods tend to have shorter life cycles due to the level of specific factors considered. This survey indicates that there is a need for decision methods within industry. This is also concluded in a study conducted within Swedish industry (Axelson et al, 2004).

7. REFERENCES

1. Abdel-Malek L, Das S. K., Wolf C. (2000), *"Design and implementation of flexible manufacturing solutions in agile enterprises"* International Journal of Agile Management Systems, Vol. 2 No. 3, pp. 187-195.
2. Borenstein D., Becker J. L., Santos E. R. (1999), *"A systematic and integrated approach to flexible manufacturing system design"*, Integrated Manufacturing Systems, Vol. 10 No 1, pp. 6-14.
3. Bourgeios F., Chiabra Z., Muth A., Neri F., Onario M., Santochi M., (2002) Assembly Net Taxonomy and Glosary, Assembly-Net Consortium.
4. Burcher P. G., Lee G. L. (2000) *"Competitiveness strategies and AMT investment decisions"* Integrated Manufacturing System. Vol. 11 No. 5, pp. 340-347
5. Boubkri N and Nagaraj S (1993), *"An Integrated Approach for the Selection and Design of Assembly Systems"*, Integrated Manufacturing System. Vol. 4 No. 1, pp. 11-17
6. Garvin, D.A (1993), *"Manufacturing strategic planning"*, California Management Review, Vol.35 No. 4, pp. 85-106.
7. Hill, T (1995), "Manufacturing strategy Text and Cases", Palgrave, New York
8. Norman and Frazier (2002), Operations Management, Southwestern college publishing, 2002, 9th Edition.
9. Pflughoeft K. A., Hutchinson G. K., Nazareth D. L., (1996) *"Intelligent decision support for flexible manufacturing: Design and Implementation of a knowledge-based simulator"*, International Journal of Management Science, Vol. 24, No. 3, pp. 347-360
10. Rao H. A. and Gu P. (1997), *"Design methodology and integrated approach for design of manufacturing systems"*, Integrated Manufacturing Systems, Vol. 8 No 3, pp. 159-171.
11. Skinner W. (1969),*"Manufacturing – missing link in corporate strategy"*, Harvard Business Review, May-June, pp.136-45
12. Axelson D. et al, (2004) Woxénrapport no 39 ISSN 1650-1888

46

A MAINTENANCE POLICY SELECTION TOOL FOR INDUSTRIAL MACHINE PARTS

Jean Khalil[1], Sameh M Saad[2], Nabil Gindy[3], Ken MacKechnie[4]

1- *School of Mechanical, Materials, Manufacturing Engineering and Operations Management, The University of Nottingham.* epxjmbk@nottingham.ac.uk

2- *School of Engineering, Sheffield Hallam University, City Campus, Sheffield, S1 1WB, UK.* s.saad@shu.ac.uk

3- *School of Mechanical, Materials, Manufacturing Engineering and Operations Management, The University of Nottingham.* Nabil.Gindy@nottingham.ac.uk

4- *Rolls-Royce PLC, Plant MTM Leader, Fan Systems.*

Industrial maintenance activities may be categorised under three strategies, preventive, corrective and predictive. It is necessary to identify domains of equipment and decide which maintenance policy suits each domain of equipment. Usually, it is assumed that maintenance managers are capable of achieving this job. This assumption however is practical, relies on the human factor, which as known to humans, could involve mistakes and hence may lead to implementing the wrong maintenance strategies. Money losses would eventually result. This paper presents a tool that deals with equipment as machine parts domains, where the domain is the group of the similar machine parts which undergo the same conditions. The suggested tool consists of a dual criteria categorisation grid, which through historical data, experts' knowledge and mathematical formulation selects the most suitable maintenance policy for each machine part (domain) individually. The implementation of this tool should guarantee the execution of the appropriate maintenance policy with each and every machine part; therefore it should result in a more economical production function and a more efficient maintenance function. Examples from industry are given to further clarify the proposed tool applications.

1. INTRODUCTION AND BACKGROUND

Equipment maintenance is a key contributor to the welfare of a production organisation. The optimisation of maintenance cost is the focus of many research works. But, before one can develop a maintenance cost optimisation model, two important questions should be answered:

- Which maintenance policy best fits the application?
- What is the level of the application; by mean would the whole industrial site adopt one maintenance policy; May each group of machines adopt a maintenance policy that suits it; or should the problem be studied on the more detailed ground of the machine **parts**?

(Wang 2002) produced a survey of the literature discussing maintenance policies. At the end some remarks were given about the optimal maintenance policy. (Chiang and Yuan 2000) and (Moustafa et al 2004) studied the deterioration of a system using a Markov chain and a semi Markov chain respectively, in order to select the best maintenance policy from: do nothing, repair and replace. Both models are discussing maintenance actions rather than maintenance policies. Their work related more to the response of the maintenance department to the life's evolution rather than the pre-planning of the optimum maintenance policy. (Bevilacqua and Braglia 2000) used the Analytical Hierarchy Process to select the optimum maintenance policy, they considered five strategies, they categorised a firm's equipment into three groups based on criticality, and hence appointed a policy to each group of machines. (Wang 2003) categorised maintenance policies into

1- scheduled
2- preventive (time periodical)
3- Condition based (C.B.M)

Fuzzy logic was then implemented to develop a model that would assist the maintenance manager in selecting the best maintenance strategy. For C.B.M the model would also assist in choosing the most suitable technique.
One common feature between the surveyed approaches is that they tackle whole *machines*; however a key point of the tool suggested in this paper is that it tackles machine *parts*. Therefore it takes the study to a more detailed level; this aspect of the proposed tool allows the flexibility of appointing different maintenance policies to different parts of the same machine. Clearly this flexibility will lead to better results in terms of the effectiveness and efficiency of the maintenance activities.

Figure 1 shows a decision making grid proposed by (Labib et al 1998) developed to recommend maintenance decisions (for *machines*) to management. "This grid acts as a map, where the performances of the worst machines are placed based on multiple criteria. The objective is to implement appropriate actions that will lead to the movement of machines towards the northwest section of low downtime, and low frequency. In the top left region, the action to implement, or the rule that applies, is OTF (operate to failure). The rule that applies for the bottom-left region is SLU (skill level upgrade) because data from breakdowns – attended by maintenance engineers – indicate that machine [G] has been visited many times (high frequency) for limited periods (low downtimes). In other words maintaining this machine is a relatively easy task that can be passed to operators after upgrading their skill levels. A machine that is located in the top-right region, such as machine [B], is a problematic machine, in maintenance words "a killer". It doesn't break down frequently (low frequency), but when it stops it is usually a big problem that lasts for a long time (high downtime). In this case the appropriate action to take is to analyse the breakdown events and closely monitor its condition, i.e.: condition based monitoring (CBM). A machine that enters the bottom right region is considered to be one of the worst performing machines based on both criteria. It is a machine that, maintenance engineers are used to seeing not working rather than performing normal operating duty. A machine of this category, such as machine [C], will need to be structurally modified and major design-out projects need to be considered, and hence appropriate rule to implement the design out maintenance (DOM).

If one of the antecedents is a medium downtime or a medium frequency, then the rule to apply is to carry on with the preventive maintenance schedules" Labib et al 1998.

The aim of this paper is to develop a classification grid tool, in order to assist the maintenance manager in appointing the right maintenance policy to every machine part. The suggested tool uses experts' information i.e. from the industrial site subject to the study, in order to assist the maintenance manager in making decisions, usually found gloomy and confusing.

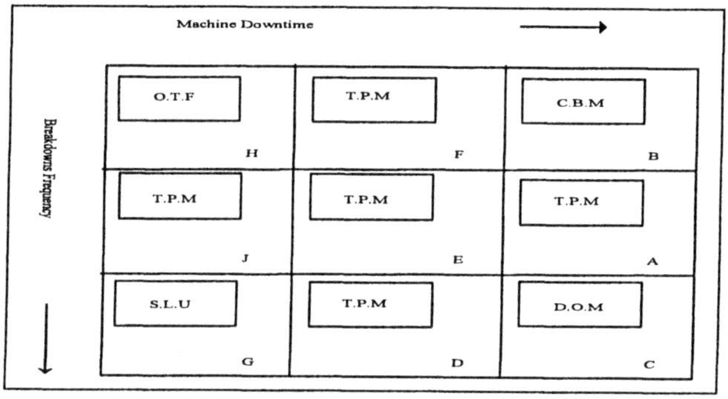

Figure 1- Factory Improvement Grid (Labib et al 1998)

2- RESEARCH METHODOLOGY AND PROPOSED GRID

The improvement grid as explained in previous section, aims to categorise a firm's *machines* with respect to two criteria: *total downtime* and *breakdown frequency,* in order to implement the right maintenance policy with every category of machines. The suggested improvement grid is modified to suit the application tackled by this paper as follows:

Instead of categorising a firm's machines; *machine parts* are categorised, in order to decide which policy would best fit every group of *machine parts*. This fact would allow different parts of the same machine to undergo different maintenance policies as appropriate. That feature, on its own should lead to significant machine-health improvement as each of its components will undergo the appropriate maintenance policy, rather than having to undergo the general policy that is labelled to the whole machine.

In Labib's paper, the judgements are called relative judgements. Relative judgements are the opposite of absolute judgements. For example, based on relative judgements, a school director could decide to gather the scientifically *worst* five pupils of every class for a meeting. Whereas based on absolute judgements, the director would decide to gather all the pupils who *failed* a certain examination. The difference is, that in the first case the pupils are judged *in comparison* with others while in the second case the students are judged with respect to an absolute scale (i.e. exam marks), against which every student is assessed individually. Within the tool suggested in this paper absolute judgements are implemented because they are believed to suit better this application.

Absolute judgements allow the experts to enter critical values rather than assuming them to be a proportion of the maximum available value. Also absolute judgement can accommodate new candidates i.e. machine parts, at any time, whereas within relative judgement the whole population i.e. ALL the machine parts must be considered right at the beginning. If one more machine part was to be added at any point of time, the whole process has to be reset.

Machine parts are assessed with respect to two criteria: the *failure frequency* and *the cost of failure* inflected on the industrial site at the event of failure. The downtime considered in Labib's grid will be one of the factors composing the *cost of failure* criterion. In the following sections the two criteria will be discussed.

2.1 The failure frequency criterion

The failure frequency criterion reflects the repetitiveness of enquiries the machine part causes to the maintenance department in an industrial firm. It expresses the load it puts on maintenance people in terms of number of jobs dispatched per unit of time. For them a job no matter how small, adds to the queue and needs response within a given time. The capture of this information should be a straight forward function of any maintenance data management system. It is only needed to know the number of failures of the given machine part per a given period of time, then divide the number of failures per the number of time units and thus get a frequency; for example if it was found that through the last six month, a certain part was reported 36 times, then the frequency is 6 failures/month. The time unit (month) is not critical, it could be a quarter or a week, the critical aspect is to compare all parts on the same basis, therefore calculate all the frequencies using the same units and plot them on the same grid. Despite the significance of the failure frequency criterion, it does not show the *impact* of a given failure, which could affect not only the maintenance department but many other parties. It is thought that the best way of reflecting the consequences of a failure is by estimating the financial losses it causes. The cost of failure criterion calculates these losses.

2.2 The Cost of failure criterion

Expressing the most considerable financial impacts of a failure resembles to a scoring technique, but in this case instead of using non-meaningful digits, currency is used. The usage of currency as a score allows the user to better feel the meaning of the output. The "cost of failure" criterion includes in its calculations four factors, crucial to the assessment of the criticality of a failure. These criteria are bottleneck penalty, production lost opportunities, possible effect of the fault on scraping production and the waste of resources in terms of labour. The price of the spare part is not included within the calculation of the cost of failure. The reason is that this criterion aims at reflecting the financial impact of the machine part's failure on the *industrial site, irrespective of its own price.* Had the spare part price being included, misleading calculations would have resulted.

The calculations of the most relevant cost factors that result at a failure event are as follows:

i- Production losses cost

$$L_p = \begin{cases} (t_{ti} + t_{su}) \times \dfrac{\alpha}{\pi} & \text{at tsu} > 0 \\ \dfrac{t_{ti}}{\pi} \times \alpha & \text{otherwise} \end{cases} \qquad [1]$$

ii- Products damage cost

$$L_d = v \times c_{pd} \qquad [2]$$

iii- Bottleneck penalty cost

$$L_b = \begin{cases} (t_{ti} + t_{su}) \times c_{dp} & \text{at } tsu > 0 \\ t_{ti} \times c_{dp} & \text{otherwise} \end{cases} \qquad [3]$$

iv- The booked labour cost

$$B_l = \begin{cases} x_1 \times t_{lc} \times s_{mh} + x_2 \times (t_{ti} + t_{su}) \times s_{oh}) \text{ at } t_{su} > 0 \\ x_1 \times t_{lc} \times s_{mh} + x_2 \times t_{ti} \times s_{oh} & \text{otherwise} \end{cases} \qquad [4]$$

The summation of these factors forms the expression presented in equation 5 $C_{failure} =$

$$[(t_{ti}+t_{su}) \times \frac{\alpha}{\pi} + v \times c_{pd} + (t_{ti}+t_{su}) \times \frac{c_{dp}}{\pi} + x_1 \times t_{lc} \times s_{mh} + x_2 \times (t_{ti}+t_{su}) \times s_{oh})] \qquad (5)$$

Table 1 shows the definition of the above mentioned symbols. Due to space limitations these symbols will be explained in details during the conference presentation.

Table 1- Variables composing the cost of failure function

Symbol	Definition	Unit
c_{dp}	Cost of production-line delay per unit of time.	£/h
c_{pd}	Value of one damaged work-piece.	£
s_{mh}	Maintenance personnel hourly rate.	£/h
s_{oh}	Operator's hourly rate.	£/h
t_{lc}	The time spent by maintenance personnel in fixing problems in case of corrective action.	h
t_{ti}	Production time loss due to a fault excluding time of set up.	h
t_{su}	Machine set up time	h
v	Number of damaged work-pieces because of a failure.	
x_1	the number of maintenance personnel involved in the repair action	
x_2	The number of operators made on stand by because of a failure	
α	Department's income due to the production of one work-piece	£
π	A product cycle time.	h

3- MACHINE PARTS CATEGORISATION GRID

The machine parts categorisation grid is a two axes plot (see figure 2)

3.1 First Axis, machine part failure frequency criterion (MPF frequency).

On the failure frequency axis, the user appoints a critical value which may be defined as:
A frequency of machine parts failure occurrences that the user considers high enough, to draw the attention, for improvement efforts.

3.2 Second Axis, machine part failure cost (MPF cost)

On the cost of failure axis the user allocates two critical values: a lower cost of failure value (CLCV), and an Upper cost of failure value (CUCV). The lower value is the maximum cost of a failure that would be considered insignificant and cheaper than the cost of an operator training course. The upper value is the maximum cost of a failure that would be cheaper to repair by the operator.

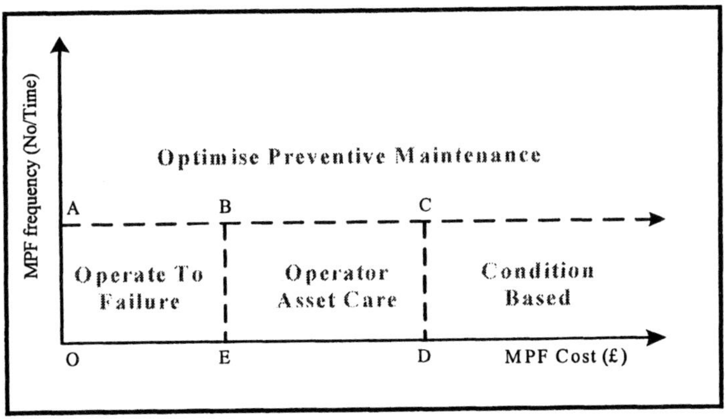

Figure 2- Modified Optimisation Grid

At this point the user should have a two axes plot with the Failure frequency criterion on one axis and the Cost of failure criterion on the other, and with one critical value for the failure frequency criterion and two critical values for the cost of failure criterion (see figure 2). The user should then start allocating the factory's machine parts subject to the study.
The factory's machine parts will therefore be allocated on the categorisation grid (see figure2) as follows:

1. A group of machine parts may fall in rectangle (OABE), this group should undergo and operate to failure policy, and hence they will run until they fail on which case a corrective action will take place.
2. Another group of machine parts may fall in rectangle (EBCD), this group should undergo an operator asset care, by mean the machine operator should be trained to repair the failure of these particular machine parts.
3. A third group will have a higher failure cost than D and a lower failure

frequency than A. The suitable maintenance policy for this group is condition based maintenance, achieved through condition monitoring or routine inspections.

4. The fourth group will have a higher failure frequency than A. For this group periodical preventive maintenance (e.g. time based) is the best policy. The usage of a preventive maintenance model is recommended with this group of machine parts.
   A preventive maintenance model aiming at optimising the total cost of maintenance for machine parts was developed in (Khalil et al 2003). The application of the model returns the cost-optimum life span of a machine part; hence a preventive maintenance action could take place and therefore avoid the occurrence of a failure. In other words, the outcome of this model assists the decision taker in the best timing with respect to cost for preventive maintenance actions. The tool proposed in this paper may be implemented prior to the application of that model or as a stand alone tool.

The operator asset care mentioned with group 2 is one of the assumptions of the T.P.M first developed by the Japanese industry. T.P.M consists briefly of involving the operator in taking care of the machine health. Within a T.P.M. policy, beside routine tasks (oiling and lubricating) the operator could also be asked to complete the repair of some faults. Usually, the list of faults that may be repaired by the operator grows up gradually. The old behaviour "this is not my job" gets replaced by "I should better take care of ***my*** equipment". But this transfer can't happen instantaneously. Actually educating the operators could be a hard task that needs time and effort. T.P.M proved in many cases to be successful and efficient however some firms prefer to adopt a different policy separating between the operators' duties and the service people duties; they thus prohibit to the operator the involvement in any equipment technical action. In industry, the latter policy is usually not particularly appreciated by the operators, as it leads to time-losses; however it is usually advocated to be safer for the operators and better for the equipment, because it only allows skilled people to work on the equipment.

5. EXAMPLES FROM INDUSTRY

The following studies were carried at a multinational industrial organisation specialised in aero-industry. Four machine parts subject to the study would be referred at, as machine part A, B, C and D. Frequency critical value is 4 failures/month, Cost lower critical value: CLCV is £300 and Cost upper critical value is CUCV: £1000. Figure 3 displays the grid for these examples and table 1 shows the collected data for these four parts.

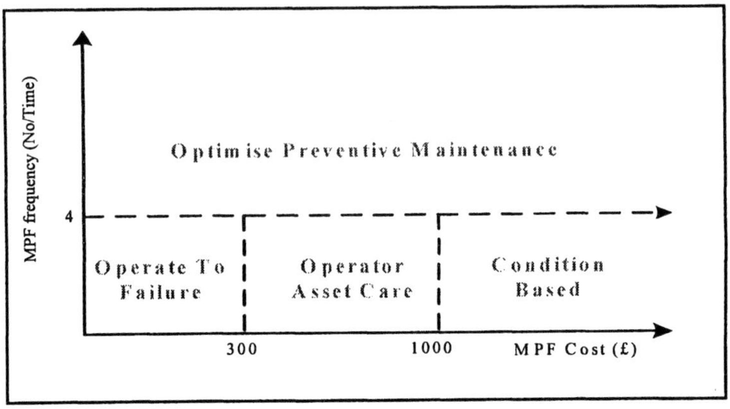

Figure 3- Illustrative Example Grid

Table 2- Real Values of the machine parts variables

Abbreviation	Unit			
	Part A	**Part B**	**Part C**	**Part D**
c_{dp}	£ 200	£ 500	£ 200	£ 150
c_{pd}	£ 900	£ 0	£ 650	£ 0
s_{mh}	£/h 25	25 £/h	£/h 25	£/h 25
s_{oh}	£/h 25	25 £/h	£/h 25	£/h 25
t_{lc}	4 h	6 h	1.5 h	6 h
t_{lp}	0.5 h	4 h	1 h	4 h
t_{ti}	8 h	6 h	4 h	6 h
t_{su}	4 h	0 h	0 h	0 h
v	1	0	1	0
x_1	2	1	1	2
x_2	1	1	1	1
α	£ 250	£ 320	£ 150	£ 50
π	1 h	1.5 h	1.5 h	0.5 h

5.1 Case 1: Machine part A

By substitution in equation 5,

$$C_{failure} = [(8+4)\times 250 + 900 + (8+4) \times 200 + 2 \times 4 \times 25 + (8+4)\times 25)] = \textbf{£6800}$$

The average failure frequency of this machine part was retrieved from the data management system and found to be **5.5 failure/ month.** Therefore, PM is recommended.

5.2 Machine part B

By substitution in equation 5,

$$C_{failure} = 6\times \frac{320}{1.5} + 6 \times \frac{500}{1.5} + 6 \times 25 + (6\times 25) = \textbf{£2630}$$

The average failure frequency of this machine part was retrieved from the data management system and found to be **1.41 failures / month.** Therefore, condition based maintenance is recommended.

5.3 Machine part C

By substitution in equation 5,

$$C_{failure} = 4\times \frac{150}{1.5} + 650 + 4 \times \frac{200}{1.5} + 1.5 \times 25 + 4\times 25 = \textbf{£ 1720.5}$$

The average failure frequency of this machine part was retrieved from the data management system and found to be **7.2 failure/ month.** Therefore, PM is recommended.

5.4: Machine part D

By substitution in equation 5,

$$C_{failure} = [6\times \frac{50}{0.5} + 0 + (6) \times \frac{150}{0.5} + 2 \times 6 \times 25 + 6\times 25)] = \textbf{£2850}$$

The average failure frequency of this machine part was retrieved from the data management system and found to be **1.1 failure/ month.** Therefore, Condition Based maintenance is recommended.

6. DISCUSSION

- One important aspect of this work is that it allows the individual consideration of each part within the machine. The previous approaches that consider the machine as the unit of the study undermine the fact that, the behaviours of the parts is usually unrelated.

- Within the calculations of the cost of failure, had the cost of the spare part been included, it would have been impossible to consider different machine parts under the same grid because $C_{failure}$ critical value could not then apply. The reason is: the machine parts' own values, would manipulate the cost of failure rather than allowing it to purely reflect the impact of its failure on the firm.

- The idea of classification and arrays is not new in maintenance science however, up to the authors' knowledge it was never used to categorise machine parts. The maintenance problem was merely ceased in a bottom up technique and thus machine parts were most of the time out of the researchers' focus.

- The development of this tool integrates to the value of preventive maintenance model presented in (Khalil at al 2003) as a decision support system, the reason being it takes away one source of potential human error; in other words substitute one human decision by a scientifically based decision. Even though the human factor contributes to the implementation of this tool, the risk of error is incredibly less than a pure human selection.

- By implementing this tool, the level of expertise required for the usage of the model presented in (Khalil et al 2003) has become less than before, which makes it more practical. After this work, the user only needs to set the critical values for both criteria (Total cost of failure and frequency of failures).

- The simplicity of the idea and its straight application should make it welcomed in industry. This paper could be taken as a good example to prove that significance and simplicity are not contradictory. It is sometimes wrongly believed that when it comes to research in industry, a limit of simplicity should be respected in order for the work to be significant.

7. CONCLUSION

This paper presented a tool aiming to assist the maintenance manager in selecting the most suitable maintenance policy for machine parts. Expert judgements are used to build a two parameters categorisation grid on which, a firm machine parts are categorised into four categories. Four maintenance strategies are considered to meet the nature of the four categories of machine parts. The suggested tool may be integrated to the previously developed preventive maintenance model developed by Khalil et al 2003 as a suitability test in which case, it eases up the usage of the model and makes it possible for a wider range of people. It also could be implemented as a stand alone tool.

8. REFERENCES

1. Bevilacqua, M and Braglia, M. The analytic hierarchy process applied to maintenance policy selection. Reliability Engineering and System safety 2000; 70: 71-83.
2. Chinag JH and Yuan J. Optimal maintenance policy for Markovian system under periodic inspection. Reliability Engineering and System safety 71 2001; 71: 165-172.
3. Khalil J, Saad S and Gindy N. A cost optimisation decision support model for preventive and corrective maintenance actions. Proceeding of the International Conference on flexible automation and intelligent manufacturing. FAIM 2003, Florida, USA.
4. Labib AW. World-class maintenance using a computerised maintenance management system. Journal of Quality in Maintenance Engineering, 1998, Vol. 4 No. 1: pp. 66-75.
5. Moustafa MS Abdel Maksoud EY Sadek S. Optimal major and minimal maintenance policies for deteriorating systems. Reliability Engineering and System safety. 2004; 83: 363-368.
6. Mechefske CK. Wang Z. Using fuzzy linguistic to select optimum maintenance and condition monitoring strategies. Mechanical Systems and Signal Processing 2003; 17(2): 305-316.
7. Wang H. A survey of maintenance policies of deteriorating systems. European journal of operational research. 2002, 139: 469-489.

PART D

MACHINE LEARNING AND DATA MINING IN INDUSTRY

47

USING DATA MINING FOR VIRTUAL ENTERPRISE MANAGEMENT

*L. Loss *, R. J. Rabelo [#], D. Luz *, A. Pereira-Klen *, E. R. Klen **
Federal University of Santa Catarina – Florianópolis (SC), BRAZIL
** {loss, luz, klen, erklen}@gsigma.ufsc.br*
[#] rabelo@das.ufsc.br

This paper presents exploratory results on how a data-mining-based tool can be used to enhance the quality of decision-making in a Virtual Enterprise environment. The developed tool is based on the Clustering mining method and implements the K-Means algorithm. The algorithm is explained, its utilization in the proposed model is introduced and the implementation results are presented and stressed in the end of the paper.

1. INTRODUCTION

Data Mining (DM) has emerged as a very powerful technique to find out patterns and relationships in large information repositories. The application of DM on several domains (e.g. marketing, investment, fraud detection, manufacturing, financial services) has increased significantly in the last years. However, its application on more volatile scenarios, like the ones represented by Virtual Enterprises (VE) is still very incipient. A VE is here considered as a dynamic, temporary and logical aggregation of autonomous enterprises that interact with each other as a strategic answer to attend a given opportunity or to cope with a specific need, and whose operation is achieved by the coordinated sharing of skills, resources and information, enabled by computer networks (Rabelo et al., 04).

Recently, many investments have been made by enterprises to support inter-enterprises communication in order to improve the information exchange among suppliers and clients as well as to enable distributed information access facilitating and enhancing the Virtual Enterprise management. The downside of this success has been information overload: how should this amount of information be used in a value added way? The fact is that there is a mass of valuable information "hidden" in the enterprises' databases which are relevant for business (Chandra et al., 2000). Examples of this include patterns of clients' behaviors, seasonal or repetitive events, suppliers' performance per product, and many others. These qualitative and quantitative unknown information correlations can be used to improve both the quality of decision-making and the formulation of successful strategies among the VE partners.

Business Intelligence (BI), Competitive Intelligence and *Market Intelligence* are examples of techniques that have been used to better organize and to properly filter the information for decision-makers (Begg et al., 2002). In spite of their potentialities, some handicaps still have to be overcome such as their application on dynamic VEs, where new suppliers and clients can enter or quit along the operation process. Supporting this requirement is extremely important as the success of VE critically depends on recognizing partners' expertise, tools and skills as marketable knowledge assets (Lavrac et al., 2002). Additionally, those techniques are not designed to be "active" tools, i.e. systems that go through information repositories in order to try to discover new information elements that can augment decision processes.

Based on that, this paper presents a hybrid approach which joins the fundamentals of DM and BI regarding the VE environment requirements. A preliminary validation of this approach was done by means of the development of an exploratory data mining tool that works together with the VE Cockpit, a BI-based VE management system (Rabelo et al., 2002).

This paper is organized as follows: Chapter 2 frames the global scenario in which the developed tool is inserted in. Chapter 3 describes the basic concepts of the data mining approach as well as explains the K-Means algorithm. Chapter 4 depicts the implemented prototype and results, and Chapter 5 provides the main conclusions.

2. GENERAL SCENARIO

In this work a VE is considered as a network of several enterprises where one of them – called VE Coordinator – has the role of managing the VE-related processes as well as of acting as the front-end with the end customer. The model presented in this section has been developed with the aim of extracting helpful information for the VE manager so that better decision-making can be taken during the VE Operation phase. The VE Manager interacts with the VE Cockpit system and is supported by its functionalities to operate the VE.

The information is stored in the VE Coordinator's database, which contains current and historical data about its suppliers, clients, and involved orders (production orders, shipment orders, sales orders and so forth). Figure 1 illustrates this model which is composed by:

- VE Cockpit system: having two main modules (Creation & Configuration, and Operation) which in turn feed the VE database during the course of the VE existence.
- Data mining tool (DM-Tool): its first module processes the database using a specific data mining algorithm (see next chapter) and sends its results to the DM Analyzer module. The second module processes these results and provides the VE manager with high-level conclusions, i.e. the envisaged information patterns.

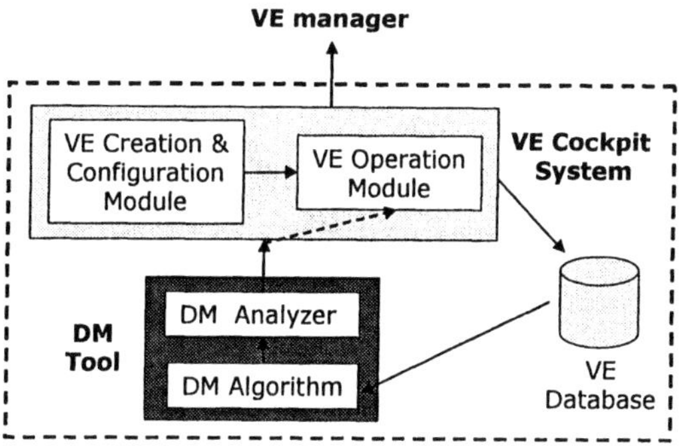

Figure 1 - General scenario of the data mining application

3. DATA MINING AND THE K-MEANS ALGORITHM

DM is the process related to the extraction of knowledge from data repositories with the aim of identify valid, new, potentially useful and understandable patterns (Fayyad et al., 1996a). DM is one of the main steps of Knowledge Discovery in Databases (KDD), generally defined as the process of automatically extracting useful knowledge from large collections of data (Adriaans et al., 1996). This is carried out by means of finding patterns in data, driven by some "rules of interest" that are defined by the user (Fayyad et al., 1996a) (Lavrac et al., 2002).

The KDD process attempts to develop technologies for automatic knowledge extraction by means of mapping low-level data (a large amount of "raw" data) into other forms that might be more compact (a short report), more abstract (a descriptive model of the process that generated the data), or more useful (a predictive model for estimating the value of future cases) (Fayyad et al., 1996a).

Based on the DM theory, two approaches can be used in the data processing. The first one is based on hypothesis tests, verifying or rejecting a hypothesis or previous ideas that are still undercover. The second one is based on the knowledge synthesis that is related to information discovering without any initial condition or supervision. Regarding the main characteristics of the VE domain and the requirements of a VE manager the second approach was chosen, as the main interest in this work is to find patterns which are not previously known.

Among a number of existing non-supervised method, *Clustering* was selected to be used considering its potentiality, simplicity and, at the same time, the facility to reach results quickly. Clustering is a common descriptive task where it is tried to identify a finite set of categories to describe data (Fayyad et al., 1996b). Examples of applications of clustering include discovering homogeneous subpopulations of potential consumers and identification of subcategories of suppliers according to some performance metrics. The clustering method is performed through an analysis of the relationships among the database's fields and tables. The similarities among attributes are in intrinsic property and it is not necessary to train pre-defined classes. Usually, it only requires an end-user to set up initial parameters and to refine them afterwards in the case a non-satisfactory result (i.e. a given configuration of data sets/patterns) is achieved. The existing clustering algorithms are based on several methods, such as (Berkhin, 2003): hierarchical methods, partitioning methods, grid-based methods, methods based on co-occurrence of categorical data, and constraint-based clustering.

The *K-Means* algorithm (MacQueen, 1967) is a widespread partitioning method that has been used in many works. In spite of some limitations, *K-Means* was the one selected to be used in this exploratory work since: it can be applied on several application domains; its implementation is relatively simple; and it works with information free of context, facilitating the search of data associations.

The K-Means Algorithm

This section will briefly illustrate the functioning of the *K-means* algorithm. As an example, consider the simple database table illustrated in Table 1. It contains records related to ten suppliers about their production capacity level and their ranking (best-delivery ranking) from the VE Coordinator point of view.

Supplier	Capacity	Ranking
S1	3	8
S2	3	6
S3	3	4
S4	4	7
S5	4	5
S6	5	5
S7	5	1
S8	7	4
S9	7	3
S10	8	5

Table 1 – Database table

Firstly, the user should indicate the value of *k*, i.e. how many clusters (grouping criteria) (s)he is interested to find information about. Assuming that Table 1 would be the only one available, up to 10 clusters could be considered. In the example showed in figure 2, two clusters are used, trying to obtain some knowledge from the suppliers' capacity and ranking. After that, a bidimensional vector / group is created to represent each supplier (Figure 2a), where, for instance, the Supplier 1 is fixed in the points (3,8).

Starting points are chosen for each group by a shuffle algorithm, after which medium points (*mps*) are calculated for each one (Figure 2b). All points are resettled according to the distance from the *mps*. Points will belong to the group that contains the closest distance to the *mp* so they can change from one group to another, i.e. new groups are created (Figure 2c). The *mps* are recalculated according to these new groups, and the process is repeated until that the new groups are equal to the previous ones, or the algorithm reach a (predefined) maximum number of iterations (Figure 2d). When a large number of database registers is involved, different final results can be reached by the algorithm. It means that different initial conditions (for instance, the number of clusters) lead to different results.

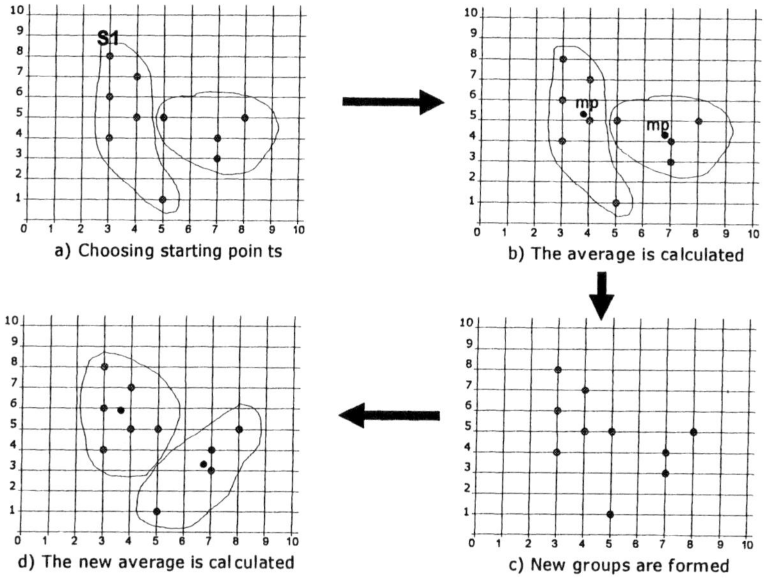

Figure 2 – Clustering steps

Summing up, on the one hand *K-Means* introduces some difficulties to decide which the most suitable solution is; on the other hand it provides other visions to the decision-maker that can enrich his/her insights.

4. PROTOTYPE

The algorithm stressed in the previous chapter has been implemented in C++ in a PC / Windows XP platform and was integrated in the VE Cockpit system (see figure 1).

The first step in that module is to indicate the database to be used as well as to select the database's tables that should be mined by the system (Figure 3 / top left).

The second step is related to the configuration of the mining system. At this point two ways are provided: *manual* and *automatic* configuration. In the *manual* way the user should define the number of clusters as well as the number of data sets (i.e. possible different/final results) to be generated (Figure 3 / bottom left). It requires a certain level of experience from the user. In the *automatic* way, the mining system generates final results automatically, taking into account four pruning parameters that the user should specify. They are: acceptable interval, standard deviation, similarity, and quantity within the interval (Figure 3 / top / inside the circle). The final results are selected according to an internal value reached by the algorithm that is related to the sum of the Euclidean distances of the clusters.

In the third step, the user should define the fields of the selected database tables that will constitute the mining sample. In figure 3, the table *TableConnectionDetails* and the fields *CD_SupplyChain*, *CD_Connection*, *CD_DPSource* and *CD_Item* were chosen. This means that the user "thinks" that useful correlations between the supply chain id, the relations among companies per item type can be revealed. The K-Means algorithm then combines these four fields trying to identify relevant correlations among them. The data set used in this prototype come from a database fed with real information from industrial partners of a research project.

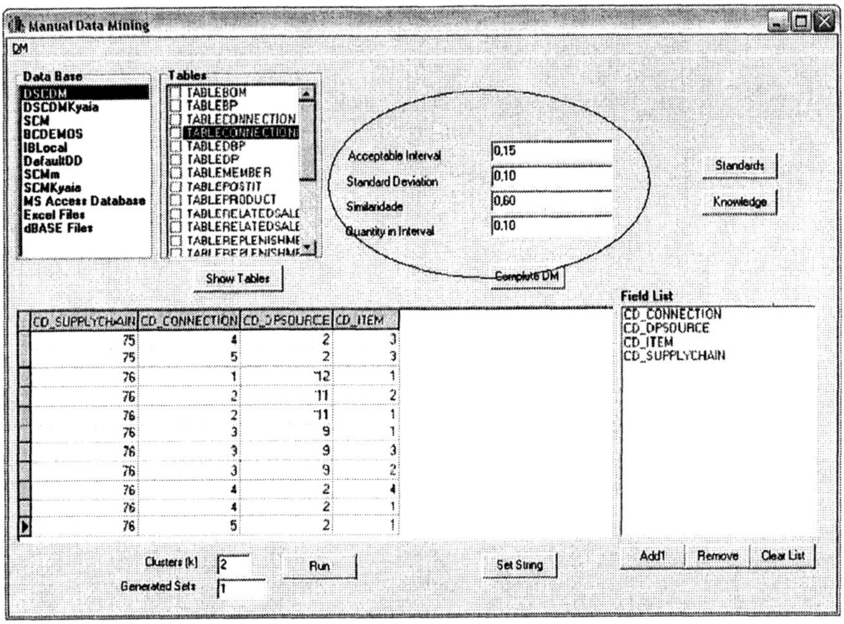

Figure 3 – Setting standards for automatic search

In the fourth step the results generated by the algorithm is shown (Figure 4), also providing the numeric association with the alphanumeric fields. In this case, *0* means *CD_SupplyChain*, *1* means *CD_Connection*, *2* means *CD_DPSource* and *3* means *CD_Item*. Results can be saved in a database or be expressed as a HTML report.

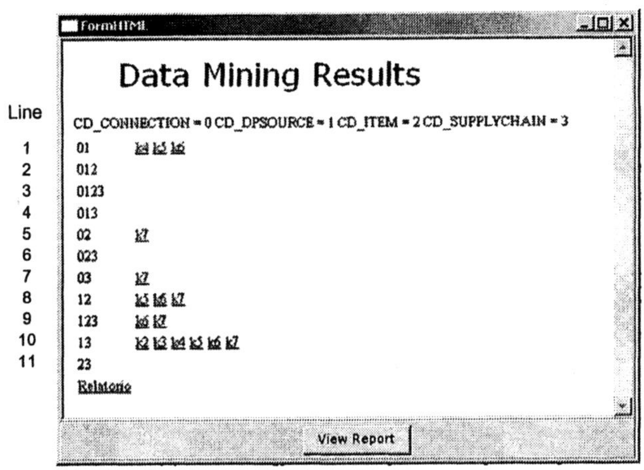

Figure 4 – Results using the automatic data mining search

From the six clusters found by the algorithm (the "lines" 1, 5, 7, 8, 9 and 10) the user would normally elect the ones which got the largest number of logical *k* occurrences (3, 3, and 6, respectively, in the lines 1, 8, and 10), i.e. the largest amount of correlations among those four fields.

The role of clustering algorithms uses to end here, at this stage and level of information processing. Nevertheless, this result is still expressed in a too low level, creating difficulties for an easier understanding and hence for an agile decision-making. For that purpose, the DM tool was extended with the DM Analyzer module in order to provide a clearer level of results description. Figure 5 shows the final results generated by the DM tool to the VE manager, i.e. the relevant correlations found out of those four database fields.

The interpretation of these results is done by the user, and each line represents *one* result. This means that *(s)he* has to elect which patterns make sense, which ones are indeed relevant and which ones deserve to be saved in a knowledge base for future use. For instance, consider that the user selects the pattern "*CD_DPSOURCE = 1.3 +- 0.5 <=> CD_ITEM = 2 +- 0 <=> CD_SUPPLYCHAIN = 73 +- 0 <=> in 9% of its values*". First of all, this can be considered a rich pattern as it comprises three of the four fields. Concretely, this means that a correlation among the fields *CD_DPSource*, *CD_Item* and *CD_SupplyChain* was found in the table's records, i.e. in *9%* of the cases the supply chain of id *73* had as a member the enterprise (source) of id *1* in the production of the item of id *2*.

Figure 5 – Report of the patterns

5. CONCLUSIONS

This paper presented exploratory results on the application of a data mining approach in a VE scenario, aiming at facilitating the decision-making process. *Clustering* technique / *K-Means* algorithm were used in this work.

Preliminary analysis from the results obtained with the software prototype have shown that data mining is a very powerful technique and can indeed support VE managers in decision-making, especially if the tool is integrated in a wider VE management system.

Three important conclusions brought up from this work:

1. Information and knowledge update: the VE data that come from the enterprises can be different from business to business, i.e. new patterns can be created and some

previous conclusions can become out of date along the time. Therefore, the user must run the system "periodically" and delete a pattern which is no longer valid. There is not a specific time to run the system. It is up to the users experience to decide when it is necessary to have a new set of patterns to be analyzed;

2 Interpretation of the results: Post-processed results (figure 5) are surely more interesting and understandable than not processed ones (figure 4). Even so, the interpretation of the patterns still remains a bit difficult, demanding an experienced VE manager to recognize their utility and validity;

3. K-Means algorithm: Databases are most usually composed by numeric and alphanumeric contents. The K-Means algorithm was designed to process only numeric data. It is then important to highlight that, in order to validate this exploratory work, only numeric data deriving from the VE scenario should be considered and used. This would enable the validation of the prototype as well as the developed approach. Efforts are currently being made in order to find a more adequate algorithm to suit the requirements of the envisaged databases and hence to better support decision-makers (VE Managers or even software agents).

Next steps refer also to a deeper validation of the results in a dynamic VE scenario where the system's knowledge base can be continuously updated with new data from the VE partners so that the VE Manager can also play the role of a Knowledge Manager.

ACKNOWLEDGEMENTS

This work was partially supported by CNPq – The Brazilian Council for Research and Scientific Development (www.cnpq.br). This work has been developed in the scope of the IST MyFashion.eu (www.myfashion.org) and IFM (www.ifm.org.br) projects. Special thanks to Solange Rezende and her team at USP-EESC, São Carlos-Brazil.

6. REFERENCES

1. Adriaans, P.; Zantige, D., Data Mining, Addison-Wesley, 1996.
2. Begg, C.; Connolly T., Database Systems: A Practical Guide to Design, Implementation, and Management. 3ed. Addison-Wesley, 2002.
3. Berkhin, P., Survey of Clustering Data Mining Techniques. Accure Software Inc. Website Accessed in November 21, 2003. http://www.accrue.com/products/rp_cluster_review.pdf
4. Chandra, C.; Smirnov, A. V.; Sheremetov, L. B.; Agent-Based Infrastructure of Supply Chain Network Management, Proc. PRO-VE'2000 Conference, pp.221-232, 2000.
5. Fayyad, U.; Piatetsky-Shapiro, G.; Smyth P.; 1996a. From Data Mining to Knowledge Discovery: an overview. In: Advances in Knowledge Discovery & Data Mining, pp. 1-34.
6. Fayyad, U.; Shapiro, G. P.; Smyth, P., From Data Mining to Knowledge Discovery in Databases. AAAIMIT Press, pp.37-54, 1996b.
7. Lavrac N.; Urbancic, T.; Orel, A., Virtual Enterprises for Data Mining and Decision Support: A Model for Networking Academia and Business, Proceedings PRO-VE'2002 Conference, pp.389-396, 2002.
8. MacQueen, J. - "Some methods for classification and analysis of multivariate observations", in LeCam, L. and Neyman, J. eds., Proceedings of the Fifth Berkeley Symposium on Mathematical statistics and probability, vol 1, pp. 281-297. Univ. of California Press, 1967.
9. Rabelo, R. J.; Pereira-Klen, A. A., Business Intelligence Support for Supply Chain Management, Proc. BASYS'2002, Kluwer Academic Publishers, pp.437-444, 2002.
10. Rabelo, R. J.; Baldo, F.; Tramontin, R. J. Jr., Pereira-Klen, A. A.; Klen, E. R.; Smart Configuration of Dynamic Virtual Enterprises, to be presented in PRO-VE'2004 Conference, Toulouse, Aug 2004.

48 MINING RULES FROM MONOTONE CLASSIFICATION MEASURING IMPACT OF INFORMATION SYSTEMS ON BUSINESS COMPETITIVENESS

Tomáš Horváth, František Sudzina, Peter Vojtáš
Faculty of Science, University Of Pavol Jozef Šafárik in Košice
thorvath@science.upjs.sk
sudzina@euke.sk
vojtas@kosice.upjs.sk
SLOVAKIA

Motivation for this paper are classification problems in which data can not be clearly divided into positive and negative examples, especially data in which there is a monotone hierarchy (degree, preference) of more or less positive (negative) examples.

We use data expressing the impact of information systems on business competitiveness in a graded way. The research was conducted on a sample of more than 200 Slovak companies. Competitiveness is estimated by Porter's model.

The induction is achieved via multiple use of two valued induction on alpha-cuts of graded examples with monotonicity axioms in background knowledge. We present results of ILP system ALEPH on above data interpreted as annotated rules. We comment on relations of our results to some statistical models.

1. INTRODUCTION

There are many theoretical articles, which argue that usage of information systems increases business competitiveness. But only a few give proofs. Our data were gathered from a sample of 206 Slovak companies, which had to submit their preference (grade) of business competitiveness and information systems usage. These data are monotonous that means that if the company is highly competitive (the grade "best") then its competitiveness is also "medium" or "low".

In the crisp framework we are restricted only to the information that is true absolutely. Since we have uncertain or noisy data, this framework is not suitable to represent this kind of information. This is a significant gap in the expressive power of the framework, and a major barrier to its use in many real-world applications. Besides probabilistic models there is an extensive study of these phenomena in manyvalued logic, especially in fuzzy logic and generalised annotated programs.

Inductive logic programming is an effective tool for the data mining in the case of non numeric data. Information is implemented in the language of predicate logic, since it is easy to understand. Unlike many data mining tools, inductive logic programming is effective for the multi relational problems, too.

After explaining data, we present a new formulation of a many-valued inductive logic programming task in the framework of fuzzy logic in narrow sense. Our construction is based on a syntactical equivalence of fuzzy logic programs FLP and a restricted class of generalised annotated programs. The induction is achieved via multiple use of classical two valued inductive logic programming. Correctness of our method (translation) is based on the correctness of FLP. The cover relation is based on fuzzy Datalog and fixpoint semantics for FLP. We present and discuss results of ILP system ALEPH on our data. Then we compare our results with a statistical method of linear regression.

The information systems usage in 206 Slovak companies was analyzed from the point of view of the most basic model, which divides information systems into enterprise resource planning (ERP) systems, management information systems (MIS) and executive information systems (EIS) (Kokles, Romanová, 2002) and (Dudinská, Mizla, 1994). Information systems cover five main areas - sales and marketing, manufacturing, finance, accounting, human resources; therefore ERP systems were divided into five categories. Data on accounting systems were not used in later analysis because all companies must do accounting therefore it cannot be the factor that might influence business competitiveness. The reason why there are companies, which do not have any accounting system, is that they outsource accounting.

We asked if the company used an information system supporting specific areas and if so, we asked how was the company satisfied with the information system. IS satisfaction was measured on a Likert-type scale 1-7 (self-assessment).

The extent of outsourcing of information systems was quite significant; therefore data on outsourcing were also used as explanatory variables. Data on outsourcing of information systems do not include outsourcing of accounting systems because managers do not consider it to be outsourcing of an accounting system but outsourcing of accounting as of a functional area.

The company size was graded. We used the same 7 intervals, which are used by the Statistical office of the Slovak Republic.

To sum up, explanatory variables include nine columns - on ERP – sales and marketing, ERP – manufacturing, ERP – finance, ERP – human resources management, MIS, EIS, partial outsourcing, total outsourcing and company size.

(Porter, 1979) suggested to analyze the value chain, all the business processes that contribute to the value added. As the main processes he identified input logistics, manufacturing, output logistics, marketing and sales, services; subsidiary processes include administration, human resource management, technological development, buying. We merged manufacturing and services in order to meet the main processes of all sectors in one value. We disagree that marketing (in contrast with selling) is a main process; therefore we included it in subsidiary processes. Each company was asked to provide self-assessment of its competitiveness in all of the dimensions. A Likert-type scale 1-7 was used to measure the competitiveness (self-assessment).

So, the dependent variables are input logistics, manufacturing and services, output logistics, sales, administration, human resource management, technological development, buying and marketing.

2. A MONOTONE GRADED ILP PROBLEM

In this section we introduce a problem of the monotone graded inductive logic programming ILP (Horváth, Lencses, Krajči, Vojtáš, 2004). We describe the problem of generalized annotated programs GAP (Kiefer, Subrahmanian, 1992), which herbrand interpretations coincides with interpretations of FLP (Krajči, Lencses, Vojtáš). After we describe our method for a multiple used classical (crisp) ILP systems to solve a monotone graded ILP problem. Our method is based on the correctness of FLP (Vojtáš, 2001).

2.1 Generalized annotated programs

Kifer and Subrahmanian (Kiefer, Subrahmanian, 1992) introduced generalized annotated logic programs (GAP) that unify and generalize various results and treatments of multi-valued logic programming. The whole theory of GAP is developed in a general setting for lattices. We restrict ourselves to the unit interval of real numbers [0,1].

In fuzzy logic programming rules had weights (or truth values) associated with them as a whole. Annotated logic, on the other hand, appeared to associate truth values with each component of an implication rather than the implication as a whole. This implication is interpreted in a "classical logic" fashion. We show how truth values in FLP can be propagated across implications to derive clauses in GAP.

Definition A function A: $[0,1]^i \rightarrow [0,1]$ is an annotation function if it is left continuous and order preserving in all variables.

The language of annotated programs consists of a usual language of predicate logic (with object variables, constants, predicates and function symbols) as in FLP and of the quantitative part of the language. The quantitative part of the language has annotation variables and a set of basic annotation terms of different arity. Every annotation term ρ is a composition of annotation functions. Notice, that $\rho\bullet$ can be considered as the truth function of an aggregation operator.

Definition If A is an atomic formula and α is an annotation term, then A : α is an annotated atom. If A : ρ is a possibly complex annotated atom and $B_1 : \mu_1, \ldots, B_k : \mu_k$ are variable- annotated atoms, then $A : \rho \leftarrow B_1 : \mu_1 \ \& \ \ldots \ \& \ B_k : \mu_k$ is an annotated clause. We assume that variables occurring in the annotation of the head also appear as annotations of the body literals and different literals in the body are annotated with different variables.

Definition Let B_L be the Herbrand base. A mapping f: $B_L \rightarrow [0,1]$ is said to be a Herbrand interpretation for annotated logic.

Note that interpretation for fuzzy logic and interpretations for annotated logic coincide.

Suppose I is an Herbrand interpretation. Then,

I satisfies a ground atom A : μ iff $\mu \leq I(A)$

I satisfies $(F \wedge G)$ iff I satisfies F and I satisfies G

(please note that iff and are metamathematical two valued connectives)

I satisfies $(F \vee G)$ iff I satisfies F or I satisfies G

I satisfies $F \leftarrow G$ iff I satisfies F or I does not satisfy G.

Definition (FLP and GAP transformations). Assume C = $A : \rho \leftarrow B_1 : \mu_1 \,\&\, ... \,\&\, B_k : \mu_k$ is an annotated clause. Then flp(C) is the fuzzy rule $A \leftarrow \rho(B_1, ..., B_k).1$, here ρ is understood as an n-ary aggregator operator.
Assume $D = A \leftarrow @(B_1, ..., B_n).r$ is a fuzzy logic program rule. Then gap(D) is the annotated clause $A{:}C_i(@^{\bullet}(x_1,...,x_n),r) \leftarrow B_1 : x_1 \,\&\, ... \,\&\, B_k : x_n$.
The satisfaction is defined differently (all variables (object and annotation) are implicitly universally quantified).
Theorem (Vojtáš, 2001) Assume *C* is an annotated clause, *D* is a fuzzy logic program rule and *f* is a fuzzy Herbrand interpretation. Then

f is a model of *C* iff *f* is a model of *flp*(*C*)

f is a model of *D* iff *f* is a model of *gap*(*C*)

This theorem is the main tool in our formal model of fuzzy ILP.

2.2 ILP system ALEPH

Since our aim is not to develop a new resp. better ILP algorithm we will not describe the used ILP systems in details – we notice just some basic properties of these systems (we are interested just in the correct transfer of the graded ILP problem to a multiple use of classical – crisp – ILP problem). In a two valued logic the Inductive logic programming (ILP) task is formulated as follows:

In ILP, given is a set of examples $E = E^+ \cup E^-$, where E^+ contains positive and E^- negative examples, and background knowledge B. The task is to find a hypothesis H such that $\forall e \in E^+$: $B \wedge H \models e$ (H is complete) and $\forall e \in E^-$: $B \wedge H \not\models e$ (H is consistent). This setting, introduced in (Muggleton, 1991), is also called learning from entailment. In an alternative setting proposed in (Džeroski, Lavrač, 2001), the requirement that $B \wedge H \models e$ is replaced by the requirement that H be true in the minimal Herbrand model of $B \wedge e$: this setting is called learning from interpretations. We will follow this in our formal model.

In order to search the space of relational rules (program clauses) systematically, it is useful to impose some structure upon it, e.g. an ordering. One such ordering is based on subsumption (clause C subsumes C' if there exist a substitution θ, such that $C\theta \subseteq C'$). Notice, that if C subsumes D then $C \models D$. The converse always not hold. Notice, that the space of clauses ordered by the subsumption is a lattice.

The ILP system ALEPH (Srinivasan, 2000, Aleph) is based on inverse entailment (Muggleton, 1995). For a given background knowledge B and examples E and the hypothesis H it must hold, that $(B \wedge H) \models E$. If we rearrange the above using the law of contraposition we get the more suitable form $(B \wedge \neg E) \models \neg H$. In general B, H and E can be arbitrary logic programs but if we restrict H and E to being single Horn clauses, $\neg H$ and $\neg E$ above will be ground skolemised unit clauses. If $\neg\bot$ is the conjunction of ground literals which are true in all models of $B \wedge \neg E$ we have $(B \wedge \neg$
$E) \models \neg\bot$. Since $\neg H$ must be true in every model of $B \wedge \neg E$ it must contain a subset of the ground literals in $\neg\bot$. Hence $(B \wedge \neg E) \models \neg\bot \models \neg H$ and so $H \models \bot$.

The complete set of candidates for H could in theory be found from those clauses which imply $\bot$. A subset of the solutions for H can then be found by considering those clauses which subsume $\perp$. ALEPH searches the latter subset of solutions for H that subsume $\perp$. $\perp$ is called saturation of example.

2.3 A monotone graded ILP problem

In a monotone graded ILP problem (Horváth, Lencses, Krajči, Vojtáš, 2004) data are not clearly divided into positive and negative examples, i.e. there is a monotone hierarchy (degree, preference) of more or less positive (negative) examples. This corresponds to fuzzy set of examples. We assume also on the side of background knowledge a monotone graded (comparative) notion of fulfilment. This corresponds to fuzzy background knowledge in the form of a definite logic program (without negation). We expect to be able to extract rules of the form

IF the satisfaction with ERP-human resources is at least 4 (or better 5, 6, 7)
AND the company size is at least 6 (more than 500 employees)
THEN the competitiveness in administration is at least 4 (or better 5, 6, 7)

Notice that we assume a positive (monotonic, increasing) influence of background factors on the degree of classification (understood in a monotonic way).

We transfer the problem of graded ILP with fuzzy (graded) background knowledge and fuzzy set of examples (graded examples) to several crisp ILP problems, so that c(B) is the knowledge acquired from B by adding an additional attribute for the truth value and for every $\alpha \in [0, 1]$ E_α^+ and E_α^- are cuts of the fuzzy set E. The fuzzy hypothesis $H \mid\!\!\!\backslash\ H_\alpha \equiv \alpha$.

Problem is that the system means the numbers like a syntactic objects and it do not distinguish the ordering between them. Therefore we must define this ordering in the background knowledge – background knowledge of ALEPH can contain rules. Since the truth value (TV) of the atoms in the background knowledge/hypotheses determines the maximum/minimum degree of compatibility ("at most"/"at least") it is convenient to define for every graded (fuzzy) predicate in the background knowledge a rule

$$\text{predicate}(X_1, X_2, ..., X_n, TV_a) \text{ :- } TV_a < TV_b, \text{predicate}(X_1, X_2, ..., X_n, TV_b).$$

This rule we rewrite to a

$\text{predicate}(X_1, X_2, ...,X_n, TV_a)$:- $\text{leq}(TV_a,TV_b)$ $\text{predicate}(X_1, X_2, ...,X_n, TV_b)$, where $\text{leq}(TV_a,TV_b)$ evaluates the relation „TV_a is less or equal than TV_b". Hereby, we add ground atoms $\text{leq}(TV_1, TV_2)$, ..., $\text{leq}(TV_{n-2}, TV_{n-1})$ such that for $i < j$ holds $TV_i < TV_j$, and TV_1 is more than the lowest grade (TV_{min}), while TV_{n-1} is the greatest grade (TV_{max}). We do this, because we need to generate the truth values (for the saturation) and not to compare them. In our case we add to the background knowledge following facts leq(1,2). … leq(5,6). leq(6,7). and for every graded background knowledge predicate rule like
sales_marketing(A,C) :- leq(C,D), sales_marketing(A,D).
human_resources(A,C) :- leq(C,D), human_resources(A,D). , etc.
Except these, the background knowledge contains facts (for every attribute and

object) like human_resources(object3,7), which means, that the company no. 3 is satisfied at the grade 7 with software for human_resources.
For example, the saturation of one example looks like

administration(A) :- company_size(A,4), company_size(A,3), company_size(A,2), company_size(A,1), manufacturing(A,6), manufacturing(A,5), manufacturing(A,4), manufacturing(A,3), manufacturing(A,2), manufacturing(A,1), finance(A,5), finance(A,4), finance(A,3), finance(A,2), finance(A,1).

One of the assets of this method is that we can define the ordering. We must tell, that in this case except the rules the background knowledge consist similar predicates (similar arity and the domain of attributes), but ILP works effectively in the case of complicated background knowledge, too.

The rules in the result hypothesis must subsume the saturations of some (all) positive and must not subsume the saturations of any negative examples. Some rules from the hypotheses evaluated by expert:

At the grade 4
marketing(A) :- sales_marketing(A,7), human_resources(A,1).
At the grade 5
marketing(A) :- sales_marketing(A,4), finance(A,4), human_resources(A,7).
buying(A) :- manufacturing(A,4), finance(A,7).
buying(A) :- sales_marketing(A,7), finance(A,7).
sales(A) :- sales_marketing(A,6), manufacturing(A,6), finance(A,5).
sales(A) :- company_size(A,2), manufacturing(A,1), human_resources(A,7).
At the grade 6
sales(A) :- sales_marketing(A,4), finance(A,6), human_resources(A,7).

Glueing hypotheses together Moreover rule obtained on the level α guarantees the result in degree α, so it corresponds to a fuzzy logic program rule with truth value α (because in body there are crisp predicates and the boundary condition of our conjunctors fulfil $C(x; 1) = x$.
The first rule corresponds to fuzzy rule
(marketing(A) :- sales_marketing(A,7), human_resources(A,1).).4
The second rule says
(marketing(A) :- sales_marketing(A,4), finance(A,4), human_resources(A,7).).5
and so on.

Here we see limitations of fuzzy logic programming in the induction, we are not able to glue them to one hypothesis. On the other side, these rules define a single annotation term for every predicate "p" in the heads of rules - a function of 9 real variables (body can contain 9 atoms) – $a_p(x_1, x_2, ..., x_9)$.
If there is no such rule then the function is the smallest monotone function extending those points, i.e.

$$a_p(x_1,x_2, ...,x_9) = \max\{a_p(y_1,...,y_9) : y_i \leq x_i \text{ for every } i=1,...,9\}$$

For example ,if the system for the predicate „sales" at grade α has induced the rule
sales(A) :- company_size(A,x_1), sales_marketing(A,x_2), manufacturing(A,x_3), finance(A,x_4), human_resources(A,x_5), mis(A,x_6), eis(A,x_7), partial_outsourcing(A,x_8), total_outsourcing(A,x_9)., then $a_{sales}(x_1, x_2, ..., x_9) = \alpha$.

Another challenging problem is to learn the function a_p, methods of (Železný, 2001) could be appropriate.

3. RESULTS AND CONCLUSION

Figure 1 represents how well can be the impact of information systems on main processes identified by linear regression and by ILP. Regression is evaluated by the coefficient of determination (R^2) because it represents the ratio of explained dispersion. Other seven bars represent the ratio of correctly classified positive instances (examples) to all positive instances for a certain grade (α = 1, 2, ..., 7) of competitiveness. Both scales are ratio scales and can be well interpreted.

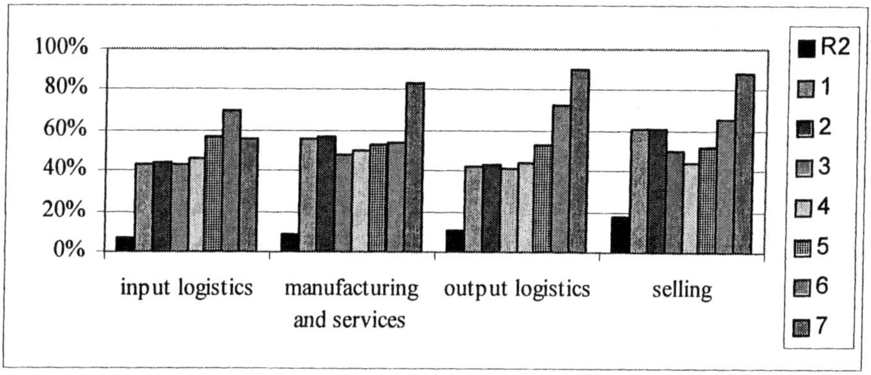

Figure 1 – Impact of information systems on main processes

Figure 2 represents how well can be the impact of information systems on subsidiary processes identified by linear regression and by inductive and logic programming.

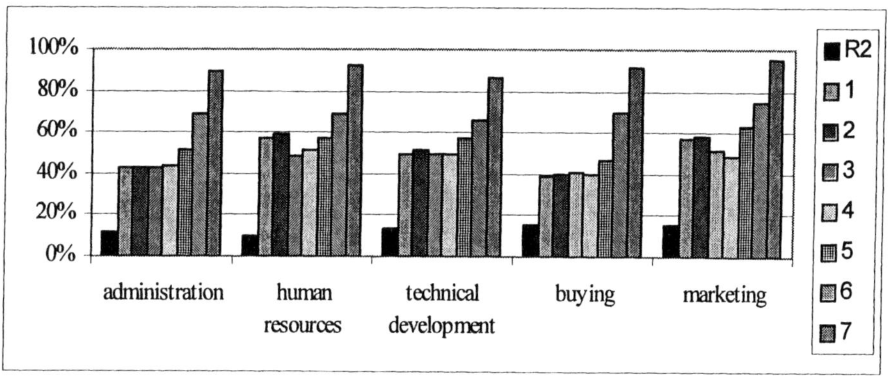

Figure 2 – Impact of information systems on subsidiary processes

To sum up, inductive logic programming on the given data yields better results in every dimension of competitiveness than regression. It could be expected that large homogenous sets of data will be in most cases better explained by inductive logic programming than by regression. As for the impact of information systems on business competitiveness, the results give evidence that business competitiveness is to some extent influenced by information systems. Differences between R^2 and other bars in Figures 1 and 2 suggest that the impact of information systems is not too linear but it is worth to invest into information systems because their impact is monotonous (non-decreasing). We plan to enhance this method in the future and compare it with other statistical methods.

4. REFERENCES

1. Doucek P. "Integrated Information Systems Improvement". In Strategic Management and its Support by Information Systems, J. Kaluža, ed. Ostrava: VŠB - University of Technology, 1999.
2. Dudinská E, Mizla M. Management information systems. Journal of Economics 1994; 42: 230-38.
3. Džecroski S, Lavrač N. "An introduction to inductive logic programming". In Relational data mining, S. Džeroski, N. Lavrač, eds. Berlin: Springer, 2001.
4. Horváth T, Krajči S, Lencses R, Vojtáš P. An ILP model for a monotone graded classification problem. Kybernetika 2004:Volume Znalosti 2003.
5. Kifer M, Subrahmanian VS. Theory of generalized anotated logic programming and its applications. J. Logic Programming 1992; 12: 335-67.
6. Kokles M, Romanová A. Information Age. Bratislava: Sprint, 2002.
7. Krajči S, Lencses R, Vojtáš P. A comparison of fuzzy and annotated logic programming. Accepted in Fuzzy Sets and Systems.
8. Lloyd JW. Foundation of logic programming. Heidelberg: Springer 1987.
9. Muggleton S. Inductive logic programming. New Gen Comp 1991; 8: 295-318.
10. Muggleton S. Inverse entailment and Progol. New Gen Comp 1995; 13: 245-86.
11. Muggleton S, Fhis C. "Efficient induction of logic programs". In Proc. Algorithmic Learning theory, Tokyo: Ohmsha, 1990.
12. Porter M. How competitive forces shape strategy. Harvard Business Review 1979; 57: 137-45.
13. Quinlan JR. Learning logical definitions from relations. Machine Learning 1990; 5: 239-66.
14. Shibata D et al. "An induction algorithm based on fuzzy logic programming". In Proc. PAKDD'99, Ning Zhong, Lizhu Zhou, eds. Beijing: Springer, 1999.
15. Srinavasan A. The Aleph Manual. Tech Rep Comp Lab Oxford Univ. 2000 at http://web.comlab.ox.ac.uk/oucl/research/areas/machlearn/Aleph/aleph.html
16. Vojtáš P. Fuzzy logic programming. Fuzzy sets and Systems 2001; 124: 361-70.
17. Železný F. "Learning functions from imperfect positive data". In Proc ILP 2001, C. Rouveirol, M. Sebag, eds. Springer, 2001.
18. ILP system ALEPH for Win http://www.comlab.ox.ac.uk/oucl/research/areas/machlearn/Aleph/aleph.pl.

49 AN APPLICATION OF MACHINE LEARNING FOR INTERNET USERS

Machová Kristína
Technical University of Košice, Department of Cybernetics and Artificial Intelligence
Kristina.Machova@tuke.sk
SLOVAKIA

This paper presents new trends in machine learning. It contains a short survey of classic methods of machine learning. Meta-learning, Boosting and Bagging are characterized in the paper as well. The paper focuses on solving the problem of Internet users' cognitive load decrease based on machine learning methods. It presents the AWS system designed to suggest Internet pages to a user on the base of his/her model. The system also offers information about visitor models for the purpose of the server content management. The AWS system is an advisory system with off-line learning capabilities, individual adaptation, and the support of global server content adaptation.

1. INTRODUCTION

Machine learning methods can be divided into supervised (with teacher) and unsupervised (without teacher) methods. A further division can be based on the type of learning task: classification or sequential task. A great deal of these methods can be found in (Mitchell, 1997) and (Machová, 2002). We will assume, that the input of a cognitive algorithm has the form of a set of training examples. The algorithm produces a concept as its output – we expect to obtain a definition of that concept as well. This definition can be of different sorts. It means, that learning methods can use various representations of knowledge to be learned. In case of the classification task, the definition can often be in the form of a classification rule. "Then" part of this rule usually represents a class (concept), to which a new example will be classified. "If" part of the rule contains a concept definition in the form of attributes with appropriate values – conditions whose meeting will be sufficient for the classification of a new example into a given class (concept). The concept definition can be of various kinds: logical conjunction, production rule, decision tree, decision list, threshold concept, criterion table, probabilistic concept and so on. Moreover, various representations of learned concepts have in common basic learning principles, for example ordered version space, hill-climbing principle, division of example space into subspaces, control with exceptions, competitive principle, score function and reduction of the number of concept versions (Machová-Paralič, 2003).

An extension of the machine learning task is represented by the field of meta-learning. Its aim is to obtain a set of rules for determining how to select the best cognitive algorithm for a given cognitive task. These rules could be incorporated into a knowledge-based system dedicated to meta-learning. (Bensuan, 2000) presents a method which is able to obtain these rules using the supervised learning approach (e.g. using an Instance Based Learning method). A task description is based on landmarks which can be of different types, for instance decision node, arbitrary selected node, the worst node, naïve Bayes, 1NN, elitist 1NN, and linear discriminator. A value of a landmark can be determined as an average error over example space. This given method employs also meta-attributes based on information theory (class entropy, average attribute entropy, mutation information, equivalent attribute number, etc.). A disadvantage of this approach is that it is based on a high number of experiments which enables to measure classification error for different combinations of learning algorithms and databases. Further research activities could be focused on uncovering general relationships between cognitive algorithms and cognitive tasks.

2. BAGGING AND BOOSTING

Bagging and boosting represent general methods for improving results of a selected classification machine learning algorithm. Both of them modify a set of training examples in order to obtain a sequence of classifiers which can be subsequently combined into a final classifier.

Bagging (Breiman, 1994) performs random selections of data from the training set. Based on these random selections, a learning algorithm produces a sequence of results – classifiers. It is possible to obtain a final classifier by selecting from this sequence of classifiers.

On the other hand, boosting (Schapire, 1999), (Quinlan, 1996) modifies the set of training data using a distribution of weights assigned to particular examples. When the first classifier is being induced, the weight distribution is uniform. For every subsequent iteration, the weights are modified. The weights of those examples which are not correctly classified by the classifier induced in the previous iteration step are increased. And the weights of correctly classified examples are decreased. The prediction of a final classifier is given as a weighted combination of predictions of particular basic classifiers. One of the surprising and recurring phenomena observed in experiments with boosting is that the test error of the generated classifiers usually does not increase as its size becomes very large, and often it is observed to decrease even after the training error reaches zero. This phenomenon is related to the distribution of margins of the training examples with respect to the generated voting classification rule, where the margin of an example is simply the difference between the number of correct votes and the maximum number of votes received by any incorrect label. The most known boosting algorithm is AdaBoost, which significantly reduces the error of any learning algorithm that consistently generates classifiers whose performance is a little better than random guessing. (Freud-Schapire, 1996) presents two sets of experiments. The first set compared boosting to Breiman's bagging method when used to aggregate various classifiers including decision trees. The performance of boosting using a nearest-neighbour classifier is studied in the second set of experiments.

Bagging and boosting can be used in any domain in which machine learning methods can be employed, for example in extracting knowledge for knowledge bases, data mining, etc. (Schapire-Singer, 2000) presents algorithms which learn from examples to perform multi-class text and speech categorisation tasks. The presented approach is based on an implementation of the boosting algorithm for text categorisation tasks, called BoosTexter. This system is applied to automatic call-type identification from unconstrained spoken customer responses.

3. AN APLICATION OF MACHINE LEARNING

3.1 Problem definition

At present, the Web represents one of the most used Internet based services. The number of accesses of various users is almost unbelievable. The Web consists of a vast number of web pages. It is not uncommon a case when a user stops its browsing through pages which seem to be uninteresting (or unattractive) for him/her although the searched information is present on these pages. Another issue is, that a user searching the Web can be currently interested in some other information than during his/her previous visits. Moreover, a huge number of links were accumulated among web pages. The basic feature of the Web – hyper-textual links representing relationships among pages – can be a source of difficulties (turning to a real nightmare) when browsing the Web. These problems related to information search and retrieval can be measured by a user cognitive load.

The problem is addressed by the AWS system striving for decreasing the cognitive load of Internet users. The focus of the system is on supporting an adaptive web. The adaptive web is able to adapt itself to its visitors – the adaptation is based on an observation of users' activities (the behaviour of users) during users' visits of the Web.

The AWS system focuses on the development of user models from users' requirements. Such model type can be used to customise the response to a user requirement – the user is provided only with those documents which are relevant to his/her profile (i.e. his/her model). User models are constructed using heuristic machine learning methods. The learning is based on logs of web servers. The AWS system represents an advisory system with off-line learning, individual adaptation (customisation for each particular user based on his/her individual model), and the support for global server adaptation (transforming pages into the form suitable for majority of visitors).

3.2 Used methods

The system employs two methods for heuristic search of concept space (namely HGS and HSG) which belong to supervised methods of machine learning and are applicable for solving the classification task. In addition, a clustering method (CLUSTER/2) belonging to unsupervised learning methods was used by this system.

Machine learning is generally based on a set of training examples and achieved results are tested using a set of test examples. Training and test examples constitute a set of typical examples. The typical examples are represented as a set of *n* attributes with their values. The last attribute can represent (in case of supervised learning) a class to which the given example belongs.

A set of typical examples is the most often given in the form of a table. An example is given in Table 1 presenting typical examples in the form used by the AWS system. This table contains typical examples characterised by an attribute *A* and belonging to a class *T*. The examples represent accesses to server pages. The attribute *A* (url) characterises those pages which were accessed (each page is stored on the server together with a set of key words which characterise the content of the page). The attribute *T* (user ID) identifies users who accessed the given pages and in this way it specifies the class to which the given accesses belong. It is quite common, that several users visit the same page – and the same typical example is classified into more than one class at the same time.

Table 1. A set of typical examples obtained from a log of a www server

Number	A (url)	T (userID)
1	/som.php	USER3eafc6cd8c98a
2	/maxnet.php	USER3eafc6cd8c98a
3	/ns_top.php	USER3eafc6cd8c98a
4	/cobweb.php	USER3eafc71274ad9
5	/id3.php	USER3eafc71274ad9
6	/c45.php	USER3eafc71274ad9
7	/pid1.php	USER3eafc7413857e
8	/psd1.php	USER3eafc7413857e
9	/plc.php	USER3eafc7413857e

Classification represents the decision on a class of a new example (with unknown class) based on definitions of available classes which were constructed using some machine learning method.

The AWS system relies on using HGS (Heuristic General to Specific) and HSG (Heuristic Specific to General) methods (Michalski, 1980) and (Machová, 2002). Both methods differ from exhaustive search of concept space – they do not search all concept space but the most promising hypotheses only. How promising particular hypotheses are can be calculated by a score heuristic function. Each algorithm iteration considers only a limited number of hypotheses (with the highest score) – this number is defined as Beam Size (BS).

Both methods (HGS and HSG) use the principle of limiting the concept space to be searched. They differ in the used direction of search. The HGS algorithm searches the concept space from more general concept descriptions to more specific (GS search direction). On the other hand, the HSG algorithm searches the concept space from more specific concept descriptions to more general ones (SG search direction).

Clustering methods can be applied when training examples do not contain any information about the class they belong to. In this case they can be grouped into natural groups or clusters using techniques of unsupervised learning (there is no feedback in the form of a class defined in advance). The clustering process starts with a set of objects – training examples. The aim is to create a set of clusters and all available training examples to distribute over the set of clusters. In general, it is possible to distinguish several different approaches to clustering: iterative, conceptual, hierarchical, and probabilistic. The AWS system employs the CLUSTER/2 clustering method (Michalski, 1983).

3.3 The AWS system

The AWS system (Machová-Klimko, 2004) was designed with the aim to enable suggestions of pages to a user based on his/her model and to carry out an individual adaptation of the content of a server. The user model is generated as a result of the heuristic search of concept space using the HGS and HSG algorithms.

At the same time, the system provides information about models of server visitors and their interests in order to support server content management. In this way it contributes to customising the server to users – it supports a global server adaptation. This feature of the system is backed up by the CLUSTER/2 clustering technique.

As depicted in Figure 1, the system consists of two parts: on-line and off-line. The on-line part is responsible for the identification of visitors and subsequent generation of suggestions based on visitors' models. The off-line part of the system is responsible for development of user models and providing information vital for the adaptation of the server content. The learning itself is performed utilising information about the content of a server and a log of the given server. The application requires the server log in the NCSA Combined Log format.

The shared part of the system is represented by a database storing user models and server logs with information about processed user requirements. The AWS system enables to identify a visitor using his/her IP address or using cookies. The identification using cookies seems to be more suitable.

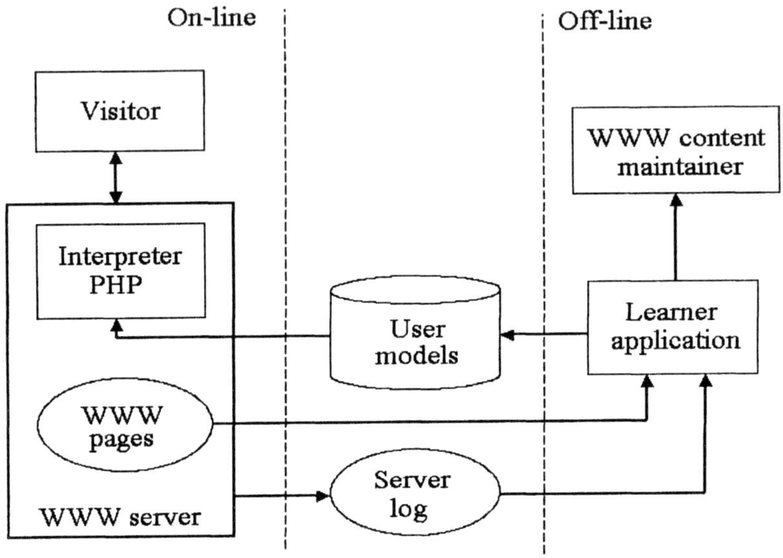

Figure 1: Structure of the AWS system

The on-line part consists of a www server, PHP interpreter, and a database of user models. A visitor sends his/her requirements on the www server. The server delivers these requirements to the PHP interpreter. The interpreter identifies the visitor and generates a query into the database. Using this query the interpreter retrieves addresses and names of pages to be suggested (based on the model of the identified visitor) and sends this suggestion to the visitor together with the page which was required by the visitor. If cookies are used to identify visitors, then in case that the visitor cannot be identified (e.g. because the user accesses the server for the first time or he/she deleted cookies in his/her web browser) a new unique identifier is generated. This identifier is sent to the client and stored on his/her disc in the form of a cookie for a subsequent identification.

The off-line part of the system represents a system mainstay. It consists of the ASW/Learner application. This application is responsible for learning (creation of user models using heuristic algorithms), populating the developed user models with relevant web pages, and for the support of the server content management based on clustering of the user models using a clustering algorithm.

3.5 A business application

The AWS system can be used in the field of advertisement as well. The system generates models of users from accesses of these users to particular web pages. These models reflect (not only) professional interests of the users. Based on this information, it is possible to recommend some product to those users who can be potentially interested in it.

For example, if a user frequently visits pages devoted to a particular software product, the AWS system can recommend him additional software packages or new versions of the product. Another example is represented by the case when the user(s) represent(s) a company. The model of this user/these users generated by the AWS system can play the role of a company profile. In consequence, new apparatus, products, or technological methods and procedures can be offered to the company representatives.

4. CONCLUSIONS

The paper focuses on machine learning and some of it's new trends, specifically meta-learning, bagging and boosting. This contribution presents also an application possibility of mentioned fields, namely solving the problem of cognitive load of Internet users by means of machine learning. A description of the AWS system is presented – the system which was designed as an advisory system with off-line learning capabilities, possibility of an individual adaptation and with the support for a global content adaptation.

The presented approach can be further extended using a method for automatic extraction of key words from documents in order to replace the manual web page description/annotation, for example using the method presented in (Paralič-Bednár, 2003).

The work presented in the paper was supported by the Slovak Grant Agency of Ministry of Education and Academy of Science of the Slovak Republic within the 1/1060/04 project "Document classification and annotation for the Semantic web".

5. REFERENCES

1. Bensusan, H., Giraud-Carrier, C.: Casa Batlo is in Passeig de Gracia or how landmark performances can describe tasks. http://www.metal-kdd.org/, 2000.

2. Breiman, L.: Bagging predictors. Technical Report 421, Department of Statistics, University of California at Berkeley, 1994.

3. Freund, Y., Schapire, R.E.: Experiments with a New Boosting Algorithm. Machine Learning: Proc. of the Thirteenth International Conference, 1996.

4. Machová, K.: *Machine learning. Principles and algorithms* (in Slovak), ELFA s.r.o., 2002, Košice. ISBN 80-89066-51-8.

5. Machová, K., Klimko, I.: Support of the adaptive WEB by means of machine learning. Proc. of the conf. Znalosti 2004, Brno, Czech republic, 2004, VSB-Technical University Ostrava, 218-225, ISBN 80-248-0456-5.

6. Machová, K., Paralič,J.: Basic Principles of Cognitive Algorithms Design. Proc. of the IEEE International Conference Computational Cybernetics, Siófok, Hungary, 2003, 245-247 ISBN 963 7154 175.

7. Michalski, R.S.: Pattern Recognition as Rule-guided Inductive Inference. *IEEE Trans. on Pattern Analysis and Machine Intelligence, 2*, 1980, 349-361.

8. Michalski, R.S., Stepp, R.: Automated Construction of Classification: Conceptual Clustering versus Numerical Taxonomy. *IEEE Transactions on Pattern Analysis and Machine Intelligence, No.5*, 1983, 219-243.

9. Mitchell, T.M.: Machine Learning. The McGraw-Hill Companies, Inc. New York, 1997, 414.

10. Paralič, J., Bednár, P.: Text Mining for Documents Annotation and Ontology Support. A book chapter in: "Intelligent Systems at the Service of Mankind", Volume 1, Ubooks, 2003, 237-247, ISBN 3-935798-25-3.

11. Quinlan, J.: Bagging, boosting, and C4.5. In Proc. AAAI 96, Portland, OR, 1996, 725-730.

12. Schapire, R.E., Singer, Y.: Improved Boosting Algorithms Using Confidence-rated Predictions. Machine Learning, 37(3), 1999, 297-336.

13. Schapire, R.E., Singer, Y.: BoostTexter: A Boosting-based System for Text Categorization. Machine Learning, 39(2/3), 2000, 135-168.

50

EVALUATING A SOFTWARE COSTING METHOD BASED ON SOFTWARE FEATURES AND CASE BASED REASONING

Christopher Irgens,
University of Strathclyde,
Department of Design, Manufacture and Engineering Management,
75 Montrose Street, Glasgow G1 1X, UK.
E-mail:chris.irgens@dmem.strath.ac.uk

Sherif Tawfik,
Arab Academy for Science and Technology
Information and Documentation Centre,
P. O. Box: 1029, Alexandria, EGYPT.
E-mail:Sherif226@hotmail.com

Lenka Landryova,
VSB-Technical University of Ostrava,
Department of Control Systems and Instrumentation,
tr.17. Listopadu 15, 708 33 Ostrava, CZECH REPUBLIC.
E-mail:lenka.landryova@vsb.cz

A common weakness of most software cost estimating models is their limited usefulness to predict the cost accurately and quickly at an early stage of the software development life cycle. This is due to the lack of data about the code quantity and code complexity at that point in the software cycle.. A method for software cost estimation has been developed to address the problem of the early and accurate establishing of cost based on Case Based Reasoning. Using the software type, functions and primitive ***(x)R(y)*** *Z-operations as parameters for characterising and handling the cases.*
The output from the model has been analysed with respect to actual recorded historical test-case cost.

1. INTRODUCTION

The software cost estimation in the early stages of the software development life cycle depends on the requirement specification of the software. The approach is to divide the software system into standard tasks, functions, and operators. The CBR process is made easier since one software system provides accumulated knowledge of a number of standard functions; each in turn providing accumulated knowledge of a number standard operators, all of which can be expressed in Z. Cases can be characterised and constructed according to standard functions and operations/operators expressed in Z. In this way the need for a large *project case-base* is diminished. Thus the main CBR disadvantage is also diminished, (Aamodt et al,1994), (Kolodner, 1993). This provides a 'featuring' mechanism that can be used to project information as has already been shown valid and useful in the product design domain (Irgens,1995).

Although there have been examples of successful CBR tools for software cost estimation (Mair, 1999), (Finnie, 1997), (Schofield, 1998), (Prietula et al, 1996). Most existing methods use the available software features to predict the software size. Then this size is used as input to one of the algorithmic models such as COCOMO or Function Point Analysis to estimate the cost. Other research concludes that in specific application domains, process control application, it is possible to estimate the software size from the user specified application features. The limitation here is that the application is restricted to a very specific domain (Mukhopadhyay et al, 1992). Sarah Jane Delay states that it appears impossible to identify features early in the life cycle that defines the size of the project (Delay, 1998).

2. THE METHOD

The new method is mimicking the successful approach taken in the product design domain (Irgens, 1995) by creating a 'feature-based' approach based upon the decomposed functionality of the software as specified during the design stage of the life-cycle. In order to provide the necessary standard notation and rigour, the use of a formal specification language was used.

Z is a formal specification language, which has been progressively developed and widely applied since its inception at Oxford University in 1980 (Sommerville, 1997). Formality implies precise, unambiguous description in the tradition of mathematics. Z is based on typed set-theory, coupled with a structuring mechanism (i.e., the schema calculus is one of its key features). A schema is a collection of variable declarations and predicates giving the relationships between the variables. This basic construct is used to structure the description of a system. The schema is divided into two main parts: declarations (or signature) and predicate. Declarative information such as object names and types are given in the signature. The predicate section provides the relationships between those objects that must hold. Preconditions and post-conditions typically are not labelled explicitly. Z describes both the state space and transitions on the states and places no restrictions on the style of specifications. The schema calculus further defines the rule of operations on schemas such as inheritance, composition, and information hiding.

1.1 The Model Mechanism

The implemented model is seen as two main parts: The first part is designed for handling the operations of adding and updating the cases in the case library. The second part is designed to enable the software project manager to determine the project effort in the early stage of the software development life cycle.

1.1.1 Part One (Adding and Updating Cases).

This part consists of programs for handling the data which have been gathered from previous historical software projects (cases) into the case library. Each of these programs has standard functions to make it simple for the project manager to handle the data in the case library.

1.1.2 Part Two (Software effort estimation process)

For determining the software project effort estimation, the project manager will go through one or more than one stage in this implemented part. These stages are as follows:

Stage (A)

The first step in this part will ask the project manager to specify the categorisation level of the software project and then to specify the system activity class that the new system belongs to.

There are two possible directions in this stage. If there are historical cases found, these projects will be displayed. Otherwise, the user will be directed to stage (B).

The user can choose one or more from the historical projects, and choose those most suitable to the new task. If the new software systems tasks are covered completely in this step, then the project manager will go directly to the next step. Otherwise, he will proceed to stage (B).

The project manager will select the appropriate project features, which affect the computing of the total estimated effort for the new project.

In this step, the model will use all the input data for estimating the new project effort and gives the result to the project manager.

In this step, the project manager could adapt the result or accept it as it is and add it to the case library.

Stage (B)

If there is no previous project for both the input categorisation and the input system activity class, or there is a new task that is not matched by any previous projects for the same categorisation and system activity class, the project manager has to divide every task into it's primary functions and specify every function by using the Z function specification to count the number of occurrences for every operator in each function in the task.

The project manager has to input the type of every function in the new task and complete by entering the number of occurrence of every operator type. Historic cases are retrieved using the nearest neighbour matching technique. The project manager can accept the effort suggested or he can modify the result by himself and store the result. The project manager will stay in step 2 until finished entering all the required new tasks of the system.

3. THE EVALUATION OF THE METHOD

The objective of the evaluation process is to measure the extent to which the model meets its predicted performance, it is necessary that the evaluation includes:

- Evaluation of the retrieval algorithm in the CBR function;
- Evaluation of the method used in the CBR function to calculate the cost;
- Evaluate the output result with respect to the actual cost.

3.1 Test Data Collection

The historical software function and cost data was needed in order to build the case library used to evaluate the prototype. The data collection was done with the help of the staff in the Information and Documentation Centre in the collaborating establishment The Arab Academy for Science and Technology and Maritime Transport (AASTMT) and was based upon historical cost information from software systems developed for internal use.

Four large completed historical software projects provide the functional and historical cost data used to evaluate the implemented model. These projects were:

- Pharmacy Inventory system;
- Clinic accounting system;
- Food Inventory system;
- Cafeteria Sales system.

3.2 Data documents

A number of documents are used in AASTMT to manage and record its software projects. The test data was collected from historical project information regarding task, functions, operations and associated spent project hours. Furthermore, the estimation method requires design information so that the software may be characterised by function type and operations profile. The number and types of Z operators are used for this purpose. Therefore a *Function Specification in Z-Notation* is also required. Each project is simplified to its basic functions. All functions are described in Z-notation and each predicate is further simplified into simple predicate in the form of ***(x) R (y)***. The number and types of the function's simple predicates composes the function's characteristic feature and can be used for the case-based search and reasoning mechanism. Every function is therefore summarised and characterised using the number of occurrences of every Z operator. This operation/operator profile forms the key feature for every basic function.

Table 1 shows a sample of Z function specification for the simple function **'add-account'**, while table 2 shows the number of occurrences of Z operators of the form ***(x)R(y)*** in **'add-account'**. These operator sets are the key *features* for every basic function forming the functions characteristics. In this manner the software project may be characterised by function type and operator density by type.

Table 1, A Sample of Function Specification in Z-Notation

add-account
HB1-mainacc : HB1mainacc
HB1-mainacc' : HB1mainacc
HB1-mainacc-record : account-key ⇸ account-data
S-mainacc-no? : mainacc-no
S-mainacc-name? : mainacc-name
S-acc-kind? :: = Madin | Dain
a-data : **P** account-data
Mess! : report
(S-mainacc-no? ∉ Dom HB1-mainacc-record
a-data = (S-mainacc-name?,S-acc-kind?)
HB1-mainacc ' = HB1-mainacc ∪{S-mainacc-no? ↦ a-data}
Mess! = "Main Account was added")
∨
(S-mainacc-no? ∈ Dom account-record
Mess! = "Main Account already exist")

Table 2, The number of occurrences of Z-operators in **'add-account'**

Operator	Number of occurrences
:	7
?	8
!	3
⇸	1
∉	1
↦	1
=	4
∈	1
∪	1
If-else	1
:: =	1
P	1
|	1
,	1

4. EVALUATION RESULTS

Using the Food Inventory system tasks as test case for the model produced the results in Table 3.

The error was measured using Mean Magnitude Relative Error (MMRE):

MMRE= $\sum_{i=1}^{n}\left(\left|(estimate_i - actual_i)/actual_i\right|\right)/n$, where $estimate_i$ is the estimated effort in hours, from the model, $actual_i$ is the actual effort, and n is the number of 'projects'. To establish whether model is biased, the Mean Relative Error (MRE) was used:

$$MRE_i = (estimate_i - actual_i)/actual_i)$$

Table 3 shows the estimated task-hours against the actual historical records, with the corresponding MRE and MMRE values.

Table 3, Estimated hours against actual recorded hours

Task number	Estimated hours based on CBR	Actual hours	MRE for CBR estimates
AS.01	6.00	6.00	0
AS.02	8.59	8.00	0.073
AS.03	7.58	7.50	0.010
AS.04	13.54	13.50	0.003
AS.05	6.00	6.00	0
AS.06	5.00	3.00	0.667
AS.07	3.00	3.00	0
AS.08	3.00	2.50	0.200
AS.09	3.00	2.00	0.500
AS.10	3.00	2.50	0.200
AS.11	2.00	2.00	0
AS.12	3.75	2.50	0.500
AS.13	2.50	2.50	0
AS.14	3.75	3.00	0.250
AS.15	8.40	8.00	0.050

Giving an MMRE=0.164.

5. CONCLUSION

The prototype was implemented in Microsoft Windows environment using Access DBMS for the working data. The results obtained were satisfactory showing acceptable variation from actual historical values. The collaborating partner has consequently advised its information centre to continue the work in order to develop the prototype for practical purposes.

The limitations can be summarised as follows:

1. The software engineers needed some time to become familiar with the model, specially the CBR function part, which depends on the good familiarity with Z notation, and the process of decomposing each complex predicate into a group of a simple predicates.
2. The implemented prototype is a first version prototype, thus it was difficult to collect and analyse more than the four historical software cases used. However, due to the nature of the method, the resultant decomposition of each software project into its constituent functions and (x)R(y) simple predicates allowed the evaluation of the method as reported above.

It is also quite clear to the authors that:

1. The strengths of both the expert judgment method and the analogy method are combined within the implemented prototype.
2. The new approach uses the software specification, which is closer to the user requirement. Moreover, instead of specifying the software project as one entity, it divides the software system into standard functions, each function is represented in standard Z notation, thus mimicking the successful feature based methods used in product design.
3. The CBR process is made effective since one software system provides knowledge of a number of standard functions; each in turn provides knowledge of a number of standard operators. In this way the need for large number of historical projects is diminished. Thus, by using a small number of historical projects the case library can be adequately populated.

6. REFERENCES.

1. Aamodt, A., Plaza, E. "Case-Based Reasoning: Foundational Issues Methodologies, Variations, and System Approaches", AI Communications. Vol. 7, pp (39-59), 1994.

2. Delay, Sarah Jane et. al. "The Limits Of CBR In Software Project Estimation". 6th German workshop on case based reasoning (GWCBR'98), eds. L.Gierl , M. Lenz, Berlin, pp. (99-108), 1998.

3. Finnie, G. R "A comparison of software effort estimation techniques: using function with neural networks, case-based reasoning and regression models". Journal of systems and software, Vol 39, pp (281-289), 1997.

4. Irgens, C. "Design Support based upon the projection of information across the Product Development life-cycle by means of Case Based Reasoning", Journal: IEE Proceedings on Science, Measurement and Technology Special issue on Manufacturing, Sept.1995, pp:345-349,Volume: 142, Issue:5, ISSN: 1350-2344.

5. Kolodner, Janet. "Case-Based Reasoning". Morgan Kaufman Publishers, California, 1993

6. Mair, Carolyn et. al. "An Investigation of Machine Learning Based Prediction System". 9 July, 1999. http://dec.buth.ac.uk/ESERG

7. Mukhopadhyay, Tridas ; Kekre, Sunder. "Software Effort Models For Early Estimation Of Process Control Applications". IEEE Transaction on software engineering. Vol. 18, No. 10, PP (915,924), August 1992.

8. Prietula M. et. al. "Software Effort Estimation With A Case Based Reasoning". Journal of Experimental and Theoretical Artificial Intelligence. 8(3-4) PP (341-363), 1996.

9. Schofield, Chris. "Non-Algorithmic Effort Estimation Techniques". Department of Computing, Bournemouth University. ESERG: TR 98-01. 1998.

10. Sommerville, Ian; Sawyer, Pete. "Requirements Enginerring A good Practice Guide". John Willy & Sons Ltd. 1997.

51

REDUCTION TECHNIQUES FOR INSTANCE BASED TEXT CATEGORIZATION

Peter Bednár, Tomáš Futej
Dept. of Cybernetics and Artificial intelligence, Technical University of Kosice, Letna 9, 042 00 Košice, SLOVAKIA

One of the most common problems in instance-based learning of text categorization is high dimensionality of feature space and problem of deciding which instances to store for use during generalization. These problems can be solved with use of reduction methods. In this paper, comparison of three reduction techniques for feature space reduction and one algorithm for reduction of storage requirements is presented. These techniques were combined with k-NN (k-Nearest Neighbors) classifier, which is one of the top-performing methods in the text classification tasks. We describe the benefit of this combination of methods and present results with the Reuters-21578 dataset.

1. INTRODUCTION

Text categorization is the problem of automatically assigning predefined categories (or classes) to text documents [3]. While more and more textual information is available, effective information retrieval is difficult without indexing of document content [1].

Document categorization is one solution to this problem. One of the top performing methods for text categorization is instance based, k-nearest neighbors, classifier. Main disadvantage of this method is high time complexity and high memory requirements. In this paper, we describe various reduction techniques, which can solve these problems.

2. INSTANCE BASED LEARNING - kNN CLASSIFIER

Example-based classifiers do not build an explicit, declarative representation of the categories, but rely on the category label attached to the training documents similar to the test document.

The kNN algorithm [3, 4, 5] is simple: given a new document, the system finds the k nearest neighbors among the training documents, and uses the categories of the neighbors to weight the category candidates. The similarity score of each neighbor document to the new document is used as the weight of the categories of the given neighbor. By sorting and thresholding the scores of candidate categories, binary category assignments are obtained. For convenience, the cosine value of two document vectors is used to measure the similarity between the documents, although other similarity measures are possible.

3. FEATURE SELECTION

One difficulty of text categorization problems is high dimensionality of the feature space. Feature space can consist of hundreds or thousands of unique terms (words or phrases) that occur in documents. We have evaluated combination of three methods, including document frequency, information gain and mutual information.

3.1 Document frequency thresholding

Document thresholding (DF) [2] is one of the simplest techniques for feature space reduction. Document frequency is the number of documents in which a term occurs. All unique terms that have document frequency in training set less than some predefined threshold were removed. The basic assumption is that rare terms are either non-informative for category prediction, or not influential in global performance. Improvement in categorization accuracy is also possible if rare terms happen to be noise terms.

3.2 Information gain

Information gain (IG) [2] measures the number of bits of information obtained for category prediction by knowing the presence of a term in a document. The information gain of term t is defined to be:

$$\begin{aligned} G(t) = & -\sum_{i=1}^{|C|} P(c_i)\log P(c_i) + \\ & P(t)\sum_{i=1}^{|C|} P(c_i \mid t)\log P(c_i \mid t) + \\ & P(\bar{t})\sum_{i=1}^{|C|} P(c_i \mid \bar{t})\log P(c_i \mid \bar{t}) \end{aligned} \tag{1}$$

where $P(c)$ is the probability of the category c, and $P(c \mid t), P(c \mid \bar{t})$ denotes conditional probability of category c given the presence or absence of term t.

3.3 Mutual information

Mutual information (MI) [2] is criterion commonly used in statistical language modeling of word associations. Mutual information can be estimated using:

$$I(t,c) \approx \log \frac{AN}{(A+C)(A+B)} \quad (2)$$

where A, B, C, and D are cells of the two way contingency table of term t and category c and $N = A + B + C + D$.

Given a training corpus, for each unique term we have computed the information gain or mutual information and removed from the feature space those terms whose IG or MI was less than some predetermined threshold.

4. INSTANCE REDUCTION

For instance selection, we have adopted algorithm called Decremental Reduction Optimalization Procudere 4 (DROP) [6]. This procedure is decremental, meaning that it begins with the entire training set, and then removes instances that are deemed unnecessary. DROP4 uses following basic rule to decide if it safe to remove an instance i from the set of training instances S:

Remove instance i from S if at least as many of its associates in T would be classified correctly without i.

To see if an instance i can be removed using this rule, each *associate* (i.e. each instance that has i as one of its neighbors) is checked to see what effect the removal of i would have on it. Instance will be removed if its removal does not hurt the classification of the instances in T (according to the F1 accuracy measure). DROP4 removes instances in the center of a category cluster and it can remove noisy instances, because a noisy instance i usually has associates that are mostly of a different category. DROP4 initially sorts the instances by the distance to their nearest enemy (i.e. nearest instance with a different category), because order of removal can have influence on instance reduction. DROP4 algorithm has additional noise-filtering pass before sorting of instances, which is based on rule similar to *Edited Nearest Neighbor* rule. It states that any instance misclassified by its k nearest neighbors is removed (if it does not hurt the classification of its associates).

5. EXPERIMENTS

We have tested described reduction techniques and their combinations on Reuters-21578 corpus. We use the ModApte version, which was obtained by eliminating unlabelled documents, and selecting the categories, which have at least one document in the training set and the test set. This process resulted in 90 categories in both the training and testing set.

The first experiment was used to setup basic parameters (such as k and c – which is the threshold for category assigning). The accuracy of the baseline k-NN classifier (defined as F1 measure) was 0.77. We can probably achieve the better result with different shareholding techniques, for example, sets threshold for each category by using cross-validation) [4].

The second experiment was oriented on selection of relevant terms based on document frequency of unique terms. According to the results for document frequency reduction (Figure. 1), useful terms have DF between 50 to 2000. Classification with selected terms does not decrease the performance of the classifier and feature space was reduced to 5.6% of the original size. The similar results are for information gain reduction (Figure. 2). If we have removed terms with lower value of IG under the 90% of terms, precision increase to 0.78. Time needed for classification was reduced 5 times. MI thresholding has different effect on performance on k-NN classifier (Figure. 3). MI is biased towards low frequency terms and this make significant accuracy loss in text categorization.

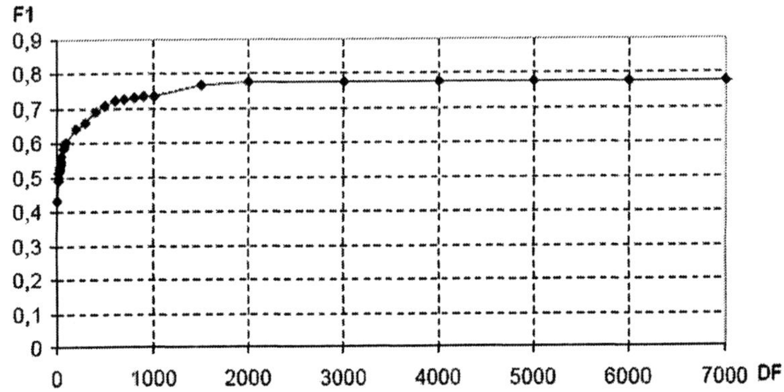

Figure 1 – Document frequency thresholding feature space reduction

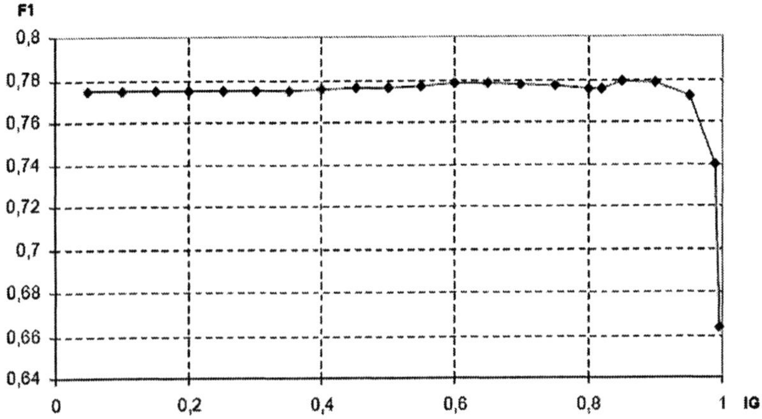

Figure 2 – Information gain feature space reduction

According to the results of experiments in (Figure. 4), the performance of classification was lower when we have used DROP4 algorithm (number of instances was reduced to 36% of original size of training set). Highest influence on this has first phase of DROP4 that is targeted to remove noisy instances. At this point, performance goes down by 2% with remove of 101 instances. Since we have categories, which have only few instances, these instances are consider being a noise and are removed by first phase of the algorithm. When we have used only basic instance of DROP algorithm called DROP1 without ENN phase the performance of classifier has risen to 0.79.

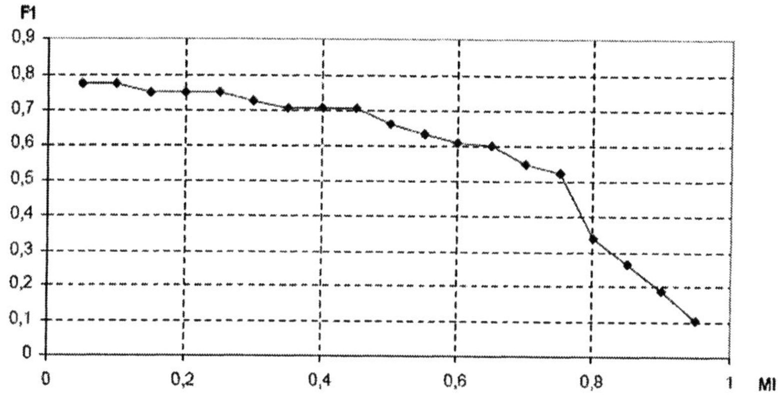

Figure 3 – Mutual information feature space reduction

6. CONCLUSION

This paper describes various algorithms for feature space and instance space reduction. In the first section of the paper, we have described some of the problems of instance based learning in text categorization task and how these problems can be solved by reduction techniques. In second section, we have introduced the results of the practical test of reduction techniques on standard Reuters benchmark. According to the results, the reduction techniques have positive influence on instance-based classification. We have achieved better performance and lower time and space complexity with DF and IG thresholding for feature space reduction and modified DROP algorithm for instance space reduction.

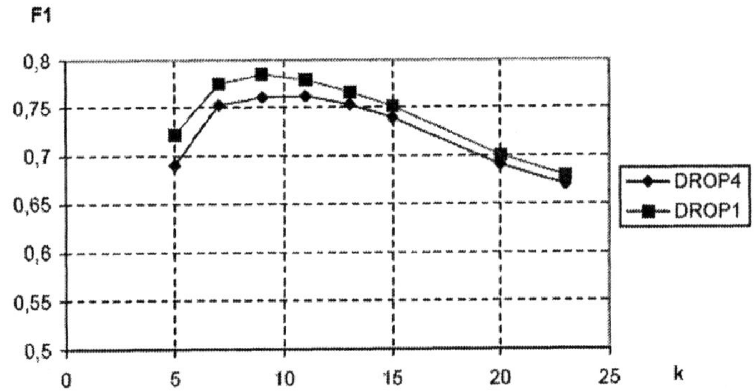

Figure 4 – F1 measure of instance space reduction techniques

6.1 Acknowledgement

This work is done within the VEGA project 1/1060/04 "Document classification and annotation for the Semantic web" of Scientific Grant Agency of Ministry of Education of the Slovak Republic.

7. REFERENCES

1. Ontology-based Information Retrieval, by J.Paralic and I.Kostial. In Proc. of the 14th International Conference on Information and Intelligent systems, IIS 2003, Varazdin, Croatia, ISBN 953-6071-22-3, 23 - 28, 2003.
2. Yang, Y., Pedersen J.P. A Comparative Study on Feature Selection in Text Categorization Proceedings of the Fourteenth International Conference on Machine Learning (ICML'97), 1997, pp412-420.
3. Yiming Yang and Xin Liu A re-examination of text categorization methods. Proceedings of ACM SIGIR Conference on Research and Development in Information Retrieval (SIGIR'99, pp 42--49), 1999.
4. Yiming Yang A study on thresholding strategies for text categorization, Proceedings of ACM SIGIR Conference on Research and Development in Information Retrieval (SIGIR'01), pp 137-145, 2001.
5. Han, E. H., Karypis G., Kumar, V. Text Categorization Using Weight Adjusted k-Nearest Neighbours.
6. Wilson D. R., Martinez T., Reduction Techniques for Instance-Based Learning Algorithms, Machine Learning 28(3):257-286, 2000.

52 APPLICATION OF SOFT COMPUTING TECHNIQUES TO CLASSIFICATION OF LICENSED SUBJECTS

Jiří Kubalík[1], Marcel Jiřina[1], Oldřich Starý[2], Lenka Lhotská[1], Jan Suchý[1]

[1] *Department of Cybernetics, CTU Prague*
Technická 2, 166 27 Prague 6, CZECH REPUBLIC
e-mail: kubalik@labe.felk.cvut.cz

[2] *Faculty of Electrical Engineering, CTU Prague*
Technická 2, 166 27 Prague 6, CZECH REPUBLIC
e-mail: staryo@fel.cvut.cz

This paper presents an application of soft computing techniques to the construction of decision support tool used for identifying the economically unstable licensed subjects. The work has been initiated by the Czech Energy Regulatory Office whose main mission is to guard the regular heat supply without significant disturbances. Thus the main goal is to develop a tool for automatic identification of the companies that could cancel the supply due to economic problems without detailed examination of each company. In order to achieve the goal two approaches have been chosen. The first one is based on development of an aggregate evaluation criterion for assessing the firms. The other one uses artificial neural networks and multivariate decision trees induced with genetic programming for classification of the firms.

1. INTRODUCTION

The presented work has been initiated by the Czech Energy Regulatory Office (ERO) whose main mission is to guard the regular heat supply without significant disturbances. Authors were involved in developing the methodology for marking the possibly problematic licensed heat and co-generation facilities that can have some problems with the financial and economic stability and therefore the energy supply could be threatened in the near future. The ERO gathers the big amount of both technical and economical data but it is difficult if not impossible to process all the information for thousands of licensed subjects. Thus the main goal is to develop a tool for automatic identification of the companies that could cancel the supply due to economic problems without detailed examination of each company.

In order to achieve the goal two approaches have been chosen. The first one is based on development of an aggregate evaluation criterion for assessing the firms.

The solved problem can be restated as a knowledge mining task, where given the existing database of firms' records one wants to extract the knowledge of what is a good and what is a bad firm (measured in terms of economic stability). If each record is assigned an indicator that expresses its stability then the task belongs to the class of supervised learning. Once a model acquiring the knowledge contained in the presented training database is built it can be used for classification of new records with unknown economic stability. In this work we use *artificial neural networks* and *multivariate decision trees* for modeling the knowledge. The multivariate decision trees are generated by *genetic programming*.

The rest of this paper is organised as follows. The next section introduces the Aggregate Evaluation Criterion, followed by sections describing the multivariate decision trees induced with genetic programming and the implementation of an artificial neural networks. We then outline the dataset and the utilized experimental methodology. The following section provides the results achieved with the decision trees and neural networks and the paper closes with conclusions.

2. AGGREGATE EVALUATION CRITERION

The original data set provided by ERO consists of raw descriptions of firms without indication of their economic stability. In order the data could be used for learning the decision tree and neural network model an economic stability value of each firm of the given training data have to be determined.

A set of five relevant risk factors and stability criteria has been selected at first. It consists of measure of long-term indebtedness, short-term financial position, operational return on assets, average equipment amortization and sales stability. All of them are real-valued attributes. Then the function called *aggregate evaluation criterion* (AEC) of financial stability that transforms the five criteria into just one has been developed, see (Beneš & Starý, 2003). This function was used to assess the stability of the given firms. The evaluated firms can be sorted using this function, the firms with the maximum value signals to ERO to focus to them. Several firms have been examined in detail in order to approve a correctness of the AEC. It turned out that AEC gives a reasonably correct ranking of firms.

The next step in the development of the decision support system was to transform the knowledge contained in the database of labeled records into the form of (1) multivariate decision trees and () artificial neural network, which will be used for classification of the firms in the future.

3. MULTIVARIATE DECISION TREES

Decision tree (Quinlan, 1986) is a tree whose internal nodes are tests (on input attributes) and whose leaf nodes are categories. Each branch (path to another node) represents a value that the attribute might take. To classify a case, the root node is tested as a true-or-false decision point. Depending on the result of the test associated with the node, the case is passed down the appropriate branch, and the process continues. When a terminal node is reached, its stored value is the answer. A decision tree is constructed top-down. In each step a test for the actual node is chosen - starting with the root node - which best separates the given examples by classes. Usually, the quality of the test is measured by the information gain. The test applied to the dataset in the given node splits the data into several subsets, each of them representing the data of the corresponding child node. Every child node is further expanded – the best split function is found, generating its descendants – until the stopping condition is fulfilled in the newly generated node. The stopping criterion is usually defined as the maximum acceptable amount of information contained in the node's data. So the process ends in a given node when the data contain samples belonging to only one class or some class significantly dominates in the node's data. The node is then assigned the identifier of that class.

In standard decision trees the test in the inner node is a test on the value of certain attribute "$attr_i < v_{ij}$", where v_{ij} is the value chosen from the domain of the i-th attribute. In this work we use *multivariate decision trees* (MDTs) with the tests of the form "$f(a_1,\ldots,a_n) < 0$", where f is an arbitrary function of the input attributes a_i using specified operations and operators (Brodley & Utgoff, 1995). Obviously such decision trees are more general than the standard decision trees, which allows to better model the given training data, see Figure 1.

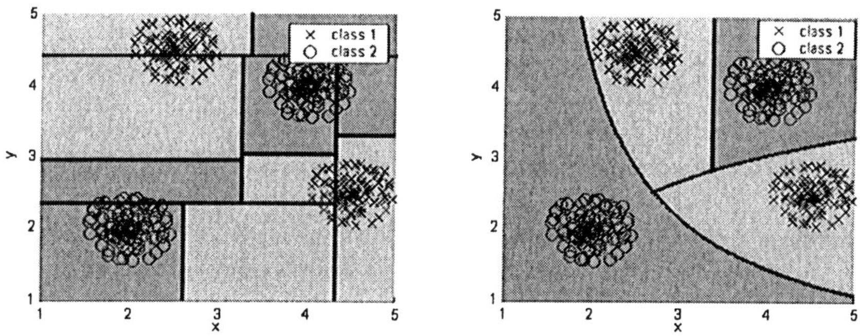

a) b)

Figure 1 – Illustration of partitioning of the pattern space of synthetic data of two classes a) by the standard decision tree and b) by the multivariate decision tree. The multivariate decision tree uses test functions composed of operators +, -, *, and /.

Utilization of genetic programming

When constructing the MDT the crucial point is to find the optimal function when new inner node is to be added into the tree. For this purpose genetic programming (GP) has been used in this work. GP is a powerful technique for automatically generating computer programs (Koza, 1992), (Bot & Langdon, 2000). GP operates on population of candidate solutions, each represented as a hierarchical parser tree – expressions representing test functions are evolved here. The complexity of evolved trees is determined by the set of operators and elementary functions. All individuals are evaluated i.e. assigned a fitness value expressing a performance measure of the represented solution. In our application the fitness reflects the quality of the split generated by given test function. A population of diverse individuals is then evolved generation by generation by means of reproduction, crossover and mutation operators. The process of evolving the population runs until some stopping criterion is fulfilled; usually it runs for some pre-specified number of generations. The best solution encountered during the whole run is then returned as the final solution.

4. ARTIFICIAL NEURAL NETWORKS

There are many paradigms of artificial neural networks. The most known and widely used is the multilayer perceptron networks (MLP), see e.g. (Bishop, 1995), (Rojas, 1996). The MLP consists of several layers of neurons called perceptron. Each perceptron calculates a post-synaptic activity (potential) as a weighted sum of its inputs and generates an output by means of an activation function. The activation function is often a logistic sigmoid or hyperbolic tangent. Frequently, the activation function of the output layer is linear, mostly the simple identity. A network of interconnected perceptrons represents powerful computational system capable of solving complex nonlinear tasks.

Thresholding. Thresholding is a process of assigning a class identifier to input pattern. Each of the five output neurons corresponding to one class can take a real value from the interval ⟨0.0, 1.0⟩. Generally, the resultant class is just the one that corresponds to the output neuron with highest value of its output.

In this work the situation is a bit different. As the boundary between adjacent classes are not sharp it may happen that for some input patterns the response of the network would be misleading in a sense that the correct class membership differs from the one that corresponds to the output neuron with the highest value. In such situations the above mentioned simple *winner-takes-it-all* strategy fails and thus is inappropriate for our purposes.

To resolve this problem we used a thresholding method that works as follows. First, two output neurons with the maximal value – the two most probable classes – are found. Then a relative difference between the two outputs is calculated. If the relative difference is less than a given threshold – indicating that the difference between the two most probable verdicts is not significant enough – then the pattern is assigned the higher (worse) class of the two ones. Otherwise, the pattern is assigned the class corresponding to the output neuron with the highest value. This thresholding strategy can be interpreted so that if any doubts of which class should be assigned to the pattern then the more pessimistic one is chosen – the worse rating is assigned to the firm. The threshold value 0.2 was used in this work.

5. REAL DATA AND EXPERIMENTAL METHODOLOGY

Classification of licensed subjects (firms) is based on five input parameters that describe different features of individual licensed subjects. To each licensed subject a category is assigned. The firms have been split into five classes according to their AEC value. The best firms with the lowest AEC value belong to class one, the worst ones belong to class five. Such a categorization is required by the ERO. The primary goal of the classification task is to identify the most unstable firms as reliably as possible in order not to miss any incompetent licensed subject.

First, the raw data were preprocessed so that each input parameter was saturated to minimal and maximal values and then the range of each input parameter was linearly scaled to <0.0, 1.0>. This adjustment is generally appropriate to eliminate large differences in the parameters.

The database provided by ERO consists of 704 records with highly unbalanced distribution of the classes, see Table 1. Classes 2 and 3 strongly dominate in the data whilst class 5 has only 16 records, i.e. this class represents only 2.3 per cent of the database. Obviously such a distribution of the classes is very bad.

Table 1 - Numbers of patterns for individual classes

Class	1	2	3	4	5
#patterns	16	286	332	54	16
% of patterns	2.3	40.5	47.2	7.7	2.3

Multivariate Decision Trees

When learning a model describing the data the whole dataset is split into training and test data so that the training data are used for training purposes and the test data are used for evaluating the final model. It is evident that applying such a concept on our data would hardly yield some general description of the training set that would correctly classify records from the test set as well. Due to this fact we decided to decompose the classification task into two parts. In the first step a *classifier I* that classifies to the following four classes is used:

- class A (corresponds to the original class 1),
- class B (corresponds to the original class 2),
- class C (corresponds to the original class 3),
- class D (union of original classes 4 and 5).

In the second step the records labeled as class D will be processed by *classifier II* which will separate class 4 records from class 5 ones. In both steps the distribution of classes in processed data is better than in the original dataset. In case of the classifier I the most important class D receives 10% of records in the data and in case of the classifier II the most important class 5 has 23% of records in the data.

In addition each of the classifiers consists of several MDTs in order to increase the robustness of the classifier. In order to determine the final verdict of such an *ensemble classifier* a simple majority rule is used – the most frequent class identifier out of the four returned answers is considered the final output of the classifier. In case of a draw the higher class identifier is taken as the final output of the classifier. This set up causes the classification to work in favor of the higher classes so a firm is always assigned the worse rating when it is on the edge.

The concept of the ensemble classifiers requires that the individual MDTs are as distinct as possible. Otherwise there would be no profit from combining multiple trees. The simplest way to obtain a set of unique trees is to use different training data for induction of each tree. The whole data set has been split into four equally sized disjunctive parts, each of them with the same proportion of classes as in the original dataset. For the learning purposes three data partitions were used the last one was used for testing the generated tree. This leads to *four different learning scenarios* that would hopefully generate four different MDTs. The division of data into four parts has been chosen with respect to the number of records of class 5, which is 16. Thus each of the learning scenarios has 12 records of class 5 in the training set and 4 records of class 5 in test set. A number of MDTs were generated for each *learning scenarios* and best representatives of each scenario were used in the final ensemble classifier. This should ensure that each MDT is to some extent unique so the final classifier should generalized well on new unseen data.

Artificial Neural Networks

When used the neural network approach the original data with 704 records were split by random into three disjunctive sets: training, validation and test sets. The database was divided into these classes in the proportion 2:1:1.

We tested three and four layered MLPs. Each MLP has five inputs and five outputs. The classification to individual classes is thus performed in code one-from-*N*. The number of hidden neurons varied from 15 to 30. The best results were obtained by the three layered MLP with 30 hidden neurons.

The post-synaptic activity is calculated by means of weighted sum of inputs in both hidden and output neurons. The activation function for the hidden layer is hyperbolic tangent and for the output layer is calculated by means of the *Softmax* rule, i.e. normalized exponential function along all outputs. This ensures that the individual outputs are in the range 0-1. Therefore, values of the outputs can be interpreted as probabilities of membership to the individual classes.

The combination of methods of the *backpropagation* (100 epochs) and conjugate gradient descent (20 epochs) was used for training of the MLP. The learning rate was 0.01 and momentum term 0.3.

6. RESULTS

This section presents results achieved with the classifier based on MDTs and provides a comparison with results of the ANN classifier. The experiments were performed using 4-fold cross-validation in order to get four different learning scenarios as described above. For each of the four training-test data configurations ten experiments were carried out and the success rates averaged over the 40 experiments are shown in tables.

Table 2 provides results achieved with the trees generated for classifier I. The average number of inner nodes of the generated trees was 12. Average accuracies on both training and test datasets show that the induced trees classify classes B (2) and C (3) with much better accuracy than the classes A (1) and D (4&5), which is given by the distribution of the classes in the data. Column "D/5" says that on average 83% and 78% of data belonging to class 5 were correctly classified as class D in training and test datasets, respectively. In other words, 83% (78%) of data belonging to class 5 proceeded to the next step where the classifier II is involved.

Table 2 – Success rates of MDTs generated for classifier I

Class	A	B	C	D / 5	Total
Training data [%]	22	89	88	**65 / 83**	85
Test data [%]	9	80	82	**56 / 78**	77

Table 3 – Success rates of classifier II

Class	4	5	Total
Training data [%]	100	**100**	100
Test data [%]	85	**38**	74

Table 3 shows average success rates of trees generated for classifier II. The average number of inner nodes (test functions) of the generated trees was 7. The trees were trained only on data labeled by classifier I as D. It shows that the trees were perfectly trained to classify the training data. In contrast the performance on the test data drops to 38%. Rather poor generalization ability results from an insufficient number of training data of class 5 – learning from 12 positive samples in 5-dimensional space can hardly be successful. However, the average number of correctly classified samples of class 5 is $1.00 \times 12 + 0.38 \times 4 = 13.5$. This is sufficient accuracy with respect to the fact that the final classifier II as well as the classifier I will be assembled from four trees, each trained on partially different data (different learning scenario).

Ensemble classifiers I and II were assembled from trees chosen according to the following criteria:

- Simple trees measured by the number of inner nodes were preferred. This criterion reduces the possibility the tree is over-learned to the training data.
- Trees best classifying data of class 5 were preferred.

Performance of final ensemble classifiers I and II as well as the classification accuracy of the compound classifier are presented in Table 4. We observe a considerable improvement in the accuracy of classification of class 3, 4 and 5 when compared to the accuracy achieved with single trees. The most important observation is that classifier I correctly classifies 80% of data of class D (classes 4&5). Among the data labeled as D all records of class 5 are present, which are perfectly identified by classifier II. On the other hand, data belonging to class 1 are not classified at all. This is because we did not make any special arrangement with the aim to correctly learn class 1 as we did for class 5.

Table 4 – Success rates of the ensemble classifiers and the compound classifier

Class	1	2	3	4	5	Total
Ensemble classifier I [%]	0	81	95	80 / **100**		**85**
Ensemble classifier II [%]	-	-	-	97	**100**	**≈100**
Compound classifier [%]	0	81	95	70	**100**	**85**

Table 5 - ANN success rate on training, validation, and test set

Class	1	2	3	4	5
Training set [%]	89	96	92	85	**78**
Validation set [%]	100	95	91	29	**75**
Test set [%]	75	94	85	69	**67**

Results achieved with ANN are summarized in Table 5. It shows results on training, validation and testing sets. The use of a simple neural network like MLP with 30 hidden neurons seems to be sufficient. The pattern space can be separated properly by hyper-planes and their combinations. The achieved overall quality of the classification is very good with respect to the given real problem and the available data. A drawback of the task is that the last fifth class contains only 16 patterns so it is difficult do make serious conclusion about classification to this class. Fortunately, utilizing the suggested thresholding can improve classification to this class. In practice, only some licensed subjects from the class 4 are classified to the class 5 but no licensed subjects from classes lower than 4 are wrongly classified to the class 5.

7. CONCLUSIONS

This paper presents an application of soft computing techniques to the construction of decision support tool used for identifying the economically unstable licensed subjects. The original data consists of raw descriptions of subjects without indication of their economic stability. First, the *aggregate evaluation criterion* has been developed for assessing the firms. Then the data labeled by AEC were used learning the classifiers based (1) on the *multivariate decision trees* and (2) on the *artificial neural network*. Achieved results show the classifiers work well on the given data. Moreover the proposed compound classifier based on multivariate decision trees is very robust so it is expected to work well on new previously unseen data as well.

It is evident that both the multivariate decision tree model and the neural network model generated using the data labeled by the AEC can only approximate the AEC. From this point of view the utilization of the models might seem useless. On the other hand, the AEC is a static model with its own parameters that are hard to tune for particular data. In particular the AEC is tailored to the currently available data and as such it might become irrelevant for the task when the situation for which it was developed changes - new data are provided or new evidence about the current firms is revealed. In such case the multivariate decision trees and neural networks would be preferred for the following reasons:

- Both types of models can be easily regenerated in order to fit well the training data when they change. As this is a long-term project new and updated data will be provided each year so the ability to adapt the model is very important.
- New factors characterizing the performance of firms can be used and easily incorporated into the models. Example of which might be the trends in attributes as the history of existing firms will be available after few years.
- Ensemble classifiers can be constructed. As the results achieved with ensemble decision tree based classifiers show a proper combination of a number of individual classifiers leads to robust classifier.

Moreover, the MDT and ANN models can be used for validating the AEC on the current data so that if both models return a different economic stability value than the AEC does for given firm it might signal the AEC is not well formed for the data.

8. ACKNOWLEDGMENTS

This research work was supported by the Grant Agency of the Czech Republic within the project No. 102/02/0132.

9. REFERENCES

1. Beneš, M., Starý, O.: Risk Ranking and Selection of Licensed Subjects. In: New Challenges for Energy Decision Makers [CD-ROM]. Cleveland: IAEE, vol. 1, 2003.
2. Quinlan, J.R. Induction of decision trees. *Machine Learning*, 1, pp. 81-106, 1986.
3. Brodley, C.E., Utgoff, P.E. Multivariate decision trees. *Machine Learning*, 19, pp. 45-77, 1995.
4. Koza, J. Genetic Programming: on the programming of computers by means of natural selection. Cambridge, MA: The MIT Press, 1992.
5. Bot, M.C.J., Langdon, W.B. Application of genetic programming to induction of linear classification trees. *Proceedings of EuroGP 2000*, LNCS 1802, pp. 247-258. Springer, 2000.
6. Bishop, C. M. *Neural Networks for Pattern Recognition*. Oxford University Press, New York, 1995.
7. Rojas, R. *Neural Networks: A Systematic Introduction*. Springer-Verlag, Berlin, Heidelberg, 1996.

53

ONE-CLASS LEARNING FOR HUMAN-ROBOT INTERACTION

QingHua Wang
IEETA, University of Aveiro,
Campus Universitário Santiágo, 3810-193, Aveiro, PORTUGAL
qhwang@ieeta.pt

Luis Seabra Lopes
IEETA/Department of Electronics & Telecommunication,
University of Aveiro,
Campus Universitário Santiágo, 3810-193, Aveiro, PORTUGAL
lsl@det.ua.pt

A Suitable learning and classification mechanism is a crucial premise for Human-Robot Interaction. To this purpose, several one-class classification methods have been investigated using wavelet features (parameters of Hidden Markov Tree model) in this paper. Only target class patterns are used to train class models. Good discrimination over outlier (never seen non-target) patterns is still kept based on their distances to class model. Face and non-face classification is used as an example and some promising results are reported.

1. INTRODUCTION

Robots are expected to exist extensively in many areas of our future daily life. So, a basic requirement arises: they must understand what people mean, e.g., instructions from us. To this end, our work focuses on basic natural language concept learning mainly through visual information in the context of Human-Robot interaction (HRI). Let's consider the task of teaching a robot to recognize an object, say, "apple", through its camera in the context of HRI. How can the teaching be conducted? To investigate some state of the art statistical approaches in literature, e.g., Hidden Markov models (Meng, 2000; Zhu & Schwartz 2002), Bayesian networks (Pham *et al*, 2002), naïve Bayes classifier (Schneiderman & Kanade, 2000), PCA based classifier (Turk & Pentland, 1994), and other state of the art methods described in (Yang *et al*, 2002), basically quite a lot of apples and enough non-apples, which is itself an ambiguous concept, must be collected to estimate the class distributions precisely. One might wonder whether these requirements are realistic in the context of HRI.

Different from other learning tasks, the learning of object recognition suitable for HRI has several constraints such as supervised, online, real-time and interactive. To learn a new object (concept), there may be only quite a limited number of samples available. Moreover, the conventional learning mechanisms mentioned above require the preparation of both target and non-target training data in order to learn the correct decision boundary between these two classes. However, in a HRI scenario, typically only positive samples are available. Thus, a so-called one-class classifier might be useful in this situation. These classifiers can learn target class models based on only target patterns, keeping good discrimination for never seen non-target patterns.

Following this idea, an approach based on the combination of the wavelet domain Hidden Markov Tree (HMT) and Kullback-Leibler Distance (KLD) was presented in (Wang & Seabra Lopes, 2004). In this approach, only target samples are used to train an object model in terms of parameters of HMTs. After that, for each sample to be recognized, its KLD to this model is computed. If its KLD is smaller than a certain threshold, obtained during the training session, it is recognized as an instance of the target class; otherwise, it is rejected. This approach has some similarity to the face detection method presented in (*Yang et al*, 2001). There, multimodal density models were used to capture the class (face) distribution only from target patterns. Based on these models, a probability for certain new patterns to be classified as target or non-target class can be computed. The problem of this HMT/KLD based approach is that it can't derive robust object models if there are big in-class variations among the training patterns.

In this paper, some methods described in (Tax, 2001) are investigated to solve this problem. The rest of the paper is organized as follows. The background and motivation to find suitable learning and classification mechanisms for natural language concept grounding for HRI is presented in section 2. In section 3, a brief description of HMTs and one-class learning is provided. The experimental setup and results are given in section 4. Conclusion is provided in section 5 with some discussion and future work.

2. BACKGROUND

Carl, a prototype of an intelligent service robot, designed having in mind such tasks as serving food in a reception or acting as a host in an organization, was developed by our group (Seabra Lopes, 2002) [i]. One of the long-term research goals of this project is to assign robots a capacity for self-development, based on rich sensory and motor capabilities. The idea is to start with minimal initial knowledge, rather than to assign robots rich knowledge in advance (Seabra Lopes & Wang, 2002). After this, robots can learn new skills, explore their environments themselves or under the human guidance.

As a basic requirement of HRI, Carl is able to enter a spoken language conversation with a person. Speech recognition is currently based on IBM NUANCE (Seabra Lopes, 2002). The recognition grammar being used can accept over 12000 different sentences. Carl is able to learn facts about the world through dialog with humans. For instance, Carl can learn that "*Peter is in France*". Later, hearing the question "*Where is Peter?*" Carl can provide an appropriate reply "France". However, Carl has no idea of what/who France or Peter are. The words (symbols) used in communication to refer to objects still have to be grounded in the robot's own sensory data.

Generally, every word in a conversation should be grounded but, currently, our focus is on grounding of symbols (nouns) that might be a cornerstone to this end, particularly symbols that refer to physical objects in the environment. Therefore, Carl's learning capabilities are being extended to visually recognize objects in a normal office environment. Thus, one can think that we ground the corresponding concepts using visual information or features extracted from visual information. In a learning phase, human tutors teach Carl these concepts. The first stage of grounding turns out to be supervised object learning.

3. LEARNING APPROACH

3.1 One-Class Learning

The design of one-class classifiers is motivated by the fact that patterns from a same class usually cluster regularly together, while patterns from other classes scatter in feature space. One-class learning and classification was first presented in (Moya et al, 1993), but similar ideas also already appeared, including outlier detection (Ritter & Gallegos, 1997), novelty detection (Bishop, 1994), concept learning in the absence of counter-examples (Japkowicz, 1999) and positive-only learning (Muggleton & Firth, 2001). In two-class approaches, information of both target class and non-target class is available. Generally, in multi-class approaches, samples are provided for all classes. Based upon this information, one can precisely capture class descriptions, and therefore find effective decision boundaries between the classes. In contrast, in one-class approaches, only samples of the target class are required. A very natural method for decision-making under this condition is to use some distance-based criterion. If the measurement of an unknown pattern x is smaller than the learned threshold, it can be accepted as the target class pattern; otherwise, it should be rejected. This can be formulated as follows.

$$Class(x) = \begin{cases} \text{target}, & if \;\; Measurement(x) \leq threshold; \\ non\text{–target}, & otherwise. \end{cases} \tag{3}$$

In some sense, this is similar to the famous Bayesian decision rule. The only difference is that, here, the threshold is learned only from target class patterns, while in Bayesian decision rule it's determined based both on target and non-target class patterns. If an appropriate model of the target class (and thus a proper threshold) is found, one can find that most patterns from this target class are accepted and most non-target class patterns are rejected. Of course, the ideal model is one that can accept all target patterns and reject all non-target patterns. But this is usually not easy to find under realistic conditions.

Several methods were proposed to construct models for one-class classification. A very natural method is to generate artificial outlier data (Roberts & Penny, 1996), and thus conventional two-class approaches can be applied. This method severely depends on the quality of artificial data and often works not well.

Some statistical methods were also proposed. One can estimate the density or distribution of the target class, e.g., using Parzen density estimator (Bishop, 1994), Gausian (Parra et al, 1996), multimodal density models (Yang et al, 2001) or wavelet-domain HMTs (Wang & Seabra Lopes, 2004). The requirement of well-sampled training data to precisely capture the density distribution makes this type of methods problematic. In (Moya et al, 1993; Tax, 2001) some boundary-based methods were proposed to avoid density estimation of small or not well-sampled training data. But a well-chosen distance or threshold is needed. Tax provides a systematic description of one-class classification in (Tax, 2001), where the decision criteria are mostly based on the Euclidean distance. The work reported in this paper is based mainly on these methods.

The best results were obtained with SV-DD (Support Vector Data Description (Tax, 2001)), a one-class classification method inspired by Vapnik's Support Vector Machines. SV-DD tries to find a sphere boundary with minimal volume, not a hyperplane as in SVM, containing all or most of objects in a data set. The sphere decision boundary is defined by the so-called support objects or support vectors (Tax, 2001). For classification, objects outside this sphere decision boundary are regarded as outliers (objects from other classes). To obtain a good and compact data description, some remote data points may be discarded although they are not real outliers.

For comparison we also investigate some other one-class classifiers, namely *PCA-DD*, *NN-DD*, *KMEANS-DD*, *KNN-DD* and *GAUSS-DD*. The *PCA-DD* uses subspace composed by Principal Components as data description. The *NN-DD* and *KNN-DD* use simple nearest neighbor and *k*-nearest neighbor data description. The *KMEANS-DD* uses *k*-clusters as data description. In each of these *k* clusters the average distance to its cluster center is minimized. The *GAUSS-DD* assumes data follows simple Gaussian distribution. For more details please refer to (Tax, 2001).

3.2 Hidden Markov Trees

Much work has shown that class distributions can be modeled by modeling the distributions of their wavelet coefficients. Hidden Markov Trees (HMTs) were proposed in (Crouse *et al*, 1998) to precisely characterize these wavelet coefficient distributions, especially the key inter-scale dependencies among parent and children coefficients. In a HMT model, after applying a wavelet transform on an image, any coefficient of it arises from a 2-state zero-mean Gaussian mixture model (Crouse *et al*, 1998). It means the magnitude of a wavelet coefficient is either "large" or "small". Given its state, it is conditionally independent from all other random variables (states of other coefficients).

More specifically, an HMT model for one of three subbands LH, HL, and HH can be fully characterized by the following parameters:

• The pmf (probability mass function) for the root S_1, $\pi = \{\pi_m\}$, $m \in \{1,2\}$. π_m is defined as

$$\pi_m = P(S_1 = m) \tag{1}$$

• The state transition probability matrix A = { $a_{i,\rho(i)}^{rm}$ }, i∈ {2, ... , n}, r, m ∈ {1, 2}. Each $a_{i,\rho(i)}^{rm}$ is the probability when the node *i* is in state m and it parent *ρ(i)* in state *r*. Thus it's defined as

$$a_{i,\rho(i)}^{rm} = P(S_i = m \mid S_{\rho(i)} = r) \tag{2}$$

• The parameters of the zero-mean Gaussian mixture, $\sigma_i^2(m)$, for the state m of each w_i, ∀ i ∈ {1, 2, ..., n}. Here, *n* is the number of coefficients.

To make the HMT model practical, it's assumed that coefficients in each wavelet subband have the same variances (Crouse *et al*, 1998). This assumption, called "tying", reduces HMT parameters to 6 for each subband of each wavelet transform level: 2 for variances and 4 for state transitions. These parameters can be learned using EM algorithm in the sense of maximum likelihood and grouped together as a parametric model, denoted as $\Theta = \{\pi, A_j, \sigma_j^{(m)}\}$, j=1,..., *L* and m=*1, 2*. Here L is the wavelet transform level. For a whole image, three independent HMTs, corresponding to subbands LH, HL and HH respectively, can be used to fully characterize it. For further information on HMTs please see (Choi & Baraniuk, 2001; Crouse *et al*, 1998; Do, 2002; Durand & Gonçalvès, 2002; Fan, 2001; Romberg *et al*, 1999).

4. EXPERIMENTS

4.1 Evaluation approach

The purpose of this paper is to investigate some one-class classifiers available from (Tax, 2001) on HMT features (parameters). The whole evaluation procedure is depicted below in Figure 1 where the specific classifier varies. In a training session, the EM algorithm is applied on target patterns to estimate HMT parameters. Then these HMT parameters are used to train one-class classifiers. For a new pattern, its corresponding HMTs are first estimated and the parameters of these HMTs are fed to learned classifiers for classification.

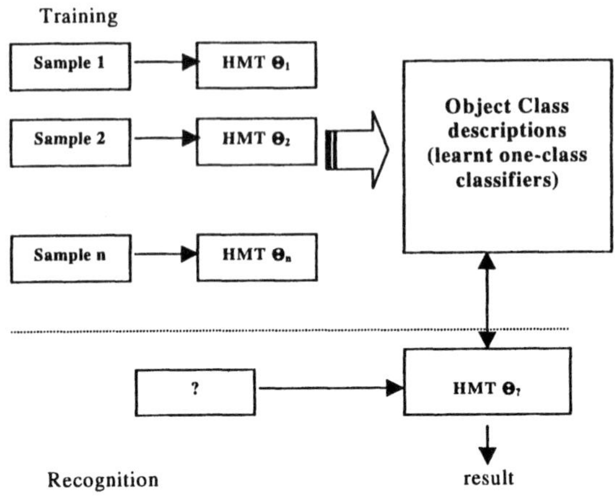

Figure 1-the evaluation scheme

4.2. Experimental Results

All the six one-class classifiers are investigated on the dataset used in (Wang & Seabra Lopes, 2004). This dataset contains two parts. There are 400 pictures from AT&T/ORL face database (AT&T, 2002) and 402 non-face pictures from (Seabra Lopes & Wang, 2002). There are some examples from each part shown in Figure 2 and 3 respectively. It should be noted that all patterns were resized as 32×32. The wavelet transform used in the following experiments is Daub4 (Wang & Seabra Lopes, 2004). Currently, experiments are run into a simulation environment which is based on PRTOOLS (Duin, 2004) and DDTOOLS (Tax, 2001).

Figure 2- Sample images from ORL face database

Figure 3-Some samples in our data set

A series of experiments are conducted in which face is the target class and face patterns are used for training. Only a part of face data is used for training, and then the rest of face data and all non-face data is used for independent testing. There is no non-face data introduced into training session. To know how well the amount of training patterns affects the final classification for each classification method, the number of training patterns is increased from 10% of the face data (40 faces randomly chosen) gradually up to 90% of the face database (360 faces randomly chosen). For a certain amount of face data (10% to 90% of the whole face database), experiments are repeated ten times and the average error rate is used as the final classification score. Thus, a total of 540 experiments were run (6 algorithms, 9 data configurations and 10 trials).

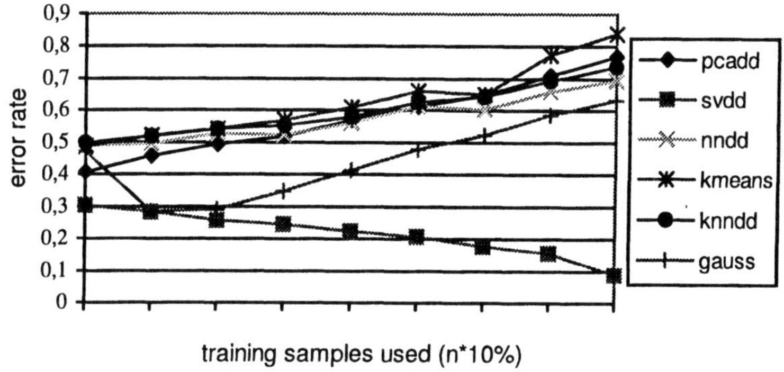

Figure 4-Error rate evaluation of all six one-class classifiers

As one can find from Figure 4, SV-DD always outperforms the other 6 methods in each of those 9 data configurations. The best error rate it can obtain is 8.8% (91.2% accuracy) when 90% faces are used into training. The worst score it achieves is around 30% in error rate when only 10% faces are involved into training. Apparently, for SV-DD can attain good results when more target patterns are involved into training.

The other five one-class classifiers don't show this characteristic. In fact, with more faces used in training, their error rates increase! This looks very unreasonable. With further analysis, some clues are found to explain this behavior. As shown in Table 1, these five classifiers generally work well with face patterns, but they don't work well with non-face patterns. They generally recognize around 80% of non-face patterns as face patterns. Thus, as the number of target patterns used for training increases, the proportion of non-target patterns in the test set also increases, resulting in higher error rates. It can be expected that with some lager data set, with more target class patterns, this might not happen again.

In some sense, several hundreds of training patterns are not easy to prepare in HRI. For the SV-DD method, it shows it can achieve reasonable performances (around 80%) with a small training set (40 to 120 training patterns, 10% to 30%). Since robot learning should be treated as a life-long learning process (Seabra Lopes & Wang, 2002), this moderate performance may form a not so bad starting point.

Finally, a brief comparison is made between the HMT/SV-DD approach and the HMT/KLD based approach described in (Wang & Seabra Lopes, 2004). Applying on the same data set used in this paper, it obtained an average accuracy about 80% on face data. At the same time, it could reject 99% non-face patterns on average although it never sees them before. The main problem of this HMT/KLD based approach is being very data dependent. Over some part of the ATT&ORL face database, its accuracies were as good as 100%; but over some other part, its accuracies were as low as some 45%. The main cause was the simple average method used to compute the overall class model. When there were great in-class variations, it usually failed since the simple average over parameters of individual HMTs smeared the true class model. When the number of training patterns increased, this problem was more serious. This method only worked well when training patterns were very much similar in pose and scale, as under this assumption the parameters of individual HMTs tended to be similar too, thus the average was closer to the true class model.

Table 1- False Negatives (FN) and False Positives (FP) when 40, 120, 240 and 360 faces were used in training

		40 faces	120 faces	240 faces	360 faces
PCA-DD	FN	0,11	0,02	0,01	0,03
	FP	0,66	0,86	0,86	0,84
SV-DD	FN	0,59	0,39	0,26	0,33
	FP	0,09	0,19	0,23	0,16
NN-DD	FN	0,02	0,02	0,02	0,04
	FP	0,89	0,88	0,84	0,77
KMEANS-DD	FN	0,05	0,01	0,01	0,01
	FP	0,86	0,92	0,92	0,92
KNN-DD	FN	0,02	0,01	0,01	0,00
	FP	0,92	0,91	0,86	0,81
GAUSS-DD	FN	1,00	0,17	0,03	0,04
	FP	0,00	0,40	0,66	0,69

In contrast, the SV-DD based approach tries to find a sphere boundary with minimal volume containing all or most of objects in a data set (Tax, 2001). Basically, when more training patterns are involved into training, this sphere decision boundary is more precise. Experiments have demonstrated this. In all experiments, SV-DD shows very steady performances, deviating in the range of ±5% to the average that is used as the final scores. This feature shows that the SV-DD based approach is less data dependent and more robust than KLD based approach.

5. CONCLUDING REMARKS

In this paper, several one-class classification methods were investigated on HMT features. Some of these classifiers can be further integrated into the context of HRI for natural language concept learning (grounding). Constraints such as only limited (target) training patterns available, interactive and flexible learning (several targets to learn simultaneously) require fast one-class classification methods. This is a very crucial step towards flexible concept learning for robots since it can relieve us from data preparation, especially outlier data. Therefore one may focus more on target class representation. In the reported experiments, only target patterns were used to train target class models. And the learned models still have good discrimination with respect to outlier (never seen non-target) patterns.

Six one-class classification methods were evaluated on HMT features. Face and non-face classification is used as an example to demonstrate their effectiveness. It can be found from the experiments that some of such one-class classifiers, particularly SV-DD, can attain very nice performance (>90%), at least on the used data. It also provides a kind of fast learning capacity. For example, when 40 faces were used for training and all the other 762 patterns were used for test, the whole process could be done in less than 1 minute in current MATLAB implementation with a laptop having P4 1.6 GHZ CPU and 256 MB RAM. It still has room to improve, e.g., to implement the algorithm in C. All other five one-classifiers performs not well or very badly on our data. It can be concluded that SV-DD forms a promising foundation for developing a learning and classification method suitable for HRI, since not only can it obtain reasonable performance with a (relative) small amount of training patterns, but also it can achieve very good results when a larger amount of training patterns are available. From a viewpoint of lifelong learning in robotics, this potential of SV-DD can be further utilized.

Obviously it's necessary to further study these one-class learning and classification methods, for example, using other data set and/or feature extraction methods. More importantly, it's interesting to integrate some of these methods into a real robot, Carl (Seabra Lopes, 2002). How to precisely connect this kind of SV-DD to certain target class representation, i.e., natural language concept grounding, should be also further studied. Furthermore, the relation between the class representation for certain target class and specific object representation of that class in terms of such SV-DD should also be considered.

6. ACKNOWLEDGEMENTS

This work is funded by IEETA (Instituto de Engenharia Electrónica e Telemática de Aveiro), Universidade de Aveiro, Portugal, under a PhD fellowship to Q. H. Wang. We thank DSP group of Rice University to let us use their HMT source code (partially). We also thank Dr David M. J. Tax of Delft University of Technology for help on using his tools.

7. REFERENCES

1. AT & T, *The Database of Faces,* formerly "The ORL Database of Faces", at http://www.uk.research.att.com/facedatabase.html, 2002.
2. Bishop C. Novelty detection and neural network validation. In: IEE Proc. Vision, Image and Signal Processing, Special Issue on Applications of Neural Networks 1994; 4: 217-222.
3. Choi H. and Baraniuk RG. Multiscale Image Segmentation using Wavelet-Domain Hidden Markov Models. IEEE Transaction on Image Processing 2001; 9:1309-1321.
4. Crouse MS, Nowak RD and Baraniuk RG. Wavelet-Based Statistical Signal Processing using Hidden Markov Models. IEEE Transaction on Signal Processing 1998; 46:886-902.
5. Do MN. Rotation Invariant Texture Characterization and Retrieval using Steerable Wavelet-domain Hidden Markov Models. IEEE Transaction on. Multimedia 2002.
6. Duin R, PRTOOLS 4.0, Delft University of Technology, The Netherlands.

7. Durand JB and Gonçalvès P. Statistical Inference for Hidden Markov Tree Models and Application to Wavelet Trees. IEEE Transaction. On Signal Processing 2002.
8. Fan Guoliang. Wavelet-Domain Statistical Image Modeling and Processing, Ph.D. dissertation, University of Delaware, 2001.
9. Japkowicz, N.: Concept-Learning in the absence of counter-examples: an autoassociation-based approach to classification, Ph D thesis, New Brunswick Rutgers, The State Univ. of New Jersy, 1999.
10. Meng LM. "An Image-based Bayesian Framework for Face detection", In Proc. of IEEE Intl. Conf. On Computer Vision and Pattern Recognition, 2000.
11. Moya M, Koch M and Hostetler L. One-class classifier networks for target recognition applications, In: Proc. World congress on neural networks (1993), pp. 797-801.
12. Muggleton, S. and J. Firth. CProgol4.4: a tutorial introduction. In S. Dzeroski and N. Lavrac, editors, Relational Data Mining, pages 160-188. Springer-Verlag, 2001.
13. Ritter G and Gallegos M. Outliers in statistical pattern recognition and an application to automatic chromosome classification. In: Pattern Recognition Letters 1997; 525-539.
14. Roberts S and Penny W. Novelty, confidence and errors in connectionist systems. Technology report, Imperial College, London, TR-96-1(1996).
15. Romberg JK, Choi H. and Baraniuk RG. "Bayesian Tree-Structured Image Modeling using Wavelet-Domain Hidden Markov Models", In Proc. of SPIE, Denver, CO, vol. 3813, pp. 31--44, 1999.
16. Parra L, Deco G And Miesbach S. Statistical independence and novelty detection with information preserving nonlinear maps. Neural Compiutation 1996; 260-269.
17. Pham TV, Arnold MW and Smeulders WM. Face Detection by aggregated Bayesian network classifiers. Pattern Recognition Letters 2002; 4:451-461.
18. Schneiderman H and Kanade K. "A Statistical Method for 3D Object Detection Applied to Faces and Cars". In Proc. CVPR 2000, pp. 746-751.
19. Seabra Lopes L. "Carl: from Situated Activity to Language-Level Interaction and Learning", In Proc. IEEE Intl. Conf. on Intelligent Robotics & Systems; pp. 890-896, Lausanne, 2002.
20. Seabra Lopes L and Wang QH. "Towards Grounded Human-Robot Communication", In Proc. IEEE Intl. Workshop on RO-MAN, pp. 312-318, Berlin, Germany, 2002.
21. Tax DMJ. One-class classification. Ph D dissertation, Delft University of Technology, The Netherlands, 2001.
22. Turk M and Pentland A. Eigenfaces for recognition. Journal of Cognitive Neuroscience 1994, 1:71-86.
23. Wang QH and Seabra Lopes L An Object Recognition Framework Based on Hidden Markov Trees and Kullback-Leibler Distance. In Proc. 6th Asian Conf. on Computer Vision: pp. 276-281, Korea, 2004.
24. Yang MH, .Kriegman D and Ahuja N. Detecting Faces in Images: A Survey. IEEE Transaction on PAMI 2002.; 1: 34-58.
25. Yang MH, Kriegman D, and Ahuja N. Face Detection Using Multimodal Density Models. Computer Vision and Image Understanding 2001; 284-304 .
26. Zhu Y, Schwartz S. "Efficient Face Detection with Multiscale Sequential Classification". In Proc. IEEE Intl. Conf. on Image Processing, pp. 121-124, Rochester, New York, USA, 2002.

---

[i] http://www.ieeta.pt/carl

54

KNOWLEDGE ACQUISITION FROM HISTORICAL DATA FOR CASE ORIENTED SUPERVISORY CONTROL

Alexei Lisounkin*, Gerhard Schreck*, Hans-Werner Schmidt**

(*) Fraunhofer-Institute for Production Systems and Design Technology
Pascalstraße 8-9, D-10587 Berlin, GERMANY
e-mail: {alexei.lisounkin, gerhard.schreck}@ipk.fhg.de

(**) ELPRO Prozessindustrie- und Energieanlagen GmbH
Marzahner Straße 34, D-13053 Berlin, GERMANY
e-mail: hans-werner.schmidt@elpro.de

This paper presents a knowledge acquisition procedure based on data mining methods and its integration within a SCADA environment for decision support of plant operators. The results shown are part of investigations using real historical data of water treatment plants.

1. INTRODUCTION

Even for high level automation of process supervision, diagnostics, and control, the facility operator role is continually increasing. Although local automation tasks are covered by an installed control system, the facility operation staff take precaution with high level functional, technological, strategic objectives. Here, human experience plays an unique role. The aim of the knowledge-based methods is to connect objective information from the facility – available historical trends and data – with experience-based evaluation and assessment through an operator team.

Use of data mining approach will support elaboration of case characteristic information and ensure objectivity of the decision making procedure. The elaboration of a generic data mining approach for the process control knowledge acquisition is the focus of this paper. The corresponding functional sequence was developed and applied for the implementation of high level control and supervision tasks for water treatment plants.

The main aim of the procedure is to extract valuable information – knowledge – from a historical data series. The analysis of the water consumption profiles is subordinate to the middle-term and long-term facility control tasks, as well as simulation based operator training. The following chapters give a description of the relevant steps and give numerical examples as illustration.

2. CASE ORIENTED SUPERVISORY CONTROL

The increasing complexity of modern process plant and the demands for energy conservation, product quality, environment protection, safety and reliability make new approaches to process automation necessary. Beside decision-making and control procedures based on mathematical models, and solvers for multi-criteria optimization tasks, knowledge based systems involving the experience of human operators are a promising approach.

SCADA systems (supervisory control and data acquisition) are usually introduced for high level process control and automation. They play an important integration function in distributed, multilevel control environments, and provide the common view to the system and according operator panels required for process operation. Typical functionalities include activation of control actions, monitoring of process states, recording of alarms and events, emergency shutdown, etc.

The human operator team is responsible for the high level process management which includes tasks like

- Assessment of process situations,
- Selection of operating points,
- Consideration of different modes of working and use of resources,
- Reaction to changing requirements / demands,
- To ensure a continuous and smooth running of the system.

Operator decisions are based on its knowledge on process situations, process trends, and process control. Hereby the experience gained by the past – historical data & situations – plays an important role to reach high level performance. Case oriented supervisory control means a mapping of known process situations to approved control strategies & actions. Therefore the identification of operation profiles of processes or sub-processes can be identified as a basic function of a decision support system. Figure 1 presents the basic concept of decision support and knowledge acquisition based on data mining. It considers the long term development of a knowledge base on approved control patterns as well as the short term decision support on actual process situations. This results to an adaptive system with high flexibility and active involvement of the human operators.

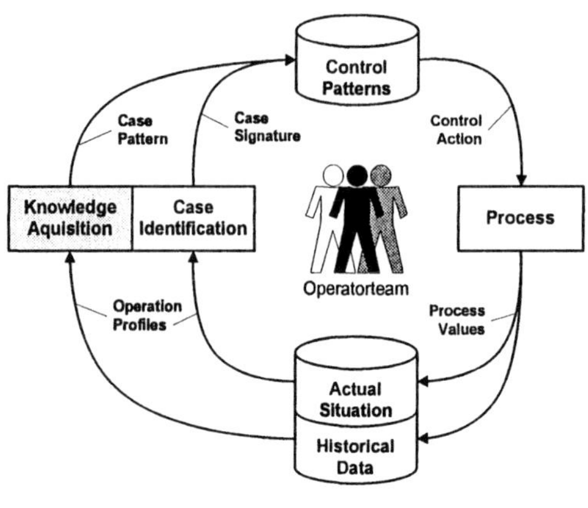

Figure 1 – Knowledge acquisition from operation profiles.

3. KNOWLEDGE ACQUISITION PROCEDURE

3.1 General Aspects

Traditionally, data analysis techniques consist of a sequence of data processing algorithms which investigate general characteristics of data series – such as minimum maximum, and mean values, statistics, images with respect to a given convolution operator, etc. – and then leave semantic data interpretation to a human. The data exploitation aspects – pragmatics – impact semantics of the entire data analysis procedure dramatically, which is reflected in the choice and parameterization of data processing algorithms. For this reason, the data analysis procedure is considered to be data driven knowledge acquisition, and our objective is to emphasize the semantic aspects for all steps of the data processing chain.

The principle steps of the data driven knowledge acquisition chain are summarized in the following list:

- data acquisition,
- data conditioning (validation, regularization and pre-processing),
- definition of semantics,
- knowledge extraction.

In Figure 2, the steps of the knowledge acquisition chain are associated with corresponding general data processing methods. This scheme defines a framework for configuration of the chain with respect to semantics and operational aspects.

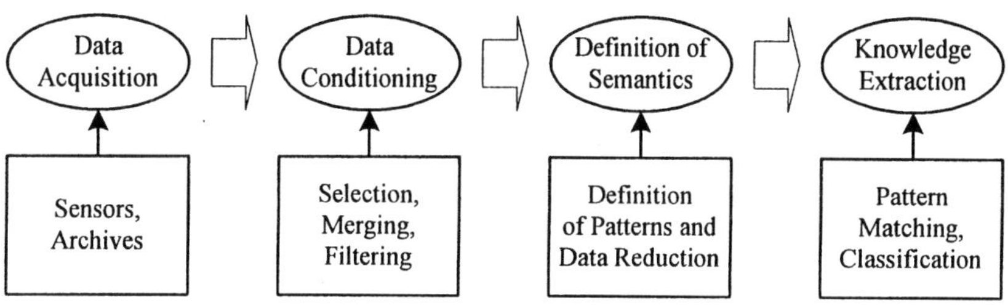

Figure 2 – Principle chain of data driven knowledge acquisition.

Special attention must also be paid to the characterization of the time enlargement of the process. Here, a continuous data flow must be separated into characteristic units possessing, in some sense, common information features. Thus, identification of repeatable operating sequences plays an important role with respect to the exploitation of data analysis results for case oriented process supervision and control.

Further, the process profile denotes a set of chronologically ordered data samples enlarged within the specified time interval. For the investigation of water consumption profiles, we set the profile time enlargement to 24 hours. This time interval corresponds to the sleep-wake living cycle of people and, moreover, it is explicitly recognizable in the process data flow.

3.2 Data Acquisition

Data samples consist of practical measurements, status information, and control signals available for the analyzed process – so-called process values. In cases where a water supply and distribution facility is controlled by a SCADA system, these process values are supplied by instruments installed in the physical system and by the logic of the control components itself. The SCADA system is usually equipped with a database which saves the process values. The archiving can be event-oriented – when a certain condition is being met – or raster-oriented – cyclic with respect to a defined time period.

For example, in Figure 3a, a single water consumption trend for a small German city (ca. 200,000 inhabitants) is depicted. Such trends collected over the period of two years have been analyzed by the authors and have offered the required input for the investigations.

Additionally, new data items within the data sample can be generated with an algorithmic analysis of the measured data sample components. Such a procedure is known as data transformation. The linear and non-linear composition of components, numerical integration and derivation are examples of data transformation. Complimentary to the transformation, data selection should also be mentioned at this point. It is reasonable to use only a subset of process values for further consideration. The other archived data items (components) will be omitted as irrelevant.

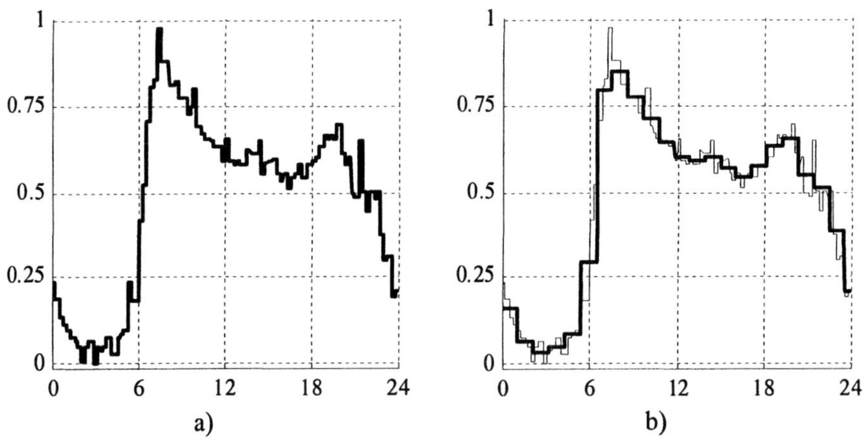

Figure 3 – a) Water consumption profile (abscise – time in hour, ordinate normalized); b) Profile after data filtering.

3.3 Data Conditioning

The objective of this step of the data analysis has is to validate data profiles and standardize their time raster, as well as to change some data properties (e.g., disturbances and noise reduction). In the beginning, the practical data profiles must be checked with respect to gaps and inaccurate data. The data validation is based on intuitive knowledge of the process – possible minimal/maximal values, expected data sampling time, etc. Inaccurate data can be set to some neutral values, gaps in the data time series can be closed by interpolation or by involving statistical information. An alternative option is to exclude profiles with gaps or inaccurate data from being considered further.

The standardization of data means the obtaining of a common time lattice (common sampling cycle) for all considered data profiles. The common time lattice is important for the profile classification procedures which operate on sets of profiles by converting them into matrix form.

Any measurement is a raw noisy data, which, in a lot of cases, must be prepared for the next steps of data analysis. Traditionally, noise reduction is proceeded by regression filters, such as low-pass or band-pass filters. The regression filters are profitable for cases in which the regression model is adequate for forecasting the process dynamics. In the described application domain, we can not expect an adequate modeling from the regression. Moreover, the use of regression filters for smoothing out the noise has the disadvantage of stretching and flattening the data: valleys and peaks become wider and their magnitude becomes smaller.

Recently, wavelet based filters have extensively been used for data processing. Here, the multi-resolution analysis is applied for the identification of meaningful data and to "smooth" out the measurement noise (Lisounkin, 2003). The shape of the mother-wavelet must be chosen in accordance with profile interpreting and archiving pragmatics. In Figure 3b, the result of water consumption profile filtering by means of the wavelet filter is depicted. Here, the Haar-mother-wavelet was selected in order to ensure minimal number of system switches and steady-state conditions in the intervals between the switch points.

3.4 Definition of Data Semantic

The objective of the data semantic analysis is to elaborate such data characteristics and a criterion which allow the interpretation of actual process data with respect to data exploitation procedure. Rules for the interpretation of the information hidden in data profiles must be defined by a human. Simultaneously, process information which is declared unimportant, will be omitted. The definition of data semantic is indeed the process of data abstraction.

An abstraction of the data is usually connected with the definition of data semantics. Thus, a semantic sensible reduction of data is applied in order to emphasize relevant features of the profile and to substitute the non-relevant information with neutral values with respect to the comparison procedure. This procedure can be characterized as morphologic processing of the data. The morphologic processing completely changes the nature of the data and represents a semantically conditioned data reduction.

With respect to water facilities supervision and control, such data which represents dynamic behavior of the water consumer is highly instructive for facility operation. For this reason, high and low amounts of water consumption were investigated, and a procedure for water consumption forecasting was developed. From this point of view, cases of high and low water consumption in water profiles are the semantic payload of the data. For the analysis and forecasting of high and low water consumption, the data model based on profile string coding and approximating string matching approach has been applied.

3.5 Extraction of Knowledge

Knowledge extraction from data series can be characterized by one of the following approaches (Müller, 2000):

- deviation detection and change measure – discovering the most significant changes in the data from previously measured or normative values,
- clustering – identifying a finite set of categories that describe the data,
- regression analysis – learning a function that maps a data item in a prediction variable,
- dependencies modeling – finding a model that describes significant dependencies between variables.

For the middle-term and long-term facility simulation and control tasks, the clustering approach has been considered. Here, the main objective of the knowledge extraction is identification of representative process data profiles and mapping of them onto the process and its context characterizations. The guides for the mapping procedure must be provided by facility staff.

It was assumed that the basis for the data driven knowledge extraction should be a set of data profiles, which possess high versatility with respect to possible facility (process) conditions.

The results achieved by semantic data analysis – a set of one-dimensional (string, sequential) patterns, or multi-dimensional patterns – is subject domain knowledge which is concentrated in a set of data signatures.

Further exploitation of the knowledge could include:

- classification – to identify in which known situation the process is operating,
- cluster verification – to check whether an existing classification is still valid,
- forecasting – to obtain a process trend for the future.

When considering water supply, data-guided knowledge is expected to be used in order to provide standard control and training scenarios. The cluster analysis approach was mainly used to produce a set of typical patterns – clusters – which represents variants in water consumption behavior. As previously mentioned, the water consumption profile analysis involves the sleep-wake cycle as an indivisible piece of information. The profiles available over a period of several months were assumed to possess information about the customer's – the water user's – behavior.

The clustering approach consists of two tasks: *first*, to structure the raw data into clusters; *second*, to map the clusters onto an a priori defined set of water user behavior scenarios. If the mapping is bijective, the clustering is successful. If the separability of the classes is high (which is given by the membership function), the used set of data profiles is informative and adequate for modeling the chosen set of water user behavior scenarios.

In Figure 4, the results of clustering with respect to the *a priori* expected behavior scenarios "workday" and "weekend" are depicted. The classification of the profiles was obtained by the *c-means* algorithm (the MATLAB software with Fuzzy Logic Toolbox (The MathWorks, Inc.)). The cluster set (Figure 4a) elaborated from a given set of data profiles will denote the signature of this set of profiles. In Figure 4b, the "workday" and "weekend" cluster centers as well as water consumption profiles over 2 months are shown as gray code matrices.

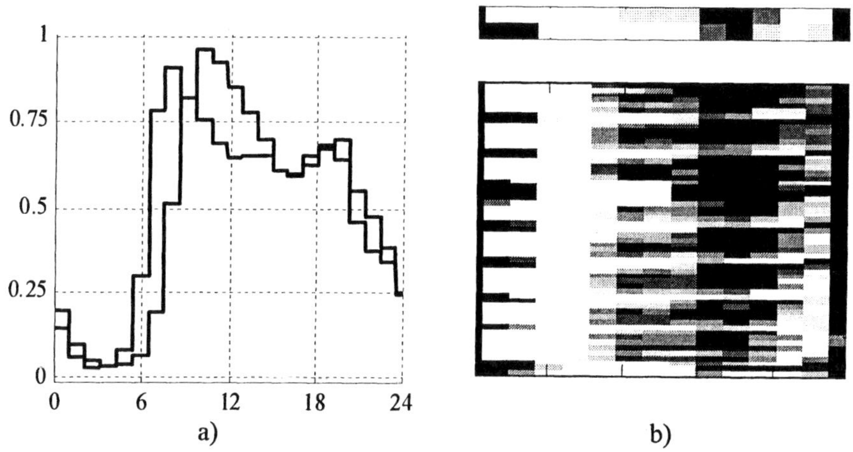

Figure 4 – Cluster centers for water consumption profiles over time axis:
a) "Workday" (solid line) and "weekend" (dashed line);
b) Above: "workday" and "weekend" cluster centers as gray code lines; Below: gray code lines for sleep-wake cycles over 2 months.

Here, matrix lines represent partial data profiles within the sleep-awake cycle. The superposed lines give impression about similarity of sleep-awake patterns over the workdays and weekends. In Figure 4b, the lines which correspond to weekends begin with a black zone which puts on view a delay in water consumption.

Considering the results of the performed data analysis, we recognize that the data signature has some relatively stable characteristic profiles for two typical situations: "workday" and "weekend". The operation scenarios of the technical facility can be elaborated for these two profiles respectively. The question of the initial operation scenario setup is solved using an a priori information – calendar. This represents the first exploitation scenario of the elaborated subject domain knowledge.

The deviation detection is based on analysis of distances between current process profile and cluster centers. Here, the membership function can be exploited. Moreover, also semantic aspects becomes an importance. By means of semantics reasoned weighting, a set of deviation functions can be provided. For calculus, different deviation functions may emphasize different properties of the data series.

Thus, the deviation identification for water consumption profiles has special semantic aspects which are listed below:

- profiles which are sequential patterns are being compared,
- time and duration of dynamic events (accelerations) are of importance;
- time and duration of maximal and minimal workload are of importance.

Moreover, the comparison of sequential patterns must have a highly robust results and be tolerant of the deviations in the range of intermediate workload.

Comparison of the codes proceeds using an algorithm for approximate string matching (Melichar, 1995). Obviously, an exact matching of two different patterns is impossible due to stochastic disturbances. Therefore, the procedure must reasonably allow, and also interpret, some "small" differences between the patterns which are being compared. The deviation calculus builds a measure for profiles similarity ("measure of confidence").

5. CONCLUSIONS AND OUTLOOK

The technique developed here was applied for analysis of water consumer behavior in German middle-sized cities. The examples shown here substantiate the approach for the modeling of the water consumers' behavior. The elaborated urban area specific set of signatures supports the facility operator as well as the manufacturer with new information which will be implemented in the control logic of the SCADA system. Moreover, this technique may build the functional interface for a data warehouse. In this context, the data warehouse model will provide the semantic background for the knowledge acquisition procedure (compare with (Kouba, 2002)).

The new application domain knowledge is already being used for the optimization of the facility operation. Thus, a simulation-based operator training with respect to typical operational situations is already available (Schreck, 2002). A model core for the water distribution facilities has already been implemented at the Fraunhofer IPK and can be supplied by data derived from the signature.

Further research activities are devoted to methods for data model adaptation at runtime, when new data samples are being imported. Herewith, a high sophisticated diversification of process control use cases should be achieved.

6. ACKNOWLEDGEMENTS

The study represented here was performed in the framework of an R&D project "Akquisition, Management und Integration von Prozesswissen in eine modellbasierte Prozessführung als Repräsentant einer neuen Generation von wissensbasierten Systemen in der Leittechnik" (http://amaryl.ipk.fraunhofer.de) partially funded by the Federal Ministry of Education and Research, Germany. We thank our industrial partner Elpro Prozessindustrie und Energieanlagen GmbH, Berlin, for extensive contacts with companies of the Water industry.

7. REFERENCES

1. Lisounkin A. "Semantic characterization of Data Series with Application to facility Control". In Proceedings of the IASTED International Conference on Signal Processing, Pattern Recognition, and Applications, June 30 – July 2, 2003, Rhodes, Greece, pp. 113-118.
2. Müller J.-A., Lemke F. Self Organizing Data Mining. Extracting Knowledge from Data. Libri Books on Demand, Dresden, Berlin, 2000.
3. Melichar B. "Approximate String Matching by Finite Automata." In Proceedings of the 6th International Conference on Computer Analysis of Images and Patterns (CAIP'95), Prague, Czech Republic, September 6-8, 1995.
4. Kouba Z., Matousek K., Miksovsky P.: "On-line analysis of utility networks", In Knowledge and Technology Integration in Production and Services: Balancing Knowledge and Technology in Product and Service Life Cycle. Edited by Vladimir Marik, Luis M. Camarinha-Matos and Hamideh Afsarmanesh, Kluwer Academic Publisher, 2002, p. 469-476.
5. Schreck G. "Simulation Services for Training of Plant Operators". In Knowledge and Technology Integration in Production and Services: Balancing Knowledge and Technology in Product and Service Life Cycle. Edited by Vladimir Marik, Luis M. Camarinha-Matos and Hamideh Afsarmanesh, Kluwer Academic Publisher, 2002, p. 79 – 86.

55 CEPSTRAL ANALYSIS IN TOOL MONITORING

Igor Vilcek, Jan Madl
Czech Technical University in Prague, Faculty of Mechanical Engineering, Dept. of Manufacturing Technology, Technická 4, 166 07 Prague 6 , CZECH REPUBLIC
phone: +420 2 24352611
e-mail: vilcek@fsid.cvut.cz, madl@fsid.cvut.cz.

Utilization of force signals to a chieve on line d rill wear monitoring is presented in this paper. After consulting the available literature it is obvious, that only some features of force component are proposed for drill wear monitoring .The really important task in drilling operations is to avoid catastrophic failure. It is desirable to make the most economic use the cutting tool without reaching catastrophic failure. Traditionally, the usual approach to tool monitoring the drilling was to the detect breakage as fast as possible and avoid overloads in the machine tool. These strategies are not enough to ensure the optimum economic performance of the machining process.

Therefore, the primary objective of this lecture assesses the feasibility of using force signature analysis as means for monitoring tool wear.

Keywords: cepstral analysis, monitoring, machining, cutting forces, harmonics

1. INTRODUCTION

Requirements for flexible manufacturing have been increasing in the last years. In order to insure effective operation of expansive manufacturing equipment, which has to run automatically and unattended, tool monitoring is important. Therefore, the essential problem to be overcome to achieve the full potential of unmanned machining is the development of effective and reliably sensors systems to monitoring the process and corrective action in case abnormal operation. With increasing wear in the twist drill margin wear causes the increase of the frictional forces between the margin and machining hole wall and leads to tensional vibrations in the cutting tool. This in turn will cause further tool wear and vibrations. If the cyclic process continuous catastrophic failure will occur at a short time. At the moment when these tensional vibrations appear, it is the appropriate time for drill bit change, since from this point on, wear increases rapidly due to the phenomenon of tensional vibrations.

Quante et al [4] recognized the importance of sensing vibrations in the twist drill for wear monitoring as a mean to overcome the difficulties of the slight sensitivity of the static component of the thrust force to wear. They proposed the use of the distance sensors without contact measuring deflection of the drill in a plane normal to the drill axis. A synchronization device was attached to the spindle emitting 256 pulses per revolution. The signal of the distance sensor was hight pass filtered at 60 Hz to avoid the effect of the spindle speed frequency at 12 Hz. An increase ranging to 5 to 8 times in the signal for a worn drill with respect to the initial value when sharp was reported . The advantage of the system is that since it senses without contact at the tool shank, as was the case with eddy current torque sensors proposed Brinksmeier et al [5], it can by applied to almost any existing machine tool without structural changes. The sensors are expensive and do not interfere with machining process.

The proposed approaches of cutting force signature analysis to be investigated and compared are the following:

- **Static component of cutting forces signal**
  One of the problems observed in the literature on the use of signal features from the static component is the occurrence of false alarms due to the stochastic character of the cutting process and especially due to variations in hardness along the workpiece. Subramian and Cook, 1977 et al. [13] established 5 percent as the maximum allowable variation in workpiece hardness for the static component of torque and thrust force to be used successfully as variables for drill wear sensing. To overcome this limitation and solve the possible false alarm problem, the thrust force-torque ratio is proposed as a method for detected the wear by means of the DC component of the signal.
- **Dynamic component of cutting forces signal: Frequency analysis**
  Analysis of the dynamic component of the cutting force signal has been neglected hitherto in most approaches to tool condition monitoring in drilling. This method is expected to be sensitive for detecting tool wear. Frequency domain methods will be applied and several analysis techniques will be explored such as the power spectrum, the power cepstrum, cross spectrum.
- **Tool failure prediction**
  It has been observed that violent and sudden oscillations occur in force signal when the tool is reaching the end of its life and is about to fail. This phenomenon is thought to be produced by a certain wear mechanism occurring at the end of tool life, when severe wear is already present, and leading invariably to catastrophic failure. To detect this phenomenon and thus predict tool failure the derivative of the cutting force signal are thought to be sensitive. Other ways of detecting the wear mechanism leading to failure will be explored; by means of frequency domain methods such as cepstral analysis and coherence function among both thrust force and torque signals.

2. SOFTWARE USED FOR EXPERIMENTS

Matlab software with a Real Time Toolbox was used for data acquisition in order to sample the cutting force signals from the dynamometer. The sampled data was saved on the hard disk of the computer for further processing and analysis. For this analysis of sampled data the Editace program, developed at the Department of Mechanical

Technology of the Czech Technical University in Prague, was used. This program is based on Matlab and is able to import the sampled data.

The measured signal can be displayed and edited, unwanted parts of the signal can be deleted, and zero can be adjusted. The basic statistical parameters of the selected part of the signal can be computed regression analysis can be performed and a curve fitted (Novak and Madl 2000) [2].

The advantage of this program is that it is very user friendly and easy to work with.

This program is used to determine empirical data for dynamic measurements of the forces and torques. This data can be utilised for optimisation of the machining process.

The results of the program should be:

- A graph of measured values
- A graph of the gradient of measured values
- A file of statistical parameter(s)
- Values of dependent parameters in the selected regression model

This software is used for drilling, milling and turning.

3. EXPERIMENTAL DEVICE

A series of drilling and milling experiments were carried out.

Cutting conditions :

HSS twist drill of diameter 2,2 and 4,5 mm.

Work material - ISO 683/XIII-86 and 11Cr 16Ni2 Mo1 Feed: 0,1mm per rev.

Number of revolutions: 2500 per min.

Dry drilling.

The thrust force and torque produced by the drilling process were measured by means of a type 9272 Kistler four-component piezoelectric dynamometer.

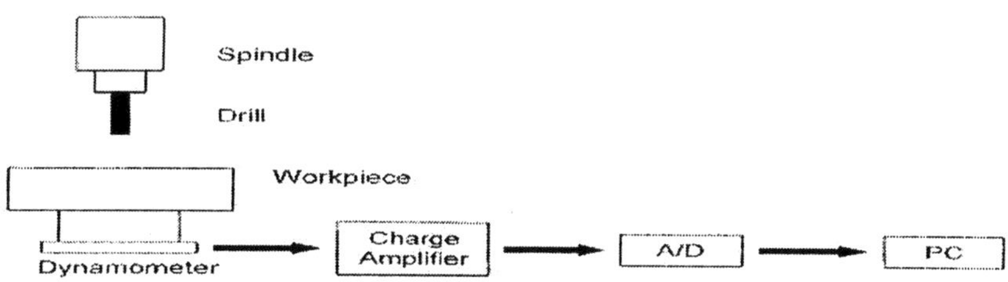

Figure 1

The signals were amplified by Vibrometer A.G type TA-3/C charge amplifiers with two individual channels. The amplified signals were sampled using a data acquisition card on the hard disk of the computer for further analysis. A schematic diagram of the experimental set-up is shown Figure1. For data acquisition Matlab software with a Real Time Toolbox are used in order to sample the cutting force signals from the dynamometer. The sampled data was saved on the hard disk of the computer for further processing and analysis. The Editace program was used to analysis the sampled data.

4. FAILURE PREDICTION BY MEANS OF CEPSTRAL ANALYSIS

The harmonics of the basic rotational frequency in fact dominates the spectrum when instability prior to failure is reached. The stronger the unstable wear phenomenon the larger the effect of harmonics.

Therefore, an efective strategy for twist drill failure prediction would be to detect the occurrence of the harmonics and to determine their presence by cepstral analysis.

Power spectrum density estimate of the thrust force for the first hole (left) and for the last hole (right) is shown in Fig. 2

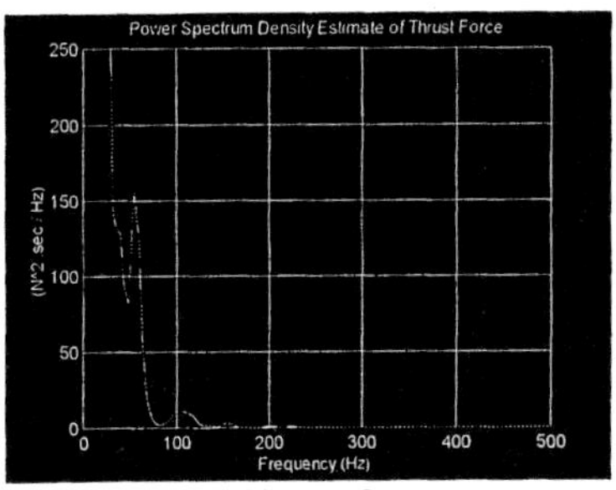

Figure 2

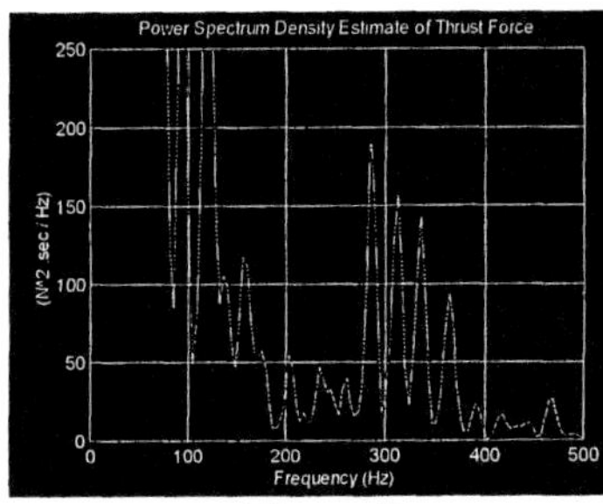

Figure 3

The cepstrum, strictly speaking the power cepstrum, is defined as the power spectrum of the logarithm of the power spectrum (Randall 1977) [1]. If the forward Fourier transform of a time function fx(t) is denoted by:

$$Fx(t) = F\{fx(t)\} \tag{1}$$

then the power spectrum may be represented by

$$Fxx\,(f) = |Fx(f)|^2 \tag{2}$$

And the cepstrum

$$Cx(\tau) = |F\,\{log\,[Fxx(f)]\}|^2 \tag{3}$$

The independent variable τ of the cepstrum is called the Quefrency, though it has the dimensions of time. The characteristic feature of the cepstrum is its ability to detect and give a measure of the phenomena, which exhibit periodicity in the spectrum, such as the harmonics. In particular, in complex signals containing a mixture of different families of harmonics, separation of the various periodicities is greatly facilitated by performing the second Fourier transform to obtain the cepstrum. A point about which different opinions exist is whether the second Fourier transform should be a forward or an inverse transform. However this is not important, since the result is identical except for a scaling factor. If an inverse Fourier transform is used, for the sake of consistency, the cepstrum is represented by

$$Cx(\tau) = |F^{-1}\{log[Fxx(f)]\}|^2 \tag{4}$$

5. FAILURE PREDICTION BY MEANS OF THE COHERENCE FUCTION BETWEEN THRUST FORCE AND TORQUE SIGNALS

An approach is put forward for catastrophic failure prediction, based on the detection of the above mentioned specific wear mechanism which occurs in the third stage of tool life, the stage of final accelerated wear (catastrophic failure).

The coherence function indicates the extent to which two signals are correlated with each other. In other words, it can be said that the coherence function gives a measure of the validity of the assumption that both signals result from the same particular generating mechanism or source.

The coherence function $\gamma_{xy}(f)$ is defined by:

$$\gamma_{xy}{}^2(f) = \frac{|F_{xy}(f)|^2}{F_{xx}(f) \cdot F_{yy}(f)}; \quad 0 \leq \gamma_{xy}(f) \leq 1 \tag{5}$$

where $F_{xy}(f)$ and $F_{yy}(f)$ are the power spectra of each signal, also often referred to as auto spectra where $F_{xy}(f)$ is the cross spectrum. The cross spectrum $F_{yy}(f)$ of $f_x(t)$ and $f_y(t)$ is the forward Fourier transform of the cross correlation function $R_{xy}(\tau)$, which is, in turn, defined by the equation:

$$R_{xy}(\tau) = \lim_{T \to \infty} \frac{1}{T} \int_{-T/2}^{-T/2} f_x(t) \cdot f_y(t+\tau).dt \tag{6}$$

This gives a measure of the extent to which two signals correlate with each other as a function of the time displacement between them. The cross spectrum can alternatively be obtained from the individual Fourier spectra $F_x(f)$ and $F_{y(f)}$ as follows:

$$F_{xy}(f) = F_x^*(f).F_y(f) \tag{7}$$

where $F_x{}^*(f)$ is the complex conjugate of $F_x(f)$.

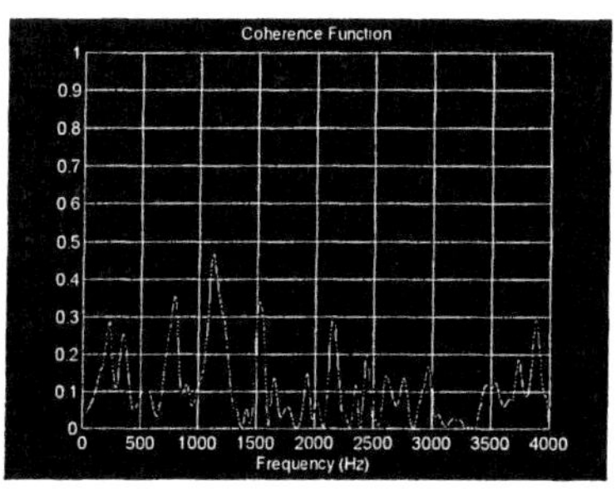

Figure 4

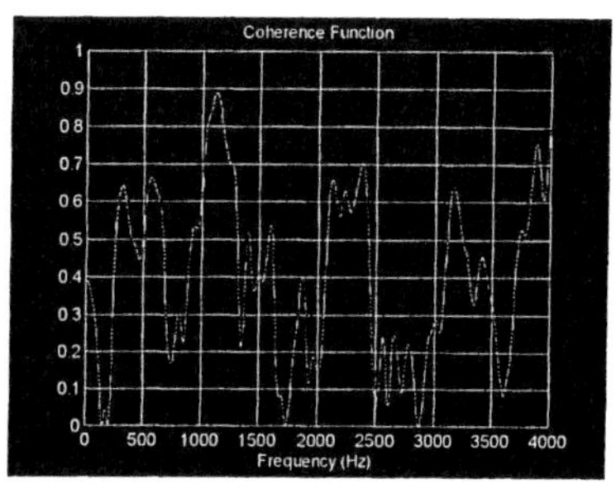

Figure 5

Therefore, three cases are possible. The coherence function can be zero, one or greater than zero and less than unity. If $g_{xy}^2(f) = 0$ for all frequencies the two signals are completely uncorrelated. In case $g_{xy}^2(f) = 1$ for all frequencies the two signals are completely correlated. If $g_{xy}^2(f)$ is between zero and one for all frequencies, one or more of the following conditions exist: a) even when the two signals y(t) and x(t) are caused partially by the same phenomenon or generating source, each is also caused in part by the other phenomena which affect it individually but do not affect the other signal, b) extraneous noise is present in the measurements, c) bias errors are a spectral estimation.
The coherence function for the first hole is shown Figure 4 or the last hole is shown in Figure 5

6. CONCLUSION

Some methods are presented for predicting of catastrophic failures in drilling, based on the detection of a specific wear mechanism operating at the end of tool life when severe wear is present and leading unavoidably to catastrophic failure. The basic characteristic of this wear mechanism is that it excites strongly tensional vibrations of the cutting tool. The proposed method relies in detecting the rise of harmonics in the spectrum of the torque signal when the wear mechanism begins to operate by means of the cepstral analysis (coherence function, power spectrum density estimate of the thrust force and torque)

7. REFERENCES

1. Randall, R.B., Application of B & K Equipment to Frequency Analysis, Bruiel and Kjaer information, Denmark, 1977
2. Novak, P. & Madl, J., Effective Evaluation of Measured Dynamic Values of Cutting Forces and Torques, Manufacturing Technology, 2001, ISSN 112386
3. El –Wardany, T.I. & Gao, D. & Elbestawi, M.A, Tool Condition Monitoring in Drilling Using vibration Signature Analysis, International Journal of Machine Tools and Manufacture, Vol. 36 No. 6, 1996, pp. 687-711
4. Quante, F. & Fehrnbach, H. & Meir, H.,-E., Automatische Uberwachung Rotierender Werkzeuge mit Abstands- und Schwingungssensoren in der Spanabhebenden Fertigung, Technisches Messen, 50. Jahrgang Heft 10, 1983, pp. 367-371
5. Brinksmeier, E., Prediction of Tool Fracture in Drilling, Annals of the CIPR, Vol.39, No. 1, 1990, pp 97-100
6. Cook, N.H., Tool Wear Sensors, Wear, Vol. 62, 1980, pp. 49-57
7. Bendat, Julius S. & Piersol, Allan G., Random data: Analysis and measurement procedures, New York, Wiley-Interscience, 1971
8. Boston O.W. and Gilbert, W.W., The Torque and Thrust of Small Drills Operating in Varius Metals, Transaction of the ASME, 1936, pp 79-89
9. Armarego, E.JA. & Brown, R.H., The Machining of Metals, Prentice-Hall, New Jersey, 1969
10. Merritt, H.E., Theory of Self-Excited Machine Tool Chatter, Transactions of the ASME, Series B, Journal of Engineering for Industry, Vol. 87, 1965, p. 447
11. Braun, S. & Lenz, E. & Wu, C.L., Signature Analysis Applied to Drilling, Transactions of the ASME, Journal of Engineering for Industry, Apr., 1982, pp. 268-276
12. Conrad, Charles J. & McClamroch, N. Harris, The Drilling Problem: A Stochastic Modeling and Control Example in Manufacturing, IEEE Transactions on Automatic Control, Vol. AC-32, No. 11, 1987, pp. 947-958
13. Subramanian, K. & Cook, N.H., Sensing of drill Wear and Prediction of Tool Life, Transactions of the ASME, Journal of Engineering for Industry, May. 1977, pp. 295-301

56 INTELLIGENT DIAGNOSIS AND LEARNING IN CENTRIFUGAL PUMPS

Jiří Kléma[1], Ondřej Flek[2], Jan Kout[1], Lenka Nováková[1]
[1]*Department of Cybernetics, CTU Prague, Technická 2*
166 27 Prague, CZECH REPUBLIC
{klema,kout,step}@labe.felk.cvut.cz
[2]*Rockwell Automation Ltd., Research Center Prague,*
Pekařská 695/10a, 155 00 Prague 5, CZECH REPUBLIC
oflek@ra.rockwell.com

This paper addresses the problem of on-line diagnosis of cavitation in centrifugal pumps. The paper introduces an application of the Open Prediction System (OPS) to cavitation diagnosis. The application of OPS results in an algorithmic framework for diagnosis of cavitation in centrifugal pumps. The diagnosis is based on repeated evaluation of a data scan providing full record of input signals which are observed for a fixed short period of time. Experimental verification of the algorithmic framework and the proposed methodology proved that a condition monitoring system built upon them is capable of diagnosing a wide range of cavitation conditions that can occur in a centrifugal pump, including the very early incipient cavitation.

1. INTRODUCTION

Pump is probably the most widespread variety of machinery in the world. Pumping systems, either directly or indirectly, are an essential part of every business throughout the world. As an ultimate illustration, a typical chemical plant utilizes on average one pump per each employee (Hennecke, 2000). It has been estimated that nearly 20% of the energy generated globally is devoted to driving pumping systems (Hart, 2002). Probably the highest percentage of pumps used in industry is accounted to centrifugal pumps. They are relatively simple, inexpensive and generally very reliable pieces of equipment. Nevertheless, the consequences of their unexpected failure include costly machinery repair, extended process downtime, health and safety problems, increased scrap levels, and loss of sales. For this reason, an increasing interest in improved maintenance regimes can be noticed among pump operators.

The traditional machinery maintenance approaches include *reactive* and *preventive* regimes (Day, 1996). The former relies upon allowing the machine to break down before being maintained. In spite of the fact that it has proved to be the most expensive option, it is still widespread among many industries. The preventive maintenance mode, whereby the maintenance is based upon specific time intervals, can lead to savings over the reactive regime, however, it is not an effective use of maintenance resources as work is undertaken regardless of the condition of the equipment. *Predictive* or *condition-based maintenance* (CBM) is based upon monitoring of condition of the equipment and determining whether corrective action is needed. By detecting the early stages of a fault, maintenance can be scheduled in advance to coincide with planned production stops. The condition-based maintenance approach relies on techniques of machinery diagnostics. Many data acquisition and analysis techniques have been developed for machinery diagnostics. Some of them rely on periodic data collection carried out by means of portable instruments and subsequent batch-mode data interpretation. However, the focus has been on schemes that provide on-line, continuous monitoring and diagnosis of equipment. Improvements in sensor technologies and mass production of a wide variety of sensors have enabled application of on-line machinery diagnostics to a wide range of equipment where such techniques would be thought too expensive just a couple of years ago.

Although the techniques of CBM rely on the ability to detect early stages of a possible failure of equipment, further benefits would certainly emerge with the ability to control the working regime of the equipment in such a way that the potential for occurrence of failure is minimized. In the case of pumps, the most frequent failures are bearing and seal failures (Marscher, 2002). One of the most important root causes of bearing and seal failures is presence of *cavitation* in a pump. The word cavitation refers to formation of vapor bubbles in regions of low pressure within the flow field of a liquid (Brennen, 1994). In the context of turbomachinery, cavitation is generally considered undesirable. Besides of the effect on life of bearings and seals, cavitation causes gradual erosion of internal surfaces of a pump. This may result in an unexpected pump failure with possible disastrous consequences.

This paper is organized as follows. Section 2 gives a brief overview of principal approaches to cavitation diagnosis. It distinguishes two sensor categories that determine whether the diagnosis will be intrusive or non-intrusive. Section 3 introduces a non-intrusive way of cavitation diagnosis based on vibration sensors' data. It describes an experimental setup we have used, defines a structure of measured data and outlines a way in which the phenomenon of cavitation may reflect in the data. A summary of principal questions to be answered by data mining is provided at the end of the section. Section 4 introduces Open Prediction System – the tool used for processing and evaluation of the measured data. The section also theoretically discusses methods relevant to domains with ordinal classifications and possibly dependent samples. Section 5 summarizes reached results and tries to answer questions raised earlier in Section 3. In conclusion, the proposed algorithmic framework for diagnosis of cavitation in centrifugal pumps is recapitulated.

2. PRINCIPAL APPROACHES TO CAVITATION DIAGNOSIS

Quite recently, first commercially available systems have appeared that reflect the trend of integration of pumping system control and condition monitoring (Stavale, 2001). Here, the ability to estimate the presence of cavitation in a pump is based on the knowledge of parameters of the pumping process, especially the pressure in pump suction. Since the phenomenon of cavitation is closely connected with pressure field in the pump, the decision to use such information for estimation of cavitation is logical. The major disadvantage of the use of pressure information is the fact that pressure sensors are considered intrusive, i.e., come into touch with the pumped fluid. Tapping such sensors into the pipework of the pumping application increases the potential risk of leakage. For that reason, such sensors must be avoided in some applications involving pumping of dangerous fluids. This disqualifies the solutions using such sensors from universal application.

The attention of the industry points towards non-intrusive sensors. Vibration sensors play a dominant role among them. Besides of being a source of information for detection of cavitation, they are capable of providing information that can be used for diagnosis of a wide range of rotating machinery faults (White, 1998). Understanding relationship between pressure pulsation in pumps and mechanical oscillation of solid pump parts creates a background for the use of pump casing vibration as a source of information for cavitation diagnosis. This paper studies possibility of indirect detection of cavitation from mechanical vibrations, the resulting diagnosis scheme must rely exclusively on information from non-intrusive sensors of this type.

3. VIBRATION SENSORS' DATA

3.1 Cavitation Research Setups

The research of cavitation diagnosis methods requires a representative amount of experimental data. The data must cover a wide range of operating conditions of the pump and various degrees of cavitation. Although data collected on a real-world system would be of highest value, there are numerous reasons why this kind of data is normally unreachable: (1) economical and safety reasons, (2) controlling the real-world system deliberately in order to cover a wide range of operating conditions is usually not allowable, (3) fitting the necessary instrumentation to the real-world system may be difficult or even impossible, (4) the cavitation condition present in the pump during experiments is usually not known with a sufficient accuracy as this mostly requires a specially modified pump that allows visual observation of the inside of the pump.

Consequently, in many cases, purpose-built experimental setups are employed to provide the data needed for research. The setup can be divided into four main subsystems: the pump (modified by a transparent material allowing visual observation of cavitation), the motor (energized either directly by mains power line or by variable frequency drive), the flow loop and the data acquisition equipment. Following signals were sensed and recorded: flow rate, pressures in both suction and discharge, temperature of the pumped fluid, shaft rotation frequency. These signals were used to control the experiments and they could not be used during a diagnosis phase. Vibration signals were sensed by accelerometers attached to the casing of the diagnosed pump. For all experiments, two accelerometers were used simultaneously. They were positioned on the casing of the respective experimental pump, in mutually perpendicular directions. One of the sensors was adjusted in the direction of the pump shaft axis (denoted as axial), the other in the direction radial to the pump shaft axis (denoted as radial). A detailed description of the setup employed in presented experiments is given in (Flek, 2002).

A typical experiment, performed with the aim of obtaining data relevant to the phenomenon of cavitation, establishes various levels of cavitation in a system operating under certain conditions. It is important to investigate cavitation in a system working at a (full) range of operating points. The operating point of the system is described by the flow rate Q, the total head rise H and the shaft rotating frequency Ω. It is set by a throttling valve in the discharge pipe. Severity of cavitation in the pump is set either by a throttling valve in the suction pipe (open tank setup) or by modification of pressure above liquid surface in the tank (closed tank setup).

3.2 Data, Preprocessing Phase, Feature Extraction

The vibration of a centrifugal pump casing is governed by numerous excitation forces acting at different frequencies: pump shaft imbalance, misalignment, impeller blade passing pulsation, bearings, cavitation, etc. Many different techniques are used to help the evaluation of vibration signals. The simplest possibilities include evaluation of amplitude information in time-domain signal. However, signals of periodical nature typically require analysis in the frequency domain. The emergence of digital signal processing techniques, especially the Fast Fourier Transform (FFT) algorithm, became the driving force behind the wide spread of frequency-domain analysis techniques.

(Flek, 2002) proposes distinguishing features of following types: (1) power spectral density of frequency band *y* (denoted as psd*y*), the number of bands depends on frequency resolution of the periodogram, in this paper we use mainly 65 bands of $\Delta f \approx 234$Hz, corresponding to a 128-line FFT with sampling frequency ≈ 30kHz, but other settings were also tested, (2) amplitude of the first 2z harmonic components of shaft rotation frequency, where z is number of impeller blades (denoted as rpm*n*), in pumps with 4 blades 8 features, (3) frequency of rotation of pump shaft (denoted as fr).

Five ordinal cavitation classes can be distinguished: 0 – normal condition (no bubbles), 1 – incipient cavitation (very first bubbles), 2 – tip vortex cavitation (a tiny stream of bubbles), 3 – moderate cavitation (a continuous stream of bubbles), 4 – severe cavitation (severe bubbles, blade cavitation). The classes are assigned on basis of visual observations during the experiments.

3.3 Data Understanding - Visualization

The influence of the phenomenon of cavitation on vibration of pump casing can be better understood with aid of visualization. Having the excitation forces and their effects decomposed into a periodic component (related to the shaft rotation frequency) and a random component (e.g., cavitation), the following visualizations can be carried out.

Figure 1 shows examples of periodograms of pump casing vibration under varying cavitation conditions. Although particular frequencies and the scale of the phenomena necessarily differ among different pumps, the figures demonstrate influence of cavitation on psd*y* attributes at different operating points. Note that a constructed classifier has to detect cavitation while not knowing the actual operating point.

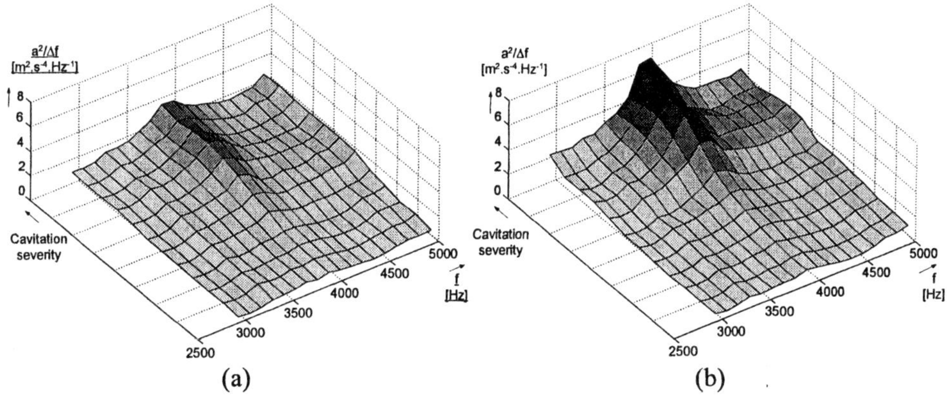

Figure 1 – Comparison of periodograms of pump casing vibration under varying cavitation conditions. Figures (a) and (b) show periodograms for two different operation points (below the flow rate of the best efficiency point and at the best efficiency point).

We have also used Radviz technique of multidimensional visualization implemented in the preprocessing tool SumatraTT (Stepankova et al., 2003). Radviz (Hofman et al., 1997) attaches to each data point fixed springs each of which is also attached at points around a circle. The springs represent dimensions of the data, the spring force for each spring is the value of the data point for that dimension. The data points are displayed at the position where the sum of the spring forces is zero, i.e., points which have one or two coordinate values greater than the others lie closer to those dimensions. The results can be seen in Figure 2.

3.4 Principal Questions To Be Answered

The main goal of the intelligent decision-making system design is to provide a tool allowing reliable and non-intrusive on-line diagnosis of cavitation in centrifugal pumps. Within this process, following principal questions regarding specific task characteristics should be answered:

- What is the optimal placement of the vibration sensors? How many of them one has to use (a minimum number of sensors should be used to save equipment and installation cost)?
- What is the influence of number (and thus resolution) of the power spectral density features? Can we deal with a large number of features having only a limited number of training examples?
- How should we deal with the measured data? Can we increase a number of training instances by generating more examples from a single (longer) signal measured under constant conditions? What is the dependence among signals measured under similar conditions (similar operating points)?
- Class values are ordered. Can we benefit from this ordering?
- How should we evaluate the resulting system and what is an optimal scoring function when developing a model? Shall we use classification accuracy only,

distinguish severity of misclassifications or rely on regression criteria (e.g., mean squared error (MSE))?

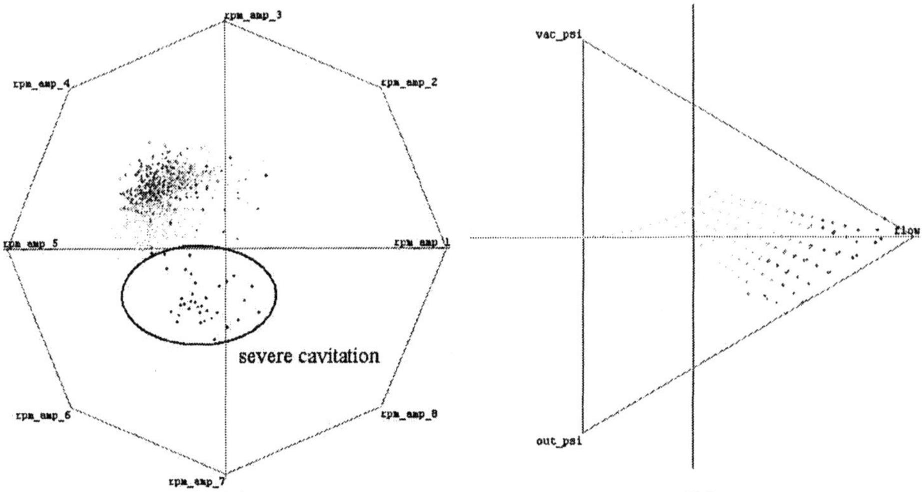

(a) (b)

Figure 2 – Radviz – class distribution in a multidimensional attribute space – the more severe cavitation the darker points. Figure (a) demonstrates ability of rpm*n* features to distinguish severe cavitation. Figure (b) shows strong dependence among cavitation, flow and pressures (these sensors are not available in a real operation).

4. CAVIATION DIAGNOSIS SCHEMES

4.1 Open Prediction System – Experimental Environment

Open Prediction System (OPS) is a prediction tool offering a solution to a wide range of prediction problems. Its main focus is on multivariate time series prediction problems with practical applications bringing cost reduction in utility companies (gas, water, heat, electricity). However, its predictive methodology (Kout et al., 2004) can be understood as fully general and able to solve both regression and classification tasks. The implemented data management structures enable data compression, data filtering, special built-in transformations, problem definition separated from preprocessing of individual models or meta-learning. These features make OPS a suitable tool for processing and evaluation of the pump data.

4.2 Available Data, Applied Methods

The analysis and results presented in this paper deal with measurements performed by Ondřej Flek on Durco Mark III 1K1.5×1-8 pump in Cleveland, Ohio, USA. In order to answer the questions summarized in subsection 3.4, a number of datasets based on these measurements were generated and processed. Several dimensions that distinguish between the datasets can be identified: (1) the data source is axial, radial or combination of both sensors, (2) the number of features varies with the selected power spectral density bandwidth – 32 and 65 bands were used, (3) the three-phase asynchronous electric motor can be energized either by direct mains power line or variable frequency drive (VFD). It follows that 3×2×2=12 different datasets were classified and compared.

The OPS offers several classifiers and predictors to be applied. It contains decision tree (DT) and random forest (RF) classifiers as well as neural networks (NN) and support vector machines (SVM). DTs and their visualization give a basic understanding of feature importance and problem complexity, while RF and SVM are techniques well-known to deal with datasets described by a large number of features regarding a number of available instances. Application of NNs gives a chance to reference the results performed in (Flek, 2002) on other two pumps. These four algorithms were applied to the datasets described in the previous paragraph.

The following two subsections describe fundamental approaches to a final task definition considering apparent class ordinality and dependence among measured instances.

4.3 Evaluating Dependent Samples – Dealing with Blocks and Series

The most often used evaluation methods (hold out, N fold cross validation, leave-one-out, bootstrapping) suppose a representative dataset on their input as well as independent instances. At the same time, they suppose that all the instances are drawn with a constant distribution and that the future instances will keep this distribution. These assumptions are violated in many practical domains. A motivational example which tries to find out whether the dataset is representative is as follows. Let us have a company producing pumps. The company produces three different pumps and wishes to equip them with a diagnostic tool that is able to detect their fault states. The fault state model is based on the data measured on pumps, but measurements are expensive. Is it necessary to measure all the pumps? An advisable approach to find it out could be to measure two pumps only. The first model can be trained on the data from the first pump and tested on the data from the second pump, the second model vice versa. When both the models give satisfactory error rates there is a solid chance that the resulting model (based on data from both the pumps) can be valid also for the third pump. Other approach, which first mixes measurements from both the pumps and then splits them randomly between training and testing subsets, is probably not a good indicator of performance on a future data from a different pump.

Under our design of experiments, a representative training data set has to be measured for each type of pump to be diagnosed. Differences between pump types are indispensable, a general cavitation model would be inaccurate. The proposed method relies on a standard pump performance testing procedure carried out by pump manufacturers with virtually every type of pump produced. It has been shown in (Flek, 2002) that such performance testing is capable of producing data covering the whole range of operating conditions.

The dataset can also contain dependent instances. This phenomenon often appears when dealing with time-series measurements, where dependence (expressed in terms of covariance) between ith and $i \pm j$th instance depends only on j and approaches 0 as $j \to \infty$. (Burman et al., 1994) proposes a modification of cross-validation called h-block cross-validation. For each instance *i* it is necessary to remove it from the training data along with *h* instances on each side of the *i*th observation. The model is trained on the reduced set and tested on the *i*th instance. This approach reduces to leave-one-out method when h=0. A slightly modified hv-block cross-validation can be found in (Racine, 2000).

Within pump measurements, it is not possible to set and measure an arbitrary number of operating conditions (time and cost reasons, limited number of valve positions, etc.). The measurements usually consist of several series, in which we start in the normal state and gradually stimulate more severe cavitation levels. In order to increase a number of training examples, we generate several instances (6) for every single operating point. A longer scan is subdivided into time slots that are processed separately. These training examples are surely dependent and must be treated in an approach analogical to the block cross-validation described above (we denote it one-block-out, OBO). Dependence among neighboring states in a series can be studied in an approach where a single series makes a single fold in the frame of cross-validation (denoted as one-series-out, OSO).

4.4 Ordinal Classification

Standard classification algorithms assume that the class values are unordered, i.e., do not exhibit any natural order. They treat the class attribute as a nominal quantity – a set of unordered values. Consequently, they cannot make use of the ordering information. (Frank and Hall, 2001) proposed a simple method that enables standard classification algorithms to make use of ordering information in class attributes. In their approach a task is transformed to a series of binary class subtasks that encode the ordering of the ordinal class. The data is first transformed from a k-class ordinal task (class attribute C with values $v_1, \ldots, v_k$) to k-1 binary subtasks, where i-th binary attribute represents the test $C>v_i$. This coding is similar to a thermometer code used in neural networks, which encodes into k binary attributes (Smith, 1996). In the next step, k-1 probabilistic models are learnt and used to estimate probabilities of the k original ordinal classes ($P(v_1)=1-P(C>v_1)$, $\forall i=2, \ldots, k-1$: $P(v_i)=P(C>v_{i-1})(1-P(C>v_i))$, $P(v_k)=1-P(C>v_{k-1})$). The class with maximum probability is assigned to the instance.

Learning to predict ordinal classes can also be rephrased as the regression problem. The class labels defined in subsection 3.2 can be either used immediately or they can be preprocessed prior to learning. We followed the second option, the *grades* transformation (Kramer et al., 2001), which depends also on fractions of values belonging to the individual classes, was carried out. In this paper, we apply the ordinal approach (denoted as ORD) altogether with the standard unordered classification (denoted as 5C) and regression (denoted as REG).

5. REACHED RESULTS

The design of experiments presented in the previous chapter results in 22 different ways of processing of each of the proposed datasets (algorithm x 5C, ORD or REG x OBO or OSO). In order to be able to evaluate and compare all the approaches, the outputs of the models were always transformed into 5 crisp classes first and then the classification accuracy was calculated. Significance of differences among the individual approaches was evaluated by McNemar's test. There were two different types of classification errors defined. The first type simply involves all misclassifications. Error occurs whenever the desired class does not agree with the generated classification. The second type pays attention to so called hard errors only. These hard errors do not occur whenever the model misclassifies an instance into its neighboring class. The detection rate of each model was also considered (ability to distinguish the normal state form an arbitrary cavitation level).

Table 1 shows a fragment of the final result table. The results proved that the axial placement of the vibration sensor gives a slightly better ability to diagnose cavitation. Moreover, combining the axial and radial data does not help to significantly increase this ability. The framework does not seem to be sensitive to the tested changes of power spectral density bandwidth, we recommend to deal with 65 psd*y* features as better portability to various pumps is assumed. The application of VFD brings another degree of freedom into the system and slightly decreases reliability of fault diagnosis. RFs proved to be the most suitable learning algorithm. Surprisingly, SVMs gave worse results than both RFs and NNs. Utilization of the ordering information significantly increases classification accuracy of DTs and SVMs, but do not help to improve RF classifiers. As for NNs, ORD approach helped to increase overall accuracy, REG approach was better considering hard errors only. OBO cross-validation results in reasonably more optimistic accuracy estimates than OSO. The results were discussed with a domain expert who regards the first sort of estimates as optimistically biased while the second one is biased pessimistically as it leaves out a certain part of the plane where a pump operating point can be set.

Table 1 – An overview of results reached by RFs on the *mains axial* and *radial* datasets with *65* power spectral density bands. The individual cells show error of classification [%]. # denotes a difference on 0.05 level of significance between corresponding tests on axial and radial data.

Method\Data	axial			radial		
	all	hard	det	all	hard	det
5C, OBO	16.3	0	7.1	16.7	0.9	8.2
5C, OSO	21.5	1.9	10.6	24.5#	6.2#	14.7#
ORD, OBO	18.2	0.1	8.0	19.2	1.8#	9.1
ORD, OSO	21.9	1.9	11.2	25.5#	5.2#	15.7#

5. CONCLUSIONS

The presented approach results in an algorithmic framework for diagnosis of cavitation in centrifugal pumps. The diagnosis is based on repeated evaluation of a data scan based on an axial vibration sensor sampled for a fixed short period of time. Signal is decomposed into periodic and random components, the methodology can deal with a large number of power spectral density features which guarantees its general applicability to various pump types. The proposed diagnosis scheme consists of the following operations: signal sensing, signal pre-processing, feature extraction, classification and presentation of diagnosis. The diagnosis can be presented either as a crisp classification (5C or REG approach), estimated as a real number (REG approach again) or in a form of a probability vector (ORD approach).

Experimental verification of the algorithmic framework and the proposed methodology suggested that a condition monitoring system built upon them is capable of diagnosing a wide range of cavitation conditions that can occur in a centrifugal pump, including the very early incipient cavitation. It can be tuned to individual pump types by means of a standard pump performance testing procedure only. This represents no extra effort since this procedure is a part of common practice exercised throughout the pump manufacturing industry. The future work lies in a verification of the proposed framework on a wider range of pump types including the extension to pump varieties other than centrifugal pumps.

Acknowledgments

This research work was supported by the research program Decision Making and Control in Manufacturing (MSM 212300013) funded by the Czech Ministry of Education.

6. REFERENCES

1. Brennen, C. E. Cavitation and Bubble Dynamics. Oxford University Press, 1994.
2. Burman, P., Chow, E., Nolan, D. A Cross-Validatory Method For Dependent Data. Biometrika 84, pp. 351-358, 1994.
3. Day, M. J. Condition Monitoring of Fluid Systems – The Complete Approach. In proceedings of the Fifth International Conference on Profitable Condition Monitoring – Fluids and Machinery Performance Monitoring. Harrogate, U.K., BHR Group, pp. 243-256, 1996.
4. Flek, O. Diagnosis of Cavitation in Centrifugal Pumps. PhD. Thesis, Czech Technical University, Department of Cybernetics, 141 p., 2002.
5. Frank, E., Hall, M. A Simple Approach to Ordinal Classification. Proceedings of the European Conference on Machine Learning, Freiburg, Germany. Springer-Verlag, pp. 145-165, 2001.
6. Hart, R. J. Pumps and Their Systems – A Changing Industry. Proceedings of 19th International Pump Users Symposium, Houston, TX, U.S.A., pp. 141-144, 2002.
7. Hennecke, F. W. Reliability of Pumps in Chemical Industry. Proceedings of Pump Users International Forum, Karlsruhe, Germany, 2000.
8. Hoffman, P., Grinstein, G., Marx, K., Grosse, I., Stanley, E. DNA Visual and Analytic Data Mining. IEEE Visualization '97 Proceedings, pp. 437-441, Phoenix, AZ, 1997.
9. Kout, J., Kléma, J., Vejmelka, M. Predictive System for Multivariate Time Series. To appear at European Meetings on Cybernetics and Systems Research (EMCSR), Vienna, 2004.
10. Marscher, W. D. Avoiding Failures in Centrifugal Pumps. Proceedings of 19th International Pump Users Symposium, Houston, TX, U.S.A., pp. 157-175, 2002.
11. Kramer, S., Widmer, G., Pfahringer, B., de Groeve, M. Prediction of Ordinal Classes Using Regression Trees. Fundamenta Informatica, 47(1-2): 1-13, 2001.
12. Open Prediction System, http://ops.certicon.cz.
13. Racine, J. A Consistent Cross-Validatory Method For Dependent Data: hv-Block Cross-Validation. Journal of Econometrics, November 2000.
14. Smith, M. Neural networks for Statistical Modeling. Boston: International Thomson Computer Press, 1996.
15. Stavale, A. E. Smart Pumping Systems: The Time is Now, ITT Industries, Fluid Technology Corporation, Industrial Pumps Group, http://www.gouldspumps.com/download_files/ /Technews/time_is_now.pdf, 2001.
16. Štěpánková, O., Aubrecht, P., Kouba, Z., Mikšovský, P. Preprocessing for Data Mining and Decision Support. In: Data Mining and Decision Support: Integration and Collaboration. Dordrecht : Kluwer Academic Publishers, pp. 107-117, 2003.
17. White, G. D. Introduction to Machine Vibration. Predict DLI, 1998.

AUTHOR INDEX